Edited by
Ernst Huenges

Geothermal Energy Systems

Related Titles

Cocks, F. H.

Energy Demand and Climate Change

Issues and Resolutions

267 pages with 30 figures
2009
Softcover
ISBN: 978-3-527-32446-0

Wengenmayr, R., Bührke, T. (eds.)

Renewable Energy

Sustainable Energy Concepts for the Future

120 pages
2008
Hardcover
ISBN: 978-3-527-40804-7

Kruger, P.

Alternative Energy Resources

The Quest for Sustainable Energy

272 pages
2006
Hardcover
ISBN: 978-0-471-77208-8

Kutz, M. (ed.)

Mechanical Engineers' Handbook, Energy and Power

1104 pages
2005
Hardcover
ISBN: 978-0-471-71988-5

Edited by Ernst Huenges

Geothermal Energy Systems

Exploration, Development, and Utilization

WILEY-VCH

WILEY-VCH Verlag GmbH & Co. KGaA

The Editor

Dr. Ernst Huenges
GeoForschungsZentrum
Potsdam
Telegrafenberg
14473 Potsdam

1st Reprint 2011
2nd Reprint 2012
3rd Reprint 2014

Library of Congress Card No.: applied for

British Library Cataloguing-in-Publication Data
A catalogue record for this book is available from the British Library.

Bibliographic information published by the Deutsche Nationalbibliothek
The Deutsche Nationalbibliothek lists this publication in the Deutsche Nationalbibliografie; detailed bibliographic data are available on the Internet at http://dnb.d-nb.de.

Cover Design Adam Design, Weinheim
Typesetting Laserwords Private Limited, Chennai, India
Printing and Binding betz-druck GmbH, Darmstadt

Printed in the Federal Republic of Germany
Printed on acid-free paper

ISBN: 978-3-527-40831-3

Contents

Geothermal Energy Systems. Edited by Ernst Huenges

ISBN: 978-3-527-40831-3

Preface

The book presents basic knowledge about geothermal technology for the utilization of geothermal resources. It helps to understand the basic geology needed for the utilization of geothermal energy and describes the methods to create access to geothermal reservoirs by drilling and the engineering of the reservoir. The book describes the technology available to make use of the earth's heat for direct use, power, and/or chilling, and gives the economic and environmental conditions limiting its utilization. Special emphasis is given to enhanced or engineered geothermal systems (EGS), which are based on concepts that bring a priori less productive reservoirs to an economic use. These concepts require the geothermal technology described here. The idea of EGS is not yet very old. Therefore, this book aims to provide a baseline of the technologies, taking into account the fact that due to a growing interest in EGS, a dynamic development may increase the specific knowledge to a large extent in the near future.

The book begins with a large-scale picture of geothermal resources, addressing expressions of the earth's heat sources and measured heat flow at different places world wide. This leads to conceptual models with a geological point of view influencing geothermal reservoir definitions based on physical parameters like porosity, permeability, and stress distribution in the underground, indicating that geothermal applications can be deployed anywhere, but some locations are more favorable than others.

The second chapter addresses the characterization of geothermal reservoirs and the implications of their exploration. A best practice for the exploration of EGS reservoirs is still to be determined and the different methods in geology, geophysics, and geochemistry have a strong local character. Some methods are successful in exploring conventional geothermal reservoirs like the magnetotellurics, whereas for EGS, seismic methods become more and more important. An overall conceptual exploration approach integrating the geophysical measurements into a geological model taking into account the earth's stress conditions is addressed in this chapter, but it has to be further developed in future contributions.

The baseline know-how of EGS drilling given in the third chapter, is based on a few case studies and therefore, somewhat different from hydrocarbon drilling

Geothermal Energy Systems. Edited by Ernst Huenges

ISBN: 978-3-527-40831-3

with reference to issues like large diameter holes, deviated wells, and mitigation of formation damage. The latter is also important for drilling conventional geothermal reservoirs, which to a great extent follow standards in operation and completion. The knowledge of underground physical conditions, especially the magnitude and direction of the local stress, is important for reliable drilling into EGS reservoirs. Awareness of the stress conditions is also a prerequisite for starting hydraulic fracturing treatment which is addressed in a following chapter.

In the fourth chapter, techniques and experiences from several EGS sites are described providing a set of methods available for addressing the goal of increasing well productivity. The case studies cover several geological environments such as deep sediments and granites. Significant progress was made in the last few years in recovering enhancing factors in the order of magnitudes. Chances and risks of companion effects of the treatments, such as induced seismicity, are addressed and will be a subject of forthcoming research.

In the fifth chapter, the state-of-the-art numerical instruments used to simulate geothermal reservoirs during exploitation are given in different case studies. Different coupled processes such as thermal–hydraulic or hydraulic–mechanical, including coupled chemical processes, are discussed. The development of the coupling of thermal, hydraulic, mechanical, and chemical processes is ongoing, hence the chapter provides the basics.

The benefits of using geothermal energy technologies for the direct use and conversion of the earth's heat into chill or power (as required), are described in the sixth chapter. Technical solutions for all tasks within the goal of energy provision exist, and approaches for improving the performance of system components are given. Special emphasis is given to techniques that can assure reliable and efficient operation at the interface of underground fluids with technical components. Processes like corrosion and scaling have to be addressed and they are still a subject of future research.

The economic learning curve is shown in the seventh chapter that provides some methods to analyze the risks of a project. A decision-making methodology is given for several stages of the project. Environmental aspects are discussed, and results of life cycle assessment with illustrations of greenhouse gas emissions are reported in the chapter.

The final chapter discusses the possibility of geothermal deployment as a part of future energy provision and an important contribution to the mitigation of CO_2 emissions. The technological, economic, and political factors controlling such deployment are discussed and should provide some assistance for decision makers.

The book was compiled by the authors, but also significantly improved by competent reviewers. Therefore, we like to thank Magdalene Scheck-Wenderoth, Albert Genter, Dominique Bruel, Claus Chur, Ron DiPippo, Wolfram Krewitt, and Harald Milsch for their excellent comments on the different chapters. In addition, we acknowledge the funds received from the EU commission, for example, for the projects ENGINE and I-GET, and the German government, especially, the Federal

Ministry for the Environment, Nature Conservation and Nuclear Safety (BMU). Special thanks go to the coworkers of the International Centre for Geothermal Research at the Helmholtz Centre in Potsdam. These colleagues assisted the development of the book with fruitful discussions over the last two years.

Potsdam, Germany *Ernst Huenges*
December 2009

List of Contributors

Mando G. Blöcher
Helmholtz Centre Potsdam GFZ
German Research Centre for
Geosciences
Reservoir Technologies
Telegrafenberg A6 R. 104
14473 Potsdam
Germany

Wulf Brandt
Helmholtz Centre Potsdam GFZ
German Research Centre for
Geoscience
International Centre for
Geothermal Research
Telegrafenberg
14473 Potsdam
Germany

Dominique Bruel
Ecole des Mines de Paris
Centre de Géosciences
35 rue Saint-Honoré
77300 Fontainebleau
France

David Bruhn
Helmholtz Centre Potsdam GFZ
German Research Centre for
Geoscience
International Centre for
Geothermal Research
Germany

Christoph Clauser
Applied Geophysics and
Geothermal Energy
E.ON Energy Research Center
RWTH Aachen University
Mathieustr. 6,
E.ON ERC Gebäude
52074 Aachen
Germany

Hans-Jörg G. Diersch
WASY Gesellschaft für
wasserwirtschaftliche Planung
und Systemforschung mbH
Walterdorfer Straße 105
12526 Berlin-Bohnsdorf
Germany

Kemal Erbas
Helmholtz Centre Potsdam GFZ
German Research Centre for
Geoscience
International Centre for
Geothermal Research
Telegrafenberg
14473 Potsdam
Germany

Geothermal Energy Systems. Edited by Ernst Huenges

ISBN: 978-3-527-40831-3

James Faulds
University of Nevada
Nevada Bureau of Mines
and Geology
Mackay School of Mines
Reno, NV
USA

Stephanie Frick
Helmholtz Centre Potsdam GFZ
German Research Centre for
Geoscience
International Centre for
Geothermal Research
Telegrafenberg
14473 Potsdam
Germany

Laurent Guillou-Frottier
Bureau de Recherches
Géologiques et Minières (BRGM)
Mineral Resources Division
3 av. C. Guillemin
BP36009
45060 Orléans Cx 2
France

Ernst Huenges
Helmholtz Centre Potsdam GFZ
German Research Centre for
Geoscience
International Centre for
Geothermal Research
Telegrafenberg
14473 Potsdam
Germany

Reiner Jatho
Federal Institute for Geosciences
and Natural Resources (BGR)
Stilleweg 2
30655 Hannover
Germany

Ralf Junker
Leibniz Institute for Applied
Geophysics
Stilleweg 2
30655 Hannover
Germany

Martin Kaltschmitt
Technische Universität
Hamburg-Harburg
Institute for Environmental
Technology and Energy Economic
Eißendorfer Straße 40
21073 Hamburg
Germany

Thomas Kohl
GeoWatt AG
Dohlenweg 28
8050 Zürich
Switzerland

Olaf Kolditz
Helmholtz Centre for
Environmental Research
Department of Environmental
Informatics
TU Dresden, Environmental
Systems Analysis
Permoser Str. 15
04318 Leipzig
Germany

Stefan Kranz
Helmholtz Centre Potsdam GFZ
German Research Centre for
Geoscience
International Centre for
Geothermal Research
Telegrafenberg
14473 Potsdam
Germany

Michael Kühn
Helmholtz Centre Potsdam GFZ
German Research Centre for
Geoscience
International Centre for
Geothermal Research
Telegrafenberg
14473 Potsdam
Germany

Patrick Ledru
AREVA Business Group Mines
KATCO
Av. Dostyk 282
050000 ALMATY
Kazakhstan

Adele Manzella
National Research Council
Institute of Geosciences and
Earth Resources
Pisa
Italy

Chris McDermott
University of Edinburgh
School of GeoSciences
UK

Inga Moeck
Helmholtz Centre Potsdam GFZ
German Research Centre for
Geoscience
International Centre for
Geothermal Research
Telegrafenberg
14473 Potsdam
Germany

Sandrine Portier
Centre de recherche en
géothermie (CREGE)
University of Neuchâtel
Emile-Argand 11, CP 158
2009 Neuchâtel
Switzerland

Simona Regenspurg
Helmholtz Centre Potsdam GFZ
German Research Centre for
Geoscience
International Centre for
Geothermal Research
Telegrafenberg
14473 Potsdam
Germany

Ali Saadat
Helmholtz Centre Potsdam GFZ
German Research Centre for
Geoscience
International Centre for
Geothermal Research
Telegrafenberg
14473 Potsdam
Germany

Gerd Schröder
Leipziger Institut für Energie
GmbH
Torgauer Strape 116
04347 Leipzig
Germany

Thomas Schulte
Helmholtz Centre Potsdam GFZ
German Research Centre for
Geoscience
International Centre for
Geothermal Research
Telegrafenberg
14473 Potsdam
Germany

Axel Sperber
Ing. Büro A. Sperber
Eddesser Straße 1
31234 Edemissen
Germany

Torsten Tischner
Federal Institute for Geosciences and Natural Resources (BGR)
Stilleweg 2
30655 Hannover
Germany

Jan Diederik Van Wees
TNO
Business Unit Geo-Information and Geo-Energy
Princetonlaan 6, Postbus 80015
3508 TA Utrecht
The Netherlands

and

Vrije Universiteit Amsterdam
De Boelelaan 1085
1081 HV Amsterdam
The Netherlands

Francois Vuataz
Centre de recherche en géothermie (CREGE)
University of Neuchâtel
Emile-Argand 11, CP 158
2009 Neuchâtel
Switzerland

Wenqing Wang
Helmholtz Centre for Environmental Research–UFZ
Environmental System Analysis
Germany

Norihiro Watanabe
Helmholtz Centre for Environmental Research–UFZ
Environmental System Analysis
Germany

Günter Zimmermann
Helmholtz Centre Potsdam GFZ
German Research Centre for Geosciences International Centre for Geothermal Research
Telegrafenberg
14473 Potsdam
Germany

1 Reservoir Definition

Patrick Ledru and Laurent Guillou Frottier

1.1 Expressions of Earth's Heat Sources

1.1.1 Introduction to Earth's Heat and Geothermics

Scientific background concerning the heat flow and the geothermal activity of the earth is of fundamental interest. It is established that plate tectonics and activities along plate margins are controlled by thermal processes responsible for density contrasts and changes in rheology. Thus, any attempt to better understand the earth's thermal budget contributes to the knowledge of the global dynamics of the planet. Information on the sources and expressions of heat on earth since its formation can be deduced from combined analyses of seismic studies with mineral physics, chemical composition of primitive materials (chondrites), as well as pressure–temperature–time paths reconstituted from mineralogical assemblages in past and eroded orogens.

Knowledge of heat transfer processes within the earth has greatly improved our understanding of global geodynamics. Variations of surface heat flow above the ocean floor has provided additional evidence for seafloor spreading (Parsons and McKenzie, 1978), and improved theoretical models of heat conduction within oceanic plates or continental crust helped to constrain mantle dynamics (Sclater, Jaupart, and Galson, 1980; Jaupart and Parsons, 1985). When deeper heat transfer processes are considered, thermal convection models explain a number of geophysical and geochemical observations (Schubert, Turcotte, and Olson, 2002). It must be, however, noted that at a smaller scale (closer to the objective of this chapter), say within the few kilometers of the subsurface where water is much more present than at depths, a number of geological and geothermal observations are not well understood. As emphasized by Elder (1981), crustal geothermal systems may appear as liquid- or vapor-dominated systems, where physics of water–rock interactions greatly differs from one case to the other. Actually, as soon as hydrothermal convection arises among the active heat transfer processes, everything goes faster since heat exchanges are more efficient than without circulating water.

Geothermal Energy Systems. Edited by Ernst Huenges

ISBN: 978-3-527-40831-3

It is thus important to delineate which type of heat transfer process is dominant when geothermal applications are considered. Examples of diverse geothermal systems are given below.

Within the continental crust, a given heat source can be maintained for distinct time periods according to the associated geological system. Hydrothermal fields seem to be active within a temporal window around 10^4–10^5 years (Cathles, 1977), whereas a magma reservoir would stay at high temperatures 10–100 times longer (Burov, Jaupart, and Guillou-Frottier, 2003). When radiogenic heat production is considered, half-lives of significant radioactive elements imply timescales up to 10^9 years (Turcotte and Schubert, 2002). At the lower limit, one can also invoke phase changes of specific minerals involving highly exothermic chemical reactions (e.g., sulfide oxidation and serpentinization) producing localized but significant heat excess over a short (10^3–10^4 years) period (Emmanuel and Berkowicz, 2006; Delescluse and Chamot-Rooke, 2008). Thus, description and understanding of all diverse expressions of earth's heat sources involve a large range of physical, chemical, and geological processes that enable the creation of geothermal reservoirs of distinct timescales.

Similarly, one can assign to earth' heat sources either a steady state or a transient nature. High heat producing (HHP) granites (e.g., in Australia, McLaren *et al.*, 2002) can be considered as permanent crustal heat sources, inducing heating of the surrounding rocks over a long time. Consequently, thermal regime around HHP granites exhibits higher temperatures than elsewhere, yielding promising areas for geothermal reservoirs. On the contrary, sedimentary basins where heat is extracted from thin aquifers may be considered as transient geothermal systems since cold water reinjection tends to decrease the exploitable heat potential within a few decades.

Finally, regardless of the studied geological system, and independent of the involved heat transfer mechanism, existence of geothermal systems is first conditioned by thermal regime of the surroundings, and thus by thermal boundary conditions affecting the bulk crust. Consequently, it is worth to understand and assess the whole range of thermal constraints on crustal rocks (physical properties as well as boundary conditions) in order to figure out how different heat transfer mechanisms could lead to generation of geothermal systems.

The following subsections present some generalities on earth' heat sources and losses in order to constrain thermal boundary conditions and thermal processes that prevail within the crust. Once crustal geotherms are physically constrained by the latter and by rock thermal properties, distinct causes for the genesis of thermal anomalies are discussed.

1.1.2
Cooling of the Core, Radiogenic Heat Production, and Mantle Cooling

The earth's core releases heat at the base of the mantle, through distinct mechanisms. Inner-core crystallization, secular cooling of the core, chemical separation of the inner core, and possibly radiogenic heat generation within the core yield estimates of core heat loss ranging from 4 to 12 TW (Jaupart, Labrosse, and Mareschal, 2007). Precise determinations of ohmic dissipation and radiogenic heat production should improve this estimate. Independent studies based on core–mantle interactions tend to favor large values (Labrosse, 2002), while according to Roberts, Jones, and Calderwood (2003), ohmic dissipation in the earth's core would involve between 5 and 10 TW of heat loss across the core–mantle boundary. The averaged value of 8 TW (Jaupart, Labrosse, and Mareschal, 2007) is proposed in Figure 1.1.

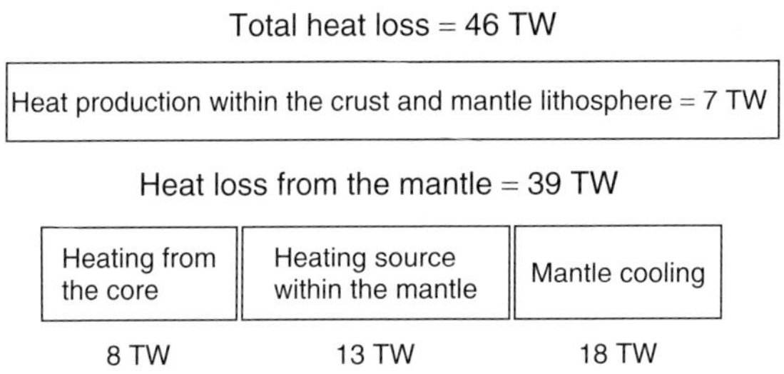

Figure 1.1 Heat sources and losses in the earth's core and mantle. (After Jaupart, Labrosse, and Mareschal, 2007.)

The earth's mantle releases heat at the base of the crust. Radiogenic heat production can be estimated through chemical analyses of either meteorites, considered as the starting material, or samples of present-day mantle rocks. Different methods have been used; the objective being to determine uranium, thorium, and potassium concentrations. Applying radioactive decay constants for these elements, the total rate of heat production for the bulk silicate earth (thus including the continental crust) equals 20 TW, among which 7 TW comes from the continental crust. Thus, heat production within the mantle amounts to 13 TW (Figure 1.1, Jaupart, Labrosse, and Mareschal, 2007). Since total heat loss from the mantle is larger than heat input from the core and heat generation within it, the remaining heat content stands for mantle cooling through earth's history.

Mantle cooling corresponds to the difference between total heat loss from the mantle (39 TW) and heat input (from the core, 8 TW) plus internal generation (13 TW). This 18 TW difference can be converted into an averaged mantle cooling of 120 $^{\circ}$C Gy^{-1}, but over long timescales, geological constraints favor lower values of about 50 $^{\circ}$C Gy^{-1}. Knowledge of the cooling rate enables one to draw a more accurate radial temperature profile through the earth (Jaupart, Labrosse, and Mareschal, 2007). However, as it is shown below, precise temperature profile within the deep earth does not necessarily constrain shallow temperature profiles within the continental crust.

1.1.3
Mantle Convection and Heat Loss beneath the Lithosphere

Heat from the mantle is released through the overlying lithosphere. Spatially averaged heat flow data over oceans and continents show a strong discrepancy between oceanic and continental mantle heat losses. Among the 46 TW of total heat loss, only 14 TW is released over continents. In terms of heat losses, two major differences between continental and oceanic lithospheres must be explained. First, oceanic lithosphere can be considered as a thermal boundary layer of the convective mantle since it does participate in convective motions. Actually, oceanic heat flow data show a similar decrease from mid-oceanic ridges to old subducting lithosphere as that deduced from theoretical heat flow variation from upwelling- to downwelling parts of a convecting system (Parsons and Sclater, 1977). Second, heat production within the oceanic lithosphere is negligible when compared to that of the continental lithosphere, enriched in radioactive elements. It follows that the oceanic lithosphere can be considered as a "thermally inactive" upper boundary layer of the convective mantle. In other words, the appropriate thermal boundary condition at the top of the oceanic mantle corresponds to a fixed temperature condition, which is indeed imposed by oceanic water.

Contrary to oceanic lithosphere, continental lithosphere is not directly subducted by mantle downwellings and behaves as a floating body of finite thermal conductivity overlying a convective system (Elder, 1967; Whitehead, 1976; Gurnis, 1988; Lenardic and Kaula, 1995; Guillou and Jaupart, 1995; Jaupart *et al.*, 1998; Grigné and Labrosse, 2001; Trubitsyn *et al.*, 2006). Even if atmospheric temperature can be considered as a fixed temperature condition at the top of continents, it does not apply to their bottom parts (i.e., at the subcontinental lithosphere–asthenosphere boundary) since heat production within continents create temperature differences at depths. Depending on crustal composition, heat production rates can vary from one continental province to the other, and lateral temperature variations at the conducting lithosphere–convecting asthenosphere boundary are thus expected. It follows that thermal boundary condition at the base of the continental lithosphere may be difficult to infer since thermal regime of continents differs from one case to the other. However, as it is suggested below, some large-scale trends in thermal behavior of continental masses can be drawn and thus a subcontinental thermal boundary condition may be inferred.

1.1.3.1 Mantle Heat Flow Variations

Since radiogenic heat production is negligible in oceanic lithosphere, heat flow through the ocean floor corresponds to mantle heat flow at the bottom of the oceanic lithosphere. This suboceanic heat flow varies from several hundreds of milliwatts per square meter at mid-oceanic ridges to about 50 mW m^{-2} over oceanic lithosphere older than 80 Myr (Lister *et al.*, 1990). When thermal effects of hydrothermal circulation are removed, this variation is well explained by the cooling plate model.

Beneath continents, mantle heat flow variations do not follow such simple physical consideration since large contrasts exist for both crustal heat production and lithospheric thickness. However, at the scale of the mantle, heat loss is mainly sensitive to large-scale thermal boundary conditions at the top of the convecting system, and not to the detailed thermal structures of the overlying lithospheres. Beneath continents, the earth's mantle is not constrained by a fixed temperature condition as is the case beneath oceanic lithosphere (see above), and thus large-scale temperature and heat flow variations are expected at the top surface of the subcontinental convecting system.

Surface heat flow measurements over continents and estimates of associated heat production rates have shown that mantle heat flow values beneath thermally stable (older than about 500 Myr) continental areas would be low, around $15 \pm 3\,\mathrm{mW\,m^{-2}}$ (Pinet *et al.*, 1991; Guillou *et al.*, 1994; Kukkonen and Peltonen, 1999; Mareschal *et al.*, 2000). On the contrary, mantle heat flow would be significantly enhanced beneath continental margins (Goutorbe, Lucazeau, and Bonneville, 2007; Lucazeau *et al.*, 2008) where crustal thickness and heat production rates decrease. Old central parts of continents would be associated with a low subcontinental mantle heat flow while younger continental edges would receive more heat from the mantle. The so-called "insulating effect" of continents is described here in terms of heat transfer from the mantle to the upper surface, where most of mantle heat flow is laterally evacuated toward continental margins and oceanic lithosphere. The term *insulating* should in fact be replaced by *blanketing* since thermal conductivity values of continental rocks are not lower than that of oceanic rocks (Clauser and Huenges, 1995).

1.1.3.2 **Subcontinental Thermal Boundary Condition**

A fixed temperature condition applies to the top of oceanic lithosphere while a low subcontinental heat flow is inferred from surface heat flow data over stable continental areas. As shown by laboratory experiments, this low mantle heat flow beneath continents cannot be sustained if continental size is small (Guillou and Jaupart, 1995). Indeed, a constant and low heat flux settles beneath a continental area for continental sizes larger than two mantle thicknesses. For smaller sizes, subcontinental heat flow is increased.

In the field, it was shown that mantle heat flow beneath stable continents may be as low as $10\,\mathrm{mW\,m^{-2}}$ (Guillou-Frottier *et al.*, 1995), whereas beneath continental margins, values around $50\,\mathrm{mW\,m^{-2}}$ have been proposed (Goutorbe, Lucazeau, and Bonneville, 2007; Lucazeau *et al.*, 2008). Beneath young perturbed areas, similar elevated values have been suggested, such as the mantle heat flow estimate of $60–70\,\mathrm{mW\,m^{-2}}$ beneath the French Massif Central (FMC) (Lucazeau, Vasseur, and Bayer, 1984).

At large scale, one may infer a continuous increase of mantle heat flow from continental centers to continental margins, but laboratory and numerical simulations of thermal interaction between a convecting mantle and an overlying conducting continent have shown that the mantle heat flow increase is mainly focused on

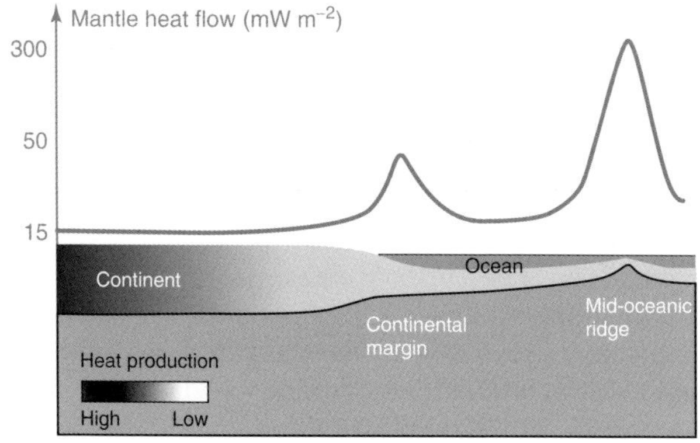

Figure 1.2 Sketch of mantle heat flow variations from continental center to mid-oceanic ridge, emphasizing a low subcontinental heat flow with a localized increase at continental margin, corresponding to a lateral decrease in crustal heat production.

continental margin areas (Lenardic *et al.*, 2000). In other words, the low and constant heat flow beneath the continent can be considered as the dominant large-scale thermal boundary condition applying above the subcontinental mantle (Figure 1.2).

1.1.4
Fourier' Law and Crustal Geotherms

Heat transfer within the continental crust occurs mainly through heat conduction. Heat advection may occur during magmatism episodes (arrival of hot magma at shallow depths enhancing local temperatures), intense erosion episodes (uplift of isotherms), and periods of hydrothermal convection. All these phenomena can be considered as short-lived processes when equilibrium thermal regime of the crust is considered. In steady state and without advective processes, the simplest form of Fourier law, with a constant thermal conductivity, a depth-dependent temperature field, and with appropriate boundary conditions for continental crust, can be written as

$$\begin{cases} k\left(\dfrac{\mathrm{d}^2 T}{\mathrm{d}z^2}\right) + A = 0 \\ T(z=0) = T_0 \\ k\dfrac{\mathrm{d}T}{\mathrm{d}z}(z=h) = Q_\mathrm{m} \end{cases} \tag{1.1}$$

where k is the crustal thermal conductivity, A heat production, T_0 surface temperature, h the thickness of the crust, and Q_m the mantle heat flow. Temperature profile within the crust thus can be written as

$$T(z) = \frac{-A}{2k}z^2 + \frac{(Q_\mathrm{m} + Ah)}{k}z + T_0 \tag{1.2}$$

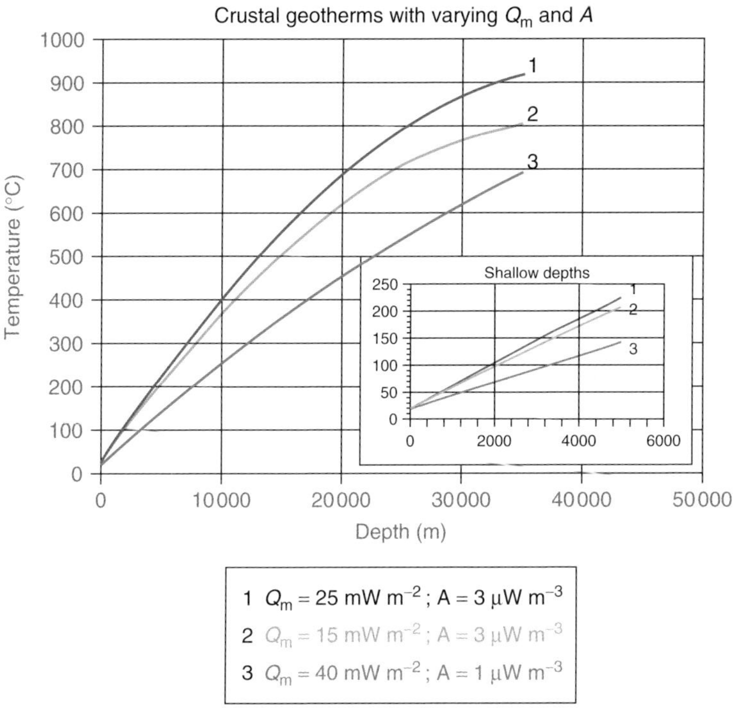

Figure 1.3 Synthetic simple temperature profiles as inferred from Equation 1.2, where mantle heat flow and bulk crustal heat production are varied.

This kind of geotherms show parabolic profiles where the curvature is controlled by A/k value. Temperature difference at depth greatly depends on both A and Q_m values. Figure 1.3 shows three crustal geotherms for a 35-km-thick crust, where surface temperature equals 20 °C, with an averaged thermal conductivity of 3 W m^{-1} K^{-1} and with different (A, Q_m) values.

As emphasized by curve 3 in Figure 1.3, a high mantle heat flow does not necessarily involve high crustal temperatures. Curvature of geotherms is indeed controlled by bulk heat production of the crust, as shown by curves 1 and 2. However, this curvature is not visible at shallow depths since the fixed temperature condition at the surface forces linear variation. These simple examples demonstrate that construction of crustal temperature profiles is strongly dependent on both estimates of mantle heat flow and of bulk crustal heat production.

At shallow depths, temperature anomalies may thus be due to anomalous HHP rocks. Likewise, other lateral heat transfer effects such as those due to thermal conductivity contrasts may lead to strong temperature differences at a given depth. In the following sections, a series of synthetic temperature profiles are built and discussed based on the geological examples.

1.1.5 Two-dimensional Effects of Crustal Heterogeneities on Temperature Profiles

1.1.5.1 Steady-state Heat Refraction

The two-dimensional heterogeneity of the upper crust is outlined by geological maps, where, for example, each rock composition is assigned one color. However, it must be emphasized that thermal properties are not necessarily correlated with rock composition, except for extreme cases (Clauser and Huenges, 1995). On one hand, one may record similar temperature profiles through distinct areas where small-scale lithological differences are observed, because the averaging effect of heterogeneities smoothes out small-scale variations. On the other hand, when large bodies with significantly distinct thermal properties are present, temperature profiles may differ by several tens of degrees at shallow depths. In other words, the horizontal geometry of anomalous bodies shall play a significant role in the establishment of temperature differences at depth.

As far as surface heat flow is concerned, small-scale lithological contrast may create large differences. For example, subvertical mineralized bodies can be rich in highly conducting minerals (e.g., volcanic massive sulfides deposits, Mwenifumbo, 1993), which may result in large surface heat flow variations, whereas

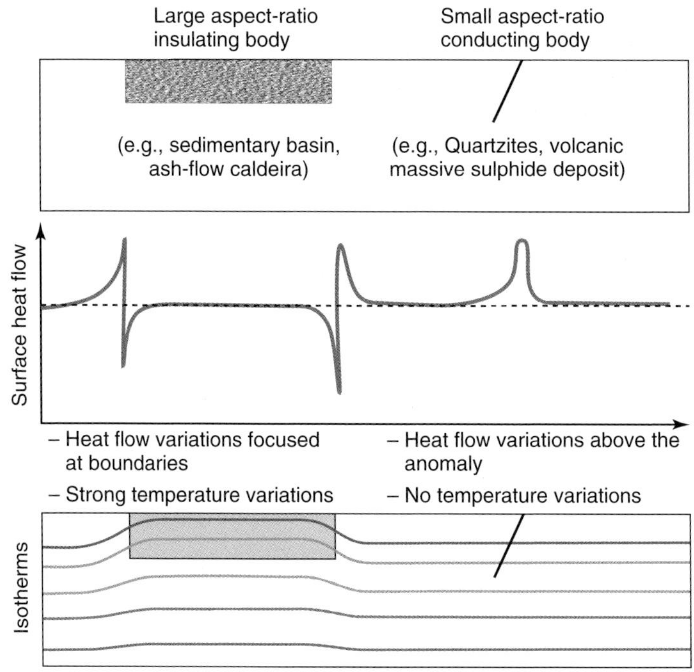

Figure 1.4 Heat refraction in two dimensions, leading to opposite effects according to the conductivity contrast or the anomaly geometry.

differences in subsurface temperatures may be negligible. These subtle effects are illustrated in Figure 1.4, where two scenarios of heat refraction effects are illustrated. The objective here is to show that high surface heat flow variations do not necessarily correlate with large subsurface temperature differences since geometrical effects have to be accounted for. Indeed, above a large aspect-ratio insulating body, isotherms are uplifted so that surface heat flow above the anomaly center corresponds to the equilibrium one. On the contrary, isotherms cannot be distorted for a small aspect-ratio conducting body but the resulting surface heat flow is enhanced.

In sedimentary basins, presence of salt may also induce heat refraction effects since thermal conductivity of halite may be four times greater than surrounding sediments (e.g., 1.5 W m^{-1} K^{-1} for sediments and around 6–7 W m^{-1} K^{-1} for rock salt and halite, according to Clauser, 2006). Consequently, temperature gradient within a thick evaporitic layer may thus be decreased by a factor of 4, leading to a cooling effect of several tens of degrees centigrade for a 2–3-km-thick layer.

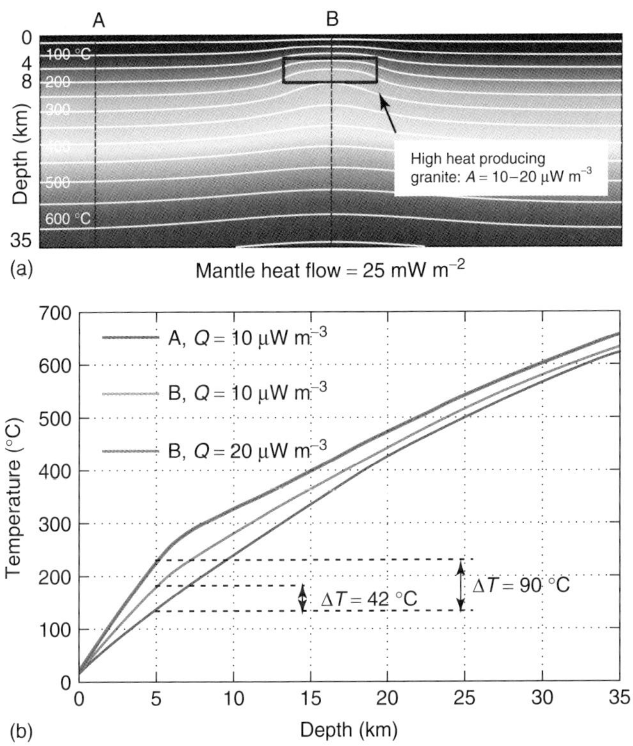

Figure 1.5 Two-dimensional effect of a high heat producing granite on temperature field (a) and geotherms (b). Here, a fixed mantle heat flow of 25 mW m^{-2} is imposed, as well as an averaged thermal conductivity of 3 W m^{-1} K^{-1} and a bulk crustal heat production of 1 μW m^{-3}.

Apart from thermal conductivity contrasts, heat production rates may also vary by a factor of 10 or more between two lithologies (Sandiford, McLaren, and Neumann, 2002; McLaren *et al.*, 2002). In the case of HHP granites, radiogenic content is so high that heat production rates may reach 10–20 $\mu W\ m^{-3}$, as it is the case of the synthetic example of Figure 1.5, where embeddings have an averaged heat production rate of 1 $\mu W\ m^{-3}$. At 5 km depth, a temperature difference of 42 °C (90 °C) is obtained for a high heat production of 10 (20) $\mu W\ m^{-3}$.

In the case of Figure 1.5, the obtained temperature anomaly depends on several other parameters such as the emplacement depth of the anomalous body. For example, same granite of Figure 1.5 emplaced at 10 km depth would involve a 30 °C anomaly at 5 km depth. This temperature difference also corresponds to the case of a shallow emplacement of 500 m below the surface. This nonobvious result can be explained by detailed analysis of geotherm curvature, as presented by Sandiford, Fredericksen, and Braun (2003).

1.1.5.2 **Transient Effects**

A number of studies have demonstrated the role of transient geological processes on crustal temperatures. Large-scale tectonic processes (thrusting events, erosion, and sedimentation) can result in temperature differences reaching several tens of degrees centigrade at a few kilometers depth (England and Thompson, 1984; Ruppel and Hodges, 1994). Magma emplacement or presence of hydrothermal convection at shallow depths may also explain disturbed temperature profiles (Cathles, 1977; Norton and Hulen, 2001).

Because thermal diffusivity of rocks is low, transient thermal evolution of rocks undergoing conducting processes is very slow, and return to equilibrium temperatures may last several tens to hundreds of million years. Figure 1.6 illustrates some examples of large-scale thermal evolution of the crust undergoing tectonic events. One may note that in the case of a thrusting event, the equilibrium thermal field (with a maximum temperature of 820 °C) is reached 120 Myr after the onset of thrusting. On the contrary, when convective processes are involved around intrusive bodies, heat transfer mechanisms through fluid circulation are accelerated, and typical timescales are lower than 1 Myr (Cathles, 1977). When smaller scale systems are considered, thermal equilibrium is reached faster. For example, serpentinization of oceanic crust may result in large amplitude thermal signatures lasting less than a few thousands of years (Emmanuel and Berkowicz, 2006).

1.1.5.3 **Role of Anisotropy of Thermal Conductivity**

Apart from steady-state heat refraction due to thermal conductivity contrasts or variations in heat production rates, other subtle effects affecting thermal properties may trigger thermal anomalies. Temperature dependence of thermal conductivity is one example, as shown in Clauser and Huenges (1995). In the case of sedimentary basins, porosity dependence of thermal conductivity is also significant, as shown in several studies (Beziat, Dardaine, and Gabis, 1988; Waples and Tirsgaard, 2002). Sedimentary basins correspond to interesting geothermal targets all the more that numerous temperature measurements may be available. When thick clayey

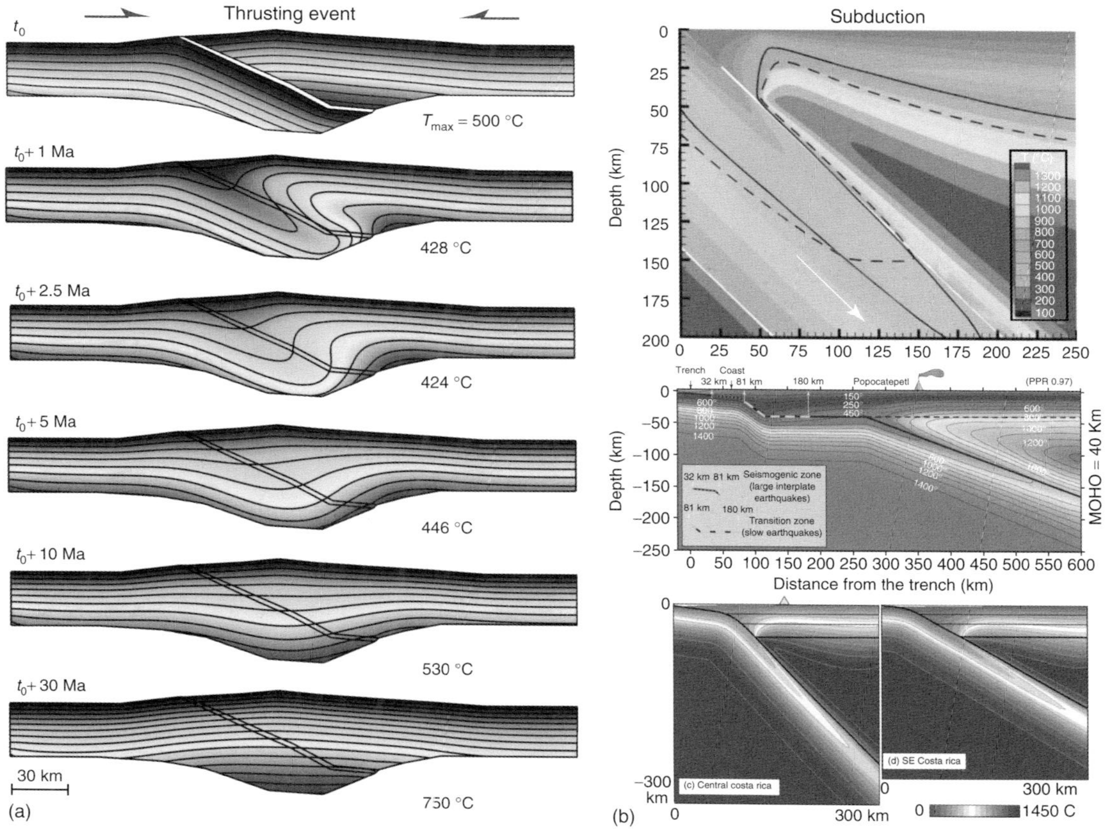

Figure 1.6 Examples of large-scale transient and steady-state thermal perturbations: (a) thrusting event resulting in a thickened and more radiogenic crust and (b) distinct models of thermal fields around subduction zones, where slab dip angle and plate velocities differ from one case to the other. (After, from top to bottom, Cagnioncle, Parmentier, and Elkins-Tanton, 2007; Manea *et al.*, 2004; Peacock *et al.*, 2005.) (Please find a color version of this figure on the color plates.)

formations are present in a basin, the role of compaction has to be accounted for since porosity decreases with compaction pressure and particles' orientation becomes horizontal with increasing pressure (Vasseur, Brigaud, and Demongodin, 1995). Both effects together with temperature-dependence effect induce important changes in thermal conductivity. First, the decreasing porosity (and thus amount of water) with depth tends to increase thermal conductivity, while temperature dependence tends to decrease it (see Harcouët *et al.*, 2007 for details). Second, the horizontal orientation of individual clay particles develops anisotropy, favoring lateral heat transfer and hindering vertical heat flow.

An example of the effect of thermal conductivity anisotropy on thermal field is illustrated in Figure 1.7, where the Paris basin is modeled according to Demongodin *et al.* study (1991)). Anisotropy ratio is increased with depth and thermal boundary conditions enable to reproduce measured surface heat flow values. Figure 1.8b shows horizontal temperature profiles at 1500 m depth, with and without anisotropy effect. When anisotropy is accounted for, heat accumulates more efficiently within the basin and a 20 °C difference with the isotropic case is reached at basin boundaries. Obviously, the importance of the anomaly critically depends on thermal conductivity values and anisotropy ratios. Measurements on representative core samples, and scaling with *in situ* conditions are thus of major importance when thermal modeling of a sedimentary basin is performed (Gallagher *et al.*, 1997).

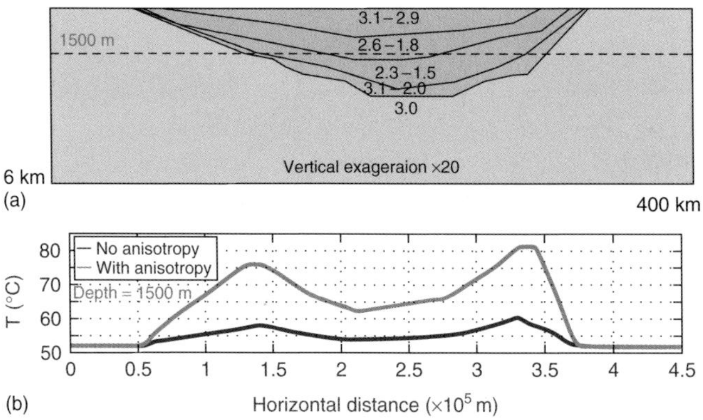

Figure 1.7 (a) Chosen model for the Paris basin (after Demongodin *et al.*, 1991), with thermal conductivity values indicated as follows: "horizontal component–vertical component." When anisotropy is not accounted for, the first value is considered as homogeneous. (b) Horizontal temperature profiles at 1500 m depth across the basin (see text).

1.1.6
Fluid Circulation and Associated Thermal Anomalies

In previous sections, heat transfer was described by pure conduction, where no heat transfer by fluid motion could occur. However, shallow geological systems are sometimes characterized by sufficiently porous layers (sedimentary units) or highly permeable areas (fault zones) in which crustal fluids may freely circulate. Depending on fluid velocity, permeability, or thickness of the porous layer, heat from several kilometers depth may be entrained by fluid circulation and thus create temperature anomalies.

Several studies have demonstrated the possibility to detect fluid motion by temperature measurements within boreholes (Drury, Jessop, and Lewis, 1984; Pribnow and Schellschmidt, 2000). Small-scale water flows through a fracture crossing the borehole may disturb locally the measured geotherm by a few degrees centigrade (Vasseur *et al.*, 1991), and large-scale fluid circulation (convective flows) may lead to cooling or warming effects exceeding tens of degrees centigrade (Lopez and Smith, 1995; Bächler, Kohl, and Rybach, 2003; Wisian and Blackwell, 2004). In some cases, the measured temperature anomalies cannot be explained by purely conductive processes and one must account for free convection in highly permeable fault zones (Rühaak, 2009; Garibaldi *et al.*, 2010).

1.1.7
Summary

Crustal temperatures are controlled by thermal boundary conditions and thermal properties of rocks. In the pure conductive regime, knowledge of mantle heat flow and crustal heat production enables to determine a probable averaged geotherm, but the natural heterogeneity of crustal composition may lead to local variations reaching several tens of degrees centigrade at a few kilometers depth. When available thermal data are used to infer deep temperatures (as it is the case in the next section), similar uncertainties can be assigned to extrapolated data.

Despite the fact that crustal temperatures are not easy to estimate, it is shown in Section 1.1 that models of geothermal reservoirs depend on several other parameters, which may be less constrained than thermal properties. In particular, the presence of fluids, which is important in the development of geothermal energy, may completely distort temperature field as soon as rock permeability is high enough (Manning and Ingebritsen, 1999). Within sedimentary basins, permeability can vary by approximately four orders of magnitude, thus allowing or preventing fluid circulation. The detailed knowledge of temperature field in an area is probably not sufficient to characterize a geothermal reservoir.

Before defining the concept of geothermal reservoir, heat flow data from Europe are reviewed and presented. The objective of this second section is to illustrate how surface heat flow and deep temperatures are not necessarily correlated, and how significant errors in deep temperature estimates can be made when shallow measurements are extrapolated at depth.

1.2
Heat Flow and Deep Temperatures in Europe

Independent of numerical modeling of heat transfer within geological systems, the best way to search for thermal anomalies in the shallow crust consists first in compiling available thermal data that correspond to the boundary conditions, as well as petrophysical parameters controlling heat transfer. Surface temperature is well known, but may however show significant spatial variations as detailed below. The few direct measurements of surface heat flow in Europe are also shown together with temperature gradients and thermal conductivity values. At mantle depths, indirect evidence for temperature variations in Europe has been evidenced.

1.2.1 Far-field Conditions

In order to constrain thermal regime of the shallow crust, one need to constrain far-field thermal boundary conditions, say at the surface and at the base of the lithosphere. Even if spatial distribution of heat producing elements within the crust is of major importance, it is necessary to estimate the amount of heat supplied at the base of the lithosphere (which is also the heat supplied at the base of the crust) and that lost at the surface.

At the surface, the ground surface temperature has been measured for centuries and can be considered as constant over length scales of several hundreds of kilometers (Hansen and Lebedeff, 1987). In Europe, ground surface temperature increases from north to south France ($\sim$1000 km) by about 5 °C, and by $\sim$10 °C from Denmark to south Italy (Figure 1.8) separated by a distance of 2000 km (Haenel *et al.*, 1980). If thermal regime of the crust is to be studied, such large-scale variations can thus be neglected. Apart from the effect of latitude, ground surface temperature can be locally disturbed by surface heterogeneities such as topography (Blackwell, Steele, and Brott, 1980) or the presence of lakes. These permanent disturbances should be theoretically considered when subsurface temperatures are studied, especially if representative length scale of surface features compare with the studied depths (e.g., warm water outflows in tunnels of Switzerland, Sonney and Vuataz, 2008). Transient changes in surface conditions such as those induced by forest fires may also affect subsurface temperatures but only for a short period. Long-period surface temperature changes such as climatic warming or cooling periods affect underground temperatures as it can be deciphered through measured temperature profiles (Guillou-Frottier, Mareschal, and Musset, 1998), but associated thermal disturbances are damped with depth, and basically cancelled at several hundreds of meters.

Contrary to the upper surface, there is no reason to consider the base of the crust as an isotherm. Seismic tomography studies have indicated that this is indeed not the case. Even if seismic velocities vary with temperature and composition, Goes *et al.* (2000) suggested that the inferred variations at 100 km depth revealed temperature differences (Figure 1.8). At shallower depths (Moho depth +20 km), Figure 1.9 shows possible large-scale temperature differences as deduced from the shear velocity model of Shapiro and Ritzwoller (2002). Local studies of seismic tomography also suggested anomalous hot zones at the base of the European crust, such as beneath the FMC and beneath the Eifel area in Germany (Granet, Wilson, and Achauer, 1995; Ritter *et al.*, 2001). These anomalously hot zones and

Figure 1.8 Mean annual surface (air) temperature (in degrees centigrade) in Western Europe as published in Haenel *et al.* (1980), after a Climatic Atlas published in 1970 by UNESCO.

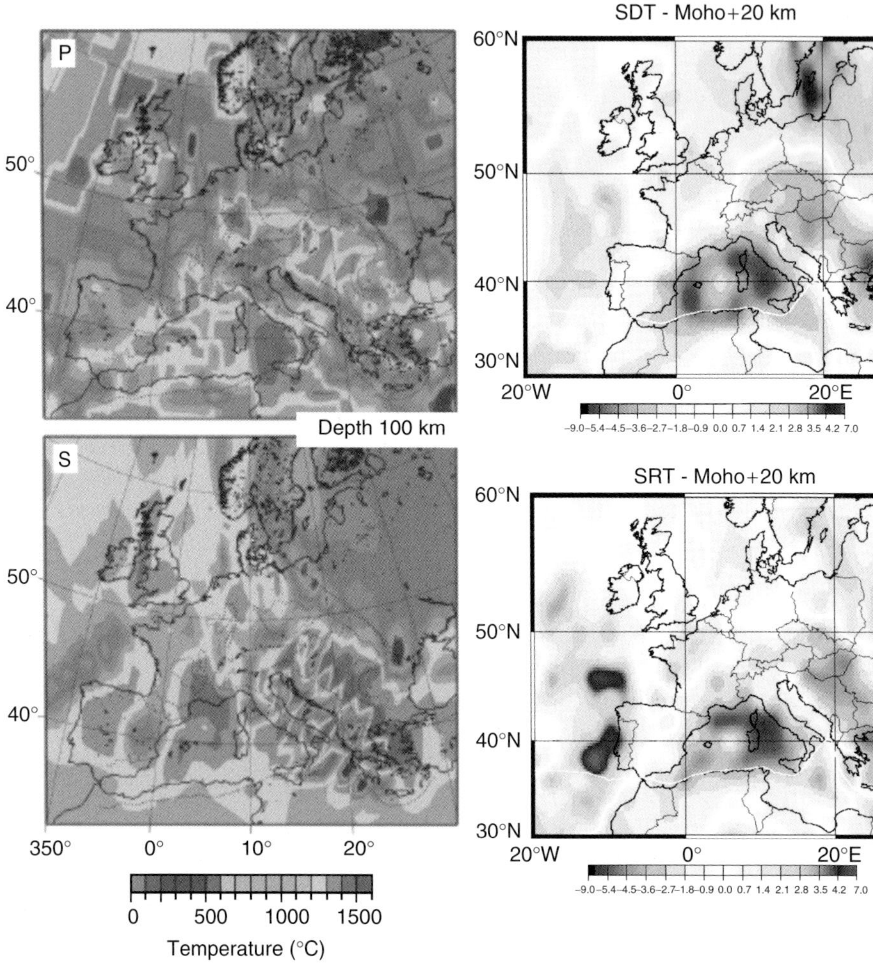

Figure 1.9 Left: Temperatures at 100 km depth estimated from the P and S velocity anomalies. (After Goes *et al.*, 2000.) Right: Tomographic models extracted from an upper mantle shear velocity model (Shapiro and Ritzwoller, 2002); top: diffraction tomography, bottom: ray tomography. (Please find a color version of this figure on the color plates.)

their surface signatures would be associated with local mantle upwellings (Goes, Spakman, and Bijwaard, 1999; Guillou-Frottier *et al.*, 2007), thus reinforcing the possible increase in underlying heat flow. For the FMC area, Lucazeau, Vasseur, and Bayer (1984) used distinct geophysical data to build a thermal model and concluded that an additional heat flow contribution from the mantle of 25–30 mW m^{-2} can explain surface heat flow data. While a mantle heat flow of 40 mW m^{-2} is present in the vicinity of the FMC, it would locally reach 70 mW m^{-2} beneath parts of the FMC where a thin crust is seismically detected.

1.2.2
Thermal Conductivity, Temperature Gradient, and Heat Flow Density in Europe

The global heat flow database (International Heat Flow Commission, IHFC; http://www.heatflow.und.edu) contains almost all published heat flow measurements that were available at the time of its publication (Cermak, 1993; Pollack, Hurter, and Johnson, 1993). Each heat flow data is provided with supplementary information such as thermal conductivity, heat production rates, and temperature gradients that were used to estimate surface heat flow. However, there is no information on data quality, which depends on several independent factors (precision of measurements, depth of boreholes, stability of temperature gradient, etc.). In order to improve the IHFC database quality, the few new published thermal data in Europe have been added (Cermak *et al.*, 1996; Nemcock *et al.*, 1998; Demetrescu and Andreescu, 1994; Aydin, Karat, and Kocak, 2005), related data have been clustered, and a quality criterion has been applied. In addition, numerous anomalous values have been removed (e.g., those lower than 25 mW m^{-2}). Because the quality criterion accounts for the number of individual boreholes used, data close to each other (separated by less than 15 km) have been affected a single mean value. The quality criterion accounts for (i) the number of individual boreholes used for heat flow estimate, (ii) the standard deviation of the estimate (s.d. in Table 1.1); (iii) the minimal depth where temperature measurements are accounted for; and (iv) the depth range where estimate is performed. These last two criteria enable to retain only stable and undisturbed temperature profiles for heat flow estimates. High, medium, and low quality criteria are detailed in Table 1.1.

This process of data treatment provided 1643 heat flow data, whereas 3520 original data were present in the IHFC database. More than 1000 data from Russia had to be removed or were clustered with neighboring ones. In Austria, the hundreds of data presented by Nemcock *et al.* (1998) decreased to 36 of quality 3. Numerous heat flow data in the IHFC database are deduced from individual

Table 1.1 Quality criteria assigned to heat flow data shown in Figure 1.10.

Site characteristics	Quality[a]
Several boreholes, good s.d. (<10%), depth range of estimate >200 m	1
Several boreholes, good s.d. (<10%), unknown or shallow (<200 m) depth range	2
Several boreholes, medium s.d. (between 10 and 30%), depth range >200 m	2
Several boreholes, medium s.d. (between 10 and 30%), unknown or shallow (<200 m) depth range	3
Several boreholes, bad s.d. (>30%)	3
One borehole, depth range >200 m	2
One borehole, depth range <200 m	3

[a]When one major information is missing (temperature gradient or thermal conductivity), then quality is decreased by one unit.

measurements where temperature gradient or thermal conductivity is not indicated, thus involving removal of the data. At last, a total of 257 data of high quality, 869 of medium quality, and 517 of low quality were obtained. Figure 1.10 illustrates the newly obtained heat flow map together with thermal conductivity and temperature gradient values.

Figure 1.10 does not present any contours of thermal data, which are simply gathered according to their range of values. Indeed, as previously indicated, there are many reasons to explain short-scale variations in temperature gradient and heat flow data. It is thus somehow dangerous to assign a given thermal regime in a geological area since even lithological changes may lead to significant local variations. In addition, one can see that thermal conductivity data appear surprisingly constant in some areas (Ukraine, Belorussia), revealing that the same value was probably assumed for tens of boreholes. These maps also indicate that heat flow values are sometimes estimated with no thermal conductivity data (e.g., Spain).

1.2.3
Calculating Extrapolated Temperature at Depth

Temperature measurements in mining or petroleum boreholes were used in the last decades to construct temperature maps at different depth levels (Haenel *et al.*, 1980; Hurtig *et al.*, 1992). Bottomhole temperature (BHT) measurements in petroleum boreholes are not necessarily representative of the equilibrium temperatures, and some corrections are needed (Goutorbe, Lucazeau, and Bonneville, 2007). However, because of the lack of information, a statistical method is often used to infer possible equilibrium temperatures. When temporal history of BHT measurements is well documented, appropriate corrections for transient disturbances can yield temperature estimates at a few kilometers depth with uncertainties of $\pm 10\,^\circ$C (Bonté *et al.*, 2010).

In the last decade, the extrapolation of European temperatures to 5 km depth was performed by petroleum industry using such BHT measurements (which are unavailable), and where any uncertainty propagates and increases with depth through a linear extrapolation. This map was reviewed and analyzed by Genter *et al.* (2003). Using the deepest equilibrium temperature gradients inferred from measurements in mining boreholes (present in the IHFC database) and from recent studies (Fernàndez *et al.*, 1998), the authors performed a critical analysis of the temperature map presented by Hurtig *et al.* (1992), which was then modified by the "heat mining Economic Interest European Group" but only available as an unpublished map. Results of color validation are shown in Figure 1.11. It must be emphasized that the objective of this work was simply to use available thermal data to check the interest of some areas which was deduced from confidential data. When one color code is not confirmed, a temperature difference greater than 20 °C is obtained. When it is partly confirmed, it means that extrapolations are coherent for only a restricted area. The analysis

(a)
(b)
(c)

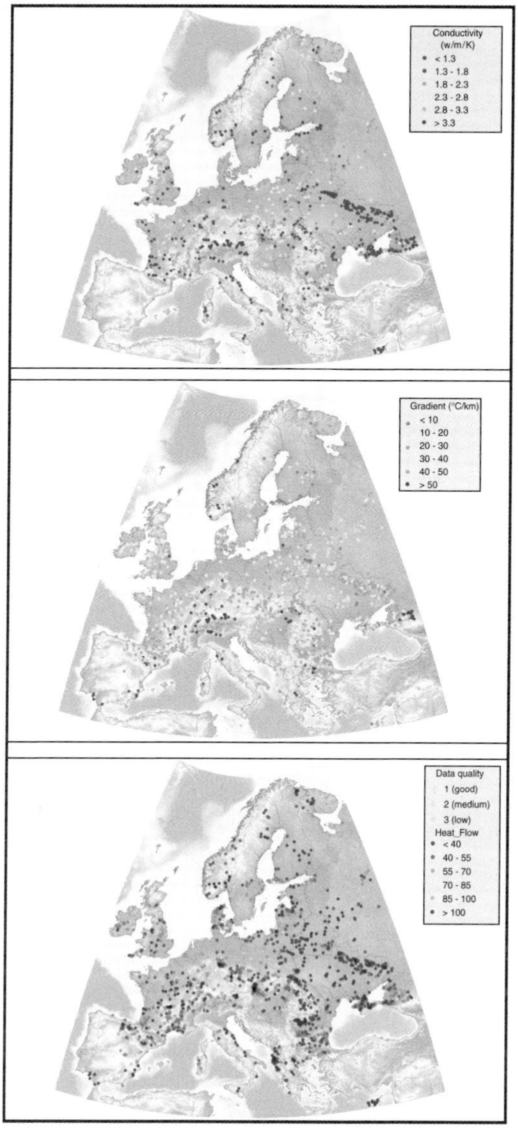

Figure 1.10 Thermal conductivity (a), temperature gradient (b), and heat flow data (c) as compiled from this study. Each color is assigned a range of values, and for heat flow data, a quality criterion is added (see text). (Please find a color version of this figure on the color plates.)

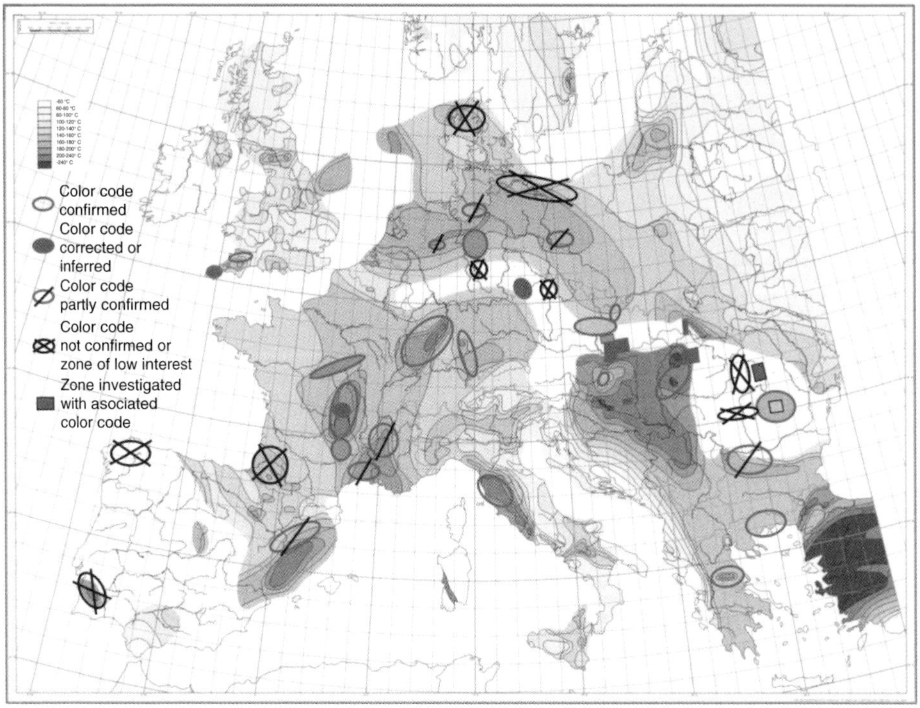

Figure 1.11 Map of temperature at 5 km depth, as inferred from unavailable (confidential) BHT measurements (Hurtig *et al.*, 1992; EIEG, 2000) and critical analysis by Genter *et al.* (2003) from published thermal data (see text). (Please find a color version of this figure on the color plates.)

was not exhaustive but it shows that differences of several tens of degrees centigrade at 5 km depth may be easily reached when two extrapolation methods are investigated.

1.2.4 Summary

These different data sets were used for a predictive survey to evaluate potential zones of high heat flow where enhanced geothermal systems could be experimented. This approach takes only the thermal aspect of the geothermal systems into account without any geological a priori. At the scale of Europe (Figure 1.11), it reveals large wavelength positive anomalies in Italy, Central-Eastern Europe, and Turkey, which correspond to well-known geothermal systems located in extensional settings within active geodynamic systems and to which Iceland could be associated although it is not represented on the map.

In Italy, since Miocene, the Northern Apennine fold belt has been progressively thinned, heated, and intruded by mafic magmas. In Tuscany, this evolution is the source of a granitic complex that has been emplaced between 3.8 and 1.3 Ma. A long-lived hydrothermal activity is recorded in this area by both fossil (Plio-Quaternary ore deposits) and active (Larderello geothermal field) systems (Dini *et al.*, 2004). In Central-Eastern Europe, the Pannonian basin is characterized since Middle Miocene by an upwelling of the asthenosphere and thinning of the lithosphere, responsible for coeval rifting in the basin and compression in the flanking Carpathian and Dinaric belts (Huismans, Podladchikov, and Cloetingh, 2001). In Turkey, the collision between the Arabian and Eurasion plates has induced the westward escape of the Anatolian block, which is accommodated by the right-lateral movement of the Anatolian fault network. Much of the geothermal activity appears to be focused along kinematically linked normal and strike-slip fault systems most commonly within E-W-trending grabens (Sengör, Gorur, and Saroglu, 1985; Ercan, 2002).

Besides these active tectonic zones, other positive anomalies are mainly distributed along a series of intracontinental grabens that cut the western European platform, corresponding to the west European rift system (Dèzes, Schmid, and Ziegler, 2004). These rift structures, the upper Rhine graben, the Limagne system, the Rhône valley, a part of Provence, the Catalonia, and the Eger grabens, were created in the Oligocene as a result of the thinning of the continental crust. Among these structures, the Rhine graben has been intensively studied over the last 10 years for its potential. It is about 300 km long, with an average width of 40 km, limited by large-scale normal faults. The post-Paleozoic sediments of the western European platform have overlain the Hercynian basement, which is made of granite, granodiorite, or other related basement rocks (Edel and Fluck, 1989). This area – characterized by a thin continental crust and a Moho at 25 km depth – shows a Tertiary volcanism that occurred in the form of isolated volcanoes of alkaline composition related to a mantle magmatic activity (Wenzel and Brun, 1991).

This preliminary analysis shows that the thermal aspect of the geothermal systems is directly linked and controlled by the past and present geodynamic context. This framework provides a first-order constrain on the location of favorable and unfavorable geodynamic sites for the exploration of potential geothermal reservoirs. In order to define conceptual models, these different contexts will be reviewed and complemented by an evaluation of the main properties of the potential reservoir in terms of porosity, permeability, fluid flow with respect to the stress field.

1.3 Conceptual Models of Geothermal Reservoirs

From a geological point of view, geothermal reservoirs are heated and pressurized water and/or vapor accumulations from which heat can be extracted from the underground to the surface. From a technical, environmental, and economic approach, the geothermal reservoir can be defined by the cost-efficiency of this extraction depending on the temperature, depth and size of the accumulation, the fluid flow, and the industrial process under which it will be processed. Although this approach is highly dependent on economical indicators that are not linked to geology (price of energy, incentives politics for access to renewable energies, etc.), reference to present-day parameters will be provided for the different types of reservoirs.

1.3.1
The Geology of Potential Heat Sources

To get heat is the first condition for defining a geothermal reservoir. How can we explore potential heat sources? It has been shown that thermal boundary conditions (the mean annual surface temperature, temperatures at depth estimated from the P and S velocity anomalies) and thermal properties of the main lithologies and structure at depth enable the first calculation of extrapolated temperature at depth and thus the delineation of potential zones of high-thermal gradient.

Such zones can also be determined through a geological empiric approach. Heat is transferred within the crust through two mechanisms:

- The main active and permanent phenomenon at the scale of the continental crust is the conduction of heat. In conduction, heat moves through the material from a hotter to a cooler zone. The feasibility and intensity of such transfer is directly linked to the thermal properties of the mineral constituting the rock that is evaluated as the thermal conductivity. As continental crust is heterogeneous and a result of the superposition of layers with different conductivity properties (stacked allochthonous units over autochthonous cover sequence or basement in orogenic zones, sedimentary basins over basement within intracratonic zones, etc.), conduction will not be homogenous at the scale of the whole continental crust. Highly conductive zones such as fractured granites will be explored with interest while refractory units such as mafic units will be considered as potential thermal insulator.
- In convection, heat is transported by the movement of hot material. The ascent and emplacement of a granitic body or of a volcanic dyke network is a typical example of convection where heat is transferred from deep source and then dissipated by conduction in the host rocks at shallow level. Contact metamorphism is a direct expression of the elevation of temperature with respect to extreme geothermal gradients reaching 500 °C for granites emplaced at around 5 km depth. Globally, convection leads to anisotropic diffusion of heat; the movement of hot material being, most of the time, controlled by the permeability system of the continental crust, mainly fracture network.

The past or present geodynamic context gives a first-order constrain on the location of favorable and unfavorable geodynamic sites for high geothermal gradients. Conduction is directly controlled by the thickness, heterogeneity, and composition of the continental crust, whereas convection processes are mainly located within active zones of magmatism and metamorphism.

Rift in accretionary systems are characterized by thinned crust and lithosphere, in relation with asthenospheric doming and upwelling. This definition covers both mid-oceanic ridges and back arc extensional systems, immerged or emerged, and to less extent pulls part systems developed along strike-slip faults. The geological setting of such rift zones is then the most favorable context because of the high mantle heat flow (Figure 1.2), the shallow depth of the mantle crust boundary, and the periodic magmatic activity (emplacement of stocks, sills, and dykes) and volcanic flow of hot mafic lavas. Moreover, convection of heat is enhanced by fluid – rock interaction – intense fracturing related to extensional tectonics favoring exchange between fluids of superficial and deep origin in the vicinity of magma chambers. Numerical modeling of rifting processes illustrates the shift of the isotherms toward surface depending on rifting velocities, presence of strain softening, and time (Huismans and Beaumont, (2002), Figure 1.12).

Iceland is the best case history for illustrating this first-order parameter for location of high geothermal gradients (Figure 1.13). A large active volcanic zone, corresponding to the mid-oceanic ridge, is running SW–NE, and displays various heat sources (dikes and magma chambers). Seawater, meteoric water, and volcanic fluids are mixed in pressurized water-dominated reservoirs, often associated with young tectonic fractures, carrying heat from several kilometers depth toward the surface (Flóvenz and Saemundsson, 1993; Arnórsson, 1995). The regional temperature gradient varies from 50 to 150 $^\circ$C km^{-1} and the highest values are found close to the volcanic rift zone.

Active margins related to subduction are sites of intense convection with respect to magmatic activity. A computed model at the scale of the lithosphere (Figure 1.6) shows that the subduction of cold lithosphere is accompanied by a raise of hot lithosphere just above the main plate boundary. Thus, large crustal zones have temperatures greater than 300 $^\circ$C at very shallow depth and undergo melting conditions at few kilometers depth. Generated calc-alkaline magmatism is responsible for the intrusion of voluminous granitic suites at shallow depth and related volcanism of intermediate to felsitic composition. This convective phenomenon at the scale of the lithosphere is also responsible of the concentration of U, K, and Th radioelements in the upper crust, which will contribute to the thermal budget of the continents over a long period.

Active margin settings are zones with almost infinite source of fluids from meteoric origin, as generated high relief is bordered by oceanic areas, and from deep source, in relation to magmatic and metamorphic processes.

Presently, New Zealand and Philippines are the zones where the exploitation of geothermal energy is the most advanced within these subduction-related contexts.

Collision zones and convergent plate boundaries may also be sites of high geothermal gradients. Collision is responsible for the development of large thrust systems that lead to a crustal thickening of several tens of kilometers. Zones situated at mid-crustal depth within the underthrust slab will then be buried and will undergo an immediate increase in pressure and a progressive increase in temperature. As discussed previously (Figure 1.6), the equilibrium thermal field is reached several ten million years after the thrusting event has ceased. This evolution is

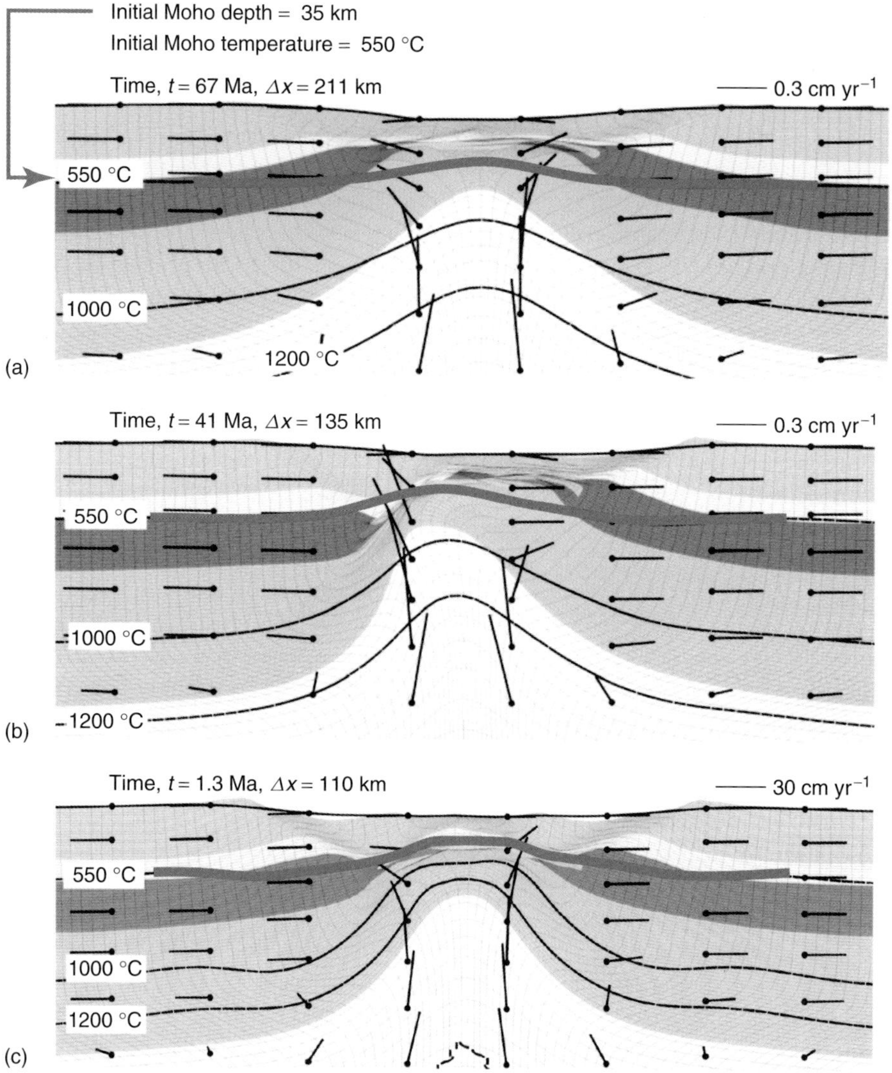

Figure 1.12 Uplifted isotherms (thick grey line) created by lithospheric extension, after Huismans and Beaumont, 2002. Crust and lithosphere are rheologically stratified. Lateral boundary conditions reproducing extension correspond to the imposed rifting velocities given in centimeters per year. (a) Case of no strain softening, after 211 km of extension; (b) asymmetric extension obtained with the introduction of strain softening and 41 Myr after rifting initiation; and (c) case of a fast rifting velocity, involving a large zone (~40 km) of hot middle crust.

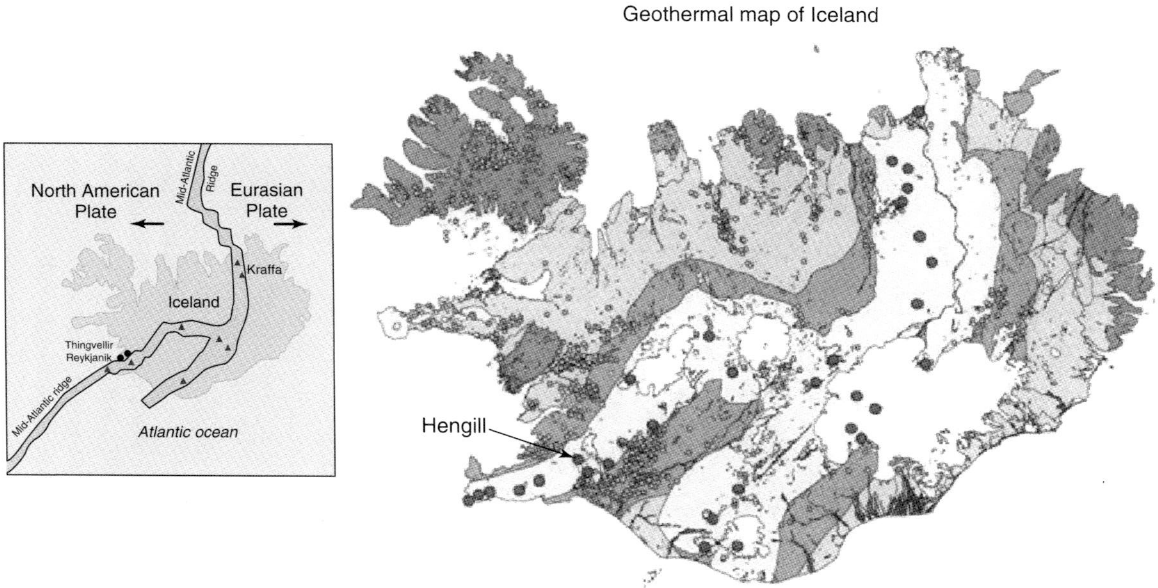

Figure 1.13 Geothermal map of Iceland. The main geothermal fields are located within prehistoric and historic lava centers and interglacial lavas, within the active rift zone.

well documented in the European Variscan belt where high paleogradients determined from mineral assemblages show that a regional geothermal system, responsible for many ore deposits (Au, U, etc.), has been generated during the late orogenic evolution of this collision belt (Bouchot *et al.*, 2005). The melting of large mid-crustal zones has been enhanced by the fertility of the crust rich in radioelements and hydrated minerals generating large volume of migmatites and granites over a long period, from 360 to 300 Ma (Ledru *et al.*, 2001). This situation reflects probably what is occurring within the Tibet Plateau – crustal thickening resulting from the collision between Asia and India being responsible for the development of migmatitic layers at depth. Taking this time delay related to the progressive re-equilibration of the isotherms in the thickened crust, such collision plate boundaries can be considered as favorable zones for high geothermal gradients. Moreover, like in the case of active margins, the concentration of radioelement-rich geological units (differentiated granites, uranium-bearing sedimentary basins, volcanic ash flows, overthrust Precambrian radiogenic granites, etc.) in the upper crust contributes to the thermal budget of the continents over several hundreds of million years.

The location of high geothermal gradients in the vicinity of transform margins and of thermal anomalies along continental-scale strike-slip faults can be related to thickening processes inherited from an early stage of collision, or linked to zones of pull-apart extension (that can be assimilated to the general case of rift systems), or a combination of both processes. In the case of the San Andreas Fault and its satellites in Nevada, it seems that the dominant feature for exploration at the regional scale is the presence of structural discontinuities bordering such pull-apart basins (Figure 1.14, Faulds, Henry, and Hinz, 2005; Faulds *et al.*, 2006).

Within plates, out of these plate boundaries, the lithosphere is considered as stabilized and the main mechanism of heat transfer is conduction. Depending on its composition (i.e., conductivity of its main lithologies) and thickness, geothermal gradients vary between 15 and 25 $^{\circ}C\,km^{-1}$. The main source of thermal anomalies is the presence of highly radiogenic lithologies such as alkaline and aluminous granites, uranium-bearing sedimentary basins, or highly conductive materials (massive sulfide). The radioactive decay is the cause of heat anomalies in the vicinity and at the apex of these radiogenic bodies, generally of small to medium amplitude and wavelength (Figure 1.5). This is the model on which exploration of deep geothermal resources is done presently in the Southern Australian craton (McLaren *et al.*, 2002; Hillis *et al.*, 2004). Highly radiogenic Precambrian granites ($\sim$16 $mW\,m^{-3}$), outcropping in large ranges and found laterally at the base of a Paleozoic sedimentary basins resting unconformably over this basement, are considered as the source of local thermal anomalies that are superposed to a regional anomaly know as the *south Australian heat flow anomaly* (SAHFA) (McLaren *et al.*, 2003; Chopra and Holgate, 2005). Paralana hot springs are observed along the main faulted contact between the basement and cover sequences and uranium-bearing sediments deposited during the erosion of the radiogenic Precambrian granites are presently exploited by *in situ* recovery (Berveley mine). The company Petratherm

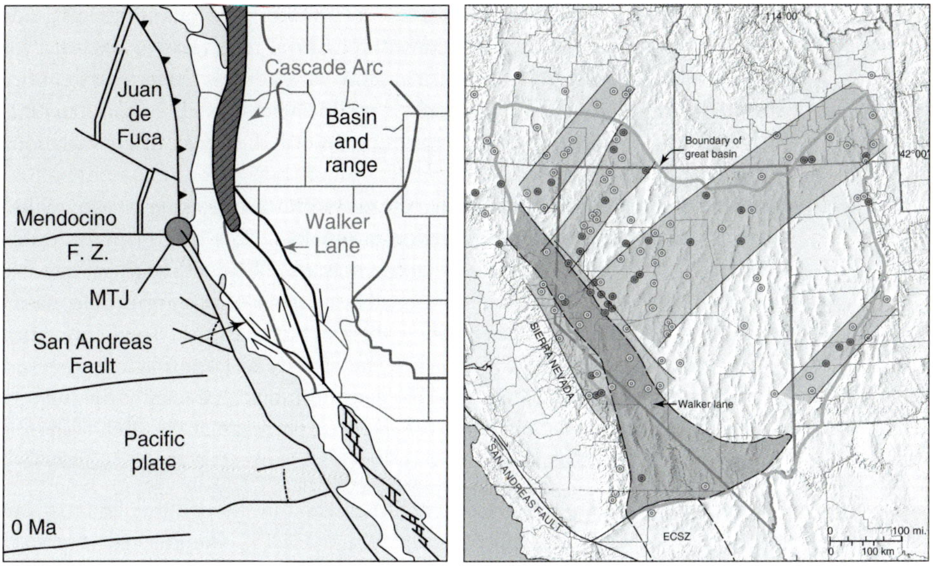

Figure 1.14 Geothermal fields in the Great Basin, western United States. Most of the activity is concentrated in the transtensional northwestern Great Basin within NE-trending belts oriented orthogonal to the extension direction and radiating from the northwestern terminus of the Walker Lane dextral shear zone (dark grey). Black spots, high temperature geothermal systems (>160 °C); open circles, low temperature systems (<160 °C); ECSZ, eastern California shear zone.

has drilled on Paralana 1B down to 1.8 km, which suggests 200 °C at 3.6 km, that is, more than 50 °C km^{-1}. In that case, the reservoir should be the Infracambrian detrital sedimentary sequences.

1.3.2
Porosity, Permeability, and Fluid Flow in Relation to the Stress Field

The permeability of the continental crust is defined by the capacity of the geological medium to transmit fluid. It constitutes a critical geological parameter for the definition of the geothermal reservoir as it plays a fundamental role in heat and mass transfer (Manning and Ingebritsen, 1999). This parameter is related to two basic properties of the rocks:

1) The porosity is the ratio of pore volume to the total volume. The intrinsic permeability is the measure of the fluid flow through the pore network of the rock and will be directly correlated to the porosity. These parameters are directly linked to the packing of the minerals within the rocks, which is a result of the nature, size, sorting of the minerals and elements, and of the compaction and diagenetic history. Sedimentary rocks such as limestone, sandstone, or

conglomerate are generally porous and can store large quantities of fluids within their pore network. They constitute natural reservoirs in the crust for all kind of fluids. The intrinsic permeability parameter is the primary control on fluid flow as it will vary from 10^{-23} m^2 in intact crystalline rocks to 10^{-7} m^2 in detrital porous sediments; meaning 16 orders of magnitude variations (Manning and Ingebritsen, 1999).

2) The fracture permeability is linked to the discontinuities that are present within the rock along which fluid circulation is possible. This type of permeability is generally well developed in crystalline massifs. Thus, although granite is a nonpermeable rock, a granitic massif will be considered as a permeable massif as a whole – fluid circulating along the fracture network. Implicitly, such permeability will be well developed in the vicinity of large fracture systems, whether active or fossil. Because of the discontinuous character of the fracture and their geometrical complexity, the intrinsic permeability of such system is more difficult to evaluate compared to stratified permeable layers.

The range of permeability observed and measured in the continental crust can be illustrated by a one-dimensional graph (Figure 1.15). Geothermal reservoirs are characterized by rather large permeability, higher than 10^{-13} m^2. Analyses of coupled groundwater flow and heat transport in the upper crust infer permeabilities in the range of 10^{-17}–10^{-14} m^2 with a mean value greater than 10^{-16} m^2 (Manning and Ingebritsen, 1999).

The identification of potential geothermal reservoir will then focus in priority on the exploration of both types of permeability related to the intrinsic and fracture permeability properties. A good knowledge of the geometry of the geological units and their physical properties, determined *in situ* or by geophysical methods, and of the structural pattern is a key for successful exploration. The building of a 3D

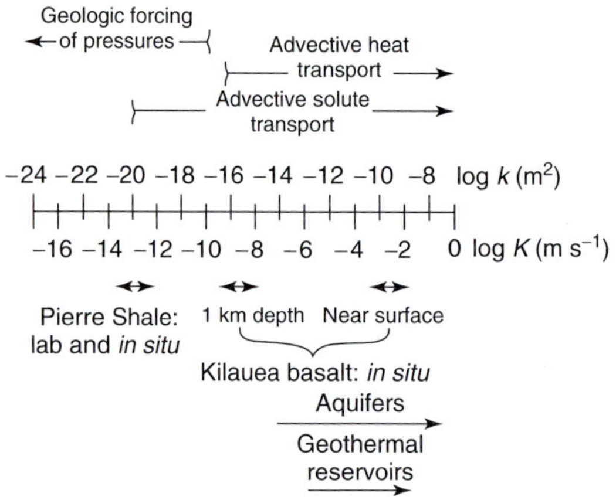

Figure 1.15 Range of permeabilities observed in geologic media: permeability (k, m^2) and hydraulic conductivity ($K = k\rho f g/\mu$, m s^{-1}) in relation to water density ρ_w and viscosity μ_w at 15 °C.

geometric model and the realization of 2D or even 3D seismic surveys are presently strategies that are promoted by many exploration companies, such as ENEL in Larderello or in the Rhine graben.

The main source of fluids is water in the ocean and meteoric water in the continents. The infiltrated part of meteoric and sea water will migrate down within permeable lithologies and/or fracture systems. Depending of the tectonic sites and of the geothermal gradient, they will be heated to temperature that will already be sufficient for direct use or even power generation. This is the case of the basin located in thinned crust such as the Rhine graben. What is the lower boundary of this infiltration?

Besides the first-order variation related to the intrinsic permeability of the systems, variation of permeability is also recorded with depth. Permeability of fractured crystalline rocks decreases with increasing pressure or effective stress. The porosity will be also controlled by the pressure–temperature conditions, and an increase in pressure with depth will reduce the porosity as the compaction will be greater. Surveys performed during deep drilling programs have demonstrated this decreasing permeability at depth, between 10^{-17} and 10^{-15} m^2 above 4 km depth and 10^{-18}–10^{-16} m^2 below (Huenges *et al.*, 1997), and a paper on the variation of permeability as a function of depth based on geothermal data and metamorphic systems was published in 1999 by Manning and Ingebritsen (Figure 1.16). With respect to this aspect, the brittle–ductile transition is a major decoupling surface at the scale of the crust that marks the lower limit that meteoric water can reach at depth and a change in fluid flow processes as it has been shown in natural analog (Famin *et al.*, 2004).

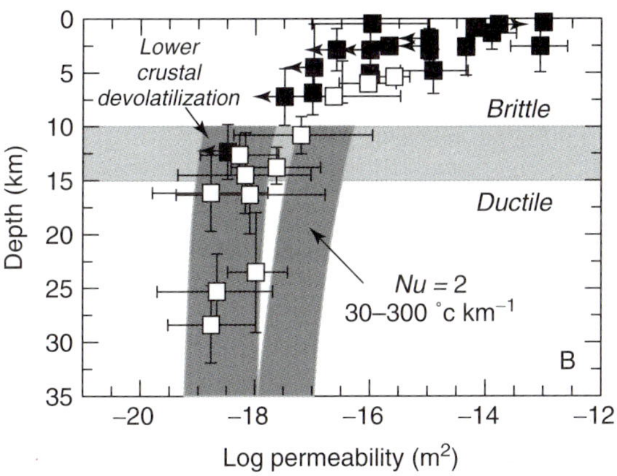

Figure 1.16 Permeability as a function of depth in the continental crust based on geothermal data (solid squares) and metamorphic systems (open squares). (After Manning and Ingebritsen, 1999)

If the intrinsic permeability of the continental crust is low below the brittle–ductile transition zones, fluid flow will occur through pervasive flow in zones of low strain rates and homogeneous lithologies while higher strains and a more heterogeneous rheology will favor a channeled flow (Oliver, 1996). In the case of the geothermal reservoirs, attention will be paid on the channeled flow that will occur along active tectonic zones or on emplacement of magmatic or volcanic suites. In such environments, hot fluids with temperature generally greater than 250 °C will move upward and finally reach the surface if they are not trapped by a cap nonpermeable layer or blocked within the fracture network. They ascend through the crust and will get connected, at the brittle–ductile transition zone, with fluids that have been infiltrated from surface. Many of the active geothermal fields are in fact resulting from the mixing between these ascending and descending fluids within the permeability network of the continental crust. A loop is created at the level of the reservoir, corresponding to a convective transfer of heat toward surface. Geysers from Iceland or from western United States are the most visible trace of this phenomenon.

This link between the permeability of the continental crust, the potential infiltration of meteoric fluids and seawater, the brittle–ductile transition at depth, and the potential connection to fluid of deep origin illustrates the need to have a global approach for the exploration of geothermal reservoirs. Thus, numerous studies dedicated to tectonic processes during prograde and retrograde metamorphism or to transfer and trapping of hydrothermal ore deposits could provide a lot of information about the variation of permeability in the continental crust.

Other variations of permeability are observed laterally within one single geological medium depending on heterogeneity, anisotropy, and time. These parameters constrain the efficiency of the reservoir and its sustainable use. The intrinsic permeability is a function of the heterogeneity and anisotropy of the medium. Fluid flow will tend to be greater parallel to the main layering of the sedimentary or volcanic rocks and foliation of metamorphic rocks rather than across them. Permeability is also a time-dependent process as fluid–rock interactions will provoke permanent dissolution and recrystallization phenomenon that will modify the permeability network. The intensity and orientation of the stress field will exert a direct control on this process by determining zones of compression and extension, in relation to the relative position of the main stress axis and resulting strain.

1.3.3 Summary

The review of the phenomena that control the distribution of heat and fluid at depth shows that conventional reservoirs for high enthalpy geothermal energy are located in zones of active volcanism or magmatism while low- to medium enthalpy can be found in varied environment. The identification of potential reservoirs for developing a heat exchanger is linked to our ability to evaluate the coincidence

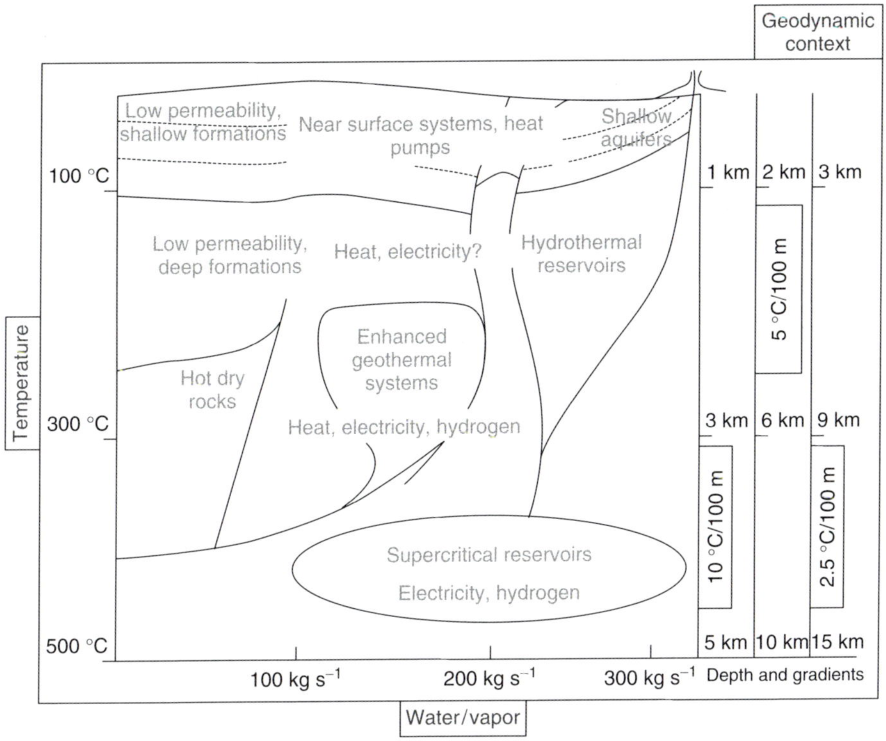

Figure 1.17 Sketch section showing the variety of reservoirs that can be used for heat extraction and the different uses of the geothermal energy.

of the following four independent parameters: heat, fluid flow, permeability, and appropriate orientation of the stress field in relation to the permeability network. Among these parameters, only fluid flow and permeability can be enhanced by engineering. These parameters are summarized in one sketch section that illustrates the variety of reservoirs that can be used for heat extraction and the various uses of the geothermal energy (Figure 1.17).

Following chapters will analyze the different processes of stimulation that can be applied for achieving such improvement. Lessons learned from the Soultz EGS experiment, the sustainable development of the Larderello field in Italy, and the Icelandic geothermal power network, among other case histories, show that the concept of geothermal reservoir must not be too restricted. Experiences of stimulation can be realized by extending active geothermal fields and the development of binary plants makes possible the exploitation of geothermal reservoirs with minimum temperatures of 85–100 °C for power generation.

A reservoir is also defined by its economic viability. Many other parameters such as the price of the steel involved in drilling, cost of drilling depending on the availability of drilling companies active on the "exploration market," and feed in tariffs supporting the development of geothermal energy must then be taken into account to evaluate the feasibility and viability of the project. These questions will be reviewed in the following chapters.

References

Arnórsson, S. (1995) Geothermal systems in Iceland: structure and conceptual models – I. High-temperature areas. *Geothermics*, **24**, 561–602.

Aydin, I., Karat, H.I., and Kocak, A. (2005) Curie-point depth map of Turkey. *Geophysical Journal International*, **162**, 633–640.

Bächler, D., Kohl, T., and Rybach, L. (2003) Impact of graben-parallel faults on hydrothermal convection – Rhine graben case study. *Physics and Chemistry of the Earth*, **28**, 431–441.

Beziat, A., Dardaine, M., and Gabis, V. (1988) Effect of compaction pressure and water content on the thermal conductivity of some natural clays. *Clay and Clay Minerals*, **36**, 462–466.

Blackwell, D.D., Steele, J.L., and Brott, C.A. (1980) The terrain effect on terrestrial heat flow. *Journal of Geophysical Research*, **85**, 4757–4772.

Bonté, D., Guillou-Frottier, L., Garibaldi, C., Lopez, S., Bourgine, B., Lucazeau, F., and Bouchot, V. (2010) Subsurface temperature maps in French sedimentary basins: new data compilation and interpolation. Submitted to *Bulletin de la Société Géologique de France in press.*

Bouchot, V., Ledru, P., Lerouge, C., Lescuyer, J.L., and Milesi, J.P. (2005) Late Variscan mineralizing systems related to orogenic processes: the French Massif Central. *Ore Geology Reviews*, **27**, 169–197.

Burov, E., Jaupart, C., and Guillou-Frottier, L. (2003) Ascent and emplacement of buoyant magma bodies in brittle-ductile upper crust. *Journal of Geophysical Research*, **108** (B4), 2177. doi: 10.1029/2002JB001904.

Cagnioncle, A.M., Parmentier, E.M., and Elkins-Tanton, L. (2007) Effect of solid flow above a subducting slab on water distribution and melting at convergent plate boundaries. *Journal of Geophysical Research*, **112**. doi: 10.1029/2007JB004934.

Cathles, L.M. (1977) An analysis of the cooling of intrusives by ground water convection which includes boiling. *Economic Geology*, **72**, 804–826.

Cermak, V. (1993) Lithospheric thermal regimes in Europe. *Physics of the Earth and Planetary Interiors*, **79**, 179–193.

Cermak, V., Kresl, M., Kucerova, L., Safanda, J., Frasheri, A., Kapedani, N., Lico, R., and Cano, D. (1996) Heat flow in Albania. *Geothermics*, **25**, 91–102.

Chopra, P. and Holgate, F. (2005) A GIS analysis of temperature in the Australian crust. Proceedings World Geothermal Congress 2005, April 24-29, 2005, Antalya, Turkey.

Clauser, C. (2006) Geothermal energy, in *Landolt-Börnstein, Group VIII: Advanced Materials and Technologies*, Energy Technologies, Subvol. C: Renewable Energies, Vol. **3** (ed K.Heinloth), Springer-Verlag, Heidelberg, Berlin, pp. 493–604.

Clauser, C. and Huenges, E. (1995) Thermal conductivity of rocks and minerals, in *Rock Physics and Phase Relations: A Handbook of Physical Constants*, AGU Reference Shelf, Vol. **3** (ed T.J. Ahrens), AGU, Washington, DC, pp. 105–126.

Delescluse, M. and Chamot-Rooke, N. (2008) Serpentinization pulse in the actively deforming central Indian basin. *Earth and Planetary Science Letters*, **276**, 140–151.

Demetrescu, C. and Andreescu, M. (1994) On the thermal regime of some tectonic units in a continental collision environment in Romania. *Tectonophysics*, **230**, 265–276.

Demongodin, L., Pinoteau, B., Vasseur, G., and Gable, R. (1991) Thermal conductivity and well-logs – a case study in the Paris basin. *Geophysical Journal International*, **105**, 675–691.

Dèzes, P., Schmid, S.M., and Ziegler, P.A. (2004) Evolution of the European Cenozoic Rift System: interaction of the Alpine and Pyrenean orogens with their foreland lithosphere. *Tectonophysics*, **389**, 1–33.

Dini, A., Gianelli, G., Puxeddu, M., and Ruggieri, G. (2004) Origin and evolution of Pliocene–Pleistocene granites from the Larderello geothermal field (Tuscan Magmatic Province, Italy). *Lithos*, **81**, 1–31.

Drury, M.J., Jessop, A.M., and Lewis, T.J. (1984) The detection of groundwater flow by precise temperature measurements in boreholes. *Geothermics*, **13**, 163–174.

Edel, J.B. and Fluck, P. (1989) The upper Rhenish shield basement (Vosges, Upper Rhinegraben and Schwarzwald): main structural features deduced from magnetic, gravimetric and geological data. *Tectonophysics*, **169**, 303–316.

Elder, J. (1967) Convective self-propulsion of continents. *Nature*, **214**, 657–660.

Elder, J. (1981) *Geothermal Systems*, Academic Press, London, 508 pp.

Emmanuel, S. and Berkowicz, B. (2006) An experimental analogue for convection and phase separation in hydrothermal systems. *Journal of Geophysical Research*, **111**, B09103. doi: 10.1029/2006JB004351.

England, P.C. and Thompson, A.B. (1984) Pressure-temperature-time paths of regional metamorphism. 1. Heat transfer during the evolution of regions of thickened continental crust, *Journal of Petrology*, **25**, 894–928.

Ercan, A. (2002) Petroleum and geothermal model of Aegean Sea and western Anatolia. *Turkey: Second Balkan Geophysical Congress and Exhibition*, 158–159.

Famin, V., Philippot, P., Jolivet, L. and Agard, P. (2004) Evolution of hydrothermal regime along a crustal shear zone, Tinos Island, Greece. *Tectonics*, **23**, TC5004. doi: 10.1029/2003TC001509.

Faulds, J.E., Coolbaugh, M.F., Vice, G.S., and Edwards, M.L. (2006) Characterizing structural controls of geothermal fields in the northwestern great basin: a progress report: *Geothermal Resources Council Transactions*, **30**, 69–76.

Faulds, J.E., Henry, C.D., and Hinz, N.H. (2005) Kinematics of the northern Walker Lane: an incipient transform fault along the Pacific –North American plate boundary. *Geology*, **33**, 505–508.

Fernàndez, A., Marzán, I., Correia, A., and Ramalho, E. (1998) Heat flow, heat production, and lithospheric thermal regime in the Iberian Peninsula. *Tectonophysics*, **291**, 29–53.

Flóvenz, O. and Saemundsson, K. (1993) Heat flow and geothermal processes in Iceland. *Tectonophysics*, **225**, 123–138.

Gallagher, K., Ramsdale, M., Lonergan, L., and Morrow, D. (1997) The role of thermal conductivity measurements in modelling thermal histories in sedimentary basins. *Marine and Petroleum Geology*, **14**, 201–214.

Garibaldi, C., Guillou-Frottier, L., Lardeaux, J.M., Bonté, D., Lopez, S., Bouchot, V. and Ledru, P. (2010) Relationship between thermal anomalies, geological structures and fluid flow: new evidence in application to the Provence basin (south-east France). Submitted to *Bulletin de la Société Géologique de France in press*.

Genter, A., Guillou-Frottier, L., Feybesse, J.-L., Nicol, N., Dezayes, C., and Schwartz, S. (2003) Typology of hot fractured rock resources in Europe. *Geothermics*, **32**, 701–710.

Goes, S., Loohuis, J.J.P., Wortel, M.J.R., and Govers, R. (2000) The effect of plate stresses and shallow mantle temperatures on tectonics of northwestern Europe. *Global and Planetary Change*, **27**, 23–38.

Goes, S., Spakman, W., and Bijwaard, H. (1999) A lower mantle source for central European volcanism. *Science*, **286**, 1928–1931.

Goutorbe, B., Lucazeau, F., and Bonneville, A. (2007) Comparison of several BHT correction methods: a case study of an Australian data set. *Geophysical Journal International*, **170**, 913–922.

Granet, M., Wilson, M., and Achauer, U. (1995) Imaging a mantle plume beneath the French Massif Central. *Earth and Planetary Science Letters*, **136**, 281–296.

Grigné, C. and Labrosse, S. (2001) Effects of continents on Earth cooling: thermal blanketing and depletion in radioactive elements. *Geophysical Research Letters*, **28**, 2707–2710.

Guillou, L. and Jaupart, C. (1995) On the effect of continents on mantle convection. *Journal of Geophysical Research*, **100**, 24217–24238.

Guillou, L., Mareschal, J.-C., Jaupart, C., Gariépy, C., Bienfait, G., and Lapointe, R. (1994) Heat flow, gravity and structure of the Abitibi belt, Superior Province, Canada: implications for mantle heat flow. *Earth and Planetary Science Letters*, **122**, 103–123.

Guillou-Frottier, L., Burov, E., Nehlig, P., and Wyns, R. (2007) Deciphering plume-lithosphere interactions beneath Europe from topographic signatures. *Global and Planetary Change*, **58**, 119–140.

Guillou-Frottier, L., Mareschal, J.-C., Jaupart, C., Gariépy, C., Lapointe, R., and Bienfait, G. (1995) Heat flow variations in the Grenville Province, Canada. *Earth and Planetary Science Letters*, **136**, 447–460.

Guillou-Frottier, L., Mareschal, J.-C., and Musset, J. (1998) Ground surface temperature history in central Canada inferred from ten selected borehole temperature profiles. *Journal of Geophysical Research*, **103**, 7385–7397.

Gurnis, M. (1988) Large-scale mantle convection and the aggregation and dispersal of supercontinents. *Nature*, **332**, 695–699.

Haenel, R. and 16 co-authors (1980) Atlas of Subsurface Temperatures in the European Community, The Commission of the European Communities, 36 p., 43 plates.

Hansen, J. and Lebedeff, S. (1987) Global trends of measured surface air temperatures. *Journal of Geophysical Research*, **92**, 13345–13372.

Harcouët, V., Guillou-Frottier, L., Bonneville, A., Bouchot, V., and Milesi, J.-P. (2007) Geological and thermal conditions before the major Paleoproterozoic gold mineralization event at Ashanti, Ghana, as inferred from improved thermal modelling. *Precambrian Research*, **154**, 71–87.

Hillis, R.R., Hand, M., Mildren, S., Morton, J., Reid, P., and Reynolds, S. (2004) Hot dry rock geothermal exploration in Australia, application of the in situ stress field to hot dry rock geothermal energy in the Cooper Basin. Eastern Australian Basins Symposium II Volume. Petroleum Exploration Society of Australia (PESA) Eastern Australian Basins Symposium, Adelaide, pp. 413–421.

Huenges, E., Erzinger, J., Kuck, J., Engeser, B., and Kessels, W. (1997) The permeable crust: geohydraulic properties down to 9101 m depth. *Journal of Geophysical Research*, **102**, 18,255–18,265.

Huismans, R.S. and Beaumont, C.B. (2002) Asymmetric lithospheric extension: the role of frictional plastic strain softening inferred from numerical experiments. *Geology*, **30**, 211–214.

Huismans, R.S., Podladchikov, Y.Y., and Cloetingh, S. (2001) Dynamic modeling of the transition from passive to active rifting: application to the Pannonian basin. *Tectonics*, **20**, 1021–1039.

Hurtig, E., Cermak, V., Haenel, R., and Zui, V. (eds) (1992) *Geothermal Atlas of Europe*, Hermann Haack Verlagsgesellschaft mbH, Germany.

Jaupart, C., Labrosse, S., and Mareschal, J.-C. (2007) Elsevier, Temperatures, heat and energy in the mantle of the Earth, in *Treatise on Geophysics*, Mantle Dynamics, Vol. **7** (eds D. Bercovici and G. Schubert), pp. 253–303.

Jaupart, C., Mareschal, J.-C., Guillou-Frottier, L., and Davaille, A. (1998) Heat flow and thickness of the lithosphere in the Canadian Shield. *Journal of Geophysical Research*, **103**, 15269–15286.

Jaupart, C. and Parsons, B. (1985) Convective instabilities in a variable viscosity fluid cooled from above. *Physics of the Earth and Planetary Interiors*, **39**, 14–32.

Kukkonen, I.T. and Peltonen, P. (1999) Xenolith-controlled geotherm for the central Fennoscandian Shield: implications for lithosphere-asthenosphere relations. *Tectonophysics*, **304**, 301–315.

Labrosse, S. (2002) Hotspots, mantle plumes and core heat loss. *Earth and Planetary Science Letters*, **199**, 147–156.

Ledru, P., Courrioux, G., Dallain, C., Lardeaux, J.M., Montel, J.M., Vanderhaeghe, O., and Vitel, G. (2001) The Velay dome (French Massif Central): melt generation and granite emplacement during orogenic evolution. *Tectonophysics*, **342**, 207–227.

Lenardic, A., Guillou-Frottier, L., Mareschal, J.-C., Jaupart, C., Moresi, L.-N., and Kaula, W.M. (2000) Amer. Geophys Union, What the mantle sees: the effects of continents on mantle heat flow, in *The History and Dynamics of Global Plate Motions*, AGU Monograph, Vol. **121** (eds M. Richards *et al.*), pp. 95–112.

Lenardic, A. and Kaula, W.M. (1995) Mantle dynamics and the heat flow into the Earth's continents. *Nature*, **378**, 709–711.

Lister, C.R.B., Sclater, J.G., Davis, E.E., Villinger, H., and Nagahira, S. (1990) Heat flow maintained in ocean basins of great age: investigations in the north equatorial west Pacific. *Geophysical Journalal*, **102**, 603–630.

Lopez, D. and Smith, L. (1995) Fluid flow in fault zones: analysis of the interplay of convective circulation and topographically driven groundwater flow. *Water Resources Research*, **31**, 1489–1503.

Lucazeau, F. and 10 co-authors (2008) Persistent thermal activity at the eastern Gulf of Aden after continental break-up. *Nature Geosciences*, **1**, 854–858.

Lucazeau, F., Vasseur, G., and Bayer, R. (1984) Interpretation of heat flow data in the French Massif Central. *Tectonophysics*, **103**, 99–119.

Manea, V.C., Manea, M., Kostoglodov, V., Currie, C.A., and Sewell, G. (2004) Thermal structure, coupling and metamorphism in the Mexican subduction zone beneath Guerrero. *Geophysical Journal International*. **158**, 775–784.

Manning, C.E. and Ingebritsen, S.E. (1999) Permeability of the continental crust: implications of geothermal data and metamorphic systems. *Review of Geophysics*, **37**, 127–150.

Mareschal, J.-C., Jaupart, C., Gariépy, C., Cheng, L.Z., Guillou-Frottier, L., Bienfait, G., and Lapointe, R. (2000) Heat flow and deep thermal structure near southeastern edge of the Canadian Schield. *Canadian Journal of Earth Sciences*, **37**, 399–414.

McLaren, S., Dunlap, W.J., Sandiford, M., and McDougall, I. (2002) Thermochronology of high heat-producing crust at Mount Painter, South Australia: implications for tectonic reactivation of continental interiors. *Tectonics*, **21**. doi: 10.1029/2000TC001275.

McLaren, S.N., Sandiford, M., Hand, M., Neumann, N.L., Wyborn, L.A.I., and Bastrakova, I. (2003) The hot southern continent: heat flow and heat production in Australian Proterozoic terranes, in *Evolution and Dynamics of the Australian Plate* (eds R.R. Hillis and R.D. Müller), Geological Society of Australia Special Publication 22 and Geological Society of America Special Paper 372, pp. 157–167.

Mwenifumbo, C.J. (1993) Temperature logging in mineral exploration. *Journal of Applied Geophysics*, **30**, 297–313.

Nemcock, M., Pospisil, L., Lexa, J., and Donelick, R.A. (1998) Tertiary subduction and slab break-off model of the Carpathian-pannonian region. *Tectonophysics*, **295**, 307–340.

Norton, D.L. and Hulen, J.B. (2001) Preliminary numerical analysis of the magma-hydrothermal history of The Geysers geothermal system, California, USA. *Geothermics*, **30**, 211–234.

Oliver, N.H.S. (1996) Review and classification of structural controls on fluid flow during regional metamorphism. *Journal of Metamorphic Geology*, **14**, 447–492.

Parsons, B. and McKenzie, D. (1978) Mantle convection and the thermal structure of plates. *Journal of Geophysical Research*, **83**, 4485–4496.

Parsons, B. and Sclater, J.G. (1977) An analysis of the variation of ocean floor bathymetry and heat flow with age. *Journal of Geophysical Research*, **82**, 803–827.

Peacock, S.M., van Keken, P.E., Holloway, S.D., Hacker, B.R., Abers, G.A., and Fergason, R.L. (2005) Thermal structure of the Costa Rica – Nicaragua subduction zone. *Physics of the Earth and Planetary Interiors*, **149**, 187–200.

Pinet, C., Jaupart, C., Mareschal, J.-C., Gariépy, C., Bienfait, G., and Lapointe, R. (1991) Heat flow and structure of the lithosphere in the eastern Canadian Schield. *Journal of Geophysical Research*, **96**, 19941–19963.

Pollack, H.N., Hurter, S.J., and Johnson, J.R. (1993) Heat flow from the Earth's interior: analysis of the global data set. *Review of Geophysics*, **31**, 267–280.

Pribnow, D. and Schellschmidt, R. (2000) Thermal tracking of upper crustal fluid flow in the Rhine graben. *Geophysical Research Letters*, **27**, 1957–1960.

Ritter, J.R.R., Jordan, M., Christensen, U.R., and Achauer, U. (2001) A mantle plume below the Eifel volcanic fields, Germany. *Earth and Planetary Science Letters*, **186**, 7–14.

Roberts, P.H., Jones, C.A., and Calderwood, R.A. (2003) Energy fluxes and ohmic dissipation in the Earth's core, in *Earth's Core and Lower Mantle* (eds C.A. Jones, A.M. Soward, and K. Zhang), Taylor and Francis, London, pp. 100–129.

Rühaak, W. (2009) Multidimensional modeling of the thermal and flow regime in the western part of the Molasse Basin, Southern Germany, PhD thesis, University of Aachen, Germany, 88 p.

Ruppel, C. and Hodges, K. (1994) Role of horizontal thermal conduction and finite time thrust emplacement in simulation of pressure-temperature-tile paths. *Earth and Planetary Science Letters*, **123**, 49–60.

Sandiford, M., Fredericksen, S., and Braun, J. (2003) The long term thermal consequences of rifting: implications for basin reactivation. *Basin Research*, **15**, 23–43.

Sandiford, M., McLaren, S., and Neumann, N. (2002) Long-term thermal consequences of the redistribution of heat-producing elements associated with large-scale granitic complexes. *Journal of Metamorphic Geology*, **80**, 87–98.

Schubert, G., Turcotte, D.L., and Olson, P. (2002) *Mantle Convection in the Earth and Planets*, Cambridge University Press, 956 p.

Sclater, J.G., Jaupart, C., and Galson, D.A. (1980) The heat flow through oceanic and continental crust and the heat loss of the Earth. *Review of Geophysics*, **18**, 269–311.

Sengör, A.M.C., Gorur, N., and Saroglu, F. (1985) Strike-slip faulting and related basin formation in zones of tectonic escape: Turkey as a case study, in *Strike-slip Faulting and Basin Formation*, Special Publication 37 (eds K.T. Biddle and N. Christie-Black), Society of Economic, Paleontologic, and Mineralogists, pp. 227–264.

Shapiro, N.M. and Ritzwoller, M.H. (2002) Monte-Carlo inversion for a global shear-velocity model of the crust and upper mantle. *Geophysical Journal International*, **151**, 88–105.

Sonney, R. and Vuataz, F.D. (2008) Properties of geothermal fluids in Switzerland: a new interactive database. *Geothermics*, **37**, 496–509.

Trubitsyn, V., Kaban, M., Mooney, W., Reigber, C., and Schwintzer, P. (2006) Simulation of active tectonic processes for a convecting mantle with moving continents. *Geophysical Journal International*, **164**, 611–623.

Turcotte, D.L. and Schubert, G. (2002) *Geodynamics*, 2nd edn, Cambridge University Press, 450 p.

Vasseur, G., Brigaud, F., and Demongodin, L. (1995) Thermal conductivity estimation in sedimentary basins. *Tectonophysics*, **244**, 167–174.

Vasseur, G., Gable, R., Feuga, B., and Bienfait, G. (1991) Groundwater flow and heat flow in an area of mineral springs. *Geothermics*, **20**, 99–117.

Waples, D.W. and Tirsgaard, H. (2002) Changes in matrix thermal conductivity of clays and claystones as a function of compaction. *Petroleum Geoscience*, **8**, 365–370.

Wenzel, F. and Brun, J.-P., ECORS-DEKORP Working Group (1991) A deep reflection seismic line across the Northern Rhine Graben. *Earth and Planetary Science Letters*, **104**, 140–150.

Whitehead, J.A. (1976) Convection models: laboratory versus mantle. *Tectonophysics*, **35**, 215–228.

Wisian, K.W. and Blackwell, D.D. (2004) Numerical modeling of Basin and Range geothermal systems. *Geothermics*, **33**, 713–741.

2
Exploration Methods

David Bruhn, Adele Manzella, François Vuataz, James Faulds, Inga Moeck, and Kemal Erbas

2.1
Introduction

Most of the world's geothermal fields exploited today are associated with volcanic and/or recent tectonic activity and have commonly been discovered through surface expressions, such as hot springs. Typically, springs and fumaroles are sampled, and geochemical analyses are used to estimate the maximum temperatures of fluids in the subsurface. The areas with highest geochemically derived temperatures are further explored in detail and eventually developed.

Some geothermal fields have been found even though there were no obvious surface manifestations; generally, by exploration drilling for some other resource, for example, in a sedimentary basin; or by drilling heat flow holes on a regular grid. Other areas are inferred to have a high geothermal potential on the basis of known heat flow data. To explore such areas and reservoirs in more detail and better establish their potential and extent, both geological and geophysical methods are usually applied.

In general, geothermal exploration strives to detect hot water or steam reservoirs in a hot and highly permeable environment. The concept of enhanced geothermal system (EGS) slightly modifies this concept, as fluid pathways can be enhanced or even generated by suitable stimulation methods. The focus has also broadened in terms of temperature, as attention is no longer reserved to the absolute hot spots. Instead, the proximity to the end user is a critical aspect for the profitability of a project, especially at lower enthalpy reservoirs, where geothermal district heating is a component of the overall economic calculations. In particular, the main utilization scheme for sustainable and efficient production from low enthalpy geothermal reservoirs is a combined heat and power generation with production and injection wells. Thus, the targets of exploration have become more diverse, just as the tools to characterize them have evolved in the past decades.

Nonetheless, the general geochemical and geophysical exploration approach is similar to that described, for example, by Lumb (1981), but is generally combined with comprehensive assessment of the geologic setting, especially of the tectonic

Geothermal Energy Systems. Edited by Ernst Huenges

ISBN: 978-3-527-40831-3

and structural framework. Thus, fruitful exploration strategies typically involve the following:

- assessment of the geologic and geodynamic setting;
- geochemistry including fluid and rock isotope chemistry;
- structural analysis of faults, fractures, and folds;
- determination of the regional stress field;
- potential methods, mainly gravity and magnetic surveys;
- electrical and electromagnetic methods (EMs);
- seismic methods, both active and passive.

Previous exploration efforts have focused primarily on obvious geothermal targets with good surficial expressions and relatively high temperatures. However, many high enthalpy systems are yet to be developed and others simply do not have obvious surficial expressions. Consequently, exploration efforts today are commonly focused on hidden (or blind) geothermal systems and unconventional low enthalpy resources, which commonly require enhancement in permeability (i.e., EGS).

Exploration of potential EGS reservoirs encompasses a broad spectrum of geological settings and therefore requires a wide variety of approaches, often a combination of several methods. For example, potential EGS targets now include deep sedimentary basins, which have previously been the almost exclusive domain of hydrocarbon (HC) exploitation. Because of the breadth of potential settings for geothermal activities, the geological characterization of an area becomes even more important, especially if no surface expressions indicate geothermal activity at depth.

There has also been considerable improvement in both technology and methods, often borrowed from the HC industry, and driven by increasing activity in the geothermal sector. For example, magnetotellurics (MTs), still in its infancy as an exploration method in the early 1980s, has become a primary tool for the detection and characterization of deep geothermal reservoirs. Similarly, seismic methods, often considered too expensive and not very useful in the traditional, mostly volcanic environments of conventional geothermal activity, have increased in popularity in the geothermal sector. Examples include well-established regions such as Larderello in Italy and the deep reservoirs in sedimentary basins such as the Molasse Basin in Southern Germany or the Rhine Graben, where development of EGS systems can become particularly important. Efforts to better characterize the geologic and structural settings of geothermal activity have also recently increased, including analyses of surficial geothermal features (Coolbaugh *et al.*, 2007, 2006) and favorable structural settings (Curewitz and Karson, 1997; Faulds *et al.*, 2004, 2006; Micklethwaite and Cox, 2004; Fairley and Hinds, 2004).

Enhancing a geothermal system generally involves drilling along deviated well paths and with large diameters, drilling with formation damage mitigating technologies, stimulating the reservoir by hydraulic fracturing, and/or targeting fault zones that will produce with high flow rates, which are usually higher than those in HC production (Huenges and Moeck, 2007). Thus, one of the key geological issues, especially critical for EGS development, is knowledge of the stress field and an understanding of geomechanics in the subsurface. The geological characterization must therefore also include various methods that constrain the stress field of a reservoir and elucidate the stress states along faults slated for stimulation.

In summary, geothermal exploration for EGS means, on the one hand, that a reservoir should be understood as a part of a complex geosystem and, on the other hand, it is part of a complex mechanical rock response in the subsurface reacting – either positive or negative – to all manipulations that need to be done from exploration over reservoir access to exploitation. Consequently, geothermal exploration for EGS should encompass a broad palette of approaches from geosystem analysis to reservoir characterization to reservoir geomechanics.

This chapter describes geological criteria, as far as they have been defined, and the most common geological, geochemical, and geophysical methods of geothermal prospecting, as well as trends and requirements for future developments.

2.2 Geological Characterization

Choosing a favorable location for a potential EGS site requires careful consideration of the geologic setting, including heat flow, stratigraphy, and structural framework. Although high permeability is critical for a conventional geothermal system, it is not required for an EGS site because the flow rate and productivity of a well can be increased artificially. Nonetheless, the delineation of suitable geologic settings is the most important aspect of EGS reservoir exploration. Exploration strategies must target accessible and extractable sources of thermal water in large enough quantities to promise sustainable power and/or heat extraction.

The geological settings for geothermal reservoirs can vary widely. High enthalpy systems typically occur in magmatic, extensional, or transtensional settings. Magmatic settings include arcs (e.g., Central America and parts of the Mediterranean), both continental and oceanic rifts (e.g., Basin and Range and Iceland, respectively), hot spots (e.g., Hawaii), and transtensional pull-aparts in strike-slip fault systems (e.g., Salton Trough, California). However, high enthalpy geothermal systems are also relatively common in amagmatic extensional and transtensional settings, as most evident in the Basin and Range province (USA) and western Turkey (Akkuş *et al.*, 2005), where normal fault systems are the primary control on geothermal activity. Lower enthalpy systems are also found in the above settings as well as in relatively quiescent tectonic environments, deep sedimentary basins (e.g., Paris and North German Basins), and convergent plate margins (e.g., Alpine orogenic belt; Hurter and Haenel, 2002).

Although the general tectonic settings favorable for geothermal activity are well known, the detailed lithologic and structural controls on individual systems are generally not well characterized. It is, however, crucial in geothermal exploration to identify geological units and structures that host hydrothermal fluids. These reservoirs can be governed by either pore space (i.e., a high porosity) or fractures (i.e., high fracture density). Fluid pathways are critical for the productivity of a reservoir and are commonly controlled by structures such as fractures and faults. A low permeable but highly porous reservoir can be enhanced into a productive geothermal system if the naturally missing fractures are artificially induced by stimulation. From this perspective, it is obvious that a successful development of a geothermal field strongly depends on the understanding of the geologic setting at different scales from microscopic pore space to regional fault patterns. Thus, any assessment of geothermal potential requires a comprehensive analysis of the geologic setting that aims to elucidate critical issues beyond temperature, such as porosity, permeability, spatial extent of the reservoir body, structural framework, and surficial features indicative of geothermal activity.

The first step in the assessment of the geological and geodynamic setting of hidden geothermal resources is the compilation of all available data for a region from all possible sources: the national or regional surveys, land owners, the oil and gas industry, downhole temperature data, published academic data, and remote sensing data. Such a data compilation is usually an inexpensive but potentially time-consuming part of the exploration process.

Remote sensing has turned out to become a more commonly used and immensely useful tool. It is possible to get detailed surface information even for areas with limited access prior to any on-site activity. Even though such methods have existed for quite some time, new technology using sensors to detect different wavelengths of light can now distinguish between different types of rock. The analysis of rock forming minerals associated with geothermal activity can be used to get a first idea about the orientation of geological structures that may control pathways of geothermal fluids, before an area is further explored with ground-based field work. With the introduction of hyperspectral surveys, a new, albeit expensive, tool with high spatial resolution from airborne instruments is now available. It allows detailed mapping of mineral distribution, which could allow the reconstruction of the geological history and potential changes in mineralogy caused by geothermal activity. That way, prospective zones without obvious surface expressions may be delineated and specifically targeted for ground-based exploration.

The next step in the exploration of a potential geothermal prospect is the evaluation of the regional temperature field. In this context, information from existing wells is of preeminent importance. Data on the subsurface temperature distribution is widely available from previous activities, although information becomes scarce and less reliable with increasing depth. If the data availability is not sufficient, there are several standard approaches to measure subsurface temperatures, which help in the evaluation of a geothermal prospect. Most commonly, shallow (1–3 m deep) holes are drilled or a hollow tube is hammered into the ground and the temperature is measured. These frequently applied methods are inexpensive and give a general idea about the distribution of a subsurface heat source. More important are temperature gradient holes (TGHs) that are drilled to get a better idea about the temperatures at depth at a given location. These TGHs can be a few hundred meters or even more than a kilometer in depth depending on the target depth. This method can yield much more reliable information than the shallower holes but can also become fairly expensive depending on the geological environment and the depth anomaly of interest.

If the temperature field of an area is well constrained, the characterization of the potential reservoir rock is of utmost importance. Fluid-hosting rock types with a high porosity include sedimentary facies, such as sandstone-rich alluvial fans within sedimentary basins, or some volcanic rocks, such as vesicular basalt or some ignimbrites. The largest geothermal fields currently under exploitation reside in rocks that range from limestone to shale and volcanic rock to granite. Volcanic rocks are the most common host for geothermal reservoirs. The lateral extent and deposition of sedimentary and volcanic lithologies is, however, limited and – especially in the case of sedimentary basins – structurally controlled. Highly porous sandstones with low clay content are common for alluvial fans and channel deposits in alluvial plains. Also, aeolian and barren sandstones are highly porous successions. Because the placement of alluvial fans and fluvial deposits are typically strongly controlled by syntectonic processes or paleorelief, the tectonic and geodynamic history of a basin may reflect the possible depocenters of potential reservoir rock types. As such, facies types and highly porous sandstone sequences therein are of limited spatial extent and strongly tied to structural highs and graben flanks. Generally, potential reservoir rocks are limited to the basin rim, whereas the basin center deposits encompass low porous finest grained siliciclastics, evaporites, or limestone, depending on climatic conditions and general evolution of the basin. In the case of the South Permian Basin System in Central Europe, the southern flank of the basin system more likely contains potential reservoir rocks due to erosion of volcanic rocks of the initial basin phase and proximity of the high mountain ranges of the Variscan orogeny. Primary reservoir rocks, however, can be altered or cemented during diagenesis, basin inversion, and related hydrothermal phases, or can simply be eroded. Thus, the comprehensive geodynamic history of a geological system needs to be analyzed to delineate potential reservoir rocks hosting fluids and/or potential for hot dry-rock treatment.

While the hosting of thermal fluids can be controlled by facies, as described above, the channelways for fluids are instead controlled by structures (e.g., faults and fractures) in many geothermal systems. Moreover, in tectonically active regions, such as the largely amagmatic extensional Basin and Range province in western North America, highly permeable and porous fault breccia may actually host geothermal reservoirs in highly faulted areas (Curewitz and Karson, 1997; Faulds *et al.*, 2006). It is therefore critical to determine which types of structures and which segments of faults are most favorable for providing fluid pathways in geothermal systems. Such structures must be characterized in known geothermal systems in order to guide exploration for new systems and expansion of existing systems (whether conventional or EGS). To facilitate this important characterization of structural controls, integrated field-based geological and geophysical studies are necessary. The field-based studies involve detailed geologic mapping of structures (e.g., faults, fractures, and folds), stratigraphic units, and surficial geothermal features, as well as kinematic analysis of faults, such that the geometry and kinematic evolution of controlling fault systems are defined. Geophysical methods include subsurface borehole imaging and analysis, gravity surveys, resistivity studies, and seismic reflection data (as discussed more thoroughly in geophysical section). The geological and geophysical data sets can then be synthesized to generate sophisticated 3D geological and stress/strain models of the geothermal system, which can provide critical information for targeting favorable drilling sites and ultimately developing a geothermal system. It is important to note, however, that in some settings (highly vegetated areas or areas of low relief in basin interiors, passive continental margins, etc.), field-based geologic studies yield limited results. In such areas, analysis of the stratigraphic and structural framework and delineation of geothermal reservoirs depends largely on geophysical investigations and their geologic interpretation.

Recent studies of fields in the western United States and western Turkey have revealed several favorable structural settings for geothermal activity in these young largely amagmatic extensional domains. Such settings, investigated by analysis of the 3D geometry and kinematic evolution of fault systems, include discrete steps in fault zones or belts of intersecting, overlapping, and/or terminating faults (Figure 2.1; Faulds *et al.*, 2006). In addition, most fields are associated with steeply dipping faults and, in many cases, with Quaternary faults. The structural settings favoring geothermal activity generally involve subvertical conduits of highly fractured rock along fault zones oriented approximately perpendicular to the least principal stress. Features indicative of these settings that may be helpful in guiding exploration include (i) major steps in range-fronts, (ii) interbasinal highs, (iii) mountain ranges consisting of relatively low, discontinuous ridges, and (iv) lateral terminations of mountain ranges (Faulds *et al.*, 2006). Even magmatic systems such as those in Iceland are also controlled by tensional fractures (Gudmundsson, 1999, 2000). However, dilational fault segments are not the only type of conduit for hydrothermal systems. It is known, for example, from other fields such as the EGS site in the Coso geothermal system, that critically stressed faults can control fluid flow (Sheridan and Hickman, 2004).

However, the above examples represent amagmatic or magmatic geothermal systems in which tectonic features are relatively well exposed on the surface. In these cases, surface features that constrain the location, geometry, and controls of the geothermal system can be synthesized with geophysical data and extrapolated into the subsurface to define the extent of the geothermal reservoir. This is not possible in many parts of the world with significant geothermal potential, including some tectonically active regions in relatively moist, highly vegetated areas (e.g., Larderello, Italy; Oregon, USA), and deep sedimentary basins on passive continental margins or in continental interiors (e.g., Paris and North German basins).

In areas with limited surface exposures, several methods must be combined to generate an integrated geological model. These include a comprehensive assessment of the geodynamic history of the region, various geophysical investigations, and quantification of both the geothermal gradient and stress/strain along interpreted faults or within sediment-hosted geothermal reservoirs. Critical geophysical methods include gravity, MT, resistivity, and seismic reflection surveys. The gravity and seismic reflection data would collectively indicate the location of major faults

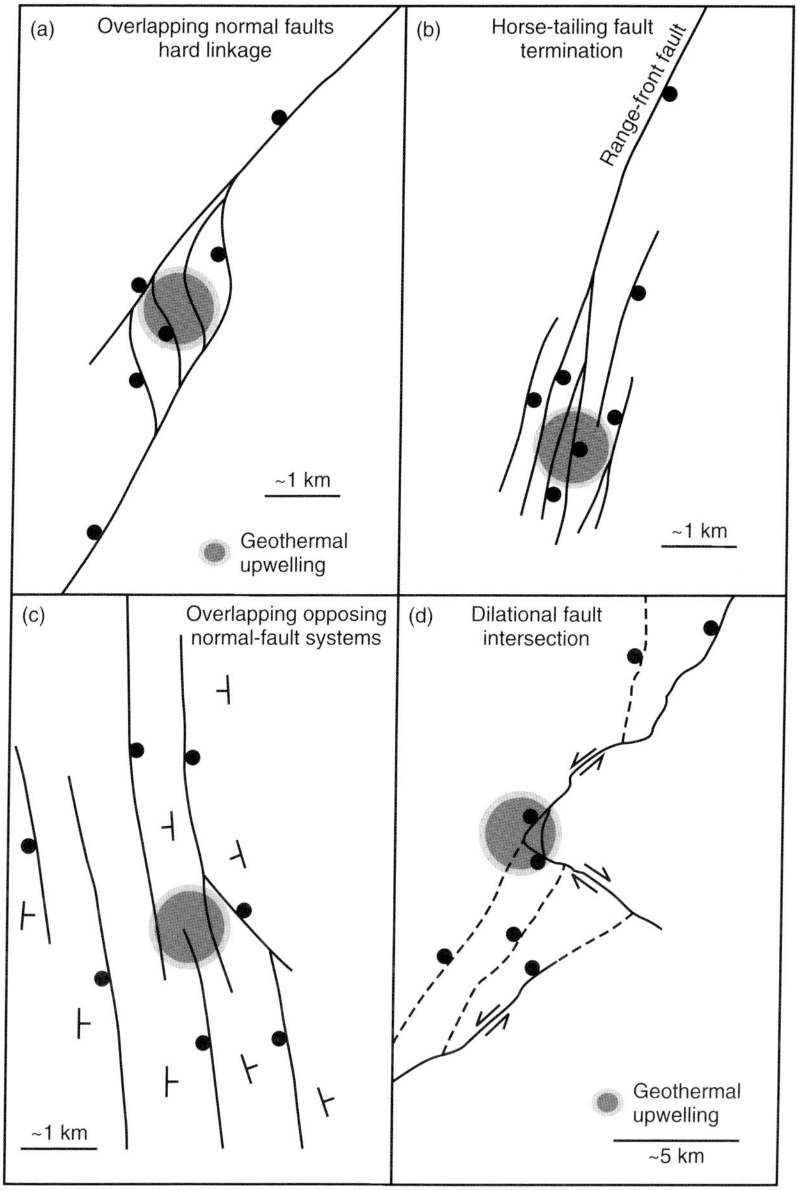

Figure 2.1 Examples of favorable structural settings for geothermal systems. Areas of upwelling geothermal fluids are shaded in dark gray. (a) Stepover between two overlapping normal fault segments with multiple minor faults providing hard linkage between two major faults. (b) Terminations of major normal faults, whereby faults break up into multiple splays or horsetail. (c) Overlapping, oppositely dipping normal fault systems that generate multiple fault intersections in the subsurface. (d) Dilational fault intersection between oblique-slip normal faults.

and areas of structural complexity, such as overlapping faults and fault intersections, where fracture density would likely be greatest. The MT and resistivity data would permit assessment of which faults or stratigraphic horizons accommodated significant fluid flow. Evaluation of the geodynamic history, GPS geodetic data, any available borehole data (e.g., breakouts and FMI (Formation Micro Image) imagery), and known regional fracture patterns would then help to quantify stress/strain along interpreted faults and fractures and thus determine which structures would likely accommodate dilation or shearing and related maximum fluid flow. The geothermal gradient would dictate at what levels these fluids would provide a viable geothermal resource. In deep sedimentary basins in tectonically quiescent areas, highly permeable sedimentary horizons, at levels deep enough to record sufficient temperatures, typically provide the most favorable geothermal reservoirs (e.g., mid-Jurassic Dogger Limestone and lower Triassic sandstone in Paris basin (Dezayes *et al.*, 2008) and lower Permian sandstone in North German basin (Moeck *et al.*, 2009)). In tectonically active regions, fault zones are commonly the most important targets as they can channel geothermal fluids from deep levels in the crust to relatively shallow reservoirs, thus providing a more accessible and more economical resource.

2.3 Relevance of the Stress Field for EGS

It is important to note, however, that exploration not only involves the delineation of systems and zones favorable to geothermal development but also requires the characterization of the subsurface to optimize the development of EGS sites. Some of the crucial aspects are determination of the stress field and the transference of lithostratigraphy to mechanical stratigraphy in terms of reservoir geomechanics. If the stress field is known and the general mechanical behavior of rock types is considered, the access and further treatments of the reservoir rock can be optimally planned.

The stress fields analyzed for a region of interest are often found to be predictable by lithospheric scale models assuming an overall control the forces exerted by the boundaries of tectonic plates and their relative movements. However, there are also examples of significant deviations from that overall trend at the regional scale (Hillis and Reynolds, 2000). In some cases, these large-scale stress orientations hold for basement level rocks; the stress regime in the sedimentary cover seems to be detached from the basement, for example, in the Central North Sea, where basement rocks show similar stress orientations to the rest of Europe, while for the sedimentary cover in the basin stress orientations varied widely (Cowgill *et al.*, 1995). Similarly, detached stress orientations have been described for the Canadian shelf (Yassir and Bell, 1994) and for the Gulf of Mexico (Yassir and Zerwer, 1997).

Generally, the regional stress field can be derived on the basis of several different types of data, as suggested by Zoback (1992) for the World Stress Map (Reinecker, Heidbach, and Mueller, 2003) and consequently used for regional studies (e.g., the

Australian Stress Map; Hillis and Reynolds, 2000); some do not require information from existing wells (1 and 2), while the others generally rely on information derived from drilling (3 and 4) and borehole engineering (5 and 6) activities in the surrounding region:

1) the determination of focal mechanism solutions from earthquakes of sufficiently high magnitudes occurring in the region, from which the principal stress orientations can be inferred;
2) geological observations such as recent fault slip and volcanic alignments can also serve as first-order stress indicators;
3) failure along the borehole walls (borehole breakouts), which occur in the direction of the minimum horizontal stress (S_h) or
4) drilling-induced tensile fractures that form parallel to the maximum horizontal stress direction (S_H);
5) hydraulic fracturing that can be induced by well and/or reservoir engineering; and
6) overcoring.

In the early stage of field development, before drilling or with no available stress magnitude data, stress models can be developed assuming that *in situ* stress magnitudes in the crust will not exceed the condition of frictional sliding on well-oriented faults. Commonly, geometrical constraints such as fault throw and fault intersections in mapped 3D fault patterns, for example from seismic surveys, indicate a limited variation of stress regimes, ranging from normal faulting ($S_V > S_{Hmax} > S_{hmin}$) to transtensional ($S_V = S_{Hmax} > S_{hmin}$) to strike slip ($S_{Hmax} > S_V > S_{hmin}$) or reverse faulting ($S_{Hmax} > S_{hmin} > S_V$), as shown in Figure 2.2a, where S_{Hmax} and S_{hmin} are the maximum and the minimum horizontal stresses, while S_V is the vertical stress.

Stress values for any given stress regime can be predicted using Equation (2.1) and assuming Andersonian fault theory (Anderson, 1951) and the Mohr–Coulomb criterion. Applying the known stresses S_V (vertical stress) and S_{hmin} (minimum horizontal stress) and Equation (2.1), the value for S_{Hmax} (maximum horizontal stress) in the reservoir can be constrained. The frictional equilibrium applicable for a geothermal reservoir is (after Jaeger, Cook, and Zimmerman, 2007)

$$\frac{\sigma_{1\text{eff}}}{\sigma_{3\text{eff}}} = \frac{(\sigma_1 - P_p)}{(\sigma_3 - P_p)} = \left[(\mu^2 + 1)^{1/2} + \mu\right]^2 \tag{2.1}$$

Parameters used in this equation include a frictional coefficient μ, ranging from 0.6 to 1.0 for most rock types, as suggested by Byerlee (1978) on the basis of experimental data, and pore pressure P_p; σ_1 and σ_3 are the maximum and the minimum principal stresses, respectively (see also Peška and Zoback, 1995; Moeck *et al.*, 2009).

The *in situ* stress tensor in a reservoir can be derived only from failure along the borehole wall, that is from borehole breakouts and tensile fractures. The opening angle of borehole breakouts can be used to determine the maximum horizontal

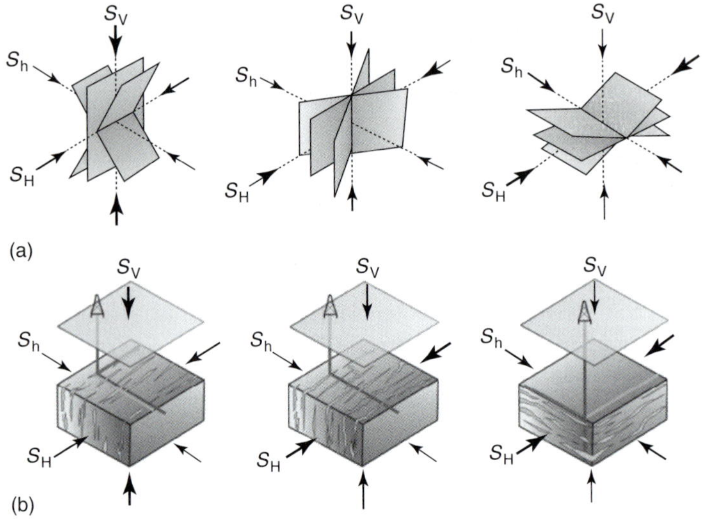

Figure 2.2 (a) Geometrical relation between stress axes, stress regimes, and fracture planes. Brown: shear fractures; blue: tensile fractures. Stress regimes from left to right: normal faulting, strike-slip faulting, and reverse faulting. (b) From left to right, orientation of tensile fractures in normal faulting, strike-slip faulting, and reverse faulting regime. Red drill path is least stable; green drill path is most stable. In strike-slip regimes, the most stable drill path depends on the stress ratios of S_V and S_H. (Please find a color version of this figure on the color plates.)

stress value if the vertical stress and minimum horizontal stress values are known (Moeck and Backers, 2007; Zoback, 2007). The magnitude of the minimum horizontal stress can be determined by hydraulically induced tensile mini-fracs or leakoff tests (LOTs), where the fracture opening pressure is nearly equivalent to the minimum horizontal stress magnitude. In critically stressed reservoirs, this value of the minimum horizontal stress might not be determinable, because a shear fracture develops prior to a tensile fracture. The orientation of the stress tensor can, however, be determined only by borehole breakouts or induced fractures in the borehole. Typical data sources for such studies are image logs such as BHTV (bore hole televiewer), FMI/UBI (formation imager), or caliper logs that measure the elongation of the borehole. Combining the methods of stress regime determination and LOT and evaluating the vertical stress, which is generally known from the overburden density and thickness, the complete stress tensor can be calculated in magnitude and direction.

Brittle failure of rock is commonly described by the Mohr–Coulomb criterion (Figure 2.3a). The Mohr circle is the illustration of acting stresses in rock. A stress field is defined by the main principal stress axes $s1 > s2 > s3$. The failure mode tensile (A, Figure 2.3a), hybrid tensile (shear and tensile; B, Figure 2.3a), and shear (C, Figure 2.3a) are dependent on the differential stress $s1-s3$. In low differential stress (near surface), tensile failure is most likely, and in depths $>$2000 m, shear failure is more likely due to high differential stresses and related high normal

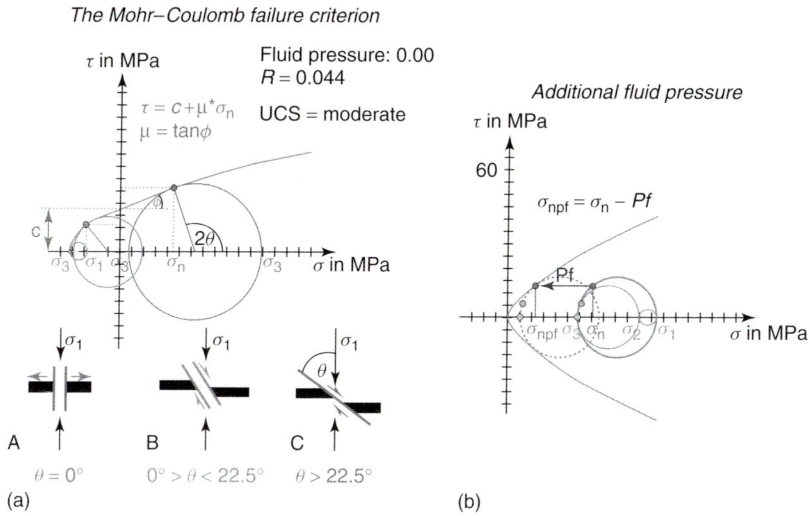

Figure 2.3 The Mohr–Coulomb failure criterion (see text). (Please find a color version of this figure on the color plates.)

stresses acting on fracture planes. Which plane will fail is dependent on the angle between failure plane and maximum principal stress axis s1. By understanding the stress field it is possible to estimate the orientation of likely failure planes in the current stress field based on Anderson's faulting theory (Figure 2.2a).

Additional fluid pressure as happened with fluid injection during reservoir stimulation has a significant effect on failure. As shown in the illustration (Figure 2.3b), additional fluid pressure decreases the normal stress and failure occurs as soon as critical shear conditions acting on a plane are reached (as defined by the green critical shear envelope). Existing fault planes have no cohesion (as illustrated in the above case), so it is easier to reactivate existing faults even when not optimally orientated to the maximum principal stress axis. Reactivation of faults is commonly shear failure because even under very low differential stresses tensile failure (the blue dot) and also hybrid shear failure (pink dot) would happen after shear failure. Only under very high additional fluid pressures all kinds of failure could occur.

An understanding of the state of stress is important for reservoir evaluation in terms of fluid flow. The stress state of faults affects the transmissivity within an often complex fault pattern. Transmissivity, which is the extent to which fractures are hydraulically conductive, depends strongly on their aperture (Cook, 1992), which in turn is primarily, yet not solely, affected by the fracture orientation within the *in situ* stress field. Preexisting faults and fractures that are critically stressed for either tensile (Gudmundsson, Fjeldskaar, and Brenner, 2002) or shear failure Barton, Zoback, and Moos, 1995) within the *in situ* stress field are most likely to be open and hydraulically conductive (Ferrill *et al.*, 1999; Talbot and Sirat, 2001). Generally, tensile failure is unlikely to be the dominant fracture reactivation mechanism below a depth of approximately 2000 m, as the high lithostatic overburden causes high differential stresses (Ferrill and Morris, 2003). In a fractured rock, shear reactivation of preexisting faults normally occurs at lower pore-fluid pressure than tensile fracturing. However, new tensile fractures can form and serve as conduits for fluid flow (Sibson, 1996, 1998) if

- rocks are intact and do not contain favorably oriented, cohesionless faults
- existing faults are not favorably oriented for shear reactivation;
- existing favorably oriented faults have become cemented and regained cohesive strength.

The last point deserves attention, as it shows that the determination of fracture orientation in space may not suffice to evaluate its potential as a fluid conduit because of mineralization due to continuous circulation of hydrothermal fluids (Morrow, Moore, and Lockner, 2001). In addition, Sibson (1998) pointed out that faults may have regained cohesive strength and thus behave like an intact rock rather than a cohesionless fault. Sufficient geochemical characterization of the reservoir, including both rocks and fluids, is essential to address this specific case adequately. In contrast, fractures that are not favorably oriented within the *in situ* stress field may well serve as fluid conduits if they are propped open by grains that prevent crack closure despite the stress orientation (Hillis, 1998).

A graphic evaluation of the orientation of fractures with respect to the *in situ* stress field, the fault rock strength, and the corresponding likelihood of the fracture to be critically stressed and hydraulically conductive is the fracture susceptibility diagram (Mildren, Hillis and Kaldi, 2002; Hillis and Nelson, 2005). It is constructed as a stereoplot (Figure 2.4), which is color coded by the amount in pore pressure ΔP_p that leads to failure of a fracture for a given failure envelope in the Mohr circle diagram.

Knowledge of fracture orientation with respect to the stress tensor is important for well planning if deviated wells are considered in a certain direction, relative to one of the principal stress axes, to cross natural tensile fractures, or to enable multiple hydraulic fractures, which are both in the plane of the maximum and intermediate stress axes (Figure 2.2a and b). The orientation of an induced tensile fracture at the wellbore wall can be predicted, given knowledge of the *in situ* stress tensor and wellbore trajectory (Peška and Zoback, 1995). A well path optimized for fracture stimulation within the stress field is, however, not necessarily the safest in terms of borehole stability; thus, changes in stress concentrations around the borehole and mechanical behavior of rocks should be considered before reservoir access (Figure 2.2b).

Fracture stimulation is used to enhance reservoir performance, particularly in low-permeability reservoirs. It is achieved by artificially increasing pore-fluid pressure (Chapter 4). This kind of human intervention can cause a modification of *in situ* stress conditions that can be significant enough to change fault behavior. Also, production and injection through wells, both important elements in sustainable reservoir management, change the stress field by modification of the pore pressure, which is also referred to as *formation pressure*. Injection causes an increase in formation pressure, which in turn causes a decrease of normal stresses acting

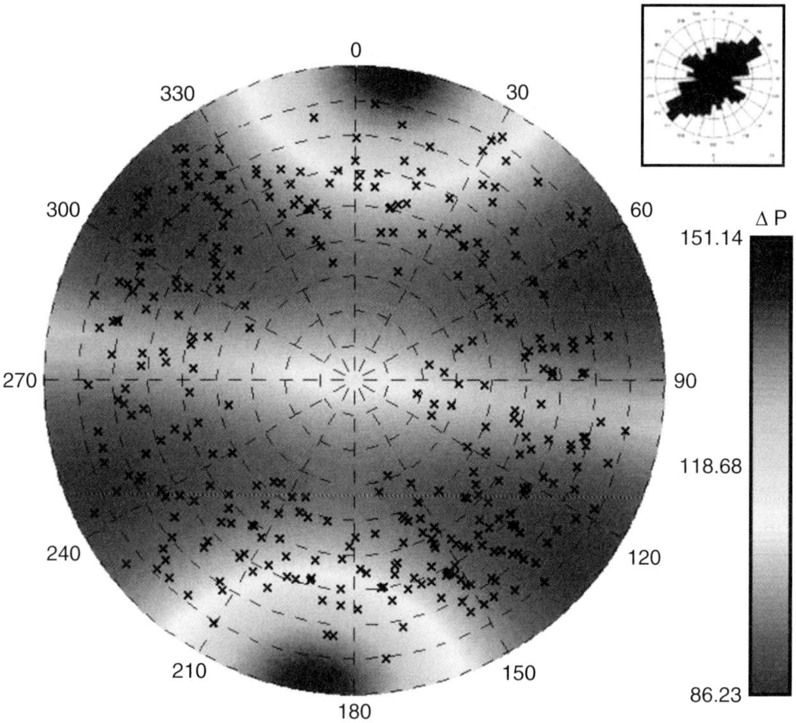

Figure 2.4 Fracture susceptibility diagram. The amount of pore pressure increase, ΔP_p needed to cause failure of a fracture with a given orientation is indicated by the color scale shown at the right edge of the image. Fracture orientations observed from image logs or oriented cores can be plotted as planes to poles. If they lie in the red areas of the diagram, ΔP_p is relatively low and fractures are more likely to fail and to be conductive than in the blue areas (after Mildren, Hillis, and Kaldi, 2002). In the example shown here, steeply dipping fractures striking NW–SE or NE–SW are much more likely to be conductive than a steeply dipping fracture striking E–W. (Please find a color version of this figure on the color plates.)

on planes. An increase in the normal stress, effectively an increase in the ratio of shear to normal stress along fracture planes, can evoke fault reactivation if the frictional strength of the fault is reached. In contrast, production means a decrease of the formation pressure causing an increase in normal stresses, which can lead to frictional blockade and closure of a fracture plane and hence to a reduced fracture transmissivity and lower production rates. It is therefore crucial to understand the fault behavior under changed stresses and to characterize the fault systems.

An approach to describe the stress state along a fault that serves as a fluid conduit is the concept of slip tendency introduced by Morris, Ferrill, and Henderson (1996). The slip tendency analysis was originally developed for fault characterization in earthquake prone areas. It is a technique that permits the rapid assessment of reactivation and leakage potential of any fault population within the stress field under initial and changing pore pressure conditions. For the EGS project at Groß Schönebeck in the Northeast German Basin, this approach was successfully applied to describe the stress state along faults under initial and modified formation pressure, and finally to assess the fault reactivation potential and to understand recorded microseismic events during massive water stimulation (Moeck, Kwiatek, and Zimmermann, in press).

The slip tendency is the ratio of resolved shear stress to resolved normal stress on a surface (Morris, Ferrill, and Henderson, 1996). It is based on Amonton's law that governs fault reactivation:

$$\tau = \mu_s{}^* \sigma_{\text{neff}} \tag{2.2}$$

where τ is the shear stress, σ_{neff} the effective normal stress ($\sigma_n - P_p$), and μ_s the sliding friction coefficient (Byerlee, 1978). According to this law, *stability or failure* is determined by the ratio of shear stress to normal stress acting on the plane of weakness and defined as slip tendency T_s (Lisle and Srivastava, 2004; Morris, Ferrill, and Henderson, 1996). Slip is likely to occur on a surface if resolved shear stress, τ, equals or exceeds the frictional sliding coefficient and slip tendency is given as

$$\tau_s = T/\sigma_{\text{neff}} \geq \mu_s \tag{2.3}$$

The shear and effective normal stress acting on a given plane depend on the orientation of the planes within the stress field that is defined by principal effective stresses $\sigma_{1\text{eff}} = (\sigma_1 - P_p) > \sigma_{2\text{eff}} = (\sigma_2 - P_p) > \sigma_{3\text{eff}} = (\sigma_3 - P_p)$(Jaeger, Cook, and Zimmerman, 2007):

$$\sigma_{\text{neff}} = \sigma_{1\text{eff}}{}^* l^2 + \sigma_{2\text{eff}}{}^* m^2 + \sigma_{3\text{eff}}{}^* n^2 \tag{2.4}$$

$$T = \left[(\sigma_1 - \sigma_2)^2 l^2 m^2 + (\sigma_2 - \sigma_3)^2 m^2 n^2 + (\sigma_3 - \sigma_1)^2 l^2 n^2\right]^{1/2} \tag{2.5}$$

where l, m, and n are the direction cosines of the plane's normal with respect to the principal stress axes, σ_1, σ_2, and σ_3 respectively. Equations (2.4 and 2.5) define effective normal stress and shear stress for compressional stress regimes, that is, σ_{1eff} is horizontal. Extensional and strike-slip regimes can be derived by changing the order of the direction cosines in these equations (Ramsay and Lisle, 2000).

Dilation of faults and fractures is largely controlled by the resolved normal stress which is basically a function of lithostatic and tectonic stresses and fluid pressure. On the basis of Equation (2.4), the magnitude of normal stress can be computed for surfaces of all orientation within a known or suspected stress field. This normal stress can be normalized by comparison with the differential stress resulting in the dilation tendency τ_d for a surface defined by

$$\tau_d = \frac{(\sigma_1 - \sigma_n)}{(\sigma_1 - \sigma_3)} \tag{2.6}$$

Slip and dilation tendency stereoplots are obtained by solving Equations (2.3 and 2.4) for all planes in 3D space, substituting in Equation (2.2) for shear stress distribution along fault planes, and by solving Equation (2.5) for normal stress distribution along fault planes plotting the results in equal area stereonets (Morris, Ferrill, and Henderson, 1996; Ferril and Morris, 2003). As such, this slip and dilation tendency analysis is a technique that permits the rapid assessment of stress states and related potential fault activity through easy visualization.

Faults with a high slip tendency are critically stressed faults with a high amount of shear stress. They have a high reactivation potential as shear fractures and are therefore prone to seismicity during stimulation of critically stressed reservoirs. Faults with a high dilatational tendency bear low shear stresses and low normal stresses (Moeck *et al.*, 2009). During the operations at the Groß Schönebeck field, a massive water stimulation lasting six days induced surprisingly low seismicity of magnitudes −1 to −2 as described by Moeck *et al.* (2009). The slip tendency analysis, however, revealed a low slip tendency of optimally oriented faults resulting from high rock strength and therefore a high frictional resistance of any faults. Thus, slip is very unlikely to occur under initial reservoir conditions and a significantly higher pore-fluid pressure of 20 MPa is needed to increase the slip tendency. Increasing the pore pressure means a reduction of the normal stress acting on a fault plane. An increasing ratio of shear to normal stress is effectively an increase in slip tendency. The visualization of slip tendency is given in the lower hemisphere projection and shows all faults prone to high slip. Figure 2.4 shows the distribution of faults with highest slip tendency in the red areas for the volcanic succession of the Groß Schönebeck reservoir. However, the slip tendency is below the value of 0.8 which is the limit of frictional resistance (Byerlee, 1978). During stimulation and consequent pore fluid increase, a fracture plane was generated as evidenced by microseismic events in the area of high slip tendency (Figure 2.5). The concert of both results indicates that the slip tendency analysis, originally developed for earthquake assessment, is an appropriate method to investigate, characterize, and understand fault behavior of engineered reservoirs.

The compilation of all available data from the surface and/or subsurface into one integrated a priori 3D geological model will facilitate a comprehensive interpretation. Depending on the geological setting and on available data, conventional geological maps can be used for 3D geological modeling (Moeck *et al.*, 2007). A priori 3D geological models are the portal to further modeling, including flow simulation as part of reservoir engineering or stress modeling, to understand the stress state and fault behavior under initial and changing stress conditions (e.g., during stimulation).

The ultimate purpose of geological exploration studies on geothermal fields is the comprehensive characterization of geological controls on the geothermal systems. A broad understanding of a geological system, including a quantitative structural geological site characterization, does not only delineates favorable areas for future geophysical exploration and drilling but also facilitates all levels of field development and utilization. Exploration geology grounded in field-based and/or

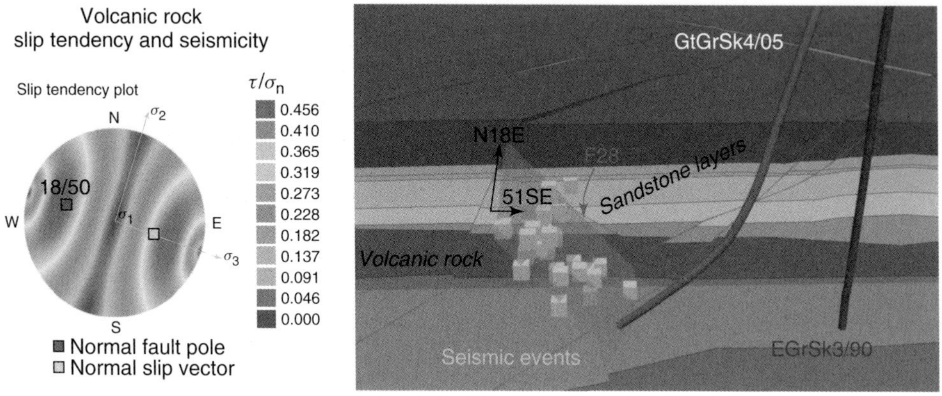

(a) (b)

Figure 2.5 (a) Slip tendency plot of the lower Permian volcanic rocks in the Groß Schonebeck field. The pole of plane represents the mean plane as derived from microseismicity. (b) Mean plane of recorded seismic events together with a spatial distribution of recorded seismicity (yellow boxes) together with least-square fitted plane (transparent yellow). The distribution of seismicity fits the orientation of the F28 fault plane within the reservoir (Moeck, Kwiatek, and Zimmermann, in press). (Please find a color version of this figure on the color plates.)

subsurface data is therefore not only the initial step in any geothermal investigation but also a crucial aspect of any EGS project.

2.4 Geophysics

Geophysical methods have played a key role in geothermal exploration for many years. Specific exploration techniques and their possible combinations for EGS applications are not well established yet, because the parameters searched for a profitable utilization are not only the physical parameters of the geothermal system itself but additionally information on the condition of the reservoir (e.g., stress, strain, and (pore-) pressure) that have to be derived from the surface.

The geophysical methods are usually aimed at yielding information about a possible geothermal reservoir, the heat source, and the hydraulic situation. In the case of exploration for possible EGS applications, the methods used should additionally help to obtain precise information about structural and tectonic setting, regional and local stress field, and many other parameters in a depth range up to several kilometers, which are critical for later stimulation procedures.

Therefore, in this chapter, today's most prominent geophysical exploration methods are outlined together with remarks on possible developments toward specific EGS exploration methodologies. These, of course, have to be most closely linked to geological, geochemical, and geophysical well logging information as well as to rock physics from laboratory experiments.

2.4.1
Electrical Methods (DC, EM, MT)

Most rocks are poor conductors, but they are usually porous and the pores are filled with fluids, which means that they are also electrolytic conductors. As the reservoirs are volumes of rock filled with hot fluids (water, vapor, and gas), electrical resistivity of subsurface rocks is the most diagnostic parameter for geothermal resources that can be measured from the surface. In addition, electrical resistivities are strongly temperature dependent. Bulk conductivity increases by more than an order of magnitude when temperatures are raised from room temperature to 200 °C (Yokoyama *et al.*, 1983). Up to the critical temperature of water, a temperature increase of the water in the pores enhances the conductivity additionally. When we approach the melting point of a rock, even more significant changes in electrical properties take place. Therefore, electrical methods have gained the same significance in geothermal exploration as seismic methods have in oil exploration. They are extensively used to obtain a first approximation of subsurface conditions, because an area of several square kilometers can be studied within a short time and the costs are relatively low.

Generally, dense volcanic, igneous, and carbonate rocks have higher resistivities than clastic sedimentary rocks, while for shales and clays resistivity values are the lowest (1–10 Ω m). Resistivities for hydrothermal reservoirs are typically lower than for the surrounding rocks and depend on several factors, for example, the porosity of the rocks and the salinity of the fluids. These interdependencies can be described by a formula empirically derived by Archie (1942):

$$\rho = \alpha \phi^{-m} S^{-n} \rho_{\mathrm{w}} \tag{2.7}$$

where ρ and ρ_w are the resistivity of the formation and of the pore water, respectively, ϕ is the porosity and S the water saturation. α, m, and n vary for different rock types. In case of sandstone, the tortuosity, α, ranges from 0.5 to 2.5, the cementation factor, m, ranges from 1.3 to 2.5, and the saturation index, n, is typically 2, while for loose sand, typically $\alpha = 0$ and $m = 2.15$. The ratio of formation resistivity to water resistivity is often referred to as the *formation factor F*, such that

$$F = \rho/\rho_{\mathrm{w}} = \alpha \phi^{-m} \tag{2.8}$$

In a porous, fluid-filled rock, conductivity is commonly composed of the conductivity of the fluid and of the surface conductivity. Assuming parallel circuit behavior, the rock conductivity σ is

$$\sigma = \sigma_{\mathrm{c}} + \sigma_{\mathrm{w}}/F \tag{2.9}$$

and as $\rho = 1/\sigma$

$$1/\rho = 1/\rho_{\mathrm{c}} + 1/F\rho_{\mathrm{w}} \tag{2.10}$$

The contribution of the fluid conductivity increases with the amount of dissolved solids, while surface conductivity becomes important with the presence of clay minerals, which often occur as a product of hydrothermal alteration and weathering.

If surface conductivity contributes significantly to the overall conductivity, as is the case in geologic formations containing clay minerals, Archie's equation is not applicable (Klein and Sill, 1982). The degree of resistivity change with clay content depends on clay mineralogy: The strongest effect is observed for montmorillonite – which can lower the resistivity by two orders of magnitude (Nishikawa, 1992) – and for sericite, whereas it is not as pronounced for kaolinite, alunite, and chlorite. The clay effect is strongest when the salinity of the fluid is low, while it becomes negligible for salinities of 0.1 mol l^{-1} KCl and more.

Methods to measure resistivity of the subsurface can basically be divided into two general groups:

- those that measure the difference in electrical potential (DC, i.e., direct current);
- those that measure electromagnetic fields, natural or artificially created.

There is one main difference between electrical potential and electromagnetic (inductive) techniques. The latter usually provide information on conductivity–thickness products of conductive layers, and, generally, only thickness information on resistive layers. In contrast, resistivity techniques usually provide information on resistivity–thickness products for resistive layers and conductivity–thickness products for conductive layers. Because in most cases the exploration target is conductive, EMs are more suitable.

Complications with electrical methods in geothermal exploration arise, if the surrounding rocks are hydrothermally altered and also display low resistivities. Such alterations, often indicative of previous hydrothermal activity, can make even dry rocks look like a promising reservoir.

2.4.1.1 **Direct Current (DC) Methods**

A common method for studying the electrical resistivity in the subsurface is to apply an electric potential to two electrodes driven into the ground separated some distance from each other. The potential field, built up between this pair of electrodes, is recorded by means of a sensitive voltmeter connected to another pair of electrodes. The depth of penetration is given by the geometry of the array used and apparent resistivities (in ohm meters (Ωm)) can be calculated for resistivity depth soundings and/or resistivity mapping. Resistivity distribution in the subsurface can then be obtained either by forward or inverse modeling.

Standard potential methods use DC depth soundings with varying electrode configurations such as the Schlumberger or the Wenner arrays. These methods are simple to use, but relatively slow in progress. The application of these soundings is limited in many geothermal areas where the lateral extent of the anomalous resistivities is small compared to the required spread between the electrodes. They are therefore often used for mapping and delineation of shallow resources.

Other DC methods are developed around dipole sources. In the dipole–dipole arrangement, two similar pairs of closely spaced electrodes are moved along a profile; all electrodes are kept in one line. This procedure is repeated with varying electrode spacings, thus yielding the so-called pseudosections of apparent resistivity. These sometimes show a good correlation with contour maps of the subsurface resistivity distribution. In the so-called bipole–dipole arrangement (or roving dipole), one pair of electrodes is kept fixed while the other pair, usually more closely spaced, is moved around, which allows the determination of anisotropies in the resistivity of the subsurface.

These methods allow reasonable resolution down to depths of 2000 m and were used quite regularly for geothermal prospecting in the past. Another application in DC electrical prospecting for geothermal anomalies is the self-potential method which was also quite common for measurements in geothermal areas where it revealed anomalous regions associated with near-surface thermal zones and faults that are thought to be fluid conduits.

More commonly applied is DC Tomography and E-Scan, which is a proprietary method. The other methods mentioned make use of electromagnetic fields.

2.4.1.2 Electromagnetic Methods

The principle behind EMs is governed by Maxwell's equations that describe the coupled set of electric and magnetic fields' change with time: changing electric currents create magnetic fields that in turn induce electric fields that drive new currents. Most EM techniques (controlled source audio magnetotellurics (CSAMT), TDEM, FDEM, GPR, and NMR) use a controlled artificial electromagnetic source as a primary field that induces a secondary magnetic field, while MT methods use the earth's natural electromagnetic field as source signal.

EM methods can be used for exploration and monitoring of circulating fluids in reservoirs or faults and thus provide important information about their activity and fluid content. As the phase change of pore fluid (boiling/condensing) in fractured rocks can result in resistivity changes that are more than one order of magnitude greater than those measured in intact rocks, EM methods can provide information of primary economic significance. In addition, production-induced changes in resistivity provide valuable insights into the evolution of the host rock and resident fluids and thus into the sustainability of a reservoir.

2.4.1.3 The Magnetotelluric Method

In the MT method, the earth's impedance to the natural EM wave field is measured to extract information about variations in the resistivity of the subsurface. The method has been used for about 30 years now and has improved continuously in both equipment and interpretation, and, despite its numerous pitfalls, it has become the standard method in geothermal exploration. The main advantage to all other electrical methods is its ability to probe depths of several tens of kilometers

In the MT method natural EM waves, generated by thunderstorm activity, provide signals with frequencies higher than 1 Hz, while frequencies lower than 1 Hz are caused by large-scale ionospheric currents created by the interaction between the solar wind and the magnetosphere. At large distances from the source, the resulting electromagnetic field is a plane wave of variable frequency (from about 10-5 Hz up to audio range at least). The subsurface structure can be studied by making simultaneous measurements of the strength of the magnetic field variations at the surface of the earth and the strength of the electric field component at right angles in the earth. Because the direction of polarization of the incident magnetic field is variable and not known beforehand, it is common practice to measure at least two components of the electric field and three components of the magnetic field variation to obtain a fairly complete representation. Another assumption in MT is that the displacement currents can be neglected since conduction currents dominate the electromagnetic behavior. The dominant diffusive process makes it possible to obtain responses of volumetric averages of the measured earth's resistivity. Measurements are usually represented as MT-apparent resistivity and phase as a function of frequency.

The investigation depth is a function of the electrical resistivity ρ of the earth and angular frequency, ω, of the EM field. Since earth is a conductor, the electromagnetic wave is governed by a diffusion process in the earth. This implies that the field strengths attenuate (decrease exponentially) with depth. A reasonable measure of the penetration scale length is the skin depth δ, which corresponds to the depth at which the amplitude of the incident electromagnetic field has attenuated by a factor of $1/e$.

A useful approximation for a uniform half-space of resistivity ρ is given as:

$$\delta \approx 500\sqrt{\rho/f} \quad D = \frac{\delta}{\sqrt{2}} = 356\sqrt{\frac{\rho}{f}} \quad \text{(meters)} \tag{2.11}$$

where D is the so-called investigation depth and f the frequency ($f = \omega/2\pi$).

The skin depth relation shows that investigation depth depends not only on frequency but also on the resistivity of the subsoil.

Depending on the measurement frequency range and thus investigation depth, MT methods are named differently. MT measures in the frequency range 1 to 10^{-6} Hz, where studies focus on imaging crustal and mantle geological targets. Natural electromagnetic source energy is usually adequate to ensure the full frequency spectrum. In the mesoscale frequency range, from 1 to 105 Hz, the method is referred to as *audiomagnetotelluric* (AMT). A controlled electromagnetic source (CSAMT, see below) is commonly used at higher frequencies to prevent low signal-to-noise ratios where cultural noise and a weak natural signal may be present. At the very shallow scale radiomagnetotellurics (RMTs) measurements in the frequency range of 15 to 250 kHz using a radiotransmitter allow detailed characterization within the first tens of meters of depth.

At each MT station, five measurements (channels) are recorded. These are the magnetic field in two horizontal directions and in the vertical direction, and the electric field in two horizontal directions, the horizontal measurements being perpendicular (e.g., north and east). A typical MT station for data acquisition consists of two pairs of electrodes set up as orthogonal dipoles with lengths between 50 and 100 m, and three magnetometers (typically flux gates or induction coils) also set up in orthogonal directions (two horizontal, the same as the electric dipoles, and vertical) as sketched in Figure 2.6. The two dipoles measure the electric field fluctuations in the horizontal directions from the potential difference

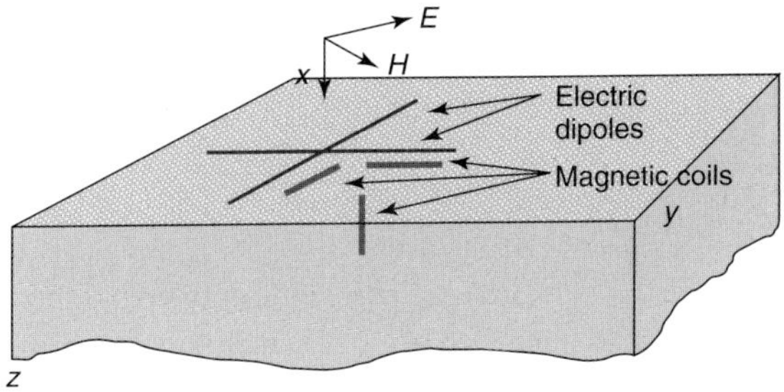

Figure 2.6 MT field setup: The directions are labeled as x, y, and z, with z being the vertical direction. The electric field is abbreviated "E" and the magnetic field is abbreviated "H", such that components of the fields measured are E_x, E_y, H_x, H_y, and H_z.

between them. The magnetic field fluctuations in the three spatial directions are measured from the electric currents induced in the magnetometers. The stations can be anywhere from a few hundred meters to tens of kilometres apart depending on the required resolution for detailed reservoir-scale mapping or a general reconnaissance.

Signals vary in strength with time. Therefore, recording times have to be long compared to the period of interest, which is time dependent on the depth to be investigated in order to get enough signal and ensure high-quality data. For a maximum period of 100 seconds, corresponding to a depth of 1–2 km, recording takes approximately one day, while for periods of 10 000 seconds and depths down to 100 km it can take several weeks. If the area is particularly noisy or the signal is low, the measurements are usually longer in order to improve the statistical properties of the data. A typical survey consists of several MT stations running in parallel and moved after the required recording. The data recorded by the sensors (time series of electric and magnetic fields) are converted to digital form and are not only stored for later spectral analysis but also usually converted immediately to spectral form and processed in real time, providing a clear idea of data quality during ongoing fieldwork.

From the acquired data, which are recorded as changes in the electric and magnetic fields with time, the values of apparent resistivity and phase versus frequency are derived (Larsen *et al.*, 1996). In the frequency domain, electric and horizontal magnetic field components are linearly related by the impedance tensor Z and the goal of data processing is to describe this relationship with the best possible accuracy. Several processing steps are usually performed to reach this goal. A crucial step is the removal of noise which is frequently recorded in the proximity of sources of artificial electromagnetic signals, such as electric pasture fences, corrosion-protected pipelines, or railway lines, especially if they are run with DC. Thunderstorms are also possible sources of noise. Noise causes the coherence that is computed as the cross-correlation between the electric and magnetic fields to deviate from unity. If the fields are linearly related, coherence is unity; if there is noise in any of the field components, the coherence will be reduced. When coherence drops below reasonable values (0.85–0.90), it is common practice to discard the apparent resistivities that are calculated (Ritter, Junge, and Dawes, 1977).

This problem of noise is usually addressed with reference to data of another site situated beyond the sphere of influence of the artificial signal (Gamble *et al.*, 1979; Clarke *et al.*, 1983), often referred to as a *remote reference site*. In areas where uncorrelated noise has been a problem in obtaining MT soundings, this procedure has resulted in significant improvements in the quality of the data, provided the electromagnetic noise at the sites is not correlated. Remote reference stations are therefore often situated on islands, where influence of cultural noise is low, for example, the stationary reference site of the Japanese and South Korean Geological Services on the Japanese island of Jeju, the island of Capraia in the Aegean Sea off the west coast of Italy, used for Italian sites in Tuscany, such as Larderello and Travale (Manzella *et al.*, 2010) or the island of Rügen in the Baltic Sea used for the MT survey in Gross Schönebeck, Germany, within the I-GET project (Muñoz *et al.*, 2010). Over a portion of the frequency range where noise is a particular problem (from 0.1 to 10 Hz, the so-called dead band, where signal is particularly low), the multiple-station approach has permitted data to be obtained where previously it had been impossible. Experience taught that a combination of local and very far remote sites – to face both local high frequency noise and far, planar noise sources – proved to be the most effective solution.

As might be expected, the spectral analysis of long data series, combined with the need for extensive tensor rotation and testing of the spectral values, results in a volume of processing that is as time consuming and as costly as the acquisition of the data. Rapid analysis in the field is necessary as the MT method does not always provide useful results, even after measurements have been made with reliable equipment and for a long time. If the natural electromagnetic field strength is unusually weak during a recording period, or if there is some phenomenon which precludes an effective analysis of the fields, it may be necessary to repeat the measurements at a more favorable time. When the analysis is done in the field, decisions about reoccupying stations and siting additional stations can be made in a timely manner that will reduce overall operating costs. After data processing, the impedance components are scaled to obtain the apparent resistivity, ρ_a, similar to that used in DC resistivity techniques, and phase, ϕ, for given frequencies. The processed data can then be used for interpretation. *Apparent resistivity* is defined as the resistivity of the homogeneous earth, which would produce the measured response at each frequency. Two data curves are defined both for resistivity and phase, which are referred to the two pairs of orthogonal electric and magnetic field horizontal components. Usually they are termed *xy* and *yx*, since they refer to E_x/H_y and E_y/H_x in a Cartesian system. For a layered earth (Figure 2.7), the apparent resistivity at high frequency is equal to the true resistivity of the surface layer, where at lower frequencies it asymptotically approaches the resistivity of the bottom layer.

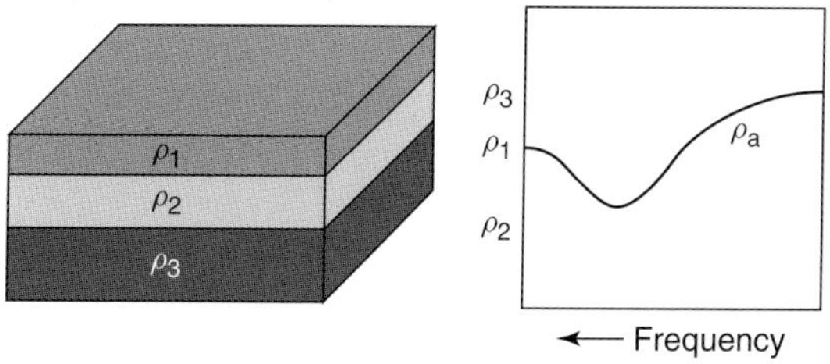

Figure 2.7 MT-apparent resistivity response for a three-layer model.

Dimensionality analysis of the impedance tensor has been proved to be highly important before multidimensional modeling process, given that many 3D environments have been approached with 2D models, which is not always satisfactory. In a stratified medium, the 1D case, resistivity changes only with depth and the impedance tensor is independent of the measurement orientation of the field components. The two polarization curves *xy* or *yx* are the same.

In the 2D case (Figure 2.8), which represents the most commonly assumed situation for MT data interpretation, geoelectrical changes occur with depth as well as in a direction perpendicular to the electrical/geological strike direction. Two different polarization modes can be defined with respect to the geological strike direction: The TE (transverse electric) mode is defined when the horizontal component of the electric field E is parallel to the strike direction and the horizontal magnetic field H is perpendicular. Conversely, the TM (transverse magnetic) mode is defined when the horizontal magnetic field H is parallel to the strike direction and E is perpendicular. When measurements are not performed along the electrical strike direction, the latter can be retrieved by the so-called decomposition analysis using several methods (e.g., Strike, Phase Tensor, and WALDIM) and by trigonometric rotation of the TE and TM curves.

The 3D case represents the most general type of geoelectrical structure where resistivity changes in all directions and the impedance tensor contains all the horizontal electric and magnetic field components independent of the measurement direction. In this case a strike direction cannot be defined. In practice, due to computational or budget limitations, many 3D environments are investigated using 2D profiles making it impossible to compute a 3D model. Since most of these 2D profiles include 3D effects that can lead to misinterpretation, the use of the determinant of the impedance tensor was proposed as a useful tool for computing routine inverse models when it is not possible to determine principal strike direction, given that the determinant is invariant under rotation.

The determinant mode reduces the distortion effects caused by shallow heterogeneities and nonfinite lateral structures, and the phase is not affected by galvanic distortions. The determinant inversion generally allows a good data fit while at the same time resolving reasonably well both resistive and conductive structures along any profile. Estimated resistivity values lie much closer to the true subsurface resistivity in between the extreme resistivities predicted by individual TE and TM

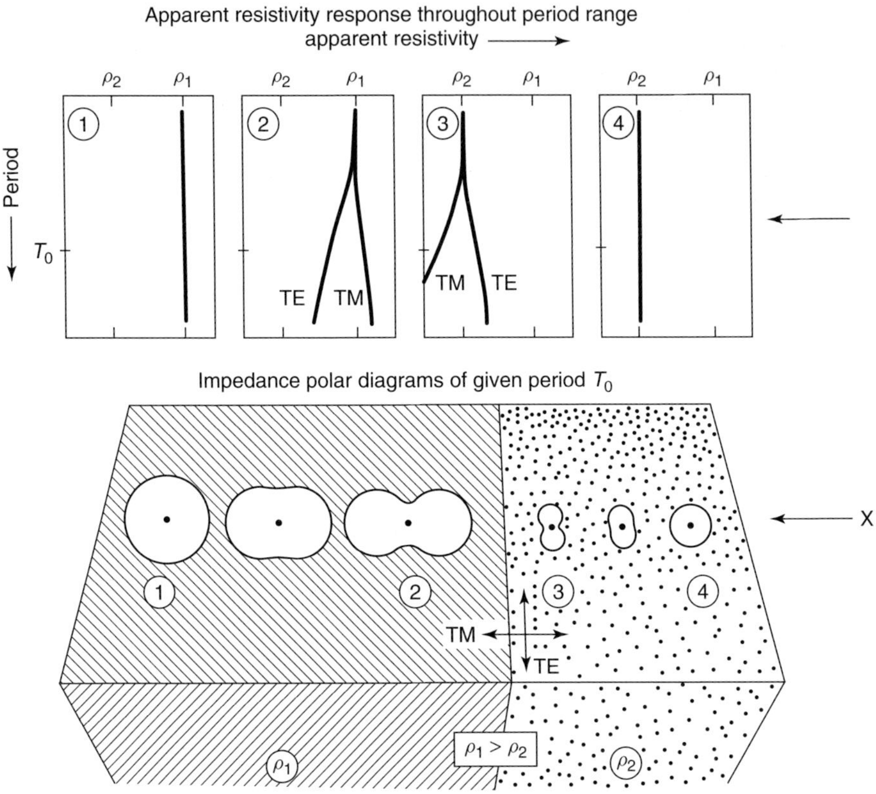

Figure 2.8 Impedance polar diagrams (at one frequency) and apparent resistivities, at four sites on a simple 2D contact model. TE and TM refer to transverse electric (*E* parallel to strike) and transverse magnetic (*H* parallel to strike) field polarizations, respectively. (From Vozoff, 1991).

mode inversions. The determinant mode has been successfully tested in several studies.

The goal of any MT interpretation is a representation of true resistivity with depth. There are two ways such a representation is achieved: forward or inverse modeling. MT multidimensional modeling techniques are well developed. Forward modeling codes can resolve 1D, 2D, and 3D structures by creating a synthetic cross section of the subsurface, computing its MT response, and then comparing it with the actual MT data, using the time-consuming trial-and-error approach. Inversion codes also exist and have been used routinely for computing 1D and 2D responses. 3D inverse codes, even though they have been available for some time (Mackie, Smith, and Madden, 1994; Newman and Alumbaugh, 2000; Zhdanov *et al.*, 2000; Siripunvaraporn *et al.*, 2005), are mostly still in the development stage (Siripunvaraporn and Egbert, 2009). Several field studies have shown promising results both in mineral exploration (Farquharson and Craven, 2009) and in the characterization of geothermal reservoirs (Uchida and Sasaki, 2006; Heise *et al.*, 2008; Newman *et al.*, 2008).

Most 3D inversion codes are based on finite difference approaches (Mackie, Smith, and Madden, 1994; Newman and Alumbaugh, 2000; Sasaki, 2004; Siripunvaraporn *et al.*, 2005), some other approaches such as the edge finite-element method (Farquharson and Craven, 2009; Han *et al.*, 2009) and the IE (Zhdanov *et al.*, 2000) have been tried to develop fast and reliable codes. While the differences in the results they yield are not as big as the step from 2D to 3D inversion, Han *et al.*, (2009) show that methods may be somewhat faster than others. Nonetheless, the main obstacle for their widespread application has been available computing power, as calculation of a 3D MT structure is very challenging. This development is likely to be accelerated with the further improvement of computing speed and power, such that 3D inversion may well become the standard interpretation in the near future.

Limitations, Problems, and Shortcomings of the MT Method The MT method has been refined considerably over the last years but problems still exist, primarily with noise in the measurements and lack of an adequate interpretation. Improvements in data collection, data processing, and three-dimensional numerical modeling continue to reduce such problems. Artifacts are inherent in every inversion algorithm due to noise, undersampling, and three dimensionality, and so inverse modeling results that provide a good data fit should not be regarded as the only possible answer. A geologically reasonable model that fits the data is still the best assurance that a model is credible.

Cultural noise may be considered as a main limitation when no filtering is possible. An alternative approach for noise removal was proposed by Weckmann *et al.* (2005), which uses a combination of frequency domain editing with subsequent single site robust processing. Nonetheless, even with a remote reference and sophisticated processing noise remains a major problem especially in industrialized areas. This problem may even occur once a geothermal field has been developed. Subsequent MT exploration and monitoring is more difficult because pipes and pumps generate a lot of electromagnetic noise that will contaminate the natural signals.

As far as resolution with depth is concerned, the deeper the unit is, the thicker it has to be in order to be mappable by MT. The MT data can be interpreted to give an estimate of resistivity variations with depth. And, because MT needs a resistivity contrast to be present in order to map a boundary, and because these units need to be fairly thick to be mapped, the sections will not have the resolution of seismic sections. A conductor below a massive salt layer – a setting that presents a challenge to seismic imaging because of the high velocity contrast between the salt and the underlying sedimentary rocks – can be detected quite successfully with MT, as salt is usually highly resistive. The opposite is the case if a weak conductor is below by a good conductor. Such a situation is difficult to resolve for MT. In high temperature reservoirs, the overlying clay cap presents such a good conductor, which may make imaging of the reservoir difficult. The clay cap can cause further confusion when a (paleo-) geothermal reservoir is exhausted. Many of these problems are discussed in detail in Pellerin, Johnston, and Hohmann (1996).

In terms of resolving power with respect to targets of interest, attention has to be given to a priori geological assumptions, mesh size, and data dimensionality. In general, a good resolution demands dense site spacing, dense meshes in the models, and use of appropriate sensors for the period range. Undersampling is often the cause for the lack of adequate resolution of the targets, because the measurement sites are located too far apart in a heterogeneous medium. The apparent target size increases with depth due to increasing recording frequency, while target resolution decreases. As seen in the skin depth section, ground resistivity can change the investigation depth and consequently the resolution of the retrieved information for the same frequency range.

Data distortion is produced by the presence of three-dimensional local scale structures, located in the shallow subsurface, producing an anomalous charge distribution over its surface area. This presents a problem often encountered in the MT method, and all resistivity methods that are based on measuring the electric field on the surface, and is usually referred to as *telluric* or *static shift*. In practice, static shift is a vertical displacement of the apparent resistivity curves, where the phase angle curve is not affected. This phenomenon is caused by inhomogeneities of the resistivity close to the electric dipoles, which often occur in areas with a heterogeneous distribution of rocks near the surface, as is usually the case in regions shaped by volcanic activity. In areas where the near-surface rocks are homogeneous, such as sedimentary layers, and with little resistivity variation at shallow depth, static shift is usually not a problem.

There are basically two phenomena that produce static shifts: (i) voltage distortion (dependence of the electric field on the resistivity where the voltage is measured) and (ii) current distortion (current channeling). Voltage distortion occurs as a variation of the voltage in the surface when a constant current density flows through domains of different resistivity. For example, when the resistivity near the dipoles is lower than that in the rocks a little further away, the electric field (or the voltage difference over a given length) is lower in the low resistivity domain. This lowering of the electric field is independent of the frequency of the current. If the dipole is closer to the higher resistivity rocks, the electric field would be higher than a little further away. Current distortion occurs when current is flowing in the ground and encounters a resistivity anomaly. If the anomaly is of lower resistivity than the surroundings, the current is deflected (channeled) into the anomaly and if the resistivity is higher, the current is deflected out of the anomaly. If the anomaly is close to the surface, this will affect the current density at the surface and hence the electric field. As for the voltage distortion, this effect is independent of the frequency of the current density.

The problem here is that the electric field at the surface is scaled by an unknown factor (shifted on log scale) by anomalies in the vicinity of the measuring dipole. Sternberg, Washburne, and Pellerin (2006) have published results of model calculations showing that the voltage distortion and current channeling can produce dramatic static shifts in MT data. There is no numerical method to correct for the static shift and it is necessary to use information from other geophysical methods such as the transient electromagnetic method (TEM or TDEM; see below) that are not affected by static shift, using the vertical magnetic field component data, or comparing all the survey responses with a priori geological or geophysical information.

Even with the most beautiful interpretation of measured MT data, it has to be kept in mind that MT models provide information on bulk resistivity alone, which in terms of interpretation cannot be directly linked to any lithology, porosity of the media, or hydraulic permeability without a priori hydrogeological information. Resistivity measurements are affected simultaneously by lithology, the presence of fluids, and structure of the pore spaces. Further research needs to address this issue with the study of petrophysical relationships in order to quantitatively convert resistivity into rock physical properties.

The single most significant disadvantage of the MT method is it provides slow coverage of a prospect area and is therefore costly – but still cheap compared to active seismic methods. While this limitation is owing to the underlying physics and thus unlikely to change, the possibilities of the method usually outweigh its shortcomings and make it the most applicable of all individual geophysical methods for the exploration of deep geothermal reservoirs.

2.4.1.4 Active Electromagnetic Methods

Active EM methods are used mainly for shallow depth resistivity studies. One of their main applications today is to support static shift corrections of MT data, for which mainly TEM is used. TEM has become the standard among all active EM measurements, as it is highly reliable and the most precise and cost effective of the resistivity techniques. In the most common central-loop TEM method (Figure 2.9), a loop of wire is laid on the ground which has a square shape, each side measuring several hundred meters. A magnetic spool is placed at the center of the square and serves as a receiver, after which DC current is applied to the loop. The current builds up a magnetic field of known strength. The current is abruptly turned of, leaving the magnetic field without its source, which induces an image of the source loop on the surface. The current and the magnetic field decay and again induce currents at greater depth. The spool at the loop's center measures the magnetic decay at the surface with time elapsed since the current was switched off. The decay rate of the magnetic field with time is dependent on the current distribution that in turn depends on the resistivity distribution. The induced voltage in the receiver coil, measured as a function of time, can therefore be interpreted in terms of the subsurface resistivity structure.

The depth of penetration is a limitation similar to most electrical methods. However, the TEM method is less expensive and its interpretation is less time consuming. It is more downward focused, has excellent resolution, and requires significantly less area than other electric methods. Both two- and three-dimensional modeling compiled from one-dimensional inversion of each TEM sounding are routinely carried out. The method has been used extensively mostly in Iceland,

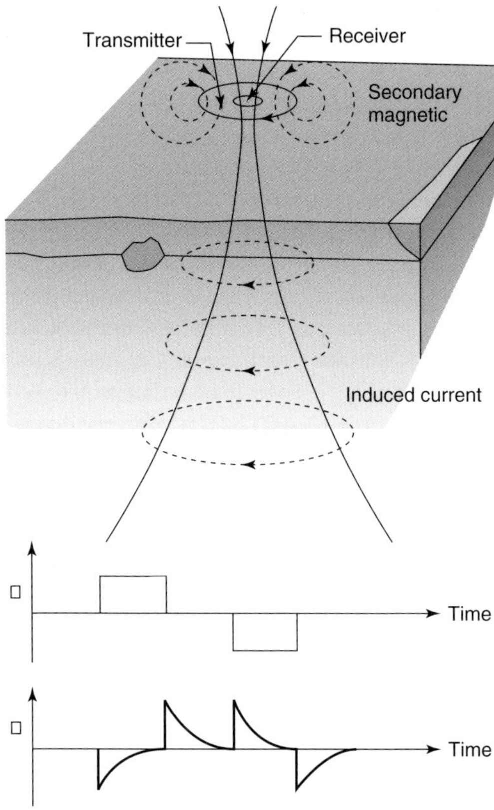

Figure 2.9 TEM sounding setup: transmitted current and measured transient voltage.

mostly to add additional information to existing geological, geochemical, and MT data or even instead of MT, as it is cheaper and has a much higher resolution at lower depths. For deep prospection in high temperature fields, however, it is mainly used to correct MT data for the static shift. Without the correction, apparent resistivities obtained by MT, which were up to a factor of 10 too low, have been observed in volcanic areas of Iceland. If the shift is not corrected for, interpretation would give 10 times too low resistivity values and about three times to shallow depths to resistivity contrasts. For this purpose, MT surveys in such areas are now routinely complemented by TEM measurements (Árnason, Eysteinsson, and Hersir, 2010).

For a joint inversion of MT and TEM, data should be collected from nearly identical places. To determine the static shift, a joint 1D inversion of MT and TEM soundings is performed. The misfit between the resistivities calculated for the two methods determines the static shift. This misfit can vary significantly within an area and between polarization directions. This was shown by a full 3D inversion of MT soundings acquired around the Hengill volcano in SW Iceland, which required correction of the static shifts separately for the two polarization directions (xy and *yx*). This correction showed that in many cases the two polarizations were shifted very differently. Thus, it does not suffice to determine static shift in one place and polarization direction alone but for every MT data point acquired.

Another active EM method that is used routinely in exploration is CSAMT, which is described in more detail, for example, by Zonge (1992). It is similar to MT, the main difference being that it uses an artificial source, and it is a method of choice, if noise is a particular problem for MT surveys. The source provides a stable signal, allowing higher precision and faster measurements than those acquired with natural-source measurements in the same spectral band. An electric dipole with a length of 1 or 2 km grounded at a distance of 4–10 km from the receiver stations in the area to be measured serves as the source. Measurements are usually made with continuous stations along a line or with individual stations in a grid to determine 2D or 3D behavior of the subsurface.

The resolution with depth is governed by the same equations (Equation 2.11) as for MT: the depth of exploration or investigation is related to the square root of ground resistivity and the inverse square root of signal frequency. These equations do not define a depth limit for the resolution; however, the maximum depth for practical use is usually between 2 and 3 km. The limiting factor on depth of exploration with all of the data in the far field is usually signal level. Both E- and H fields vary as a function of frequency and earth resistivity and decrease as $1/r^3$, where r is the separation between the transmitter and receiver, so signal strength decreases rapidly with depth. As a general rule, when sounding over a relatively homogeneous territory, transmitter and receiver should be about five times the depth of exploration apart, so for an investigation depth of 1 km a receiver–transmitter separation of about 5 km is recommended. If the distance between transmitter and receiver is less than three times the depth of interest, the far-field condition is no longer applicable and the change of resistivity with depth no longer obeys the rules summarized in Equation (2.11), and calculation of the subsurface properties becomes more complicated. Therefore, surveys are usually carried out with receiver–transmitter separations between 5 and 15 km.

Lateral resolution depends mainly on the length of the electric dipole serving as a source. Theoretically, the dipole can be reduced as much as necessary to get the desired lateral resolution, but a reduction in dipole length also reduces the strength of the signal. Received signal strength is directly proportional to the length of the dipole, such that half the dipole length results in half the signal strength.

CSAMT is often used in environments where the background noise is more than 10 times the signal level, and MT measurements are of limited use. An example of such an application is the survey for the potential EGS site near Skierniewice in Poland, which was performed within the I-GET project (Bujakowski *et al.*, 2010). The original MT measurements yielded highly noisy data, making an interpretation of the reservoir properties at 4 km depth nearly impossible, despite good quality remote reference data. Additional CSAMT measurements helped to determine the resistivity patterns of the uppermost kilometer and to put constraints on the interpretation of the MT data for the rocks below.

Advantages of EM Geophysical Methods Among the geophysical methods that sense bulk electrical and effective properties of the subsurface, EM have a greater depth capability and provide better resolution than DC electrical measurements. In terms of resolution, only reflection seismics has the potential to yield better results. EM methods are cost effective, relatively easy to operate in the field, and a variety of data processing options are available, ranging from the construction of apparent resistivity curves or pseudosections for fast subsurface evaluations to 1D and 2D forward and inverse modeling. 3D inverse modeling is not yet fully developed although research is moving forward rapidly in this field, where new codes are being tested. However, the main concerns in all EM methods are cultural noise sources such as power lines, pipelines, and DC trains among others that screen and disturb the geophysical signal.

Electromagnetic induction methods are the most widely used and versatile geophysical methods in geothermal exploration and investigation at different scale ranges. A diverse set of techniques and instruments available provides the possibility of conducting cross-scale investigations. Selection of the appropriate technique depends strongly on the objectives of the study, time, financial aspects, and computational facilities.

2.4.2 Seismic Methods

The reason for the widespread application of seismic methods in many exploration tasks is that they provide the most detailed structural information at depth. They are standard exploration methods in HC exploration and therefore highly developed in every aspect: data acquisition, logistics, and interpretation. In geothermal exploration, the focus are fluid-filled rock volumes that are not necessarily linked to specific structures but the structural setting itself (e.g., faults, dykes, and grabens) and the parameters of possible resource regions as well as underground conditions (e.g., stress, strain, and pore pressures) are also in focus of the investigations.

Body waves travel through the interior of the earth (Figure 2.10). They follow ray paths bent by the varying density and modulus (stiffness) of the earth's interior. The density and modulus, in turn, vary according to temperature, composition, and phase. Two basic types of seismic waves are of interest in exploration of subsurface resources: P-waves and S-waves. P-waves are longitudinal (compressional) waves and they are the fastest of the elastic waves (P-waves = primary waves). S-waves, also called *shear waves* or *secondary waves,* are transverse waves that travel more slowly than P-waves and thus appear later than P-waves on a seismogram. Particle motion of S-waves is perpendicular to the direction of wave propagation and they do not exist in fluids as water or in gases (air).

Seismic methods can be divided into two main subclasses:

- active seismic methods, which make use of waves created by artificial sources
- passive seismic methods, for which the sources are natural earthquakes or rupture processes, induced by for example, injection or extraction officials (e.g., hydraulic fracturing).

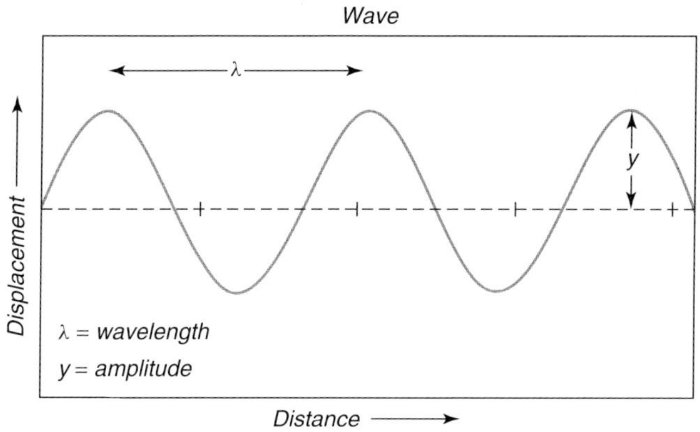

Figure 2.10 Seismic wave parameters: wavelength $\lambda = v/f$, where v *is* speed of propagation and f is frequency. The *period* T is the time for one complete cycle for an oscillation of a wave. The *frequency* f is the number of periods per unit time (for example, 1 second) and is measured in hertz. $f = 1/T$.

Seismic methods determine subsurface elastic properties influencing the propagation velocity of elastic waves: as the waves travel through the subsurface, wave velocities change depending on the density of the rock, and wave paths are reflected and refracted by elastic discontinuities such as sedimentary layering, boundaries between different rock units, and fractures. Because fractures present a considerable elastic discontinuity affecting the path and velocity of a wave, seismic methods have the potential to identify not only their presence but also fracture attributes such as orientation, density, aperture, and filling. This potential makes them particularly interesting for EGS exploration, where the orientation and distribution of faults, fractures, fissures, and cracks is of utmost importance for the access to and exploitation of the desired resource.

In terms of resolution, seismic methods provide the most detailed structural information at depth. The maximum possible resolution is between one quarter and on eight of the dominant wavelength if recent advances in the incorporation of amplitude information are applicable. For a porous rock with a velocity of 2000 m s^{-1} and a frequency of 100 Hz resulting in a wavelength of 20 m, the resolution limit would be 2.5–5 m. Seismic resolution decreases with depth as the velocities normally increase and high frequencies are lost due to absorption. So, the smallest features to be seen on a seismic diagram are still large at the surface outcrop scale.

2.4.2.1 **Active Seismic Sources**

The physical phenomenon measured with geophones during seismic surveys is the ground motion due to the elastic waves generated by explosions (shots) or weight drop (Vibroseis). These geophones record the arrival of the waves that travel along varying paths in the subsurface and thus arrive at different times and with

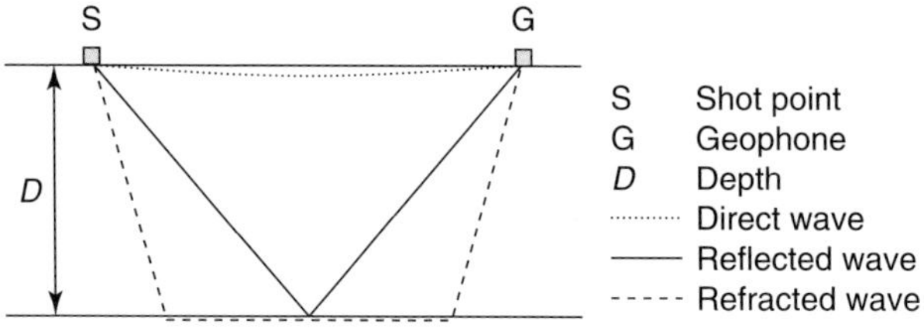

Figure 2.11 Different ray paths of seismic waves: a part of the waves travel through the air (compressional direct wave) and along the surface, while in the subsurface waves are refracted and reflected at interfaces between rock units of different elastic parameters. While reflected waves are created at boundaries with the angle of incidence, refraction process is more complex as these waves are created at the boundary and travel with the velocity of the underlying layer. They can be observed at surface only if there is an increase of seismic velocity with depth.

different incidences, generating a long and complex signal (Figure 2.11). The arrival times of the different waves generating the signal depend on the compressional wave velocities v_P and the shear wave velocities v_S. These are dependent on rock composition, density and/or the degree of fracturization, temperature, and the presence of fluids and their pressure and degree of saturation. If the seismic wave velocity in the rock is known, which is usually determined in the laboratory, the travel time may be used to roughly estimate the depth of a structure.

There are many different surface seismic methods and combinations of methods, using 2D-, 3D-, P-wave-, and S-wave sources. The different approaches and various processing techniques with all their assumptions, advantages, and pitfalls have been developed largely for the HC industry and are described in detail in numerous and often voluminous textbooks (Sheriff and Geldart, 1995; Keary, Brooks, and Hill, 2002). A very good summary of seismic approaches for the geothermal context is given by Majer (2003a).

A subsurface structure of interest can be imaged with the transformation of the acquired data from the timescale, to the depth scale. Depth conversion is ideally an iterative process. Good seismic processing, seismic velocity analysis, and, if available, information from wells in the area are required to refine a conversion. Processing involves numerous steps with the goal to suppress noise, enhance the signal and migrate seismic signals generated by subsurface structures to their appropriate locations in the *xy*-time space of the seismic data. These steps allow better interpretation of the observations, as subsurface structures become more apparent and can be located more accurately.

The analysis of seismic wave velocities provides some useful information for geothermal exploration. Fractures, higher temperatures, and the presence of fluids cause a decrease in v_P and the ratio v_P/v_S. Pressure, temperature, and saturation may tell us whether a reservoir is steam- or liquid dominated: at high temperatures and low pressures and saturation, steam will be the dominant phase and v_P and v_P/v_S will be relatively low, while at comparatively lower temperatures and/or higher pressures and saturation v_P and v_P/v_S will be higher, indicating a liquid-dominated

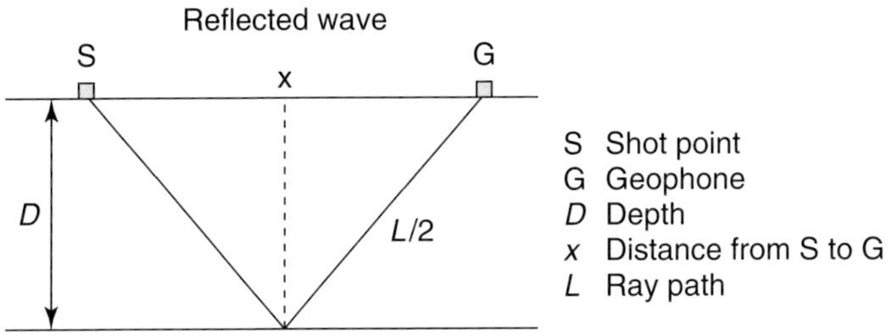

Figure 2.12 Schematic diagram of reflected wave.

system. v_S is not as sensitive to saturation, such that the ratio v_P/v_S is a very helpful indicator. In contrast, the attenuation of v_P is relatively sensitive to the presence of vapor and can therefore be indicative of zones containing steam.

Both seismic reflection and seismic refraction surveys have been used in geothermal exploration. Refraction surveys are limited to some extent because of the amount of effort required to obtain refraction profiles giving information at depths for more than a few kilometers and the difficulties caused by the generally complex geological structures in areas likely to host geothermal systems. Seismic refraction is normally restricted to cases where the densities of the rocks and thus seismic velocities increase with depth. In addition, geophone arrays for refraction measurements need profile length of at least four to five times (sometimes even eight times) the sampling depth because of the very nature of refraction. These distances require higher shot energy (i.e., more explosives) and limit the applicability of refraction methods in exploration to shallower targets or to large-scale investigations of the earth's crust and upper mantle with very energetic sources. Most of the time and also within reflection surveys it is used to get a first approximation about the velocity distribution at depth.

In general, reflection seismic methods are more commonly used in geophysical exploration, as they require much shorter profiles and therefore less shot energy and have a much higher lateral resolution. However, reflection signals are much more complex to detect and to analyze than refraction signals as they never arrive first, which implies time and labor intensive filtering and detection from a multitude of overlapping data. Moreover, the specific setup for reflection measurements requires more logistic preparation and personnel, which makes it generally a lot more expensive than refraction methods. It is nonetheless the method of choice in HC exploration, as it can resolve structural details of a reservoir.

In seismic reflection, the two way travel time is measured, which is the time it takes for a wave from its source to the reflector (some sort of mechanical discontinuity) and back to the receiver (Figure 2.12). Unless the rocks above the reflector and their seismic velocities are known, the depth of the reflector and the velocity can be determined by the use of many seismic stations and many different shot points.

The travel time T between shot point S and geophone G is given by the length of the ray path L and the seismic velocity v within the subsurface layer, such that

$$T = \frac{L}{v} \tag{2.12}$$

From Figure 2.12, we see that

$$\left(\frac{L}{2}\right)^2 = D^2 + \left(\frac{x}{2}\right)^2 \tag{2.13}$$

Such that $T = 1/v(4D^2 + x^2)^{1/2}$

This equation contains D and v as unknowns, which can be constrained if measurements of T are available for many geophones and shot points. Depending on the shot point layout and the spacing of geophones, there is usually considerable overlap of measurements over a common point on the reflector, which is referred to as *fold*. Assuming a horizontal layer as the reflector, the travel times for reflection events from a common point vary with offset (x in Figure 2.12). This variation in travel time depends only on the velocity of the subsurface layer, thus the subsurface velocity can be derived, assuming this velocity does not change horizontally.

Incidence elastic waves reflected at a single reflector and then detected at the surface are called *primary reflections*. However, in reality, many waves are reflected at multiple interfaces before they are detected and are therefore referred to as *multiple reflections* or *multiples*. Multiples generally have lower amplitudes than primary reflections as energies are split at every reflection. The correct identification of multiples is a crucial step in the interpretation of seismic traces. Travel times of multiples can be calculated from the corresponding primary reflections and can be identified and filtered by appropriate processing techniques. As signal strength decreases significantly with depth, processing will always need to involve improvement of the signal-to-noise ratio.

Each shot from the source generates not only multiple reflections but also several different primary reflections from different boundaries at various depths. The arrival times of the reflected waves vary with the depth of the reflector and with the velocities of the different layers crossed by the wave. The seismic trace resulting from a single shot at one receiver is thus composed of a series of arrivals. These "spikes" will vary considerably in amplitude depending on the attenuation within the subsurface. Generally, amplitude decreases rapidly with depth. In addition to the reflections generated at different depths and the arrival of multiples, seismic traces contain a lot of seismic noise and signals from surface waves and air waves. All these signals result in a rather complex diagram, which requires extensive processing before reflections can be recognized and interpreted. The traces recorded by all receivers resulting from an individual shot are assembled in shot gathers. Usually the traces are plotted side by side, allowing an alignment of reflection events and their correlation from trace to trace.

Reflection profiles are taken with shot points and geophones aligned and moved along lines, resulting in a 2D seismic survey. Such surveys are very common, also for geothermal exploration. They supply sufficient information of the subsurface, if structures to be determined are of uniform geometry and if the geology of the area is nearly 2D, as in some sedimentary environments. For 3D structures, several closely spaced lines are necessary to provide adequate coverage of lateral changes. Even though 2D seismic profiling is a standard procedure in exploration, for which abundant off-the-shelf software packages exist for processing and interpretation, each reflection survey needs to be designed specifically to optimize the measurements for the required information. Geologists and geophysicists need to communicate the problem to be addressed clearly to the contractors doing the profiling.

Resolution with depth was shown to depend on the wavelength and thus the frequency of the signal. As higher frequencies are lost with depth, resolution can be improved with a higher energy signal, requiring a stronger shot, which is not always feasible. Lateral resolution also depends on wavelength and thus decreases with depth. However, a crucial point which can be controlled by the layout is receiver spacing: it should be sufficiently narrow to allow reliable correlation of reflections from the reflection interfaces.

To get a 3D image of the subsurface and of a potential reservoir, 3D seismic surveys are highly desirable. When fractures are important a 3D approach is, usually, also required. With receivers arranged on and shot points moved along a grid, processing and interpretation of data is usually very time consuming and additionally complex. In result such surveys are rather expensive such that large-scale 3D surveys are rarely performed in geothermal prospecting. Perhaps, more importantly they have been developed primarily for oil exploration in sedimentary environments that usually display less structural complexity laterally than, for example, volcanic areas or other areas favorable for geothermal exploration (Figure 2.1). A rare example of such a survey was conducted in the Italian geothermal area of Travale in 2003 (Cappetti *et al.*, 2005). Despite difficult terrain, the survey generated sufficient data to significantly improve the deep geothermal reservoir of the area, although severe reprocessing was required (Casini *et al.*, 2010).

One of the limitations of seismic signals generated and detected at the surface is their restriction to horizontal or gently dipping reflectors. To detect and image more vertically situated structures, vertical seismic profiling (VSP) was developed, which takes advantage of measurements within an existing well. An array of receivers and the setup of one or more sources well adjusted to the problem not only allow resolution of vertical reflectors such as faults but also provides a highly reliable calibration tool for surface seismic measurements. VSP is also very useful when dealing with seismic anisotropy.

2.4.2.2 **Seismic Anisotropy and Fractures**

Most commonly, the reflections of P-waves are used to image the presence and orientation of fractures at depth. The underlying assumption in this approach is that fractures cause P-wave reflection anisotropy, with the fast and high amplitude direction parallel to the fractures and the slow and low amplitude direction perpendicular to the fracture. Fractures are also assumed to be the cause for P-wave attenuation. Stress can close cracks, water and/or steam can influence the crack properties as well. It is therefore important to measure as many azimuths as possible (i.e., a dense 3D grid) to detect and distinguish the potential influence of crack orientation and fluid-fill on the signals. With increasing variety of crack orientations (greater heterogeneity), it becomes more difficult to derive solutions for fracture orientation from observed signals, such that there is almost always some ambiguity in the results and careful processing and interpretation are crucial.

Out of the many possible more or less advanced processing and reprocessing steps and procedures, the amplitude variation with offset (AVO) and amplitude versus azimuth (AVA) methods deserve special mentioning, as they are often applied to address fracture anisotropy. Variations in the AVA are analyzed assuming that fractures attenuate the P-waves as they travel across the fractures. Thus, an analysis of the variation of amplitude measured from different angles can yield information about the fracture anisotropy. Similarly, analysis of AVO uses variations in amplitude as function of reflection angle to derive information about anisotropy. Approaches such as AVO assume that there is a dominating set of fractures with a certain orientation. The detection of this preferred trend would help understand the potential anisotropy in permeability and thus be of great importance for geothermal exploitation. However, AVO does not always work. Especially, older terrains with a complex geological history tend to have multiple fracture sets of various orientations, some of which may be open simultaneously despite unfavorable orientation of the stress field. So, before AVO analysis is carried out, it has to be determined if the rock physics and fluid characteristics of the target reservoir are likely to give a usable response. Such a step will include seismic forward modeling including realistic geological and petrophysical boundary conditions of the area. AVO quality is also dependent on depth, as signal-to-noise ratio gets worse and higher frequencies are more attenuated, geology often gets more complicated, making AVO less applicable with increasing depth. A detailed coverage of AVO, its strengths and pitfalls, is given by Avseth, Mukerji, and Mavko (2005).

Generally, current technology can often locate fracture trends, but it usually does not provide the accuracy in locating high permeability zones to site wells. Seismic attributes such as P-wave anisotropy, AVO, or AVA are helpful in defining overall fracture properties and the detection of fracture zones. But these approaches have not been able to define the specific fracture sets that control permeability. Theoretically, the resolution with depth would allow precise localization of productive zones. But, even with the highest theoretical data quality of today, determining the significance of the underground images obtained remains the greatest challenge; there appears to be an agreement among seismic experts that a large part of the seismogram is not yet understood and contains valuable information that may one day be retrievable. The challenge is to define seismic properties that might image flow properties in the reservoir and permeability, rather than simply geologic features. While it is theoretically possible to reach this goal with adequate conditions such as enough measurement points of sufficient quality, enough computing power, and sufficient frequency content, several practical obstacles prevent its successful attainment.

In a study for the oil and gas industry, Majer (2003b) lists a number of problems and potential ways to address them. The limitation of image resolution by the amplitude and frequency content of the seismic waves and by the level of complexity of the ambient and signal-generated noise fields is hard to overcome with current techniques and surface sources. A major cause for this problem is the heterogeneity and thickness of the weathered surface layer which can attenuate the high frequency content and the coherence of the signal severely. This problem is partly solved by VSP, as receivers and sources are placed beneath the surface layer in a vertical array within the well. An approach to reduce imaging limitations is the incorporation of S-wave properties and the converted waves (P to S and S to P) generated by the multiple reflections in the earth. Majer (2003b) also points out that including amplitude and converted waves in the analysis could even make surface methods more useful, particularly where P-, S- and converted waves can be examined directly. S-waves have already become more and more common parts of the analyzed wave spectrum in recent years, which is of specific use for the definition of anisotropy and fracture orientation of a rock. Generally, to make use of the full potential seismic methods have to offer, three-component data including P- and S-wave reflection as well as VSP is required.

2.4.2.3 Passive Seismic Methods

The passive seismic method takes advantage of naturally occurring seismicity. The energy of seismic events is high enough to be detected by standard seismometers, even if it is not felt by the population. Such low magnitude earthquakes occur quite frequently in tectonically active regions, where most geothermal reservoirs are located. Moreover, microseismicity is often associated with hydrothermal convection, thus responding directly to the resource to be detected. Thus, passive seismic studies have been found to have a promising potential in pinpointing active faults or fracture systems that are not always found on the surface, as well as their elevation and inclination.

Seismic surveys of microseismicity require a sufficiently dense network of recording stations placed around the potential reservoir and an extended period of recording time, usually several months. Several well-located events are necessary to reliably characterize an active fault. If these active faults are located, sophisticated use of recording and the recorded data can help to construct a three-dimensional image of fluid flow in the reservoir, as fluid circulation occurs in open faults and fracture systems, which are often responsible for the observed microseismicity. The frequencies associated with fluid circulation in open fractures are usually at the lower limit of the recording spectrum. This problem can be solved by the use of broadband stations that record a much broader spectrum of frequencies than standard seismometers. Since an increase in temperature results in the reduction of P-wave velocity over a large volume in the crust, the measurement of delay times from teleseismic events (distant earthquakes) have been used to locate large hot bodies that act as the source of geothermal systems. However, teleseismics far enough to be of use in P-wave delay surveys occur only rarely, and a long period of recording time and relatively huge investments are necessary.

The first target of passive seismic methods is to determine hypocenters, whose location is directly linked to those of faults – including those created by stimulation and hydrofracturing – and to the tectonic signature of the area. In addition, information about the geology and tectonics can be obtained from fault plane solutions and first motion studies of these earthquakes, which are valuable in determining whether the earthquake activity in a prospect area is anomalous or typical for the region. If there are enough microseismic events and if they are homogeneously distributed with respect to the recording stations and potential targets, a 3D distribution of seismic velocities can be constructed. As the different seismic velocities v_P, v_S, and the ratio v_P/v_S depend on various physical parameters in a geothermal environment including fluid content of a rock, mapping of v_P/v_S can be a powerful tool both in the exploration as well as for monitoring during exploitation of a geothermal reservoir. Changes in v_P, v_S, and the ratio v_P/v_S are expected when the steam volume increases, since it causes a strong P-wave attenuation and an even sharper drop of v_P/v_S. Extensive fracturing of a liquid-filled rock causes a minor reduction in the P-wave velocity and a significant reduction in the S-wave velocity, so that v_P/v_S is higher than normal. In addition, as v_S is more sensitive than v_P to anisotropies of the rocks, v_P/v_S can vary with azimuth. Such variations can contain important clues about preferred orientation of fluid circulation.

One method that takes advantage of anisotropies is the analysis of shear wave splitting (SWS), which is based on the separation of the shear wave into a fast wave traveling parallel to the fracture direction and a slow one traveling perpendicular to the fluid-filled fractures (Crampin, 1981; Hudson, 1981) (Figure 2.13). The time delay is proportional to the number of cracks per unit volume along the path of the wave. Provided polarization of the fast wave and time delay are observed, the detailed analysis of the polarization of shear waves in the seismograms allows the determination of fracture orientation and of fracture density (Rial, Elkibbi, and Yang, 2005). Tomographic inversion and the differences in arrival times can be used to map the 3D distribution of the fractures, crack geometry, and thus regions of potentially productive reservoir rocks. The SWS method is therefore highly useful for the detection and development of EGS reservoirs, particularly if one well has already been drilled and is used for stimulation procedures. The seismicity induced by these operations provides excellent sources of shear waves near the area of interest. Thus the method can be used even where natural seismicity is scarce.

The largest data sets on SWS connected to geothermal reservoirs have been collected by the University of North Carolina at Chapel Hill (J. Rial and his group). From their experience gathered so far, there is also a list of limitations for the method. One major problem can be the scarcity of detectable seismic events, which is often the case in sedimentary basins. This can be overcome by long term surveys or permanent arrays. An assumption and prerequisite for all successful SWS analysis is the mechanical isotropy of the uncracked rock volume. Any preexisting

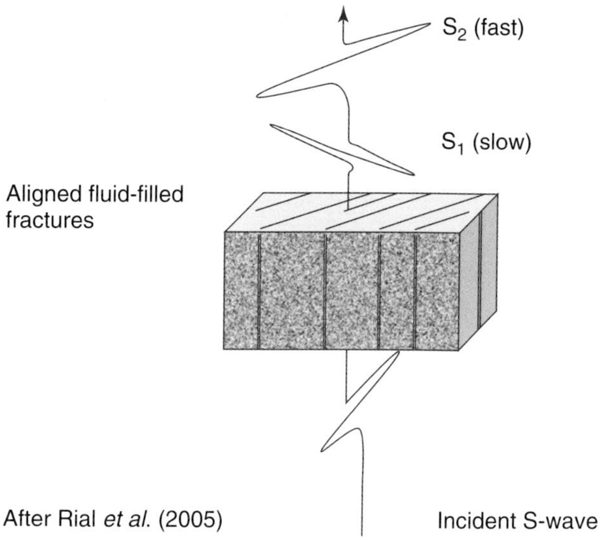

Figure 2.13 Shear wave splitting. The incident S-wave with arbitrary orientation is split into a fast S-wave oscillating parallel to the direction of the fractures, while the slow S-wave oscillates perpendicular to the fracture orientation.

lithologic anisotropy or strong heterogeneity can severely limit the usefulness of the SWS method. In addition, the volume of aligned cracks needs to be sufficiently large to produce a measurable effect at the surface. A layer of limited thickness at great depth maybe below the resolution limits of the technique, even if the fractures would present a good target for EGS operations.

Success in the application of the SWS analysis is critically dependent on the data acquisition. Fracture parameters such as density, strike, dip, fluid-fill content, and/or aspect ratio can be determined only if SWS data are collected from many different azimuths and incident angles. A dense network of stations can usually overcome this limitation. To determine where along the ray path the cracked areas responsible for the SWS are located, an even denser spacing may be required, which is of course a cost factor. Fracture orientation can most easily be determined among the desired parameters, particularly for parallel vertical cracks. If cracks are shallow dipping or more than one crack system with varying orientations exist, the analysis requires the very dense seismic arrays mentioned above. The determination of fracture dip is also more strongly dependent on ray path coverage quality.

In addition to the dense seismic arrays, a high sampling rate is necessary to not only measure fast shear wave polarization orientations but also track ray path–dependent variations in observed time delays (Rial, Elkibbi, and Yang, 2005). The high sampling rates are particularly important for the accurate determination of fracture density, as variations in time delay are subtle compared to that of polarization direction. The determination of other parameters such as aspect ratio of cracks and fluid-fill content, which can give information about the state of the fluid, cannot easily be accomplished and requires high-quality data sets. But even then, results may be non-unique, as the effect of cracks saturated with vapor on shear wave polarizations is similar to that of water-filled cracks with high aspect ratios (MacBeth, 1999). Clearly, these data acquisition requirements also need adequate computing power.

2.4.3 Potential Methods

2.4.3.1 Gravity

Gravity measurements are used to determine differences in density and their lateral extent in the subsurface. These differences are usually very small and require highly sensitive equipment to determine relative gravity anomalies. Measured data need time-dependent (e.g., drift and tidal effects) and static (e.g., elevation and topography) corrections for local and regional conditions and are then used to construct a contour map of Bouguer anomaly with lines of equal gravity anomaly. These lines are called *isogals* – gal in memory of Galileo Galilei. Positive gravity anomalies (compared to their surroundings) correspond with higher density subsurface. They can be of interest for geothermal exploration, as they are associated with mafic to intermediate intrusions, and geologically young intrusions ($<\sim$1 Ma) can provide a potential heat source. Such structures would also commonly be detectable by a positive magnetic anomaly. Positive density anomalies can also be caused by deposition of silicates from hydrothermal activities.

Negative gravity anomalies can have several causes, some of which also have promising implications for geothermal exploration. For example, lower densities can be caused by felsic intrusions such as granites, magma bodies, higher porosities, or by highly fractured parts of a rock. Highly porous or highly fractured rocks would provide potentially interesting zones of higher fluid content and/or permeability. Alteration minerals produced by circulation of hot water can also cause a negative density anomaly.

Faults can also be traced by gravimetric tools, as they usually display a distinct change in density across a more or less well-defined linear zone. These faults that might have no surficial impressions can accommodate the upwelling of hot water. Gravity anomaly maps can show the extent of the sedimentary cover in basins as negative anomalies and be used to estimate the depth of the underlying basement. Such maps can provide useful first information about heat and volume of a potential geothermal reservoir. For example, gravity surveys were performed at the classic Italian site of Larderello (Fiordelisi and Bertani, 2006; Orlando, 2005), where 23 000 stations were acquired, corresponding to 1 station per kilometer, to provide subsurface structural information. 2D/3D modeling in conjunction with experimentally determined density data pointed out deep low density bodies (6 I-GET report WP2) related to molten intrusions: the potential heat source of the system.

In a volcanic environment, the gravimetric differences depend strongly on the chemistry of the rocks and on their porosity. Generally, solid magmatic bodies are much denser than layers of pyroclastic rocks, where densities are usually low. In highly porous rocks such as rhyolitic tuffs (where porosities can be higher than 40%), densities are also strongly influenced by the fluid content. A dry rock will be much lighter than a fluid saturated rock; steam and liquid water would also be clearly distinguishable. Examples of volcanic geothermal reservoirs, where this observation was used to delineate the zones of hydrothermal alteration, are Broadlands and Ohakuri in the Taupo geothermal area of New Zealand (Hunt *et al.*, 1990).

The primary use of gravimetric measurements today is to help constrain the structural context of an area, outline trends of faults, and determine the depth to basement. Gravity surveys using surface, airborne, or even satellite data are used more and more for this purpose as one of the first steps to characterize a region of interest and to help constrain areas for further inspection. Even though they can help constrain the extent of the resource, interpretation of the results is rarely unequivocal to make economical decisions. For example, mafic to intermediate composition intrusions, commonly associated with positive anomalies, can be negative anomalies, if they ascend through a dense metamorphic basement (for example, the Denver Basin, Colorado, USA). Another example of positive anomalies would be the occurrence of hydrothermal alterations in sedimentary environments. They are usually thought to cause a densification, as observed at Salton Sea, California, USA. But, hydrothermal alterations are also used to explain local gravity lows at The Geysers steam field in California. In the case of potential EGS, the differences in gravity may be even less meaningful, as porosities and permeabilities affected by hydrothermal activities can be much smaller than those in conventional systems and with higher temperatures.

Another common use of gravity monitoring surveys in geothermal areas is to define the change in groundwater level and for subsidence monitoring. Fluid extraction from the ground which is not replaced within an adequate time span causes an increase of pore pressure and hence of density. This effect may induce subsidence at surface, whose rate depends on the recharge rate of fluid in the extraction area and the rocks affected by compaction. Repeated gravity monitoring in conjunction with weather monitoring may define the relationship between gravity and precipitation that produces the shallow ground water level change. When gravity is corrected by this effect, gravity changes show how much natural inflow replaced water mass discharged to the atmosphere. The underground hydrological monitoring of a gravity survey is an important indication of the fluid recharge in geothermal systems and the need for reinjection.

The advantages of gravimetric methods over other geophysical methods are that they are comparatively easy to use and fairly economical. They do provide a good estimate of the extent of bodies with certain density contrasts and can thus help constrain the location and extent of reservoirs. The resolution and quality of data, however, decrease considerably with depth. Gravimetric studies therefore provide a useful tool for shallow ($<\sim 2$ km) reservoirs in conventional systems and, given their often ambiguous results, in combination with other geophysical methods. For potential reservoirs in EGS, this method may be even more limited than for conventional systems.

2.4.3.2 Geomagnetics and Airborne Magnetic

Magnetic surveys measure changes of the earth's magnetic field over time and space. The latter are associated with the composition and structure of the rock formations at the subsurface. The parameter of interest here is the magnetic susceptibility, which is a material property and can be determined and calibrated using rock samples in the laboratory. The magnetic susceptibility of rocks influences the natural magnetic field, which is measured in nanotesla (1 gamma = 1 nT (nanotesla)). Measurements are performed either at the surface or airborne, if the objective is regional mapping. Various instruments can be used. The most commonly used instruments are high precision proton precession magnetometers and cesium–vapor magnetometers, which nowadays are operated together with differential GPS to record time and location accurately.

Rock magnetism is acquired when the rock forms, and it reflects the orientation of the magnetic field at the time of formation. But, rock magnetism can also change with time: if the rock is subjected to temperatures above a certain point, called the *Curie temperature*, it loses its magnetic properties. It is remagnetized once it cools down, now induced by the magnetic field present at that time.

The shape and magnitude of a magnetic anomaly depends primarily on two factors:

- The shape and orientation/position of the magnetic structure in the subsurface.
- The latitude of the location. This factor is important because of the dipolarity of the earth's magnetic field. The inducing magnetic field has a dip angle that varies from place to place over the surface of the earth: at the magnetic North Pole, it is vertical, and the pattern of magnetic anomalies is symmetrical, while the patterns of anomalies that are recorded become more complex away from the pole.

The magnetic susceptibility of a rock and the temperature at which it disappears depend strongly on the rock components, the more or less magnetic minerals. Minerals with high magnetic susceptibility are, for example, magnetite, ilmenite, hematite, and pyrrhotite. Silicate minerals, rock salt (halite), and limestones (calcite) have very low magnetic susceptibilities and are therefore not useful for magnetic measurements. Consequently, sedimentary rocks usually have much lower magnetic susceptibilities than igneous or metamorphic rocks. Thus the magnetic method has been traditionally used for identifying and locating masses of igneous rocks that have relatively high concentrations of magnetite, which is the most common of the magnetic minerals. Strongly magnetic rocks include basalt and gabbro, while rocks such as granite, granodiorite, and rhyolite have only moderately high magnetic susceptibilities. Since hydrothermal activities are often associated with plutonism, magnetic interpretation can be the first step in finding areas favorable for the existence of a potential geothermal reservoir.

However, today, high-resolution aeromagnetic (HRAM) surveys have a resolution in the subnanotesla scale, such that magnetic surveys are no longer restricted to magmatic rocks but can also be used to map intrasedimentary faults, as long as there are some layers containing elevated magnetite concentrations that generate small anomalies (>10 nT at 150 m elevation, Nabighian *et al.*, 2005). Such HRAM surveys are considered industry standard, and they are often used in HC exploration, but flight specifications for a high-resolution survey vary from one country to another. Typical exploration HRAM surveys have flight heights of 80–150 m and line spacings of 250–500 m (Millegan, 1998).

Hydrothermal activity influences the susceptibility of rocks. In the conventional volcanic environment, circulation of hydrothermal fluids causes alterations in the rock, which in turn cause a reduction in susceptibility. This reduction is a consequence of the destruction of the magnetite contained in the rocks. That way, units of volcanic rocks and lava flows can easily be distinguished from hydrothermally altered rock units, which makes geomagnetic surveys a useful tool for geothermal prospecting at high enthalpy volcanic reservoirs. Alterations are usually caused by high temperature fluids that may be related to a geothermal reservoir and structures such as faults or dykes, which allow fluid circulation. Areas where this method was used to outline such features for exploration are, for example, in New Zealand, Japan, Kenya, Iceland, or the western United States, to name but a few.

An additional potential for the geomagnetic method is its ability to detect the depth at which the Curie temperature is reached. Various ferromagnetic minerals have differing Curie temperatures, but for the two most strongly magnetic minerals, magnetite and pyrrhotine, the temperatures are 580 and 320 °C, respectively. For magnetite, the temperatures can vary with titanium content, adding a degree of uncertainty to depth estimates using the degree of magnetization. Nonetheless, keeping in mind the mentioned uncertainties, the deepest level of detectable magnetization provides a useful estimate for the temperature at the depth and thus of the temperature gradient and the heat content. For magnetic field observations made at or above the surface of the earth, the magnetization at the top of the magnetic part of the crust is characterized by relatively short spatial wavelengths, while the magnetic field from the demagnetization at the Curie point in depth will be characterized by longer wavelength and lower amplitude magnetic anomalies. This difference in frequency characteristics between the magnetic effects from the top and bottom of the magnetized layer in the crust can be used to separate magnetic effects at the two depths and to determine the Curie point depth. This approach, using the creation of a Curie point depth map as an integral part of the exploration, has been adopted for many attempts to discover new geothermal prospects (Yellowstone National Park, Cascade Range of Oregon, Japanese Islands, Northern Red Sea, Trans-Mexican Volcanic Belt, parts of Greece, etc.).

For regional exploration, magnetic measurements can be important for understanding the tectonic setting, for example, in Iceland or at Dixie Valley, Nevada, USA (Smith, Grauch, and Blackwell, 2002). With the HRAM surveys, the study of basin structure has an important economic application, especially in oil and gas exploration, but could also be applied for geothermal reservoirs and potential EGS systems. For the most part, basin fill typically has a much lower susceptibility than the crystalline basement. Thus, it is commonly possible to estimate the depth to basement and, under favorable circumstances, quantitatively map basement structures, such as faults and horst blocks (Prieto and Morton, 2003).

The magnetic method has thus expanded from its initial use solely as a tool for finding iron ore to a common tool used in exploration for minerals, HCs, ground water, and geothermal resources. The speed with which the measurements can be made and the relatively low cost for campaigns have made the method very popular during the last 30 years. Restrictions are the resolution with depth; the complexity of the interpretation, which makes it most reliable only for structures with simple geometric shapes; and the insensitivity to the actual presence of water. With these restrictions in mind, the method is not more or less useful for EGS or conventional geothermal systems. Its value is mainly the potential to determine heat at depth, a characterization of the regional tectonics and the outline of a potential heat source.

2.4.4 Data Integration

The most important objective of applying geophysical methods is to obtain quantitative information over the subsurface model space. The transformation from raw data to an estimated geophysical model is usually achieved using numerical forward modeling and inversion procedures, to provide a description of the subsurface fitting to the observed data. Joint inversion of different geophysical sets is used to constrain the possible subsurface models with multiple independent data sources, using either a deterministic approach or a probabilistic approach such as stochastic inversion methods

To perform the integration of geophysical measurements with hydrogeological and hydrothermal measurements, the scale problem, as well as the nonuniqueness and uncertainty of the geophysical and geochemical models, and which specific petrophysical relationship is most appropriate for each case study have to be considered. Thereafter, integration and estimation approaches that focus on defining the spatial distribution and the magnitude on the geothermal system can be applied.

The first step is to obtain reliable geophysical models with which to translate geophysical properties into (thermo)hydraulic parameters. The second step is the quantitative conversion of the geophysical and geochemical property to hydrogeological and hydrothermal properties that may be obtained (i) via direct mapping using a petrophysical relationship, the so-called deterministic approach, or (ii) by applying stochastic methods such as geostatistics or Bayesian techniques, the so-called probabilistic approach.

The most general way to integrate a priori information and data for nonlinear problems is to apply stochastic inversion methods where the resulting model parameters are given by a probability distribution. The probabilistic weight of each element is considered in the iterative posterior inversions to improve the models.

These geostatistical and simulating methods include Montecarlo simulation, neural networks, fuzzy logic, and Bayesan methods. Computational time and a priori distributions of model parameters are the main concerns.

A very recent example (Muñoz *et al.*, 2010) of integrated interpretation of MT and seismic data sets shows that there is a huge potential for the future in combining different methods even if fundamental laws linking the parameters investigated cannot be formulated explicitly.

2.4.4.1 Joint Inversion Procedures

Geothermal exploration research is increasingly turning to joint inversion strategies in which multiple geophysical data sets and/or geophysical–hydrothermal data sets are processed simultaneously to produce more realistic estimates of the hydrologic parameters that satisfy all the available data sets. Thus, joint inversion methods are configured either as a coupled inversion of geophysical and hydrological data or as a coupled inversion of multiple geophysical data.

When two data sets are both sensitive to the same physical property, the simultaneous inversion is achieved by minimizing the misfit of both data sets. On the other hand, if the geophysical data sets are sensitive to different physical properties final models will provide complementary information at the same location point. Joint inversion of hydrogeological, hydrothermal, and geophysical data is expected to improve the final hydrogeological model. Hydrogeological and hydrothermal data calibrate the hydrogeophysical variables based on the assumption that any relevant hydrogeological structure has a geophysical signature.

2.5 Geochemistry

2.5.1 Introduction

In any type of geothermal systems, high temperature fluid is in chemical equilibrium with the surrounding rock. This equilibrium controls the natural porosity and permeability of the fractured rocks and will shift due to forced-fluid circulation of the exploitation phase.

Geochemical techniques provide information that could support future EGS developments. During the exploration phase of an EGS resource, fluid and rock geochemistry is a major tool in the determination of the origin and the quality and quantity of the fluid resource, helping to build a conceptual model.

Fluid and gas composition, water–rock volume ratio, and reservoir temperature are important parameters when forecasting the processes taking place in an EGS reservoir.

The first task is of course to prepare a review of published research and spatial data on the fluid and rock geochemistry. Data from springs and wells (groundwater,

Table 2.1 Geochemical tools and interpretation.

Geochemical tool	Selected parameters	Information on the EGS reservoir
Secondary minerals	Mineral assemblages (silica, silicates, carbonates, sulfides, zeolites, clays)	Qualitative estimation of reservoir temperatures, reservoir heterogeneity
Fluid and gas chemical composition	Major and minor species	Rock type, chemical equilibrium, fluids mixing processes
Fluid and gas isotopic composition	Hydrogen, carbon, sulfur, strontium isotopes	Elevation of the infiltration zone, underground residence time, origin of the minerals, fluids mixing processes, crustal/mantle origin
Gas composition	N_2, CO_2, H_2S, H_2, He, Ar, Rn	Fracture monitoring during drilling
Water-stable isotopes	$D/^{18}O$	Oxygen shift enrichment
Solute geothermometers	SiO_2, Li, Na, K, Ca, Mg, F, SO_4	Reservoir temperature, fluid uprising processes, mixing processes
Isotope geothermometers	$^{18}O(H_2O)$, $^{18}O(SO_4)$	–
Gas geothermometers	CO_2, H_2S, H_2, CH_4, Ar	–
Mud and fluid logging	Major species, CO_2, He	Fracture monitoring during drilling

oil, and geothermal exploration boreholes) can reveal important features on deep fluid flow and water–rock interactions.

The natural surface manifestations of an EGS reservoir are by definition less active and visible than for hydrothermal high temperature systems. In rift or in mountainous conditions, convective fluids from great depths can rise toward the surface and discharge in hot springs or in shallow porous aquifers.

Geochemical methods, applied during the various stages of exploration and evaluation as well as exploitation, should be used simultaneously to geological and hydrogeological appraisals and are particularly important because of information they supply at costs that are relatively low compared to geophysical methods and drilling.

Geochemical methods are applied to waters, gases, and secondary minerals (Table 2.1). The parameters to be measured and analyzed concern the physicochemical parameters, the dissolved chemical species (major, minor, and trace elements), as well as the isotopes (stable and radioactive).

Interpretation of the geochemical data gives some input on the following characteristics of a deep EGS reservoir: reservoir temperature (or reservoir depth if the geothermal gradient is already known), type of reservoir rock, mixing processes, multiple fluid origin, elevation of the infiltration basin, underground transit time, behavior of fluid exploitation (Tester *et al.*, 2006; Taylor, 2007).

A sound understanding of the fluid chemistry in the exploration phase facilitates the building of a conceptual model and approaches the future exploitation conditions. For the past several decades, the water chemistry and gas chemistry of geothermal fluids have proved very effective in evaluating subsurface temperatures, determining water origin, identifying and eliminating mixing effects, and predicting scaling and corrosion (Fournier, 1977; Rybach, and Muffler, 1981; Arnorsson, Gunnlaugsson, and Svavarsson, 1983; Arnorsson, 2000; Giggenbach, 1988).

2.5.2
Fluids and Minerals as Indicators of Deep Circulation and Reservoirs

Geochemical and isotopic methods have been found to be very effective for the geothermal exploration and potential assessment of geothermal field and surrounding systems. On a regional scale, hydrothermal system fluids associated with heat sources carry geochemical and isotopic signatures that provide insights into the deeper crustal processes operating in the magmatotectonic areas from which they originate (Bowen, 1989; Giggenbach, 1997a). On a smaller scale, geochemical and isotopic signatures also provide valuable information about physical reservoir processes in geothermal reservoirs, and location(s) of source inflows into geothermal fields (Arnorsson, 2000; Goff and Janik, 2000). Such information is of course invaluable for resource exploration and development programs.

Several geochemical methods are based on the relationships between hydrothermal alteration minerals occurring in high temperature geothermal systems and fluids circulating within these systems (Table 2.1). These methods include not only the popular chemical geothermometers and other techniques focusing on water–rock interaction (or mineral–solution equilibrium) but also the evaluation of irreversible mass transfer taking place during water–rock interaction (Giggenbach, 1984), which helps the reconstruction of the thermal history of the system (Reed, 1997). Knowledge of hydrothermal alteration mineralogy developing in high temperature geothermal systems can outline the extent of the reservoir and the temperature of alteration, and is of utmost importance for fluid geochemistry.

Hydrothermal alteration minerals found in outcrops and the drill holes, which can be sampled from the cuttings and the less frequent cores, have been the subject of many investigations (Browne, 1970, 1982; Heald, Foley, and Hayba, 1987; Stober and Bucher, 2000). Observations include mineralogy, and fluid inclusions in alteration minerals can be made in the laboratory.

The increasingly sensitive instruments and techniques available to investigate fluid inclusions have proven useful in describing the evolution of geothermal systems. Increasing proficiency in analysis of the gases and solids in fluid inclusions has contributed valuable details about the genesis of fluids. The case histories developed from fluid inclusions for a limited number of geothermal systems have greatly aided in developing exploration models for additional geothermal resources.

Although the sequence of alteration minerals varies from system to system, there is a general relationship between hydrothermal alteration minerals and temperature ranges, as summarized by Henley and Ellis (1983); (Figure 2.14). Some hydrothermal minerals (such as pyrite, calcite, and quartz) are of little use for evaluating deep temperatures and permeabilities, because these minerals are stable over large temperature intervals. The most informative minerals are the authigenic feldspars that are sensitive to both temperature and permeability. The occurrence of hydrothermal minerals typical of active geothermal systems depends on several factors such as temperature, pressure, fluid composition, and permeability (Browne, 1970).

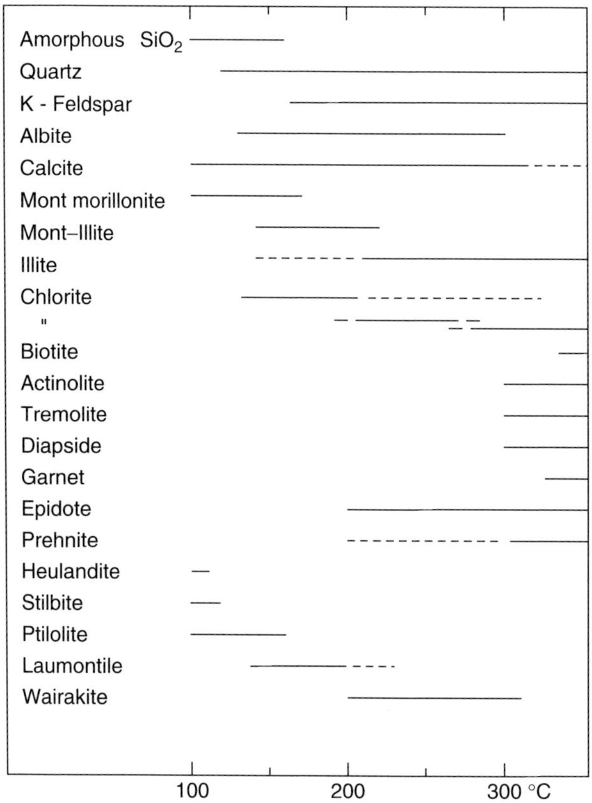

Figure 2.14 Temperature ranges for typical hydrothermal alteration minerals observed in active geothermal systems. (Henley and Ellis, 1983). Solid and dashed lines indicate the most and less frequent temperature ranges of occurrence, respectively.

Several hydrothermal minerals (e.g., epidote and chlorites) form solid solutions that can adapt to some extent to changes in rock composition by changing their composition, thus increasing their stability range. Further complications are due to the development of mixed-layer minerals, involving clays and chlorites. In spite of these complications, the geothermal systems explored through deep drilling have shown a thermal zoning of the hydrothermal alteration mineralogy, which has led to the identification of four hydrothermal alteration zones. The shallowest zone is the argillic zone, which is characterized by the presence of montmorillonite, eventually accompanied by illite, chlorites, and low temperature zeolites (e.g., heulandite, stilbite). This zone develops up to temperatures of 150–160 °C, above which montmorillonite becomes unstable. The strong increase in chlorite and illite contents and the appearance of mixed-layer clays characterize the transition to the phyllitic zone, also termed *illite–chlorite zone*, which develops up to temperatures close to 200–250 °C. The zeolite mineral typical of this zone is laumontite. The following zone, called *propylitic zone* or zone of Ca–Al–silicates, is characterized by the presence of secondary minerals, which are close to equilibrium with neutral, sodium–chloride aqueous solution. This zone develops up to temperatures of 300 °C. Epidote, the most typical mineral, can start to form in small amounts within the phyllitic zone, but it becomes abundant in the propylitic zone. Epidote is usually accompanied by abundant adularia, albite, and sulfide minerals (e.g., pyrite, pyrrhotite, and sphalerite). The zeolite mineral typical of this zone is wairakite. Chlorite and illite are also stable within this zone, but are less abundant than in the phyllitic zone. The deepest zone is the thermometamorphic zone, which is characterized by remarkable textural reorganizations of the original lithotypes and by the appearance of high temperature mineral phases, such as amphiboles (e.g., actinolite and tremolite), pyroxenes (e.g., diopside), biotite, and garnets.

The rocks affected by argillic and phyllitic alterations are characterized by extremely low permeability. In fact, the minerals typical of these two zones behave plastically under mechanic stress, acting as a reservoir cap rock. The hydrothermal minerals of the propylitic and thermometamorphic zones exhibit instead brittle behavior, permitting the development of fractures that act as high permeability pathways for geothermal fluids. Therefore, these two hydrothermal alteration zones indicate the existence of geothermal reservoirs.

The petrographic logs, which are generally carried out during deep geothermal drillings, are based on this thermal zoning of the hydrothermal alteration mineralogy.

If the geology is poorly known, it will be possible to predict the reservoir host rock from the fluid chemistry.

2.5.3
Mud and Fluid Logging while Drilling

In the rotary drilling of geothermal wells, a drilling mud is used both to transport the cuttings up to the surface and to impose hydrostatic pressure on the walls of the borehole. For these functions the mud must, for example, have an acceptable viscosity and density. It is therefore important to monitor the characteristics of the mud, e.g. by monitoring its ionic composition and to keep them within certain limits. The techniques involve separation of the mud into liquid and solid portions, and analysis of both.

The presence of geothermal reservoirs can be detected while drilling in geothermal formations by maintaining a chemical log of selected parameters in the return drilling fluid or by tracing a well drilling mud. The geochemical monitoring of drilling fluids appears an interesting method to characterize the rocks or waters of drilled formations and to forecast and quantify fluid circulation depths. The geochemistry of the formations is reflected in the composition of the drilling fluid (Aquilina and Brach, 1995). Anomalies in the gas content of the fluids could be measured when fractures are intersected. From the variations of gas composition along the drilling depth, we are able to detect some possible fluid-producing zones in different depths, which can correlate well with the logging data and drilling cores (Vuataz *et al.*, 1990).

2.5.4
Hydrothermal Reactions

Possible sources contributing major and trace elements to the discharges (Figure 2.15) include the host rocks, the magma, and the fluids circulating in the subsurface. Craig (1963) found that the deuterium content of geothermal waters was always close to that of local meteoric waters, indicating that by far the major

Figure 2.15 Borehole discharge showing evidence of hydrothermal reactions. (Photo F.-D. Vuataz).

proportions of water in hydrothermal discharges are of local meteoric origin. On the basis of these findings, the formation of hydrothermal solutions was explained largely in terms of the interaction of meteoric waters with crustal rocks at elevated temperatures, with magmatic contributions limited to the supply of heat (Ellis and Mahon, 1964).

Thermal waters can be described comprehensively by collecting samples from a reasonable number of thermal and nonthermal waters, distributed all over the investigated area, to be analyzed, for the following constituents: Na, K, Mg, Ca, alkalinity, SO_4, Cl, Li, F, B, SiO_2, and NH_3. Additional constituents useful to investigate specific problems are Al, H_2S, Rb, Cs, Br, As, and Hg. The analysis of H_2S is generally performed for environmental purposes and is not part of the standard analytical routine. If possible, the hydrogeochemical survey should be carried out at the end of the dry season to get water samples least affected by mixing with surface water. The field measurements to be carried out are temperature, pH, Eh, conductivity, and alkalinity. Sample size depends on the number of constituents to be determined and on laboratory requirements. Large water amounts are generally needed for tritium determination and when trace elements are analyzed. It is advisable to repeat the analysis of a given sample (stored in sufficient amount in the laboratory), although the concentration of some solutes may change with time. Acidification is needed to preserve cation contents of high temperature waters, which become supersaturated upon cooling, and to prevent precipitation of trace metals from both high- and low temperature waters. Dilution of filtered or filtered–acidified samples is advisable for silica determination. Unfortunately, Al concentrations are rarely measured in geothermal liquids and sometimes poor representative values are obtained, because finely dispersed aluminum oxyhydroxides pass through the membrane filters. The quality of water analysis is usually checked computing the ionic balance; however, possible errors for minor constituents (e.g., Li and F, but also Mg and SO_4 in high temperature geothermal fluids) or neutral species (e.g., SiO_2 and NH_3) cannot be detected in this way. At best, ionic balance gives an indication on the analytical accuracy of major constituents.

The dissolved constituents of geothermal waters may be subdivided into two groups according to their behavior:

- Mobile or conservative constituents are those whose activity is not limited by saturation with respect to a solid or a gas phase; comparatively mobile constituents in geothermal waters (and in most natural waters as well) are Cl, Br, B, and, to a minor extent, Li, Rb, and Cs; once they have been added to a geothermal water through a complex history, their contents along the upflow path are changed only by mixing and steam loss.
- compatible or reactive constituents are those whose activity is controlled by saturation with respect to a solid or a gas phase; they equilibrate under reservoir conditions and may respond to thermochemical changes along the upflow path of the geothermal water; Ca, Mg, Na, K, HCO_3, SO_4, F, SiO_2, and so on, usually have compatible behavior in geothermal environments.

In order to calculate the saturation index, the chemical speciation in the aqueous solution has to be reconstructed first, taking into account all ion complexes (Bowers, Jackson, and Helgeson, 1984). Due to these effects of ion complexing, total (analytical) concentration ratios of compatible cations (Na^+, K^+, Ca^{2+}, and Mg^{2+}) and compatible anions (SO_4^{2-}, F^-, and HCO_3^-) diverge, to variable amounts, from free ions activity ratios, which are uniquely fixed, at a given temperature or temperature–P_{CO2} condition, by mineral–solution equilibrium (Helgeson, Kirkham, and Flowers, 1981). This is a great step forward with respect to simple geothermometers (Guidi *et al.*, 1990). Calculations have to be carried out with the aid of a computer program, specifically implemented for this purpose (Reed and Spycher, 1984; Reed, 1982; Pitzer, 1981). Since most hydrothermal minerals are aluminum silicates, the aluminum concentration in the aqueous solution has to be introduced in the computations. However, the absence or the poor analytical quality of Al data is not a major problem and it can be circumvented assuming that aluminum concentrations are constrained by equilibrium with a given aluminum silicate (Pang and Reed, 1998).

In addition to geothermal waters, analysis of gases escaping from geothermal areas or in the fluid produced by the geothermal wells is useful in understanding subsurface conditions (D'Amore and Panichi, 1987; D'Amore and Nuti, 1977).

Geothermal gas samples are usually analyzed for H_2O, CO_2, H_2S, NH_3, He, Ar, O_2, N_2, H_2, CH_4, and CO; in addition to these constituents, HCl, HF, and oxidized S species (mainly SO_2) have to be taken into account too in high temperature volcanic gases (Giggenbach, 1996).

In general, the major component of geothermal gases is H_2O, which is followed by CO_2 and H_2S in order of decreasing importance. Other gas species present in lower concentrations are N_2, H_2, CH_4, CO, NH_3, Ar, and He. Strongly acid gases, that is, SO_2, HCl, and HF, which are typical of fluids degassed from magma bodies (Chiodini *et al.*, 1992), are virtually absent in geothermal fluids (Giggenbach, 1980). Sulfur dioxide was detectable only in some fumarolic discharges from active volcanic areas.

Similar to what is done for waters, for which a first classification step is needed before investigating mineral–solution equilibria, also for gases it is convenient to carry out an initial evaluation involving the less reactive constituents, with the aim to get information on the possible origin of fluid components, on the main processes controlling their distribution and on the secondary processes possibly interfering with gas equilibria evaluations (Chiodini and Marini, 1998). The most obvious constituents to use are N_2, Ar, and He, as suggested by Giggenbach (1991a).

2.5.4.1 **Boiling and Mixing**

Depending on pressure and temperature conditions, the main component of geothermal fluids, H_2O, can be present in different physical states. The presence of a single liquid phase in the geothermal reservoir is the most frequent situation, but it is not the only one, since either a two-phase liquid–vapor mixture or a single vapor phase can also be present in the reservoir. These possibilities can be ascertained by accurate enthalpy data or by gas geochemistry (Giggenbach, 1980; Bertrami *et al.*, 1985).

The geothermal wells, which tap a single liquid phase at temperatures above 100 °C under reservoir conditions, obviously discharge two-phase liquid–vapor mixtures, which are generated through boiling of the original single liquid phase. Sampling and analysis of both liquid and vapor phases, separated at known pressure and temperature conditions, are required to recalculate the composition of the original single liquid phase (Drummond, 1981). The two phases can be separated by means of a wellhead pressure separator. When samples of separated liquid and vapor phases are collected from a geothermal well, it is necessary to know the separation temperature and/or pressure and the well-bottom temperature and/or pressure and total discharge enthalpy (steam to water ratios).

Where boiling (steam separation) occurs there is a partitioning of dissolved elements between the steam and the residual liquid; dissolved gases and other relative volatile components concentrate in the steam and nonvolatile components become concentrated in the liquid in proportion to the amount of steam that separates. When steam separation takes place, the less soluble gases (e.g., N_2, H_2, CH_4, and CO) enter preferentially the vapor phase, while the more soluble gases (CO_2, H_2S, and NH_3) are retained in part in the aqueous phase (Truesdell, 1975).

The physical mechanisms of boiling processes are extremely complicated. Two limiting mechanisms of boiling can be recognized (Arnorsson, 2000; Tonani, 1970): single-step separation, where the steam, continuously produced by decompression of the uprising liquid, remains in contact and in equilibrium with the liquid until it is separated in a unique separation event; and continuous separation, where the steam is continuously separated from the liquid as soon as it forms. An infinite number of intermediate mechanisms can also exist (multistep separation).

However, hot waters ascending from a geothermal reservoir may cool in upflow zones not only conductively and/or by boiling due to depressurization but also by mixing in the upflow with shallow, relatively cold water. Since cold waters are most often lower in dissolved solids than geothermal waters, mixing is often referred to as *dilution*. Large variations in the temperature and flow rates of thermal springs in a particular field that can be linked with parallel variations in the concentrations of nonreactive components in the water, such as Cl, usually constitute the best evidence that mixing has occurred (Marini and Cioni, 1985).

Mixing models have been developed to allow estimation of the hot water component in mixed waters emerging in springs or discharged from shallow drill holes (Truesdell and Fournier, 1977). There are essentially three kinds of mixing model:

- the chloride–enthalpy mixing model;
- the silica–enthalpy warm spring mixing model;
- the silica–carbonate mixing model.

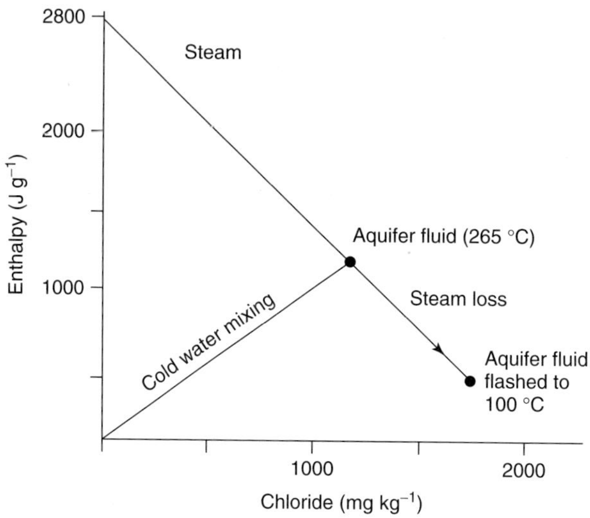

Figure 2.16 Enthalpy versus chloride plot showing the effects of boiling and dilution on a geothermal aquifer liquid at 265 °C. (Fournier, 1979a).

Linear relationships between the concentrations of conservative components such as between Cl and B or Cl and δ^2H are generally considered to constitute the best evidence for mixing.

The magnitude of the oxygen shift depends on the extent of the water–rock interaction. There appears to be a crude relationship between the temperature of geothermal waters and their ^{18}O shift. Generally, increasing temperatures enhance chemical reaction rates including reactions involving water and rock, thus increasing the ^{18}O isotope shift. Mixing of geothermal water with local cold water may manifest itself in a linear relationship between the δ values for ^{2}H and ^{18}O or between these values and the aqueous concentrations of conservative elements such as Cl.

Recognition of mixed water on the basis of chemical composition of a single sample is generally not convincing and it is necessary to establish that the sampled and analyzed waters are truly mixed before applying mixing models to estimate reservoir temperatures.

The enthalpy versus chloride plot is a suitable tool to distinguish the effects of boiling (steam loss) and mixing, since both steam and cold waters, which generally have low chloride contents, are characterized by very different enthalpy values (Fournier, 1979a). The enthalpy–chloride plot of Figure 2.16 shows that boiling moves the liquid from the point representative of the 265 °C geothermal liquid toward higher chloride contents and lower enthalpies, whereas addition of cold, dilute waters determine a decrease in both enthalpy and chloride. If discharged water is cooled mainly through conductive heat loss, the chloride concentration of the deep hot water remains unchanged.

2.5.5
Chemical Characteristics of Fluids

The chemical classification of waters is essential for a correct utilization of geochemical techniques, which can be confidently applied only to particular kinds of fluids with limited ranges of composition, reflecting the environment of provenance. For instance, most ionic solute geothermometers can be applied only to the samples representative of water–rock equilibrium at depth. Therefore, these samples have to be properly identified and selected. Furthermore, possible phenomena affecting the original characteristics of waters (i.e., addition of cold, shallow groundwaters, boiling, dissolution, or precipitation of mineral phases) have to be recognized and evaluated.

The $Cl-SO_4-HCO_3$ ternary diagram is one of the diagrams used for classification of natural waters (Giggenbach, 1988; Figure 2.17). It helps to discern

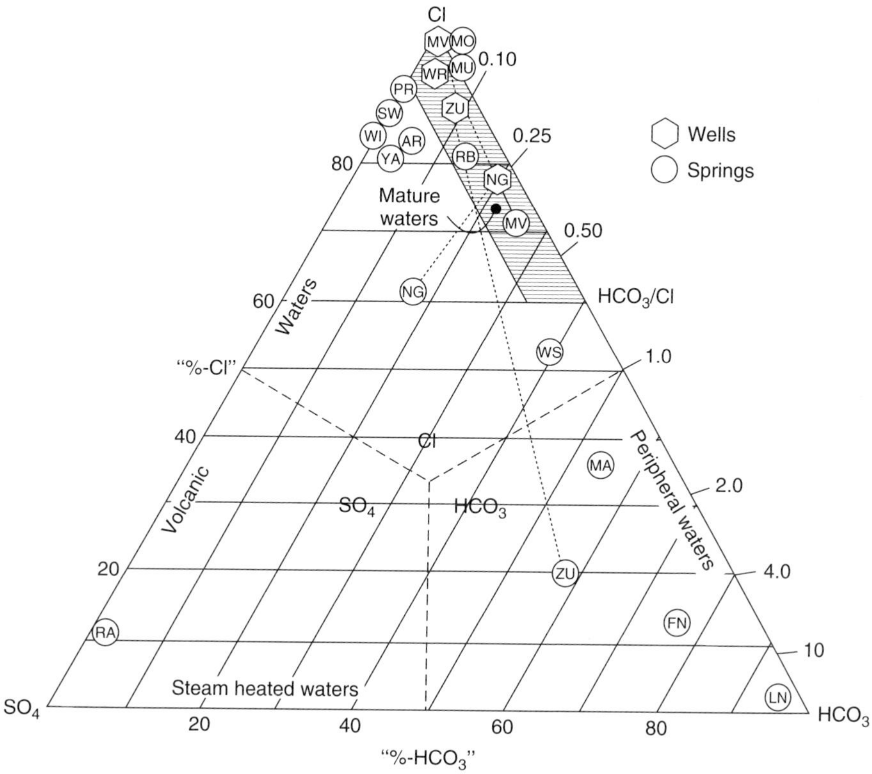

Figure 2.17 Ternary plot used to classify geothermal waters based on the relative proportions of chloride, sulfate, and bicarbonate ions. (From Giggenbach, 1988).

immature unstable waters and gives an initial indication of mixing relationships or geographical groupings. The diagram distinguishes several types of thermal water including immature waters, peripheral waters, volcanic waters, and steam-heated waters. It gives a preliminary statistical evaluation of groupings and trends.

The Cl–Li–B triangular diagram is useful for evaluating the origin of geothermal fluid. The alkali metal probably least affected by secondary processes is lithium (Li). It may therefore be used as a tracer for the initial deep rock dissolution process and as a reference to evaluate the possible origin of two important conservative constituents of geothermal waters. The boron (B) content of thermal fluids is likely to reflect to some degree the maturity of a geothermal system; because of its volatility, it is expelled during the early heating up stages. It is striking that both Cl and B are added to the Li containing solutions in proportions close to those in crustal rocks. The Cl/B ratio is often used to indicate a common reservoir source for the waters.

Four types of waters circulating in high enthalpy geothermal systems are generally described (Ellis and Mahon, 1977; Henley *et al.*, 1984; Giggenbach, 1988; Truesdell, 1991). It must be underscored, however, that each of these waters may mix with each other giving rise to hybrid water types.

2.5.5.1 **Sodium–Chloride Waters**

Waters circulating in deep, high enthalpy geothermal reservoirs usually have sodium–chloride composition and chloride contents ranging up to 10 000 mg kg^{-1}. pH of these waters is close (±1 or 2 units) to the neutral pH at depth (e.g., 5.5–5.6 at 200–300 °C). Silica, potassium, lithium, boron, fluoride are much higher than in cold waters. The high chloride waters also contain appreciable calcium. Magnesium is instead much lower than in cold waters. The main dissolved gases are CO_2 and H_2S.

In general, the waters circulating in deep, high enthalpy geothermal reservoirs are mainly of meteoric origin, but in some systems connate or other saline waters may be present. In geothermal systems with close volcanic–magmatic association and located along convergent plate boundaries, the deep, magmatic heat source may add acid gases such as HCl, HF, SO_2, H_2S, and CO_2 as well as some andesitic water. The ration of chloride to sulfate is high.

Conversion of the initially acid aqueous solutions to neutral sodium–chloride waters requires extensive rock–water interaction and virtually complete removal of magmatic sulfur species in the form of sulfates and sulfides. The deep sodium–chloride waters may flow directly to the surface and discharge from boiling, high chloride springs, whose pH ranges from near neutral to alkaline; alternatively, they may mix with shallow, low-salinity waters to give relatively diluted chloride waters.

2.5.5.2 **Acid–Sulfate Waters**

Acid–sulfate waters are typically found above the upflow part of the geothermal systems, where steam separation takes place. Boiling results in the transfer of gas species, mainly CO_2 and H_2S, into the vapor phase. This vapor phase can reach the surface without any interaction with shallow or surface waters, in the form of fumaroles and steam jets. Alternatively, separated vapor may condensate, at least partly, in shallow groundwaters or surface waters to form steam-heated waters. In this environment, atmospheric oxygen oxidizes H_2S to sulfuric acid producing acid–sulfate waters. These are characterized by low chloride contents and low pH values (0–3) and react quickly with host rocks to give advanced argillic alteration paragenesis, which are dominated by kaolinite and alunite. Dissolved cations and silica are mainly leached from the surrounding rocks, whose compositions may be approached by these acid waters. Shallow steam-heated waters may themselves boil, separating secondary steam, which reaches the surface in the form of low-pressure steaming grounds.

2.5.5.3 **Sodium–Bicarbonate Waters**

Bicarbonate-rich waters originate through either dissolution of CO_2-bearing gases or condensation of geothermal steam in relatively deep, oxygen-free groundwaters. Because the absence of oxygen prevents oxidation of H_2S, the acidity of these aqueous solutions is given by dissociation of H_2CO_3. Although it is a weak acid, it converts feldspars to clays, generating neutral aqueous solutions, which are typically rich in sodium and bicarbonate, particularly at medium–high temperature. In fact, the low solubility of calcite prevents the aqueous solution to increase in calcium content; potassium and magnesium are fixed in clays and chlorites, respectively; and sulfate concentration is limited by the low solubility of anhydrite.

Sodium–bicarbonate waters are generally found in the condensation zone of vapor-dominated systems and in the marginal parts of liquid-dominated systems. However, sodium–bicarbonate waters are also present in deep geothermal reservoirs hosted in metamorphic and/or sedimentary rocks.

2.5.5.4 **Acid Chloride–Sulfate Waters**

These acid waters do not come from separate reservoirs, but are produced through inflow of acid magmatic gases into the deepest portions of convecting neutral pH, NaCl systems. The acidity and the chemistry of the aqueous solution depends on the extent of water–rock titration, which is also a function of the amount of magmatic gases added to the water and of the availability of minerals that are able to neutralize acids.

This type of waters is commonly found in crater lakes. The chemistry of crater lake waters, especially during periods of intense volcanic activity, is obviously dominated by inflow and absorption of magmatic gases rich in HCl and S species, mainly SO_2 and H_2S, leading to the production of strongly reactive aqueous solutions with respect to cation leaching or rock dissolution, leading to deposition of alunite, anhydrite, pyrite, and kaolinite.

At greater depths, magmatic gases interact with water and masses of rocks much larger than in crater lakes, and at higher temperatures and for longer periods of time, with respect to crater lakes, thus leading to higher extents of neutralization and ultimately to the formation of neutral NaCl waters (Giggenbach, 1997a; Reed, 1997).

2.5.6 Isotopic Characteristics of Fluids

An integrated isotopic approach can provide needed information to geothermal prospecting regarding

- the sources of geothermal fluids and heat;
- the spatial distribution of fluid types;
- subsurface fluid flow directions and paths;
- the physiochemical processes affecting fluid composition, for example, water–rock reaction paths and rates;
- the temporal evolution of geothermal systems.

The isotopic compositions of elements in geothermal fluids provide a quantitative measure of material balance and can be applied to fluid samples from production wells, hydrothermal and nonthermal springs, fumaroles, and so on; fluid inclusions in rocks and minerals; and host rocks and minerals themselves (Tonani, 1970).

The isotopic compositions of elements in a fluid moving through the crust will be modified in space and time in response to varying chemical and physical parameters and/or by mixing with other fluids (Giggenbach, 1992a; O'Neil, 1986). During this process, elements will either be conserved, thus preserving isotopic information related to initial conditions and sources, or modified in a fashion that is diagnostic of chemical reactions along a flow path.

The noble gases (He, Ne, Ar, Kr, and Xe) are excellent natural tracers for heat and fluid sources, fluid origins, and reservoir processes. They are chemically inert and, therefore, conserved in water–rock systems. Because noble gases have very low solubility in fresh water, particularly at high temperatures ($T > 150\,^{\circ}$C), phase separation (e.g., boiling) will generate a residual liquid that is strongly depleted in noble gases and the relative noble gas composition will be fractionated relative to the original fluid composition. The concentrations of noble gases in the residual brines will be extremely depleted relative to the original reservoir fluid (Arnorsson, 2000). For instance, with a steam fraction of only 2.5%, the residual liquid is depleted in ^{36}Ar by a factor of ~20.

Although N_2 is reactive, in general its contents are not perturbed by chemical reactions since it is by far the predominant nitrogen species. Considering a large number of gas analyses coming from different tectonic settings, Giggenbach (1991a) showed that the relative concentrations of He, Ar, and N_2 delineate the major source components. In principle, the ^{3}He/^{4}He ratios could be used to discriminate the contributions of mantle and crustal gases.

As fluid containing mantle helium flows through the crust, the elevated ^{3}He/^{4}He ratio (elevated because of the mantle ^{3}He contribution) will become diluted with the accumulating radiogenic ^{4}He produced locally in the host rocks. The resulting gradient in helium isotopic composition along the fluid flow path will be a function of the fluid velocity and ^{4}He-production rate. Geothermal systems hosted in continental crust exhibit a wide range in ^{3}He/^{4}He ratios. There is evidence that the range may reflect variations in the contributions from different heat sources (Nicholson, 1993).

The very low solubility of noble gases in water make them very sensitive natural tracers for monitoring the return of cooler reinjected production fluids in geothermal reservoirs. Field management and production strategies need a reliable and sensitive tracer for monitoring the breakthrough of reinjected fluids (D'Amore and Panichi, 1987).

Fractionating elements (H, C, O, and N) are elements whose isotopic composition can be modified by chemical reactions involving the breakdown of chemical bonds, electron transfer, or by phase change (e.g., boiling and condensation), (Horita and Wesolowski, 1994). These elements can provide useful information regarding the source of recharge (meteoric) fluids, water/rock ratios, and chemical equilibration temperatures.

The δD and $\delta^{18}O$ values of liquids circulating in deep, high temperature geothermal systems are controlled by several processes (Figure 2.18), including (Giggenbach, 1991b) rock–water ^{18}O exchange, mixing of different waters (meteoric waters, marine waters, connate waters, magmatic waters, etc.), and boiling (steam separation). The decrease in temperature brings about a quick decrease

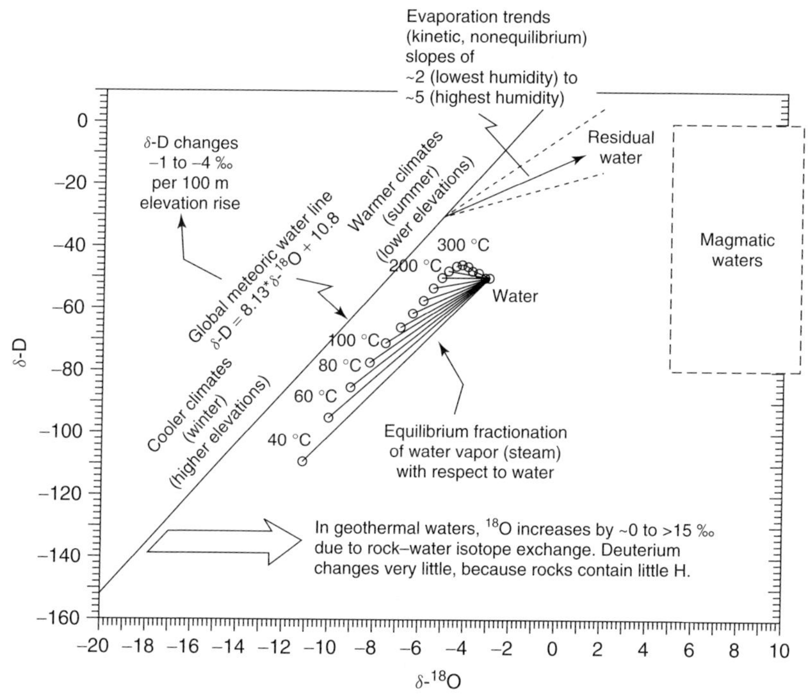

Figure 2.18 Basic processes affecting oxygen and hydrogen stable isotopes in geothermal waters. (Klein, 2006).

in the kinetics of the ^{18}O exchange between minerals and water. Therefore, ^{18}O enrichments in Na–Cl geothermal liquids with respect to meteoric waters have been traditionally considered as indicators of high temperatures (>150 °C) in the geothermal reservoirs of provenance (Truesdell and Hulston, 1980). However, the ^{18}O shifts depend also on the initial isotope composition of the liquid and solid phases involved in the exchange process as well as on the water/rock ratio, or on system dynamics (Giggenbach, 1991b). A second type of processes affecting the δD and $\delta^{18}O$ values of geothermal liquids is mixing between local meteoric waters and waters of different origin, such as marine waters; high-salinity waters, or connate waters, especially in sedimentary basins and magmatic waters (Horita, Cole, and Wesolowski, 1995). Steam separation (boiling) brings about isotopic fractionation, with heavy isotopes concentrating in the liquid (at any temperature for ^{18}O and at temperatures below 229 °C for deuterium in a pure water system), whereas the vapor phase is obviously depleted in heavy isotopes.

Deuterium, oxygen 18, and tritium contents of water are useful tracers of the meteoric origin of the water and the marine origin of the salinity in the saline thermal waters (Arnorsson, 2000). Carbon 14 is a good dating tool for geothermal systems in the granite terrain if sufficient dissolved carbon is available for analysis. The *residence time* of the geothermal fluid is defined as the time since it was last isolated from atmosphere. The estimation of fluid ages permits to constrain flow rates through the system. The presence of tritium in the thermal water is an indicator of dilution by the young shallow groundwater. However, it seems not to be a good dating method because the age of geothermal waters is usually far beyond the limit (about 100 years) of the tritium method. Sulfur and oxygen isotopes of dissolved sulfate are good tracers of the origin of the salinity in the case of marine sulfate.

The isotope ratios of dissolved elements (e.g., C, S, Sr, Nd, Pb, U) are influenced by the rock the waters pass through. The resulting isotopic contrasts give rise to spatial and/or temporal patterns in isotope ratios that contain information about fluid flow paths, water–rock interaction, and mixing relationships (Richet, Bottinga, and Javoy, 1977). This approach has been used in hydrologic studies from the catchments to the regional scale.

Nonfractionating elements (e.g., Sr, Nd, Pb, and U) are too heavy for chemical reactions or phase changes to have a significant impact on their isotopic compositions (also any natural fractionation that has occurred will be lost to normalization procedures used in the isotopic measurement). However, their isotopic compositions in fluids can be altered by fluid–rock exchange via dissolution/precipitation. The evolution of their isotopic compositions in fluids can provide insight into chemical reaction rates and paths and fluid flow directions and velocities. The isotope ratios of "nonfractionating" elements (e.g., Sr, Nd, and Pb) in solution-deposited minerals are always equal to those of the parent water; thus directly recording water conditions at the time of precipitation. On the other hand, isotope ratios of "fractionating" elements (e.g., H, C, and O) record the isotope ratios of the parent fluid, modified by temperature-dependent fractionation.

The isotopic ratios of both nonfractionating and fractionating elements in groundwater at any point along a fluid flow path will be the product of the isotopic composition of the original fluid and that of the solute acquired from the host rock. Therefore, isotope ratio measurements, either on sampled groundwater or water-deposited minerals, yield spatial or temporal patterns that contain information concerning groundwater flow and transport conditions.

Many geochemical techniques, more or less sophisticated, are effective to indicate the presence of geothermal reservoirs at depth and can be used to map their extension (Chiodini *et al.*, 1998).

2.5.7 Estimation of Reservoir Temperature

At present, the most useful geothermometers are silica, Na–K, Na–K–Ca, and $\Delta^{18}O$ ($SO_4^{2-} \cdot H_2O$). Each of these geothermometers requires special consideration in its application (Figure 2.19). In many places, some or all of these geothermometers applied to hot spring waters give good indications of deep reservoir temperature. In other places, however, these geothermometers give information only about shallow reservoirs containing more dilute and lower temperature fluids than present in

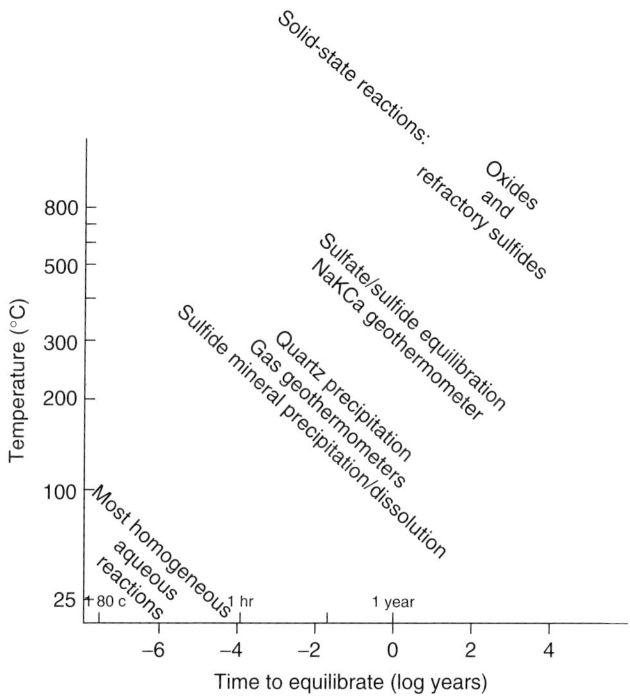

Figure 2.19 Qualitative comparison of reaction times to equilibrate. (Henley *et al.*, 1984).

deeper reservoirs. Under some conditions, mixing models may be used to estimate reservoir temperatures and salinities in deeper reservoirs. Geochemistry can be used to estimate reservoir temperatures encountered by newly drilled wells long before temperature logs (Rybach and Muffler, 1981). Fluid–mineral equilibrium modeling helped the selection and improved the use of conventional chemical geothermometers.

2.5.7.1 Geothermometric Methods for Geothermal Waters

The ratios of the cations in geothermal fluids are controlled by temperature-dependant water–rock interactions with chemical equilibrium attained in high temperature reservoirs when fluid residence times are relatively long (years). Water geothermometers can be classified into two groups (Nicholson, 1993):

- those based on temperature-dependent changes in the solubility of individual minerals, such as the silica geothermometers;
- those based on temperature-dependent ion exchange reactions, which involve at least two minerals and the aqueous solution, thus fixing the ratios of suitable dissolved constituents (e.g. the ionic solutes geothermometers).

It must be stressed that all geothermometers are used assuming that (Arnorsson, 2000)

- the geothermal liquid is in equilibrium with relevant hydrothermal minerals in the reservoir;
- the pore fluid pressure in the reservoir is fixed by coexistence of liquid and steam;
- the geothermal liquid cools either conductively or adiabatically, through steam separation at 100 °C;
- the geothermal liquid does not mix with cold, shallow waters during the ascent toward the surface;
- the geothermal liquid does not precipitate any relevant mineral phase along the upflow path.

2.5.7.2 Silica Geothermometer

The rate of quartz dissolution and precipitation depends strongly on temperature and is relatively fast at high temperatures and very slow at low temperature (Rimstidt and Barnes, 1980). This explains why in the geothermal reservoirs of constant, high temperature (generally $>180\,^\circ C$), liquids attain saturation with respect to quartz, after relatively long water–rock interaction, and little dissolved silica polymerizes and precipitates during the relatively fast ascent of geothermal waters, even though saturation with respect to quartz is largely exceeded. On the other hand, amorphous silica precipitates relatively fast when saturation is exceeded, although the lack of solubility and polymerization data at high temperatures, pH, and in multicomponent solutions limits the understanding of amorphous silica behavior (Chan, 1989).

At temperatures below 300 °C, and at depth generally attained by commercial drilling for geothermal resources, variations in pressure at hydrostatic conditions have little effect on the solubility of quartz and amorphous silica (Fournier and Rowe, 1977). The effects of added salts are significant only for concentrations greater than 2–3 wt% approximately (Marshall, 1980; Fleming and Crerar, 1979; Fournier, 1985). However, above 300 °C small changes in pressure and salinity become important. The solubility of silica is also affected by pH, for pH values above 7.8–9.3, depending on temperature. However, since pH of geothermal reservoir liquids is generally constrained at values of 5–7 by water–rock reactions, corrections for pH effects are rarely needed in geothermometric calculations. For these reasons, dissolved silica in solutions of near neutral pH from geothermal wells is a reliable geothermometer. The interpretation of dissolved silica from hot springs is somewhat ambiguous because of uncertainties about the mineral controlling dissolved silica and the amount of steam possibly separated (Fournier, 1991).

Relating the solubility of quartz to enthalpy instead of temperature has several advantages. At a given enthalpy, there is only one value of dissolved silica, while at a given temperature there are two values of dissolved silica, one for the liquid and the other for the steam. The enthalpy can be treated similar to mass, which makes the silica versus enthalpy plot a good tool to investigate isoenthalpic mixing and/or boiling.

2.5.7.3 **Ionic Solutes Geothermometers**

A general decrease in Na–K ratios of thermal waters with increasing temperatures was observed long ago (White, 1957; Ellis and Mahon, 1964). The initial attempts to derive, from these observations, an empirical Na–K geothermometer led to equations with relatively small temperature dependences due to the inclusion in the data sets of poorly equilibrated spring waters. It was recognized long ago that Mg contents of thermal waters are strongly dependent on temperature, and this relationship was early attributed to equilibration of geothermal liquids with chlorites (Ellis, 1970) or other Mg-bearing minerals, for example, montmorillonites and saponites.

The Na–K–Ca geothermometer (Fournier and Truesdell, 1973) is probably the most popular and used ionic solute geothermometer. The Na–K–Ca function is entirely empirical and assumes two different exchange reactions. The Na–K–Ca function gives erratic results below 200 °C due to high partial pressures of carbon dioxide (Paces, 1975) and due to the occurrence of exchange reactions involving also Mg. Fournier and Potter (1979) proposed a quite complex Mg correction to the Na–K–Ca function. In addition, precipitation of calcite causes an overestimation of the equilibrium temperature obtained by means of the Na–K–Ca function.

However, Reed and Spycher (1984) have suggested that the best estimate of reservoir temperature can be attained by considering simultaneously the state of equilibrium between specific components in water and many hydrothermal minerals as a function of temperature. Therefore, if a group of minerals converges to equilibrium at a particular temperature, this temperature corresponds to the most likely reservoir temperature.

2.5.7.4 Gas (Steam) Geothermometers

There are essentially three types of steam geothermometers:

- those based on gas–gas equilibria that require only data on the relative abundance of gaseous components in the gas phase;
- those based on mineral–gas equilibria involving H_2S, H_2, and CH_4, but assuming CO_2 to be externally fixed according to empirical methods;
- those based on mineral–gas equilibria that require the information on gas concentrations in steam.

Steam geothermometry is more difficult to handle than water geothermometry. Gas concentrations in geothermal reservoir fluids are affected by the ratio of steam to water of that fluid (Chiodini *et al.*, 1991a). The gas content of fumarole steam is affected by the boiling mechanism in the upflow, steam condensation, and the separation pressure of the steam from parent water (D'Amore and Panichi, 1980). The flux of gaseous components into geothermal systems from their magmatic heat source may be quite significant and influence how closely gas–gas and mineral–gas equilibria are approached in specific aquifers. The gas geothermometers are useful for predicting subsurface temperatures in high temperature geothermal systems. They are applicable to systems in basaltic to acidic rocks and in sediments with similar composition, but should be used with reservation for systems located in rocks, which differ much in composition from the basaltic to acidic ones (Arnorsson and Gunnlaugsson, 1985).

2.5.7.5 Isotope Geothermometers

Several isotope-exchange reactions can be used as subsurface temperature indicators. It may not be possible to apply gas isotope geothermometers in some liquid-dominated geothermal systems during the exploration phase because of lack of natural gas manifestations. The exchange of ^{18}O between dissolved sulfate and liquid water is most useful because of rapid equilibrium at reservoir temperatures greater than 200 °C and pH < 7, conditions that favor sulfate ion–water exchange (Arnorsson, 2000). The sulfate water ^{18}O geothermometer gives correct reservoir temperature prediction. Water isotopic composition may also be affected by evaporation or mixing with shallow groundwater.

2.5.8 Forecast of Corrosion and Scaling Processes

The solids and gases occurring in geothermal fluids are thermodynamically capable of causing corrosion by attacking metal surfaces or scaling during the utilization cycle (Vetter and Kandarpa, 1987). During geothermal reservoir exploitation, temperatures may range from 350 °C at the bottom of the well to 60 °C or so in the reinjection lines, and associated pressures can vary from 300 to 0.08 bar. Of course, geothermal fluids may contact air in the condenser and when reinjected. However, the chemical composition of these fluids will range from almost pure water to hot brine with a total dissolved solid content of $200\,g\,l^{-1}$ or more. Consequently, it is almost impossible to find a solution to corrosion and scaling problems except on a specific basis for particular sites.

The chemistry from different fields can vary substantially. Higher temperature resources with the higher water ratios have increased levels of silica that cause tremendous scaling and deposit problems. By contrast, dry steam fields do not experience the silica problems, but instead have aggressive corrosion problems associated with hydrogen chloride and hydrogen sulfide attack. Still other geothermal fields are hit with a double misfortune and encounter both scaling and corrosion problems at the same time (Figure 2.20).

Corrosion attacks occur in many geothermal operations and these result in severe equipment damage (Lichti and Wilson, 1999). Corrosion may be general, a sort of "rusting" that affects all metal surfaces in geothermal equipment, or localized. There are multiple mechanisms, for example, stress corrosion and sulfide stress cracking, of corrosion attack contributing to the failure of pipe and equipment in these systems (Corsi, 1986). The first is due to stress coupled with the existence of chloride ions in the environment and the second is caused by the presence of hydrogen sulfide in an aqueous phase. The thermal and hydraulic machines require materials that can withstand stress as well as corrosion erosion resulting

Figure 2.20 Wellhead submitted to corrosion and scaling. (Photo F.-D. Vuataz).

from high velocity geothermal fluid passing through them. It is possible to estimate pitting corrosion by a method of determining uniform corrosion rate by weight loss measurement, in which a clean sample of test metal is measured, weighed, exposed to a corroding attack for a known time interval, removed, cleaned, and reweighed. Stress corrosion can be measured by using test coupons of different types (Bridges and Hobbs, 1987). High salinity and high gas content imply specific equipments and production conditions of the EGS power plant.

The formation of scale can present challenging operating problems for geothermal plants. The major species of scale in geothermal brine typically include calcium, silica, and sulfide compounds. Calcium compounds frequently encountered are calcium carbonate and calcium silicate. Metal silicate and metal sulfide scales are often observed in higher temperature resources. Typical metals associated with silicate and sulfide scales include zinc, iron, lead, magnesium, antimony, and cadmium. Silica can present even more difficulties, as it will form an amorphous silica scale that is not associated with other cations.

There are three potential situations favoring scale accumulation (Bowen, 1989). First, deposition may take place from a single-phase fluid saturated with respect to the relevant solids (reinjection pipeline). Secondly, it may occur from flashing fluids (wells, separators, two-phase pipelines). Flashing is caused by drops in pressure or cavitation in turbulent flow and probably produces calcite scale. It enhances supersaturation by steam loss from the liquid phase, increasing concentration of the residual solutes, by temperature diminution during expansion, and by loss of stable gases such as carbon dioxide and hydrogen sulfide promoting increase in pH. Thirdly, scaling can result from steam carryover (separators, turbines, and steam pipelines). This can affect turbines badly where they are exposed only to steam. Nucleation and depositional kinetics are a function of the degree of supersaturation, pressure, temperature, and catalytic or inhibitory effects due to minor elements.

Chemical thermodynamic methodology should be used to quantitatively assess scaling tendencies from geothermal waters (Vetter and Kandarpa, 1987). Such an assessment should be routinely carried out as a part of any geothermal development program to identify optimum conditions for injection of waste geothermal fluids and, at the same time, minimize the need for using inhibitors. The rate of scale formation depends on temperature, the aqueous concentrations of the scale forming components, the degree of supersaturation and kinetics.

Because of conductive or adiabatic cooling, geothermal fluids may become supersaturated with respect to amorphous silica, which precipitates relatively fast. Occurrence of amorphous silica precipitation hinders the use of quartz and chalcedony geothermometers (Arnorsson, 2000). More importantly, the consequent deposition of amorphous silica in surface installation and in reinjection circuits is a major problem in the use and disposal of geothermal liquids for electrical production. For this reason, it is very important to evaluate the temperature at which saturation with respect to amorphous silica is attained. In the case of adiabatic cooling (boiling spring or discharge from a well), a pH increase is expected due to CO_2 loss. Boiling will cause an increase in silica and a decrease in enthalpy of the liquid. It is evident that the operating pressures of wellhead steam separators must be decided considering the pH of the liquid phase to avoid silica scaling. If the salinity of the geothermal liquid is higher than that of seawater, its effect on amorphous silica solubility has to be considered too (Fournier, 1991). Deposition of amorphous silica from supersaturated water could possibly be reduced, even inhibited, by rapid cooling of the water to 50 °C or less (Gunnarsson and Arnorsson, 2007).

Calcium carbonate scaling often occurs because almost all geothermal systems contain dissolved carbon dioxide. Prevention of calcium carbonate scaling can be achieved by alteration of either the partial pressure of carbon dioxide or the pH, and also by adding chemical scale inhibitors (Stamatakis *et al.*, 2006). Good results were obtained with the sodium polyacrylate (Yanagisawa, Matsunaga, and Sugita, 2006). Acidification is also known to lower the rate of deposition and to prevent silica scaling. Mixing of condensate and brine should always be considered as a possible means of preventing scaling in injection wells.

Safe disposal of thermally spent geothermal brines that contain environmentally hazardous constituents is commonly obtained by reinjection. During the exploitation phase, the reinjection process also serves to maintain reservoir pressure, enhance thermal recovery, and eliminate possible compactional subsidence.

References

Akkuş, I., Akilli, H., Ceyhan, S., Dilemre, A., and Tekin, Z. (2005) Türkiye Jeotermal Kaynaklari Envanteri: Envanter Serisi, Ankara, 849 p.

Anderson, E.M. (1951) *The Dynamics of Faulting*, Oliver & Boyd, London, 83 p.

Aquilina, L. and Brach, M. (1995) WELCOM (Well Chemical On-line Monitoring): evolution of chemical monitoring of drilling fluids and industrial perspectives. *Society of Petroleum Engineers, Drilling and Completion*, 158–164.

Árnason, K., Eysteinsson, H., and Hersir, G. (2010) Joint 1-D inversion of TEM and MT data and 3D inversion of MT data in the Hengill area, SW Iceland. Special volume about the European I-GET project. *Geothermics*, **39** (1), in press.

Archie, G.E., (1942) The electrical resistivity log as an aid in determining some reservoir characteristics. *Petroleum Transactions American Institute of Mining, Metallurgical, and Petroleum Engineers* **146**, 54–62.

Arnorsson, S. (ed.) (2000) *Isotopic and Chemical Techniques in Geothermal Exploration, Development and Use*, International Atomic Energy Agency, Vienna, p. 351.

Arnorsson, S. and Gunnlaugsson, E. (1985) New gas geothermometers for geothermal exploration – calibration and application. *Geochimica Cosmochimica Acta*, **49**, 1307–1325.

Arnorsson, S., Gunnlaugsson, E., and Svavarsson, S. (1983) The chemistry of geothermal waters in Iceland. III. Chemical geothermometry in geothermal investigations. *Geochimica Cosmochimica Acta*, **47**, 567–577.

Avseth, P., Mukerji, T., and Mavko, G. (2005) *Quantitative Seismic Interpretation*, Cambridge University Press, 356 p.

Barton, C.A., Zoback, M.D., and Moos, D. (1995) Fluid flow along potentially active faults in crystalline rock. *Geology*, **23** (8), 683–686.

Bertrami, R., Cioni, R., Corazza, E., D'Amore, F., and Marini, L. (1985) Carbon monoxide in geothermal gases. Reservoir temperature calculations at Larderello (Italy). *Geothermal Research Council Transactions*, **9**, 299–303.

Bowen, R. (1989) *Geothermal Resources*, 2nd edn, Elsevier Applied Science, London, p. 485.

Bowers, T.S., Jackson, K.J., and Helgeson, H.C. (1984) *Equilibrium Activity Diagrams for Coexisting Minerals and Aqueous Solutions at Pressures and Temperatures to 5 kb and 600 °C*, Springer, 397 p.

Bridges, C.E. and Hobbs, W.M. (1987) Corrosion Control in the Geothermal Drilling Industry. Material Performance, August 1987, pp. 34–41.

Browne, P.R.L. (1970) Hydrothermal alteration as an aid in investigating geothermal fields. *Geothermics*, **2**, 564–570.

Browne, P.R.L. (1982) Mapping of geothermal discharge features, in *Introduction to Geothermal Prospecting* (ed. M.P.Hochstein), University of Auckland Publications, pp. 67–69.

Byerlee, J.D. (1978) Friction of rocks. *Pure and Applied Geophysics*, **116**, 615–626.

Bujakowski, W., Barbacki, A., Czerwińska, B., Pająk, L., Pussak, M., Stefaniuk, M., and Trzeoniowski, Z. (2010) Integrated seismic and MT geophysical researches at Skierniewice test site for geothermal exploration. *Geothermics*, **39** (1), in press (Special volume about the European I-GET project).

Cappetti, G., Fiordelisi, A., Casini, M., Ciuffi, S., and Mazzotti, A. (2005) A new deep exploration program and preliminary results of a 3D Seismic survey in the Larderello-Travale geothermal field (Italy). Proceedings World Geothermal Congress, No. 759, Antalya.

Casini, M., Ciuffi, S., Fiordelisi, A., Mazzotti, A., and Stucchi, E. (2010) 3D Seismic Results in the Travale Test Site (Italy). *Geothermics*, **39** (1), 4–12. (Special volume about the European I-GET project).

Chan, S.H. (1989) A review on solubility and polymerization of silica. *Geothermics*, **18**, 49–56.

Chiodini, G., Cioni, R., Guidi, M., and Marini, L. (1991a) Chemical geothermometry and geobarometry in hydrothermal aqueous solutions: a theoretical investigation based on a mineral-solution equilibrium model. *Geochimica Cosmochimica Acta*, **55**, 2709–2727.

Chiodini, G., Cioni, R., Guidi, M., Marini, L., Raco, B., and Taddeucci, G. (1992) Gas geobarometry in boiling hydrothermal systems: a possible tool to evaluate the hazard of hydrothermal explosions. *Acta Vulcanologica, Marinelli*, **2**, 99–107.

Chiodini, G., Cioni, R., Guidi, M., Raco, B., and Marini, L. (1998) Soil CO_2 flux measurements in volcanic and geothermal areas. *Applied Geochemistry*, **13**, 543–552.

Chiodini, G. and Marini, L. (1998) Hydrothermal gas equilibria: the H_2O-H_2-CO_2-CO-CH_4 system. *Geochimica Cosmochimica Acta*, **62**, 2673–2687.

Clarke, J., Gamble, T.D., Goubau, W.M., Koch, R.H. *et al.* (1983) Remote reference magnetotellurics: equipment and procedures. *Geophysical Prospecting*, **31** (1), 149–170.

Cook, N.G.W. (1992) Jaeger memorial dedication lecture; natural joints in rock; mechanical, hydraulic and seismic behavior and properties under normal stress. *International Journal of Rock Mechanics and Mining Sciences and Geomechanics*, **29**, 198–223.

Coolbaugh, M.F., Sladek, C., Faulds, J.E., Zehner, R.E., and Oppliger, G.L. (2007) Use of rapid temperature measurements at a 2-meter depth to augment deeper temperature gradient drilling. Proceedings of the 32nd Workshop on Geothermal Reservoir Engineering, Stanford University, Stanford, CA, January 22–24, 2007, pp. 109–116.

Coolbaugh, M.F., Sladek, C., Kratt, C., Shevenell, L., and Faulds, J.E. (2006) Surface indicators of geothermal activity at Salt Wells, Nevada, USA, including warm ground, borate deposits, and siliceous alteration. *Geothermal Resources Council Transactions*, **30**, 399–405.

Corsi, R. (1986) Scaling and corrosion in geothermal equipment: problems and preventive measures. *Geothermics*, **15**, 839–856.

Cowgill, S.M., Meredith, P.G., Murrell, S.A.F., and Brereton, N.R. (1995) The orientation of stress-induced wellbore breakouts in the North Sea Basins – a case study, in *Rock Stresses in the North Sea: Proceedings ISRM International Conference, Trondheim, Norway, February 1995* (eds M.Fejerskov and A.M. Myrvang), SINTEF, Rock and Mineral Engineering, Trondheim, pp. 202–217.

Craig, H. (1963) The isotopic geochemistry of water and carbon in geothermal areas, in *Nuclear Geology in Geothermal Areas* (ed. E. Tongiorgi), Consiglio nazionale delle ricerche, Laboratoria di geologia nucleare, Pisa, Italy, pp. 17–53.

Crampin, S. (1981) A review of wave motion in anisotropic and cracked elastic media. *Wave Motion*, **3**, 343–391.

Curewitz, D. and Karson, J.A. (1997) Structural settings of hydrothermal outflow: fracture permeability maintained by fault propagation and interaction. *Journal of Volcanology and Geothermal Research*, **79**, 149–168.

D'Amore, F. and Nuti, S. (1977) Notes on the chemistry of geothermal gases. *Geothermics*, **6**, 39–45.

D'Amore, F. and Panichi, C. (1980) Evaluation of deep temperature of hydrothermal systems by a new gas-geothermometer. *Geochimica Cosmochimica Acta*, **44**, 549–556.

D'Amore, F. and Panichi, C. (1987) Geochemistry in geothermal exploration, in *Applied Geothermics* (eds J. Economides and P.O. Ungemach), John Wiley & Sons, Inc., pp. 69–89.

Dezayes, C., Genter, A., Thinon, I., Courrioux, G., and Tourliere, B. (2008) Geothermal potential assessment of the clastic Triassic reservoirs (Upper Rhine Graben, France). Proceedings of the 32nd Workshop on Geothermal Reservoir Engineering, Standford University, January 28–30, 2008, SGP-TR-185, Standford.

Drummond, S.E. (1981) Boiling and mixing of hydrothermal fluids: chemical effects on mineral precipitation. Ph.D. thesis, Pennsylvania State University.

Ellis, A.J. (1970) Quantitative interpretation of chemical characteristics of hydrothermal systems. *Geothermics*, (Special Issue 2), **2** (Part 1), 516–528.

Ellis, A.J. and Mahon, W.A.J. (1964) Natural hydrothermal systems and experimental hot water/rock interactions. *Geochimica Cosmochimica Acta*, **28**, 1323–1357.

Ellis, A.J. and Mahon, W.A.J. (1977) *Chemistry and Geothermal Systems*, Academic Press, 392 p.

Fairley, J.P. and Hinds, J.J. (2004) Rapid transport pathways for geothermal fluids in an active Great Basin fault zone. *Geology*, **32** (9), 825–828.

Farquharson, C.G. and Craven, J.A. (2009) Three-dimensional inversion of Magnetotelluric data for mineral exploration: An example from the McArthur River uranium deposit, Saskatchewan, Canada. *Journal of Applied Geophysics*, **68**, 450–458. (doi: 10.1016/j.jappgeo.2008.02.002).

Faulds, J.E., Coolbaugh, M., Blewitt, G., and Henry, C.D. (2004) Why is Nevada in hot water? Structural controls and tectonic model of geothermal systems in the northwestern Great Basin. *Geothermal Resources Council Transactions*, **28**, 649–654.

Faulds, J.E., Coolbaugh, M.F., Vice, G.S., and Edwards, M.L. (2006) Characterizing structural controls of geothermal fields in the northwestern Great Basin: a progress report. *Geothermal Resources Council Transactions*, **30**, 69–76.

Ferrill, D.A. and Morris, A.P. (2003) Dilational normal faults. *Journal of Structural Geology*, **25**, 183–196.

Ferrill, D.A., Winterle, J., Wittmeyer, G., Sims, D., Colton, S., Armstrong, A., and Morris, A.P. (1999) Stressed rock strains groundwater at Yucca Mountain, Nevada. *GSA Today*, **9**, 1–8.

Fiordelisi, A. and Bertani, R., (2006) Exploration of geothermal resources in Italy. *ENGINE – Workshop 1, Defining, Exploring, Imaging and Assessing Reservoirs for potential Heat Exchange, Potsdam* 6–8 November 2006, Germany.

Fleming, B.A. and Crerar, D.A. (1982) Silicic acid ionization and calculation of silica solubility at elevated temperature and pH. Application to geothermal fluid processing and reinjection. *Geothermics*, **11**, 15–29.

Fournier, R.O. (1977) Chemical geothermometers and mixing models for geothermal systems. *Geothermics*, **5**, 41–50.

Fournier, R.O. (1979a) Geochemical and hydrological considerations and the use of enthalpy-chloride diagrams in the prediction of underground conditions in hot-spring systems. *Journal of Volcanology and Geothermal Research*, **5**, 1–16.

Fournier, R.O. (1985) The behavior of silica in hydrothermal solutions, in *Geology and Geochemistry of Epithermal Systems*, Reviews in Economic Geology, Vol. **2** (eds B.R.Berger and P.M. Bethke), (Reviews in Economic Geology is a journal published by the University of Texas at E1 Paso, E1 Paso, Texas 79968-0637, USA.), pp. 45–61.

Fournier, R.O. (1991) Water geothermometers applied to geothermal energy, in *Application of Geochemistry in Geothermal Reservoir Development* (F. D'Amore, co-ordinator), UNITAR, pp. 37–69.

Fournier, R.O. and Potter, R.W. (1979) Magnesium correction to the Na-K-Ca chemical geothermometer. *Geochimica Cosmochimica Acta*, **43**, 1543–1550.

Fournier, R.O. and Rowe, J.J. (1977) The solubility of amorphous silica in water at high temperatures and high pressures. *American Mineralogist*, **62**, 1052–1056.

Fournier, R.O. and Truesdell, A.H. (1973) An empirical Na-K-Ca geothermometer for natural waters. *Geochimica Cosmochimica Acta*, **37**, 1255–1275.

Gamble, T.D., Goubau, W.M. *et al.* (1979) Magnetotellurics with a remote magnetic reference. *Geophysics*, **44**, 53–68.

Giggenbach, W.F. (1980) Geothermal gas equilibria. *Geochimica Cosmochimica Acta*, **44**, 2021–2032.

Giggenbach, W.F. (1984) Mass transfer in hydrothermal alterations systems. *Geochimica Cosmochimica Acta*, **48**, 2693–2711.

Giggenbach, W.F. (1988) Geothermal solute equilibria. Derivation of Na-K-Mg-Ca geoindicators. *Geochimica Cosmochimica Acta*, **52**, 2749–2765.

Giggenbach, W.F. (1991a) Chemical techniques in geothermal exploration, in *Application of Geochemistry in Geothermal Reservoir Development* (F. D'Amore, co-ordinator), UNITAR, pp. 119–144.

Giggenbach, W.F. (1991b) Isotopic composition of geothermal water and steam discharges, in *Application of Geochemistry in Geothermal Reservoir Development* (F. D'Amore, co-ordinator), UNITAR, pp. 253–273.

Giggenbach, W.F. (1992a) Isotopic shifts in waters from geothermal and volcanic systems along convergent plate boundaries and their origin. *Earth and Planetary Science Letters*, **113**, 495–510.

Giggenbach, W.F. (1996) Chemical composition of volcanic gases, in *Monitoring and Mitigation of Volcano Hazards* (eds R.Scarpa and R.I. Tilling), Springer, pp. 221–256.

Giggenbach, W.F. (1997a) The origin and evolution of fluids in magmatic-hydrothermal systems, in *Geochemistry of Hydrothermal Ore Deposits* (ed. H.L.Barnes), John Wiley & Sons, Inc., pp. 737–796.

Goff, F. and Janik, C.J. (2000) Geothermal systems, in *Encyclopedia of Volcanoes* (eds H. Sigurdsson, B. Houghton, S. McNutt, H. Rymer, and J. Stix), Academic Press, San Diego, pp. 817–834.

Gudmundsson, A. (1999) Fluid overpressure and stress drop in fault zones. *Geophysical Research Letter*, **26** (1), 115–118.

Gudmundsson, A. (2000) Active faults and groundwater flow. *Geophysical Research Letter*, **27** (18), 2993–2996.

Gudmundsson, A., Fjeldskaar, I., and Brenner, S.L. (2002) Propagation pathways and fluid transport in jointed and layered rocks in geothermal fields. *Journal of Volcanology and Geothermal Research*, **116**, 257–278.

Guidi, M., Marini, L., Scandiffio, G., and Cioni, R. (1990) Chemical geothermometry in hydrothermal aqueous solutions: the influence of ion complexing. *Geothermics*, **19**, 415–441.

Gunnarsson, I. and Arnorsson, S. (2005) Impact of silica scaling on the efficiency of heat extraction from high-temperature geothermal fluids. *Geothermics*, **34**, 320–329.

Han, N., Nam, M.J., Kim, H.J., Song, Y., and Suh, J.H. (2009) A comparison of accuracy and computation time of three-dimensional magnetotelluric modelling algorithms. *Journal of Geophysical Engineering*, **6**, 136–145. doi: 10.1088/1742-2132/6/2/005.

Heald, P., Foley, N.K., and Hayba, D.O. (1987) Comparative anatomy of volcanic-hosted epithermal deposits: acid-sulfate and adularia-sericite deposits. *Economic Geology*, **82**, 1–26.

Heise, W., Caldwell, T.G., Bibby, H.M., and Bannister, S.C. (2008) Three-dimensional modelling of magnetotelluric data from the Rotokawa geothermal field, Taupo Volcanic Zone, New Zealand. *Geophysical Journal International*, **173** (2), 740–750. doi: 10.1111/j.1365-246X.2008.03737.x.

Helgeson, H.C., Kirkham, D.H., and Flowers, G.C. (1981) Theoretical prediction of the behavior of aqueous electrolytes at high pressures and temperatures: calculation of activity coefficients, osmotic coefficients, and apparent molal, and standard and relative partial molal properties to 600°C and 5kb. *American Journal of Science*, **281**, 1249–1516.

Henley, R.W. and Ellis, A.J. (1983) Geothermal systems, ancient and modern: a geochemical review. *Earth Science Review*, **19**, 1–50.

Henley, R.W., Truesdell, A.H., Barton, P.B.Jr., and Whitney, J.A. (1984) Fluid-mineral equilibria in hydrothermal systems. *Reviews in Economic Geology*, **1**, 267.

Hillis, R.R. (1998) The influence of fracture stiffness and the *in situ* stress field on the closure of natural fractures. *Petroleum Geoscience*, **4**, 57–65.

Hillis, R.R. and Nelson, E.J. (2005) In situ stresses in the North Sea and their applications: petroleum geomechanics from exploration to development, in *Petroleum Geology: North-West Europe and Global Perspectives – Proceedings of the 6th Petroleum Geology Conference* (eds A.G. Doré and B.A. Vining), © Petroleum Geology Conferences, Ltd, Geological Society, London, pp. 551–564.

Hillis, R.R. and Reynolds, S.D. (2000) The Australian stress map. *Journal of the Geological Society, London*, **157**, 915–921.

Horita, J., Cole, D.R., and Wesolowski, D.J. (1995) The activity-composition relationship of oxygen and hydrogen isotopes in aqueous salt solutions: III. Vapor-liquid water equilibration of NaCl solutions to 350°C. *Geochimica Cosmochimica Acta*, **59**, 1139–1151.

Horita, J. and Wesolowski, D.J. (1994) Liquid-vapor fractionation of oxygen and hydrogen isotopes of water from the freezing to the critical temperature. *Geochimica Cosmochimica Acta*, **58**, 3425–3437.

Hudson, J.A. (1981) Wave speeds and attenuation of elastic waves in material containing cracks. *Geophysical Journal of the Royal Astronomical Society*, **64**, 133–150.

Huenges, E. and Moeck, I., The Geothermal Project Group (2007) Directional drilling and stimulation of a deep sedimentary geothermal reservoir. *Scientific Drilling*, **5**, 47–49. doi: 10.2204/lodp.sd.5.09.2007.

Hunt, T.M., Allis, R.G., Balkeley, M.R., and O'Sullivan, M.J. (1990) Testing reservoir simulation models for the Broadlands geothermal field using precision gravity data. *Transactions – Geothermal Resources Council*, **14** (1-2), 1287–1294.

Hurter S. and Haenel R. (2002) Atlas of Geothermal Resources in Europe. European Commission Office for Official Publications of the European Communities, Luxemburg, 93 p.

Jaeger, J.C., Cook, N.G.W., and Zimmerman, R.W. (2007) *Fundamentals of Rock Mechanics*, 4th edn, Blackwell Publishing, Oxford.

Keary, P., Brooks, M., and Hill, I. (2002) *An Introduction to Geophysical Exploration*, 3rd edn, Blackwell, 262 p.

Klein, C.W. (2006) *Exploration Geochemistry. GRC Exploration Workshop*, GeothermEx, Inc., September 2006.

Klein, D.J. Sill, W.R., (1982) Electrical properties of artificial clay-bearing sandstones., *Geophysics*, **47**, 1593–1601.

Larsen, J.C., Mackie, R.L., Manzella, A., Fiordelisi, A., and Rieven, S. (1996) Robust smooth magnetotelluric transfer functions. *Geophysical Journal International*, **124**, 801–819.

Lichti, K.A. and Wilson, P.T. (1999) Corrosion in New Zealand geothermal systems. *Corrosion Reviews*, **17**, 181–203.

Lisle, R.J. and Srivastava, D.C. (2004) Test of the frictional reactivation theory for faults and validity of fault-slip analysis. *Geology*, **32**, 569–572.

Lumb, J.T. (1981) Prospecting for geothermal resources, in *Geothermal Systems: Principles and Case Histories* (eds L. Rybach and L.J.P. Muffler), John Wiley & Sons, Ltd, pp. 77–108.

MacBeth, C. (1999) Azimuthal variations in P-wave signatures due to fluid flow. *Geophysics*, **64** (4), 1181–1192.

Mackie, R.L., Smith, J.T., and Madden, T.R. (1994) Three-dimensional electromagnetic modeling using finite difference equations: the magnetotelluric example. *Radio Science*, **29**, 923–935.

Majer, E. (2003a) 3-D Seismic Methods For Geothermal Reservoir Exploration and Assessment –Summary, Lawrence Berkeley National Laboratory, Paper LBNL-49881.

Majer, E. (2003b) Fracture Quantification in Naturally Fractured Gas Reservoirs: Research Update. GasTIPS, pp. 4–9.

Manzella, A. Ungarelli, C. Ruggieri, G. Giolito, C. and Fiordelisi, A. (2010) Electrical resistivity at the travale geothermal field (Italy). Proceedings World Geothermal Congress 2010, Bali Indonesia, pp. 25–29 April 2010.

Marini, L. and Cioni, R. (1985) A chloride method for the determination of the enthalpy of steam/water mixtures discharged from geothermal wells. *Geothermics*, **14**, 29–34.

Marshall, W.L. (1980) Amorphous silica solubilities – III. Activity coefficient relations and predictions of solubility behavior in salt solutions, 0–350°C. *Geochimica Cosmochimica Acta*, **44**, 925–931.

Micklethwaite, S. and Cox, S.F. (2004) Fault-segment rupture, aftershock-zone fluid flow, and mineralization. *Geology*, **32** (9), 813–816.

Mildren, S.D., Hillis, R.R., and Kaldi, J. (2002) Calibrating predictions of fault seal reactivation in the Timor Sea. *Australian Petroleum Production and Exploration Association Journal*, **42**, 187–202.

Millegan, P.S. (1998) *Geologic Applications of Gravity and Magnetics: Case Histories (AAPG Studies in Geology)* (eds R.I. Gibson and P.S. Millegan). Society of Exploration Geophysicists, Tulsa, OK, USA.

Moeck, I., Backers, T., and Schandelmeier, H. (2007) Assessment of mechanical wellbore assessment by numerical analysis of fracture growth, EAGE 69th Conference and Exhibition, 11–14 June 2007, extended abstracts volume, D047, London, UK. CD-ROM. ISBN: 978-90-73781-54-2. Co Productions, Houten/The Netherlands.

Moeck, I., Dussel, M., Tröger, U., and Schandelmeier, H. (2007) Fracture networks in Jurassic carbonate rock of the Algarve Basin (South Portugal): Implications for aquifer behaviour related to the recent stress field, in (eds J. Krasny and J M. Sharp), *Groundwater in Fractured Rocks*, IAH-Selected Paper Series, Vol. **9**, p. 483–497. ISBN: 978-0-415-41442-5, Taylor and Francis Group, Routledge/UK.

Moeck, I., Brandt, W., Blöcher, G., Holl, H.-G., Zimmermann, G., Huenges, E., Saadat, A., and Backers, T. (2008a) From gas to geothermal exploration: a case study from the NE-German Basin. Extended Abstracts Volume, EAGE 70th Conference and Exhibition, June 09-12, 2008, Rome, D022.

Moeck, I., Schandelmeier, H., and Holl, H.G. (2009) The stress regime in Rotliegend reservoir of the Northeast German Basin. *International Journal of Earth Science (Geological Rundschau)*, **98/7**, p. 1643–1654.

Moeck, I., Kwiatek, G., and Zimmermann, G. (2009) Slip tendency analysis, fault reactivation potential and induced seismicity in a deep geothermal reservoir. *Journal of Structural Geology*, **31** (10), 1174–1182.

Morris, A., Ferrill, D.A., and Henderson, D.B. (1996) Slip tendency analysis and fault reactivation. *Geology*, **24** (3), 275–278.

Morrow, C.A., Moore, D.E., and Lockner, D.A. (2001) Permeability reduction in granite under hydrothermal conditions. *Journal of Geophysical Research*, **106** (B12), 30,551–30,560.

Muñoz, G., Bauer, K., Moeck, I., and Ritter, O. (2010) Combining magnetotelluric and seismic models for exploration of the Groß Schönebeck geothermal site: a statistical approach. *Geothermics*, **39** (1), in press (Special volume about the European I-GET project).

Nabighian, M.N., Grauch, V.J.S., Hansen, R.O., LaFehr, T.R., Li, Y., Peirce, J.W., Phillips, J.D., and Ruder, M.E. (2005) The historical development of the magnetic method in exploration. *Geophysics*, **70** (6), 33–61.

Newman, G.A. and Alumbaugh, D.L. (2000) Three-dimensional magnetotelluric inversion using non-linear conjugate gradients. *Geophysical Journal International*, **140**, 410–424.

Newman, G.A., Gasperikova, E., Hoversten, G.M., and Wannamaker, P.E. (2008) Three-dimensional magnetotelluric characterization of the Coso geothermal field. *Geothermics*, **37** (4), 369–399. doi: 10.1016/j.geothermics.2008.02.006.

Nicholson, K. (1993) *Geothermal Fluids. Chemistry and Exploration Techniques*, Springer-Verlag, Berlin, 263 p.

Nishikawa, N. (1992) Relationships Between Clay Alteration and Physical Properties in the Hishikari Gold Mine Area (in Japanese). FY1991 Geological Survey Report, Metal Mining Agency of Japan, pp. 157–187.

O'Neil, J.R. (1986) Theoretical and experimental aspects of isotopic fractionation, in *Stable Isotopes in High Temperature Geological Processes*, Reviews in Mineralogy, Vol. 16 (eds J.W. Valley, H.P. Taylor, Jr., J.R. O'Neil), pp. The Mineralogical Society of america, 1–40.

Orlando, L. (2005) Interpretation of Tuscan gravity data. *Bollettino della Societá Geologica Italiana*, **3**, 179–186.

Paces, T. (1975) A systematic deviation from Na-K-Ca geothermometer below 75°C and above 10-4 atm PCO2. *Geochimica Cosmochimica Acta*, **39**, 541–544.

Pang, Z.-H. and Reed, M. (1998) Theoretical chemical thermometry on geothermal waters: problems and methods. *Geochimica Cosmochimica Acta*, **62**, 1083–1091.

Pellerin, L., Johnston, J.M., and Hohmann, G.-W. (1996) A numerical evaluation of electromagnetic methods in geothermal exploration. *Geophysics*, **61**, 121–130.

Peška, P. and Zoback, M.D. (1995) Compressive and tensile failure of inclined well bores and determination of *in situ* stress and rock strength. *Journal of Geophysical Research*, **100** (B7), 12,791–12,811.

Pitzer, K.S. (1981) Characteristics of very concentrated solutions, in *Chemistry and Geochemistry of Solutions at High Temperatures and Pressures* (eds D.T. Rickard and F.E. Wickman), Pergamon, pp. 249–264.

Prieto, C. and Morton, G. (2003) New insights from a 3D earth model, deepwater Gulf of Mexico. *The Leading Edge*, **22**, 356–360.

Ramsay, J.G., and Lisle, R.J. (2000) *Modern Structural Geology*, Applications of Continuum Mechanics in Structural Geology, Vol. 3, Academic Press, London, 560 p.

Reed, M.H. (1982) Calculation of multicomponent chemical equilibria and reaction processes in systems involving minerals, gases and an aqueous phase. *Geochimica Cosmochimica Acta*, **46**, 513–528.

Reed, M.H. (1997) Hydrothermal alteration and its relationship to ore fluid composition, in *Geochemistry of Hydrothermal Ore Deposits*, 3rd edn (ed. H.L. Barnes), John Wiley & Sons, Inc., pp. 517–611.

Reed, M.H. and Spycher, N.F. (1984) Calculation of pH and mineral equilibria in hydrothermal waters with application to geothermometry and studies of boiling and dilution. *Geochimica Cosmochimica Acta*, **48**, 1479–1492.

Reinecker, J., Heidbach, O., and Mueller, B. (2003) The 2003 Release of the World Stress Map, available online at *http://www.world-stress-map.org world-stress-map.org*.

Rial, J.A., Elkibbi, M., and Yang, M. (2005) Shear-wave splitting as a tool for the characterization of geothermal fractured reservoirs; lessons learned. *Geothermics*, **34** (3), 365–385.

Richet, P., Bottinga, Y., and Javoy, M. (1977) A review of H, C, N, O, S and Cl stable isotope fractionation among gaseous molecules. *Annual Review of Earth and Planetary Science Letters*, **5**, 65–110.

Rimstidt, J.D. and Barnes, H.L. (1980) The kinetics of silica-water reactions. *Geochimica Cosmochimica Acta*, **44**, 1683–1699.

Ritter, O., Junge, A., and Dawes, G.J.K. (1998) New equipment and processing for magnetotelluric remote reference observations. *Geophysical Journal International*, 535–548.

Rybach, L. and Muffler, L. J. P. (eds) (1981) *Geothermal Systems: Principles and Case Histories*, John Wiley & Sons, Ltd, Chichester, 359 p.

Sasaki, Y. (2004) Three-dimensional inversion of static-shifted magnetotelluric data. *Earth, Planets and Space*, **56**, 239–248.

Sheridan, J. and Hickman, S. (2004) *In situ* stress, fracture, and fluid flow analysis in well 38C-9: an enhanced geothermal system in the coso geothermal field. Proceedings of the 29th Workshop on Geothermal Reservoir Engineering, Standford University, January 26–28, 2004, SGP-TR-175, Standford.

Sheriff, R.E. and Geldart, L.P. (1995) *Exploration Seismology*, 2nd edn, Cambridge University Press, 592 p.

Sibson, R.H. (1996) Structural permeability of fluid-driven fault-fracture meshes. *Journal of Structural Geology*, **18**, 1031–1042.

Sibson, R.H. (1998) Conditions for rapid large-volume flow, in *Water– Rock Interaction: Proceedings of the 9th International Symposium-WRI-9, Taupo, New Zealand, March-April 1998* (eds G.B. Arehart and J.R. Hulston), Balkema, Rotterdam, pp. 35–38.

Siripunvaraporn, W., Egbert, G., Lenbury, Y., and Uyeshima, M. (2005) Three-dimensional magnetotelluric inversion: data-space method. *Physics of the Earth and Planetary Interiors*, **150**, 3–14.

Siripunvaraporn, W. and Egbert, G., (2009) WSINV3DMT: Vertical Magnetic Field Transfer Function Inversion and Parallel Implementation. *Physics of the Earth and Planetary Interiors*, **173** (3–4), 317–329.

Smith, R.P., Grauch, V.J.S., and Blackwell, D.D. (2002) Preliminary results of a high-resolution aeromagnetic survey to identify buried faults at Dixie Valley, Nevada. *Geothermal Resources Council Transactions*, **26**, 543–546.

Stamatakis, E., Chatzichristos, C., Sagen, J., Stubos, A.K., Palyvos, I., Muller, J., and Stokkan, J.-A. (2006) An integrated radiotracer approach for the laboratory evaluation of scale inhibitors performance in geological environments. *Chemical Engineering Science*, **61**, 7057–7067.

Sternberg, B.K., Washburne, J.C., and Pellerin, L. (1988) Correction for the static shift in magnetotellurics using transient electromagnetic soundings. *Geophysics*, **53**, 1459–1468.

Stober, I. and Bucher, K. (eds) (2000) *Hydrogeology of Crystalline Rocks*, Kluwer Academic Publishers, Dordrecht, p. 275.

Talbot, C.J. and Sirat, M. (2001) Stress control of hydraulic conductivity in fracture-saturated Swedish bedrock. *Engineering Geology*, **61**, 145–153.

Taylor, M.A. (2007) The State of Geothermal Technology. Part I: Subsurface Technology. Publ by the Geothermal Energy Association for the US DOE, 70 p.

Tester, J.W. *et al.* (2006) The future of geothermal energy: the impact of Enhanced Geothermal Systems (EGS) on the United States in the 21st century. Prepared by the Massachusetts Institute of Technology, under Idaho National Laboratory Subcontract No. 63 00019 for the U.S. Department of Energy, Assistant Secretary for Energy Efficiency and Renewable Energy, Office of Geothermal Technologies. (eds B.J., Anderson, A.S., Batchelor, D.D., Blackwell, R., DiPippo and E.M., Drake 358 p.

Tonani, F. (1970) Geochemical methods of exploration for geothermal energy. *Geothermics*, **2** (Part 1), 492–515.

Truesdell, A.H. (1975) Geochemical techniques in exploration. Proceedings of the 2nd UN Symposium on the Development and Use of Geothermal Resources, vol. 1, pp. 53–86.

Truesdell, A.H. (1991) Origins of acid fluids in geothermal reservoirs. *Geothermal Research Council Transactions*, **15**, 289–296.

Truesdell, A.H. and Fournier, R.O. (1977) Procedure for estimating the temperature of a hot-water component in a mixed water by using a plot of dissolved silica versus enthalpy. *U.S. Geological Survey of Journal of Research*, **5**, 49–52.

Truesdell, A.H. and Hulston, J.R. (1980) Isotopic evidence on environments of geothermal systems, in *Handbook of Environmental Isotope Geochemistry*, The Terrestrial Environment, A, Vol. 1 (eds P. Fritz and J.Ch. Fontes), Elsevier, pp. 179–226.

Uchida, T. and Sasaki, Y., (2006) Stable 3D inversion of MT data and its application to geothermal exploration. *Exploration Geophysics*, **37** (3), 223–230. doi:10.1071/EG06223.

Vetter, O.J. and Kandarpa, V. (1987) Chemical thermodynamics in geothermal operations, in *Applied Geothermics* (eds J. Economides and P.O. Ungemach), John Wiley & Sons, Inc., pp. 125–135.

Vuataz, F.-D., Brach, M., Criaud, A., and Fouillac, C. (1990) Geochemical monitoring of drilling fluids: a powerful tool to forecast and detect formation waters. *Society of Petroleum Engineers, Formation Evaluation*, **5** (2), 177–184.

Weckmann, U., Magunia, A., and Ritter, O., (2005) Effective noise separation for magnetotelluric single site data processing using a frequency domain selection scheme. *Geophysics Journal International*, **161** (3), 635–652.

Vozoff, K. (1991) The magnetotelluric method, in *Electromagnetic Methods in Applied Geophysics*, vol. 2, applications, parts A and B, Society of Exploration Geophysicists, SEG Series "Investigation in Geophysics" (ed M. Nabighian), Tulsa, Oklahoma, pp. 641–711. ISBN 1-56080-061-5

White, D.E. (1957) Magmatic, connate, and metamorphic waters. *Geological Society of American Bulletin*, **69**, 1659–1682.

Yanagisawa, N., Matsunaga, I., and Sugita, H. (2006) Scale Inhibitor Test at the Hijiori HDR SIte, Yamagata, Japan. GRC, November 2006, San Diego.

Yassir, N.A. and Bell, J.S. (1994) Relationships between pore pressure, stresses, and present-day geodynamics in the Scotian Shelf, offshore eastern Canada. *AAPG Bulletin*, **78**, 1863–1880.

Yassir, N.A. and Zerwer, A. (1997) Stress regimes in the Gulf Coast, offshore Louisiana: data from well-bore breakout analysis. *AAPG Bulletin*, **81**, 293–307.

Yokoyama, H., Nakatsuka, K., Abe, M., and Watanabe, K. (1983) Temperature dependency of electrical resistivity of water saturated rocks and the possibility of underground temperature estimation. *Journal of Geothermal Research Society of Japan*, **5** (2), 103–120 (in Japanese with English abstract).

Zhdanov, M.S. and Hursan, G., (2000) 3-D electromagnetic inversion based on quasi-analytical approximation. *Inverse Problems*, **16**, 1297–1322.

Zoback, M.D. (2007) *Reservoir Geomechanics*, Cambridge University Press, New York, 449 p.

Zoback, M.L. (1992) First- and second-order patterns of stress in the lithosphere: the world stress map project. *Journal of Geophysical Research*, **97**, 11703–11728.

Zonge, K.L. (1992) Introduction to CSAMT, in *Practical Geophysics II for the Exploration Geologist* (ed. R. van Blaricom), Northwest Mining Association, Spokane.

Further Reading

Electric Methods

Flóvenz, O.G. and Karlsdóttir, R. (2000) TEM-Resistivity image of a geothermal field in N-Iceland and the relation of the resistivity with lithology and temperature. World Geothermal Congress 2000, May 28-June 10, Kyushu-Tohoku.

Jousset, P., Haberland, C., Bauer, K., and Árnason, K. (2010) Detailed structure of the Hengill geothermal volcanic complex (Iceland) inferred from 3-D tomography of high-dynamic broadband seismological data. *Geothermics*, **39**, in press.

Mackie, R.L. and Booker, J. (1999) *Documentation for mtd3fwd and d3-to-mt*, GSY-USA, Inc., San Francisco, User documentation.

Newman, G.A., Recher, S., Tezkan, B., and Neubauer, F.M. (2003) 3D inversion of a scalar radio magnetotelluric field data set. *Geophysics*, **68**, 791–802.

Pellerin, L. and Hohmann, G.-W. (1990) Transient electromagnetic inversion: a remedy for magnetotelluric static shifts. *Geophysics*, **55**, 1242–1250.

Spichak, V. and Manzella, A. (2009) Electromagnetic sounding of geothermal zones. *Journal of Applied Geophysics*. **68** (4), 459–478.

Wanamaker, P.E., Rose, P.E., Doerner, W.M. *et al.* (2004) Magnetotelluric surveying and monitoring at the Coso Geothermal area, California, in support of the enhanced geothermal concept: survey parameters and initial results. Proceedings of the Twenty-Ninth Workshop on Geothermal Reservoir Engineering, Stanford University, January 26–28, 2004, SGP-TR-175, Stanford, pp. 287–294.

Seismic Methods

Elkibbi, M. and Rial, J.A. (2005) The Geysers geothermal field; results from shear-wave splitting analysis in a fractured reservoir. *Geophysical Journal International*, **162** (3), 1024–1035.

Ferrazzini, V. and Aki, K. (1987) Slow waves trapped in fluid-filled infinite-crack: implication for volcanic tremor. *Journal of Geophysical Research*, **92**, 9215–9223.

Hiroyuki, K. and Chouet, B. (2000) Acoustic properties of a crack containing magmatic or hydrothermal fluids.

Journal of Geophysical Research, **105**, 25,493–25,512.

Jousset, P., Neuberg, J., and Sturton, S. (2003) Modelling the time-dependent frequency content of low-frequency volcanic earthquakes. *Journal of the Volcanology and Geothermal Research*, **128**, 201–233.

Moriya, H., Niitsuma, H., and Baria, R. (2003) Multiplet-clustering analysis reveals structural details within the seismic cloud at the Soultz geothermal field, France. *Bulletin of the Seismological Society of America*, **93** (4), 1606–1620.

Tang, C., Rial, J.A., and Lees, J.M. (2005) Shear-wave splitting; a diagnostic tool to monitor fluid pressure in geothermal fields. *Geophysical Research Letters*, **32** (21), 3.

Vlahovic, G., Elkibbi, M., and Rial, J.A. (2003) Shear-wave splitting and reservoir crack characterization; the Coso geothermal field. *Journal of Volcanology and Geothermal Research*, **120** (1-2), 123–140.

Geochemistry

Economides, M.J. and Ungemach, P. O. (eds) (1987) *Applied Geothermics*, John Wiley & Sons, Ltd, Chichester, 238 p.

Gupta, H. and Roy, S. (2007) *Geothermal Energy: An Alternative Resource for the 21st Century*, Elsevier, Amsterdam, 279 p.

3
Drilling into Geothermal Reservoirs

Axel Sperber, Inga Moeck, and Wulf Brandt

3.1
Introduction

Drilling is an essential and expensive part of geothermal exploration, development, and utilization. The aim of geothermal drilling is not only to access the reservoir at its most safe but also most low priced way. Cost reduction of geothermal drilling is therefore a major issue that should be considered for the economic development of geothermal energy.

The consequences of reducing cost are often impressive, because drilling and well completion can account for more than half of the capital cost for a geothermal power project. Geothermal drilling can be even more expensive (in cost/depth) than onshore oil and gas drilling for two principal reasons:

1) **Technical challenge:** Geothermal reservoirs may host highly corrosive fluids of high temperature in great depth, which can mean that special tools and techniques are required for the harsh downhole conditions.
2) **Large diameters:** Because the produced fluid (hot water) is of intrinsically low value, large flow rates and thus, large holes and casing, are required.

Geothermal drilling cost reduction can take many forms, for example, reduction of downtime (i.e., the NPT = nonproductive time), use of optimized drilling technique and tools, higher per-well production through multilaterals and less trouble by understanding geological–technical risk before the start of drilling.

The most important part of a successful geothermal drilling project is however, to understand the complexity of drilling a well from planning to completion. Drilling means the interaction of geology and engineering, of rock material and techniques. Thus, both fields each inherently complex, need to be integrated theoretically and practically.

The main idea of this chapter is to provide an overview and a systematic approach to improve the planning and the design of geothermal wells. Since geothermal environments can differ considerably we also focus on the special demands in well design and drilling operation concerned with the specific geologic conditions of geothermal fields. The reader will be introduced to drilling techniques, drilling

Geothermal Energy Systems. Edited by Ernst Huenges

ISBN: 978-3-527-40831-3

fluid technologies, casing and cementation procedures, well planning, drilling operation, risk evaluation, well completion, and economic considerations. In particular, practical borehole stability analysis and geomechanical explanations and prognosis might be new but necessary paragraphs in a geothermal drilling chapter. Finally, we discuss some future research and development tracks for geothermal drilling with the aim to reduce drilling costs within the scope of well integrity and long-term borehole stability.

3.1.1
Geothermal Environments and General Tasks

Drilling operations are directly concerned by the subsurface geological–geothermal environment. As earlier mentioned in this book, various types of geothermal systems are defined either by temperature or by existence of high amounts of thermal fluids. Table 3.1 summarizes the characteristics and utilization of geothermal systems. The geological conditions of geothermal wells are generally more harsh compared with standard shallower geothermal or oil and gas wells and might be similar to HPHT (high pressure high temperature) wells known from HC (hydro carbon) drilling. HPHT wells are critical due to their small design margins and where well control is difficult to handle since HPHT wells are often characterized by extreme pressures and temperatures coupled with small pore pressure-fracture gradient margins. Consequently, the casing design demands maintenance of high dimensional efficiency, and quantitative risk evaluation is necessary for more complex casing and tubing design. In classical tubular designs, reduction in risk results in an increase in initial costs (e.g., thicker casing wall, higher grade

Table 3.1 Summary of characteristics of geothermal systems and their utilization.

	Geothermal system			
Characteristics	**High enthalpy**	**Low enthalpy**	**HDR**	**Aquifer**
High temperature (>150 °C)	x		x	x (also use of vapor)
Low temperature (<150 °C)		x		x
High reservoir pressure (>690 bar)	x			x
Low reservoir pressure (<690 bar)	Unusual	x	x	x
Deep reservoir (>2500 m)	x		x	x
Shallow reservoir (<2500 m)	x	x		x
Production + injection	x	x	x	(x)
Closed water circuit	x	x	x	x
Sedimentary rock	Unusual	x		x
Metamorphic rock	x	x	x	Unusual
Magmatic rock	x	x	x	x

material), and reduction in the consequence (casing or tubing failure; or other catastrophic incident) costs. The optimum design, however, is one that minimizes total costs (initial costs + consequence cost). Such an approach is necessary for geothermal wells with significantly higher temperatures and also higher pressures than conventional HC wells. Usually, the definition of the geothermal well design premises and performing the well design require several iterations because of the narrow margins. For readers, interested more in operational details of HPHT wells, the publications by Krus and Prieur (1991), French and McLean (1993), and Seymour and MacAndrew (1994) are recommended.

One of the main tasks in reservoir drilling is to keep the formation damage low. *Formation damage* is a term used throughout the industry to describe negative interaction between the drilling operation and producing formation resulting in an impaired near-wellbore permeability. Coincident with this permeability impairment is a reduction in production. In a geothermal well where economic viability is predicated on the production of prodigious amounts of heated water and/or steam, formation damage must be understood, controlled, and minimized.

3.2 Drilling Equipment and Techniques

3.2.1 Rigs and Their Basic Concepts

There are different types of drilling rigs in the international market; however, each drilling rig is last, but not the least, simply a system of technical devices to lower, pull and rotate the drillstring, circulate and clean the drill mud, and cool the mud if necessary. Additionally, it has to offer safety equipment which allows to shut in the well safely if necessary.

3.2.1.1 Hoisting System

The hoisting system is used to lower and pull the drillstring as well as casing strings. It normally consists of drawworks with a wire rope, and a pulley, consisting of crown block and traveling block which are equipped with sheaves to increase (multiply) the load capacity of the wire rope. The hook is mounted onto the traveling block, and the hook takes the swivel which allows the rotation of the drillstring without turning the traveling block, too.

Other systems use hydraulic pistons or a rack instead of a pulley to move the swivel up and down.

3.2.1.2 Top Drive or Rotary Table

These items are used to rotate the drillstring from the surface to the bit on the bottom. The "old fashioned way" is the combination of a rotary table (Figure 3.1) with a square or hexagonal kelly (Figure 3.2) and a kelly bushing fitting into the rotary table. The kelly bushing transmits the rotation onto the kelly, which allows traveling the string up and down in the rollers of the kelly bushing.

Figure 3.1 Rotary table with kelly bushing und kelly (hexagonal). (From Schlumberger Oilfield Glossary.)

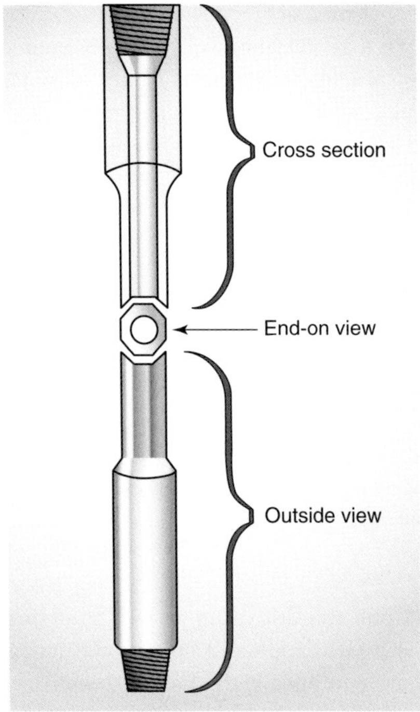

Figure 3.2 Illustration of a kelly (hexagonal).

The modern alternative is the top drive (Figures 3.3 and 3.4). This device has a built-in engine (either with an electric or hydraulic motor) to rotate the drillstring and allows traveling and rotating the string without the need of a kelly. However, very often a rotary table is installed as a backup for the top drive.

3.2.1.3 Mud Pumps

The mud pumps are used to circulate the drill mud in a kind of a "closed loop" through standpipe, swivel, drillstring, and the bit down to the borehole and in the annulus between the drillstring and borehole wall back upward to the surface.

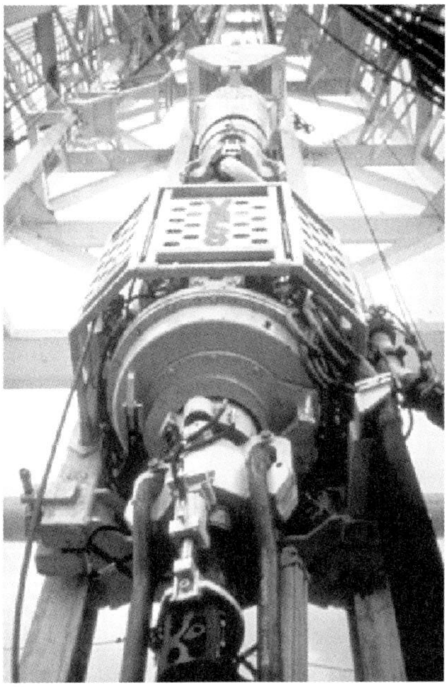

Figure 3.3 Example of a top drive engine. (figure from Schlumberger Oil field glossary 2009.)

Figure 3.4 Drill rig with top drive engine. (GFZ drill rig InnovaRig developed by Herrenknecht Vertical GmbH.)

Normally, the mud pumps are equipped with strong engines and are the main energy consuming machines of a drill rig because they have to circulate the mud with high pump rates (up to 4000–5000 l min^{-1}) and at high pressure (up to 350 bar). Usually two to three mud pumps are available to be run parallel in order to give sufficient flow rate.

3.2.1.4 Solids Control Equipment

When the mud comes back to the surface it has to be cleaned of drilled solids in order to prevent an increase in density due to accumulation of solids content and to prevent excessive abrasive wear in pumps and lines.

This cleaning is done with a chain of solids control equipment consisting of

- shale shakers (single or double deck) with screens and
- hydrocyclones (desander, desilter).

Often, centrifuges are also used to remove ultrafine solids, which otherwise will increase the viscosity of the mud excessively.

3.2.1.5 Blowout Preventer (BOP)

This is a safety device which allows to shut-in the borehole in case of an emergency. The preventer stack normally consists of two to three ram-type preventers and one annular-type preventer. The annular-type blowout preventer (BOP) is capable of closing around different diameter drillstring parts and also to shut completely. One of the ram types is dedicated to close the borehole totally even if any pipe is in the hole (and is capable to cut drillpipe and even drill collars! (DCs)), the others close around the drillpipe or DCs of specific diameters.

A special high pressure accumulator closing unit is a substantial part of each BOP stack.

3.2.2 Drillstring

The drillstring consists of the bottomhole assembly and the drillpipe (DP). All elements are hollow to allow circulation of drill mud through them (normally inside down, outside up = direct circulation).

3.2.2.1 Bottomhole Assembly

The bottomhole assembly consists of different items like drill bits, stabilizers and reamers, drill collars (DCs), a jar, a shock sub, and heavy wall drillpipe.

- **Drill bits:** Drill bits are the tools to destroy the rock; the selection is made according to rock types and properties and the applied drilling technique. Some tools are shown in Figure 3.5.

  In general, the following types are used for different applications:

- **Rollercone bits:** Rollercone bits have (mostly) three conical shaped rollers, which are equipped either with steel teeth or with special inserts made of tungsten

Figure 3.5 Rollercone- and PDC-drill bits. (From RockBit International, 2005.)

carbide (TCI). The rollers roll on bottom and crush the rock when the bit is rotated. Rollercone bits are available for all kinds of formations to be drilled. Generally, the teeth or insert size decreases for harder rock while the total number increases.

- **PDC bits:** These bits have fixed cutters distributed to three and more blades. The cutters are scraping on the bottom and are made of polycrystalline diamond compacts (PDCs) to withstand quick abrasive wear. PDC bits are used in soft to medium hard formations.
- **Diamond bits, surface set:** These are also fixed-cutter bits using scraping action for drilling. Here, natural diamonds are fixed in the nonwear resistant matrix body. When the diamonds are worn out the bit has to be pulled for change. Surface set diamond bits are used in very hard rock. They normally give a fast penetration rate in the beginning, which slows down when losing sharpness of the exposed diamonds.
- **Diamond bits, impregnated:** Here, only small diamond particles are embedded into a semiwear resistant matrix material which is exposed to slow abrasive wear. This allows drilling as long as a certain height of the matrix–diamond mixture still exists on the bit body. Impregnated diamond bits are used in ultrahard rock and give the whole bit life a relative constant rate of penetration (ROP) when the matrix wear characteristics fit to the rock to be drilled.
- **Stabilizers and reamers:** Stabilizers and reamers are tools which stabilize (centralize) the DCs in the open hole; they are equipped with smooth hard facing made of tungsten carbide in the areas of contact to the borehole wall. Reamers

are similar, but have reaming edges with rough hard facing which enables to ream tight spots in the open hole.

- **Drill collars:** DCs are thick-walled pipes (e.g., $9\frac{1}{2}$ in. × $3\frac{1}{2}$ in. = 241 mm × 89 mm; 76 mm wall) which give high weight (e.g., $9\frac{1}{2}$ in. about 310 kg m $^{-1}$; about 3t per joint!) and a high bending stiffness, so that these joints are well suited for high axial loads. Each joint is normally approximately 9.5 m long (Figure 3.6). They are used to generate the "weight on bit" (WOB) to assure penetration of the bit cutters into the formation to be drilled. The "neutral point"(transition from compression to tension) should always be in the DCs in order to prevent buckling in parts with low bending stiffness. Both ends of each joint are equipped with threads to allow quick connection and disconnection (male thread on lower end and female thread on upper end).
- **Jar:** A jar is a device which allows producing high kinetic (shock) energy onto a stuck drillstring section below the jar. There are different types (mechanically or hydraulically operated) in the market. The general working principle is the same for all types: The free part of the drillstring is pulled into tension to store energy, and the jar releases this stored energy after a time delay by an immediate elongation in a travel joint, which allows a certain upward travel until the movement is restricted again by hitting of the hammer part of the travel joint onto an "anvil" in the housing of the tool. Thus, the kinetic energy of the fast moving DCs below the jar is transmitted as a "blow" to the stuck point.
- **Shock sub:** A shock sub is a device to dampen axial vibrations of the drillstring, which are induced by the drill bit. Most of the tools use belleville springs and/or rubber elements as dampeners but some tools use hydraulic fluid dampeners.
- **Heavy wall drillpipe:** Heavy weight drillpipe (HWDP) is an "intermediate" between the extremely heavy DCs and the normal drillpipe. It is capable of withstanding some axial load due to a thicker wall than the normal drillpipe. Each joint is equipped with up-set ends on which the "tool joints" (i.e., the thicker threaded ends) are friction-welded. Additionally, it is equipped with an

Figure 3.6 Drill collars (right) and drillpipe (left). (from Schlumberger Oilfield Glossary 2009.)

up-set part in the middle of the pipe body to give some support at the borehole wall if it is run in compression.

3.2.2.2 **Drillpipe**

The function of the drillpipe is to transmit rotation and torque from the surface to the bit and to serve as a "pipeline" for the drill mud. The drillpipe normally consists of joints with an approximate length of 9.5 m; each joint is equipped with up-setted ends on which the "tool joints" are friction welded (Figure 3.7). The drillpipe is thin-walled (e.g., 5 in.-DP: 9.2 mm wall) and therefore has only low bending stiffness and is not suited for high axial compressive loads.

Figure 3.7 Tool joint of a drillpipe. (From OMSCO Industries, 1997.)

3.2.3
Directional Drilling

Directional drilling is done if a borehole has to be steered into a specific direction and/or into a target. It also allows drilling several wells from one single site into different targets. To implement directional drilling, special equipment such as those listed and described below is needed.

3.2.3.1 Downhole Motor (DHM)

It consists of a rotor which is connected via a flexible joint to the drill bit, and a stator connected to the drillstring. The stator has a "bent housing" with an adjustable bent (Figure 3.8). The rotation of the rotor is done by pumping drill mud through the motor. The downhole motor (DHM) will deviate the borehole direction into the direction of the bent as long as the drillstring is not rotated; if the drillpipe is rotated slowly no preferred direction is given anymore and the trajectory will be (more or less) straight.

DHMs are also used as an alternative to rotary/top drive for straight hole drilling if it is more economical due to the higher ROP, so that the higher daily costs are overcompensated by the ROP increase achieved with it.

3.2.3.2 Rotary Steerable Systems (RSS)

Rotary steerable systems (RSSs) are a modern alternative to DHMs; they allow steering while rotary drilling, that is, with continuous rotation of the whole drillstring (Figure 3.9). They have some advantages, for example, a more constant WOB because the drillstring can slide better in a highly deviated borehole due to

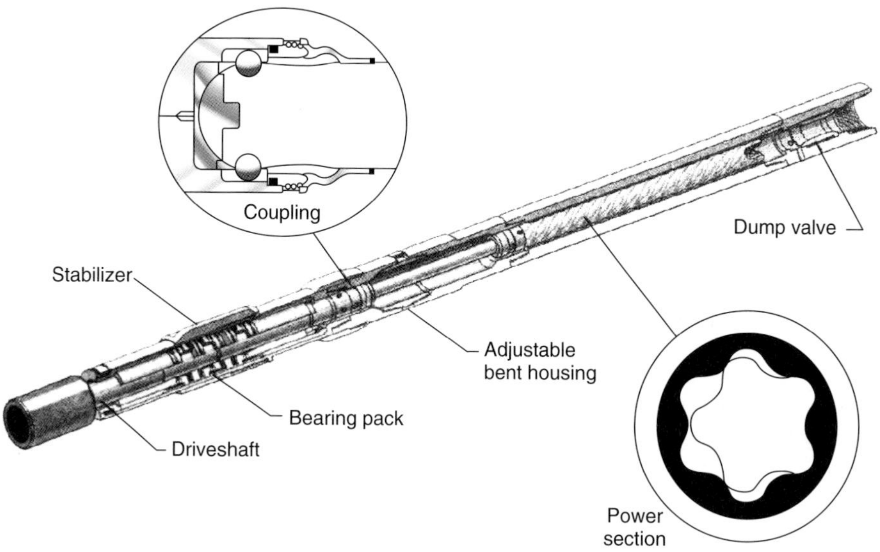

Figure 3.8 Downhole motor. (From Economides, Watters, and Dunn-Norman, 1998.)

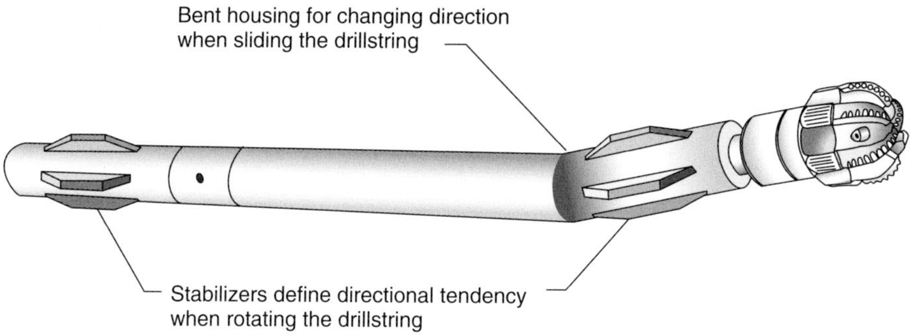

Figure 3.9 Steerable system – downhole motor and ben-thousing. (From Economides, Watters, and Dunn-Norman, 1998.)

the fact that the string is continuously in a "sliding friction mode" rather than in a "static friction mode" and therefore the friction between drillstring and borehole wall is reduced compared to directional drilling with conventional steerable systems with DHM. However, despite the advantages, a main disadvantage is the (still) higher cost of these new systems. Hence, RSSs are mainly used for extended-reach wells and offshore. The RSS is described in more detail in Section 3.11.1.2.

3.2.3.3 **Downhole Measuring System (MWD) with Signal Transmission Unit (Pulser)**

Normally (for directional drilling purpose only) the "measurement while drilling" (MWD) system is equipped with sensors for borehole inclination and borehole direction (azimuth), but other sensors are also available (e.g., for "logging while drilling" (LWD): γ ray, resistivity, conductivity, etc.). Signals are most correct when measurement is taken during short interruptions of the drilling process when the drillstring is not rotating and the mud is not flowing. Hence, it is common to make such measurements during the short standstill when a drillpipe joint is added.

The signals from the sensor are then transmitted to a signal transmission unit. This is usually a magnetic valve data pulser, where short partial reductions of the inside flow area generate pressure increase inside drillstring (pressure pulse). These pressure pulses are transmitted through the down-flowing drill mud inside the drillstring to the surface.

3.2.3.4 **Surface Receiver to Receive and Decode the Pulser Signals**

In the surface receiver, the pressure pulse pattern is analyzed for the data.

3.2.3.5 **Special Computer Program to Evaluate Where the Bottom of the Hole Is at Survey Depth**

The borehole trajectory is calculated with the special computer program. Taking into account that approximately every 10 m a single survey of inclination and azimuth is taken together with the measured depth (MD) in which the sensors are placed at the time of measurement, it is clear that no continuous trajectory is measured. So in earlier days the assumption was to connect the measured points by

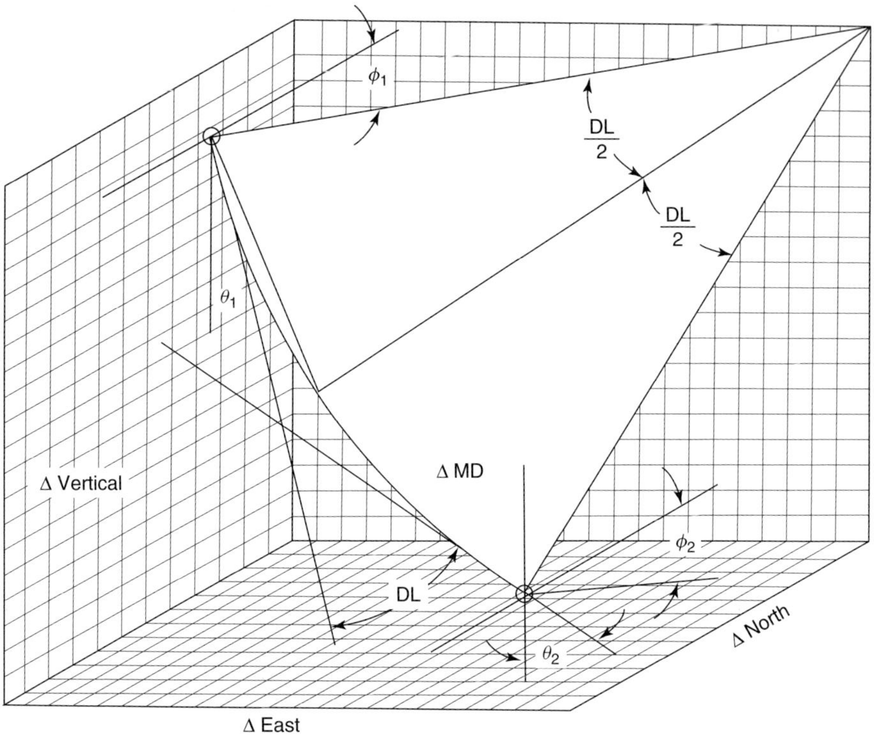

Figure 3.10 Borehole trajectory with two measuring points. (From Economides, Watters, and Dunn-Norman, 1998.)

a straight line which of course was not correct. Currently, a curve with a maximum radius fitting into the measured points is assumed as shown in Figure 3.10.

For the calculation, the following formulae are used:

$$\Delta_{\text{North}} = \frac{\Delta\text{MD}}{2}[\sin(\Phi_1) \times \cos(\Theta_1) + \sin(\Theta_2) \times \cos(\Phi_2)] \times \text{RF}$$

$$\Delta_{\text{East}} = \frac{\Delta\text{MD}}{2}[\sin(\Theta_1) \times \sin(\Phi_1) + \sin(\Theta_2) \times \cos(\Phi_2)] \times \text{RF}$$

$$\Delta_{\text{Vert}} = \frac{\Delta\text{MD}}{2}[\sin(\Phi_1) + \cos(\Theta_2) \times \cos(\Phi_2)] \times \text{RF}$$

with

$$\text{RF} = \frac{360}{\pi DL}\left[\tan\frac{\text{DL}}{2}\right]$$

$$\cos\text{DL} = \cos(\Theta_2 - \Theta_1) - \sin\Theta_1 \times \sin\Theta_2[1 - \cos(\Phi_2 - \Phi_1)]$$

With this technique, it is possible to steer a well's trajectory with an accuracy in the range of a meter. It is commonly used to drill several wells from one site into different directions, for example, from offshore platforms. It also allows to drill horizontal wells; the actual world record is held from the well BD-04A in Qatar with a total length of 12 290 m (40 320 ft) with a departure (horizontal distance of final point from start point) of 10 900 m (35 770 ft)! The entire horizontal section was drilled within a reservoir target which is only 6 m (20 ft) thick.

3.2.4 Coring

Coring – or core drilling – is a technique where not all of the rock is destroyed but only a kerfs area while the remaining "core" is recovered to the surface, depending on the applied technique either by pulling the complete drillstring with the core barrel or only the inner core barrel (which contains the core) by wireline. Normally, all core barrels consist of an outer core barrel which is connected to the core bit at the lower end and to the drillstring at its top end, and an inner core barrel which takes the core. In order to protect the core against destruction by the drilling process the inner barrel does not rotate with the drillstring due to a rotational bearing between both barrels.

The advantages of coring are that cores give an exact picture of the formation and allow direct measurements of rock permeability/porosity, and so on. In high enthalpy reservoirs, coring sometimes represents the only possible way to obtain reservoir information due to temperature limitations of logging tools.

The disadvantages are the high additional costs (due to a general time delay compared to drilling) and sometimes a higher risk for the borehole. However, the additional cost should be compared to geophysical logging runs which may be needed if no cores are taken. Sometimes coring may represent the only opportunity to obtain detailed reservoir information.

Two different techniques are to be mentioned:

- **Continuous coring (wireline coring):** The wireline technique is mainly used if a long section of the borehole is to be cored. It reduces the time to retrieve the core to the surface remarkably because there is a need to pull the whole drillstring only if the core bit has to be changed; otherwise the inner core barrel which contains the drilled core is pulled out of the hole by wireline. Of course this principle only can work if the drillstring has an inside diameter larger than the outer diameter of the inner core barrel. Therefore, normally special drillstrings – and sometimes even special drillrigs – are used in conjunction with wireline coring.
- **Spot coring:** In most of the cases "spot coring" will be done, which allows recovering cores between 9 and 27 m length per run. It allows using conventional drillpipe as it is also used for normal drilling operations.

3.3 Drilling Mud

The drill mud has a lot of different functions in a borehole. To have the proper drill mud in the hole is of high importance for its success. Hence, sometimes the mud is called to be the "blood" of a borehole!

It has to

- cool the drill bit;
- transport the cuttings (drilled solids) in the annulus to surface;
- avoid settling of cuttings if circulation is stopped;
- give cuttings (and entranced gases) free at surface;
- reduce friction between drillstring and borehole wall;
- stabilize the borehole wall;
- exert hydraulic pressure (via the hydrostatic head of the mud column) to prevent gas or fluids from entering the borehole;
- prevent mud and filtrate from entering the formation through the borehole wall;
- power DHMs;
- transport information of the formation drilled to surface (gases, cuttings, fluids).

There are several types of drill mud in use, mainly depending on the specific characteristics of the formation to be drilled and on the function of the well. Some types are given in the following sections.

3.3.1 Mud Types

3.3.1.1 Water-based Mud

Water-based mud is most common. It consists mainly of water. Depending on the formation either fresh water or brine is used, for example, when drilling through salt sections brine is more or less salt saturated.

Usually, in all water-based muds additives are added to improve mud characteristics; for example, marble or barite to increase density, polymers to reduce filtrate and to improve rheology, friction reducer, and others.

3.3.1.2 Oil-based Mud

Oil-based mud consist of a high percentage of oil, a lower percentage of water, and several additives (emulsifier, weighing material, and polymers as described above). While in the past crude oil was used it is today considered to be an environmentally friendly oil. Oil muds are mainly used in extremely water-sensitive formations to ensure good borehole stability even under adverse situations.

3.3.1.3 Foams

Foam type muds are used particularly if the formations to be drilled have a pressure gradient lower than the hydrostatic gradient. It is done to avoid severe mud (filtrate) losses into permeable formations and protect the formation against plugging of pores.

Foams are formed of water, a gas (air, nitrogen), and foaming additives.

3.3.1.4 Air

Air is used as a "drilling fluid" mainly in dry hard rock where borehole stability is not a problem. If water bearing formations are drilled high water influx into the borehole may occur. If this is encountered it can cause severe problems ("buried" bit or drillstring) because of "overload" of the carrying capacity of the air column in the borehole.

Possible advantage of air drilling is a generally higher drilling progress (ROP) due to the pressure relief on bottom. Thus, the ROP with air as drilling fluid may reach up to 10 times the ROP of a "conventional" mud.

The use of air as drilling mud in geothermal wells is clearly a desirable alternative to minimize formation damage, but not practical in all circumstances (e.g., due to borehole stability reasons).

3.3.2 The Importance of Mud Technology in Certain Geological Environments

3.3.2.1 Drilling through Plastic/Creeping Formations (Salt, Clay)

Soft plastic formations like clay, clay-rich marl, shale, and salt rock pose a special challenge in the drilling and completion processes. Creeping formations tend to decrease the caliper during drilling when a very low mud pressure is used. Washouts represent caliper increases where a very high mud flow is used. Although soft formations entice to drill with a high ROP, a moderate penetration combined with moderate WOB allows stress releases in plastic formations and prevents abnormal creeping during and specially after the drilling operation. In clay-rich formations inhibitors can minimize the swelling process whereas in salt rock overgauge or washouts (especially in potassic salt) this can almost be not avoided. Besides the mud technology, special emphasis should be given to straighten the selection of the drill bit and bit hydraulics. As stress increases with well bore depth, a formation becomes more plastic and requires different bit cutting mechanics to cut or to shear efficiently. Special emphasis on sustainable wellbore stability should be given on plastic formations that can exert abnormal stresses at the borehole wall.

3.3.2.2 Formation Pressure and Formation Damage (Hydrostatic Head, ECD)

The formation pressure is the pressure that is exerted by the formation fluids within the pores of the rock and is a key issue in well integrity (e.g., well control during operation or best casing shoe depth selection). Usually the formation pressure can be assumed as hydrostatic or as pressure exerted by a column of water from the formation's depth to ground level. During circulation, the downhole annulus pressure is always higher than that of the static column. The equivalent mud weight corresponding to this circulation pressure is called the equivalent circulation density (ECD). ECD can be interpreted as the density of a hypothetical fluid, which in static conditions and at any depth produces the same pressure as a given drilling mud in dynamic conditions. It can be calculated as follows:

$$\mathrm{ECD} = \rho_\mathrm{d} + \Delta P_\mathrm{a} / \left(g \times L/100\right)$$

where

$\rho_\mathrm{d}(\mathrm{kg\,l^{-1}})$ = density of the mud containing cuttings (mud in annulus),
ΔP_a (bar) = circulation pressure loss in the annulus,

g (9.81 kg × m s^{-2}) = acceleration due to gravity,
L (m) = vertical depth.

Drilling problems occurring due to very high mud density (e.g., differential stuck pipe, mud losses by hydraulic fracturing) can be accentuated by the ECD effect. Estimates of formation pressures made before drilling are based primarily on correlation of available data from nearby wells (leakoff tests, mud report of offset wells) and seismic data. To estimate formation pressure from seismic data, the average acoustic velocity as function of depth must be determined and the reciprocal of the velocity – the porosity-dependent *interval transit time* – is generally displayed (Bourgoyne *et al.*, 1986). During the drilling operation, the formation pressure can be determined by the pressure at the bottom of a well when it is shut in at the wellhead. If abnormal formation pressures in critical formations (overpressure due to underconsolidation or aquathermal effects in heated low permeability rocks) are presumed and threaten the drilling process, the use of FPWD (formation pressure while drilling logging tool) can resolve drilling safety (Charlez, 1997). If formation pressure increases, mud density should also increase to balance pressure and to keep the wellbore stable.

Assessing the reservoir top and during all further operations the initial reservoir conditions are disturbed by the drilling process. Effectively, problems may arise from formation damages when the well is putted on stream at a later stage (Perrin, 1999). In particular, the pay zone where the major geothermal water influx is prospected may be damaged by the fluids, the mud solids, cement slurry, and so on, causing a significant reduction in productivity. Thus, the careful and concerted to the reservoir conditions selection of mud chemistry, mud solid type and size, and mud weight belong, among others, to the critical influencing factors to minimize formation damage. For example, near-balanced mud weight minimizes the infiltration of mud solids within the pore space of siliciclastic formation. However, a low mud weight might increase hoop stresses (that are the tangential stresses) at the borehole wall initiating breakouts. Therefore, formation damage and wellbore stability are two counteracting issues and essentially need to be aligned to the well integrity.

In case of incurred or unavoidable formation damage, productivity can be restored easily in carbonate formations than in sandstone formations by acidizing. In metamorphic or granitic rock, chemical treatments can be successful, too (Tischner *et al.*, 2007). Commonly, hydraulic fracturing is finally applied to overcome the damaged zone and to increase permeability in general. In any case, it requires costly treatments in terms of rig time and treatment itself.

3.4
Casing and Cementation

Casing strings are needed to be run from time to time to secure drilled hole sections (e.g., in unstable formation), to separate hole sections/formations with different pressure gradients and to seal formations against influx of fluids (gas, oil, and water) from the previous or next hole section. Normally, the annulus between the casing and the borehole is filled with cement to achieve a sealing "connection" between both (Figure 3.11).

3.4.1 Casing and Liner Concepts

In order to be able to run the casing into a borehole the casing size (diameter) must fit into the hole which means it must be smaller than the hole diameter. Because generally more than one casing string is necessary to complete a borehole, the bit size for the next drill section has to pass through the previous casing. So, every well starts with a large diameter at the top and ends in a smaller diameter at final depth; for more details see Section 3.5.

Some casing strings are designed as "surface strings" which means that the pipe string leads from bottom to surface. Other casing strings are run as a "liner," only reaching 10–100 m into the previous casing string. Liners are run on drillpipe to the desired depth and are equipped with a special "liner hanger" which is a tool that allows hanging the weight of the complete liner into the lower part of the previous casing (Figure 3.12). Liners offer the advantage to have available a larger casing inner diameter above the liner hanger.

In geothermal wells in particular, large diameters are necessary to allow the circulation of high flow rates through the casings because otherwise the circulation pressure drop will be very high and will increase the energy costs for the production and reinjection pumps; for details see Section 3.9.

3.4.2 Casing Materials

There are different steel materials (steel grades) in use for casing, depending on strength, corrosion resistance, and so on. Generally, casing sizes, wall thicknesses, and material grades are "standardized." The most common standard used worldwide is the "API Specification 5CT" from the American Petroleum Institute; here, all material grades and properties are specified, including chemical composition of the different grades.

Common grades for carbon steel material are: H40, J55, K55, L80, N80, C95, P110, Q125; in all cases the numbers give the "minimum yield strength" of the material in "pound/square inch" $\times 1000 (= \text{psi} \times 1000)$. Other materials are available, for example, corrosion resistant materials for highly corrosive fluids.

Common sizes (outer pipe diameters) are 24 1/2, 20, 18 5/8, 13 3/8, 11 3/4, 10 3/4, 9 5/8, 8 5/8, 7 5/8, 7, 6 5/8, 5 1/2, 5, 4 1/2, 3 1/2, 2 7/8, 2 3/8 in. From these sizes a casing program for the wells can be gathered, together with the desired drill bit diameters for the corresponding hole section (Section 3.5).

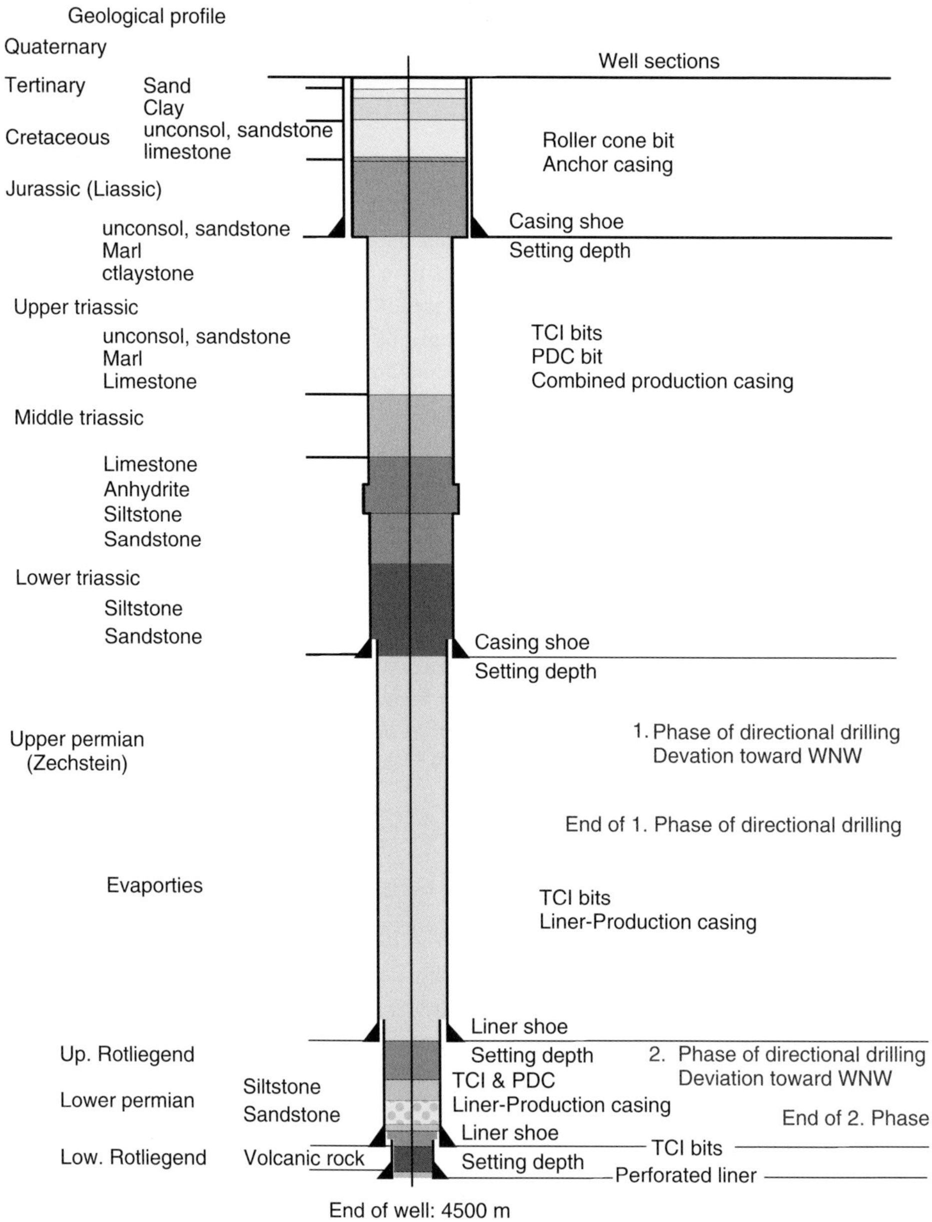

Figure 3.11 General well completion scheme. Example of the geothermal well GTGrSk4/05 (Groß Schönebeck/Germany).

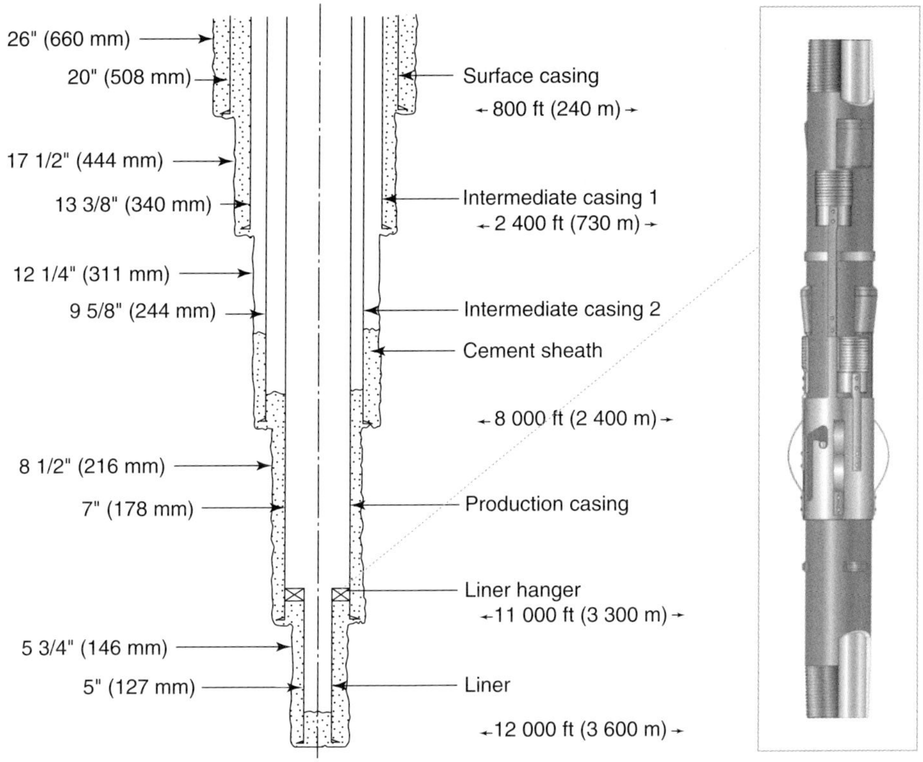

Figure 3.12 Example of liner hanger (dual cone liner hanger) as component of a well completion system. (Well completion from Perrin, 1999; liner hanger developed by Smith, *www.ed-oiltools.de/images/produkte-smith*.)

3.4.3 Pipe Centralization

A proper pipe centralization in the borehole is mandatory if the casing has to be cemented. If the pipe stands eccentric in the hole the displacement of the drill mud will be incomplete resulting in incomplete cement with mud channels. Pipe rotation can help but is not always possible (Figure 3.13).

Such an incomplete cementation with channels can neither assure a good distribution of loads from the formations to the casing, particularly not in plastic formations, nor seal the annulus properly. Therefore, "centralizers" of different shapes are normally mounted onto the casing joints at a certain distance (Figure 3.14).

In deviated and bent sections the distance between centralizers is reduced so that a sufficient load distribution is assured and the specific permissible load per centralizer is not exceeded.

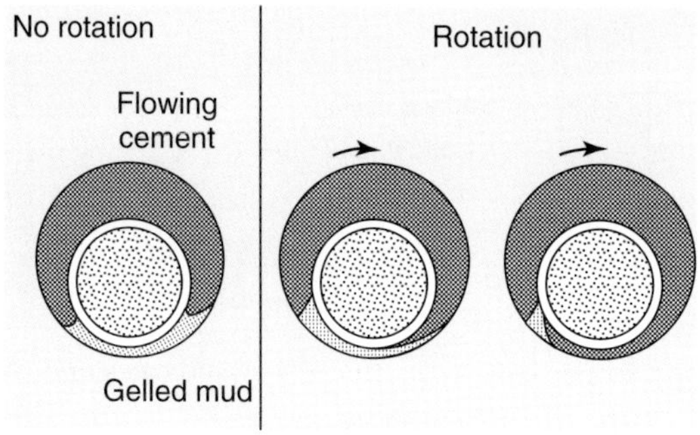

Figure 3.13 Eccentric pipe in borehole. (From Schlumberger, 1984.)

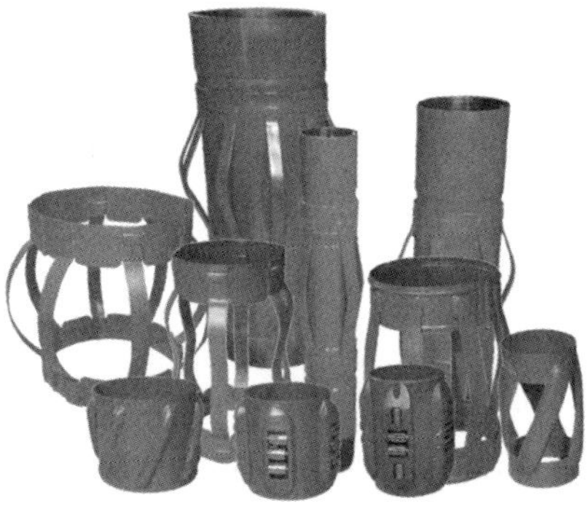

Figure 3.14 Different types of centralizers. (Courtesy of Weatherford, 2008.)

3.4.4
Cementation

According to specific needs several cementing techniques are common. However, in each case, first the cement is pumped down through the casing shoe and enters the annulus to be sealed by the cement, followed by the displacement mud which is necessary to pump the cement to the planned position in the annulus. Care has to be taken that the cement slurry is properly separated from the displacement mud because the cement slurry normally has a higher density than the mud so that a "heavy" fluid has to be lifted by a "lighter" fluid. Due to this fact also one to two valves which only allow flow through it into one direction (from top to bottom) are installed at the bottom joint of casing in order to avoid backflow of the heavy cement slurry from the annulus into the casing at the end of displacement.

Common casing cementing techniques are as follows:

- **Stinger cementation (through a drillpipe string inside the casing):** This technique is used mainly to cement large-diameter casing strings. A "stinger" is run on a drillpipe string down to the uppermost valve in the casing and latched into it. This small diameter "pipeline" allows pumping down the cement slurry quickly and with less mix-up with the mud which has to be displaced by the cement on the way down.

- **Plug cementation (two plugs are used to separate mud and cement):** In this technique a first plug is pumped in front of the cement slurry and a second plug between the end of the cement and displacement mud. Both plugs are made of rubber and act as a "floating" seal between the different fluids in order to avoid mixing of the fluids. The first plug is equipped with a membrane which bursts at a predetermined differential pressure to allow flow of the cement slurry after having opened. The second plug seals finally and indicates end of cementation.
- A schematic of a typical plug cementation (primary cement job of a surface casing string) is shown in Figure 3.15; details of the two-plug system are shown in Figure 3.16.
- **Liner cementation (with two plugs):** A liner cementation is generally similar to a two-plug cementation; the difference is a drillpipe dart (a small plug which passes through the drillpipe), latches into the liner plug and seals it while pumped down as a "unit."
- **Cementation "top-down":** While all normal cementations are done as described above (cement rises in annulus from bottom upward) sometimes it may be necessary to pump cement from top-down in the annulus. Those operations are often critical and do not allow a separation between mud and cement by plugs. This technique is sometimes used to fill the upper part of an annulus if the normal way around the casing shoe is plugged, for example, due to a partial cementation of the lower part of the annulus. Special care has to be taken not to collapse the casing because of possible high annular pressure occurrence when "squeezing" the cement from top, particularly if it is applied to large-diameter casing which have only low collapse pressure resistance!

3.4.5 Cement Slurries, ECD

Standard cement slurries consist of water, cement, and some additives. Relatively high water content is essential because the cement slurry must always be sufficiently pumped through. Hence, densities are normally ranging from approximately 1.5 to 2.0 kg l^{-1}. However, lower as well as higher densities can be mixed, for example, by adding hollow glass beads to reduce density or barite or iron oxide to increase density if necessary. Other additives – besides "density steering" additives – are fluid loss reducer, retarder or accelerator, and "friction reducer" to keep the slurry flowing even if the solids content is extremely high.

Sometimes two different slurries are used for a single cementation. Most often the reason is either a limited collapse pressure resistance of large-diameter casing or a limited fracture gradient of the rock at casing setting depth. In such cases, normally a lighter "lead" cement slurry (for the upper section) and heavier "tail" slurry (for the lowermost section) is used to seal the annulus.

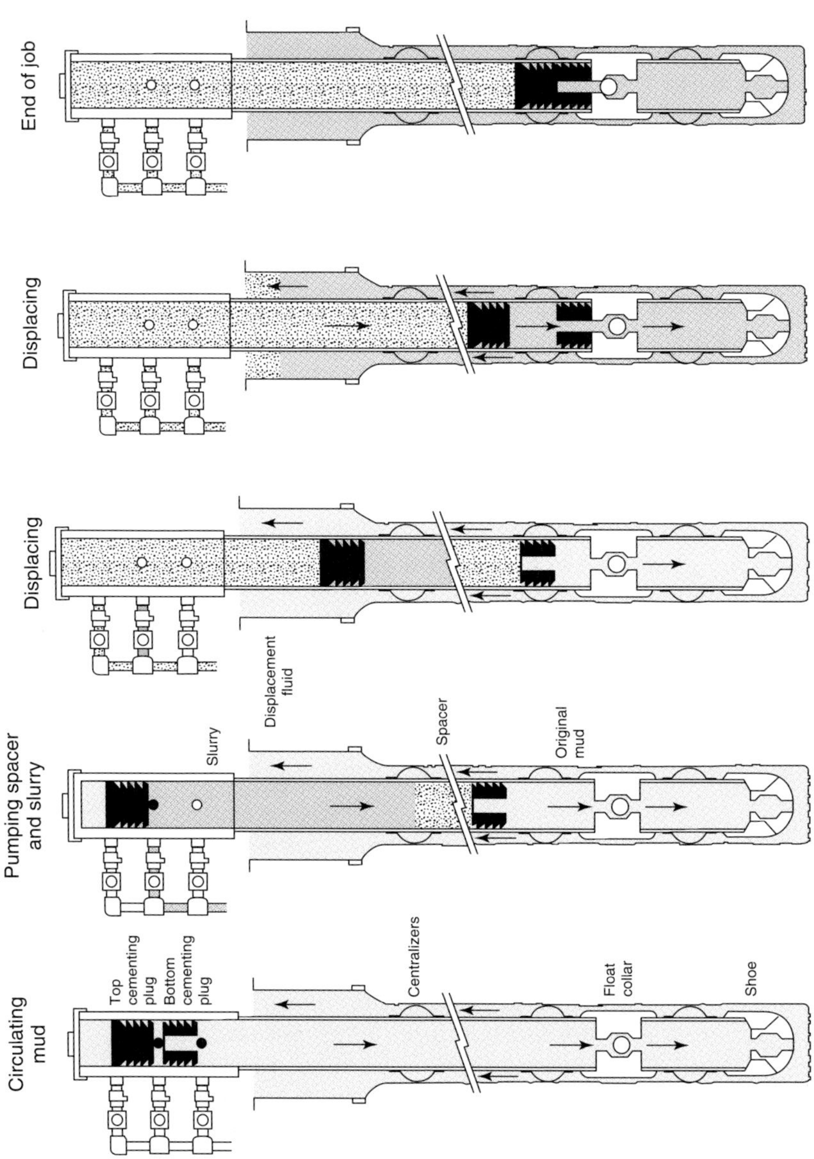

Figure 3.15 Two-plug cementation (schematic). (Courtesy: Nelson, E.B. (1990) *Well Cementing*, 2nd edn, Schlumberger Educational Services, Houston, TX.)

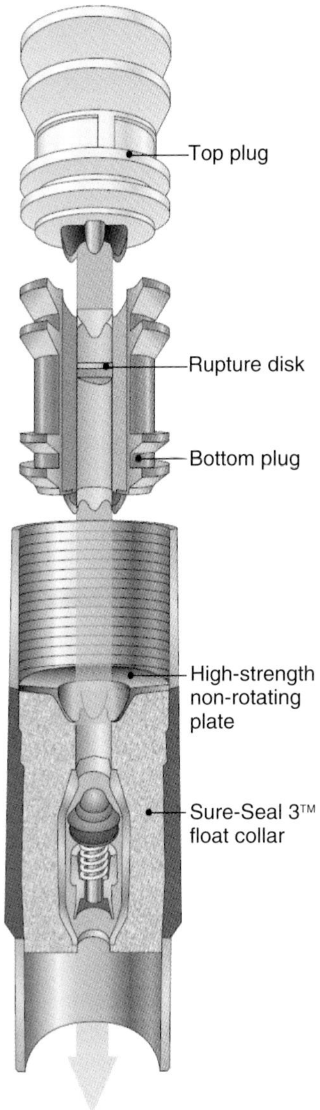

Figure 3.16 Two-plug system. (Courtesy of Weatherford, 2008. Cementing Products.)

Care also has to be taken not to fracture the rock when pumping (high viscous) cement slurries through narrow annuli, because the circulation pressure in the annulus is acting on the bottom together with the "hydrostatic head" of the mud and slurry column.

In general, it is recommended to calculate the ECD by a computer simulation of critical cement jobs, taking into account rheology of the slurries and planned pump rates.

3.4.6 Influence of Temperature on Casing and Cement

During the lifetime of a borehole, temperature changes will occur and affect basically the open hole as well as the cased hole sections. Temperature effects start even when drilling the well, because the circulating drill mud transports heat from the lower hot well section to the upper colder part. Hence, the temperature downhole will be reduced while the upper part is heated up. After having stopped circulation, for example, when tripping for a new drill bit, the temperature profile will tend to go back to the undisturbed temperature curve. The magnitude and time span of temperature changes are dependent on various factors like specific heat capacity of rock and mud, circulation rate, duration of phases with and without circulation.

If a borehole is producing hot water (or steam) the whole length will be heated up to nearly bottomhole temperature.

During well treatment, for example, acidizing, cold fluid (water) is injected, which may cool down the hole significantly, depending on the amount and pump rate of the treatment.

All temperature changes create stress and load changes on the casing strings and the borehole wall; both are discussed in Sections 3.5 and 3.6.

3.5 Planning a Well

3.5.1 Geological Forecast

The whole complexity of planning a well is fundamentally based on the forecast of a geological depth profile. Such a geological depth profile describes and evaluates the thickness and lithology of rock mass which is crosscut by the presumed well path. The geological pile should not only be provided as lithostratigraphy but also as mechanical stratigraphy. A profound geologic-geomechanical knowledge is important for several reasons: (i) the target can be defined by precise structural-geologic description, (ii) wellbore safety can be evaluated by understanding the geomechanical reaction to drilling process and (iii) design of the well path geometry and completion by integrating detailed geological-rock mechanical knowledge. The fundamental knowledge for geological forecasts comes from seismic data, offset wells, and most importantly from regional geology.

Geological depth profiles along the drill path should be provided in MD (measured depth, that is, the length of a wellbore), in TVD (true vertical depth, that is, the vertical depth below drill rig from ground elevation to depth) and TVDMSL (true vertical depth below mean sea level, that is, the vertical depth related to mean sea level – these values are negative and are needed for geological modeling because they are georeferenced to sea level). To complete a geological well section, additional information might be integrated like the planned logging and coring program, the casing shoe depths, bit and casing diameters, mud program, and specification of presumed geologically risk zones. Additionally, a 3D geological model helps to define well planning and finally serves as communication platform between the various disciplines from geoscientist, engineers, and drillers at the drill site. The geological forecast should be distributed before drilling to all participating parties and drilling services.

3.5.1.1 Target Definition

The target within a geothermal field is defined by preceding exploration efforts. If the starting point of a well (its elevation) and the target horizon is known, the well path can be defined by the starting point at ground surface, entry point in the reservoir, landing points in the pay zone and end of well. Directional drilling is often used and necessarily includes kick-off points for deviation and/or azimuthal changes. During directional drilling, the MWD tools allow to "steer" the well into the geologic target.

The decision-making process in geologic target definition is governed by 2D/3D geological/geophysical models that visualize the geometrical constrains of geological layers and their properties. The latter are generally interpreted by borehole logs from offset wells or estimated in connection with regional/local geological knowledge.

Important formation properties for the drilling process are described in the following paragraphs.

3.5.1.2 Pore Pressures/Fracture Pressure/Temperature

The pore pressure is the pressure exerted by the fluid that is in the pore space or within fractures. The often used term for the pore pressure is *formation pressure.* Usually the pore pressure is calculated by the depth of formation and the density of fluid. When the formation pressure is approximately equal to theoretical hydrostatic pressure (i.e., the compaction of water with depth is negligible due to existence of relatively permeable flow path to the surface) for a given depth, formation pressure is called *normal*, expressed by the hydrostatic gradient. Abnormal fluid pressures are pressures above the hydrostatic gradient, for example, caused by high fluid temperatures (e.g., in volcanic areas of Iceland), whereas subnormal formation pressures are fluid pressures below the hydrostatic gradient (e.g., in the Malmkarst of the South German Molasse Basin).

Generally, abnormal formation pressures are found in most of the sedimentary basins worldwide, caused by mechanisms like compaction effects, diagenetic effects, differential density effects, and fluid migration effects. Specially, compacted shales and sealed permeable formations can bear fluids that cannot escape, leading to anomalously high formation pressures. Thus, abnormal pressures that effect wellbore safety in drilling and completion need to be considered in assessing geothermal reservoirs in sedimentary basins and high temperature fields.

The fracture pressure is the pressure where a certain rock type cracks, that is, the pressure where tensile fractures occur. This pressure is an important value for drilling operation to avoid mud or cementation losses by inducing tensile fractures.

The fracture gradient can be determined by leakoff tests or mini-frac tests where a drop during mud pressure increase indicates induced mud loss through a tensile fracture. In particular, the fracture propagation pressure determined during leakoff tests is similar to the minimum principal stress value which is a key parameter for stimulation planning and treatments in enhanced geothermal systems (EGS).

3.5.1.3 Critical Formations/Fault Zones

Critical formations comprise high permeable layers with abnormal fluid pressures, and highly ductile, unstable, or swelling successions like salt rock, shale, or clay. In particular, the creeping process is often accelerated under high temperatures and threatens both drill and casing string. To avoid complications in those creeping formations, certain mud pressures and mud chemistry and thick-walled casing strings need to be considered in well planning. Especially the long-term safety of the large-diameter geothermal wells depends on the right thick-walled completion design. Special effort should be focused on the cementation process of deviated wells: in the bend sections of deviated wells the cement is not necessarily filling the annulus completely – despite use of centralizers – and the resulting abnormal hoop stresses might cause casing collapse.

Unstable formations like shale and also fault zones tend to result in break outs and cavings (as evidenced in the San Andreas Fault Zone Observatory at Depth (SAFOD) research well, drilled through the San Andreas fault, Prevedel, 2007); Karstic structures like faults in the South German Molasse Basin are geologic targets but bear high jeopardy of massive fluid loss while drilling. Drilling through seismogenic zones threaten a complete well if the fault is activated by natural or artificial influences (shearing of complete casing strings through faulting).

3.5.1.4 Hydrocarbon Bearing Formations

During the drilling operation and within the geothermal target hydro carbon (HC) bearings can delay or stop a whole geothermal project. Generally, HC bearings can be encountered in nearly all sedimentary environments. High gas contents under high pressure bear the risk of a blow out, thus, a BOP needs to be installed below the rig even in geothermal drilling projects. Large-scale crossing faults can be pressurized by gas contents leading to unexpected gas kicks even in presumed non-HC bearing or impermeable formations. High gas contents finally can be mixed up with the geothermal fluid and need to be considered in the well completion of a production well.

3.5.1.5 Permeabilities

Formations that transmit fluids readily, such as sandstones, are described as permeable and tend to have many large, well-connected pores. As such, sandstones are typical geothermal targets in sedimentary basins. Special mud chemistries and mud weights allow an optimization in reservoir assessment to mitigate formation damage. If the mud pressure is kept slightly overbalanced, the risk of differential sticking is high in permeable sandstones. Impermeable formations such as sandstones of mixed grain size, diagenetically cemented sandstone, shale, siltstone, and limestone, and metamorphic or magmatic rocks can be affected by high fracturization especially in tectonically stressed zones. Fractures themselves typically do not have much volume, but by joining preexisting pores or interconnected faults, they may enhance permeability and hence productivity, significantly.

3.5.2
Well Design

Borehole design means to plan the trajectory of the well, select casing setting depths, select sizes of casing and corresponding borehole (drill bit) diameters, and calculate the necessary wall thicknesses and material grades for the casing strings.

3.5.2.1 Trajectory

The borehole trajectory is the "connection" from the spud point at surface through the target(s) to final well depth. Designing the well path the casing diameters have to be taken into account particularly for the sections where changes are planned in inclination and azimuth in order to avoid sharp bends with high impact on casing collapse resistance.

3.5.2.2 Casing Setting Depths

Setting depths have to be chosen according to the geological profile forecast. Generally, casings have to be set to separate formations with strongly different pressure gradients from each other in order to (i) to separate formations which need different drill mud types (e.g., fresh water mud against saturated salt mud), (ii) to seal formations against influx of fluids (gas, oil, and water) from the previous or the next section, and (iii) to secure drilled hole sections (e.g., in unstable formation).

3.5.2.3 Casing Sizes

As shown above the needed number of casing/liner strings depends mainly on the geology. To find the fitting casing sizes, the whole well planning process should always start at final depth with the desired diameter there and move upward stepwise (casing size by casing size, Figure 3.17). Casing diameters depend on bit diameters, and bit diameters on casing diameters. Casing (pipe) diameters and corresponding bit diameters are – more or less – "standardized" as already pointed out in Section 3.4. Typical diameters of hole and casing are shown below. Starting from the smallest casing/liner size at final well depth the corresponding hole size can be chosen by either following a full line (conventional scheme) or a dotted line (feasible but uncommon) to the hole size, from there to the next size of casing, and so on, until the planned well scheme has reached the starting diameter.

A "golden rule" for designing a casing scheme is "drill as small as possible, but as large as necessary!"

The reason for this is the simple fact that drilling a large-diameter hole is normally more expensive than drilling a smaller one. However, this should not lead to a well with insufficient diameters of casing strings, because this may adversely affect the cost of the operational phase of the power plant for the whole lifetime!

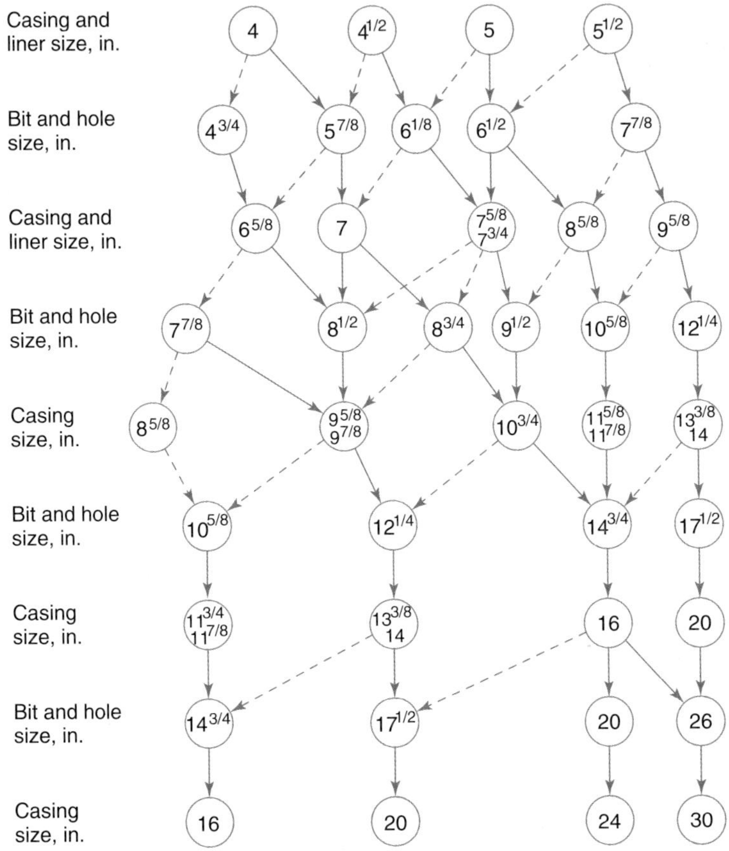

Figure 3.17 Open hole- and casing sizes (possible combinations). (Barker, 1998.)

3.5.2.4 **Casing String Design**

Casing string design is the process to decide by calculation which casing material is necessary to meet the particular needs of the casing string in terms of material grade, wall thickness, and connection type.

This is in the first step done by the calculation of

- the expected external and internal pressures to act on the casing string, taking into account formation pressure gradients and drill mud densities;
- the expected load- and stress situations on the casing string, under consideration of string weight in mud, bending (due to "doglegs," deviated wells), drag (when running casing in/out), and additional tension or compression due to temperature changes.

In the second step, the matching casing types for each string and the desired diameter are selected by the calculation of pressure and load capacities of different casing grades and wall thicknesses against

- collapse pressure,
- burst pressure,
- tensile load of casing.

This has to be done for the pipe body as well as for the thread-connection.

By preselecting a certain wall thickness (which are also "standardized" for each size) the yield strength of the pipe material is the most important material parameter for the calculation of load and pressure capacities. Additionally, the tensile strength is used to calculate the connection strength.

In these calculations the different temperatures are also to be considered, because temperature has an influence on the material's strength; it is reduced by elevated temperatures, causing a reduction of the pressure capacities of the pipe as well as of load capacities of casing pipe body and connection (Figure 3.18).

Always several load cases have to be calculated for each casing string, for example,

- Calculation of different pressure situations for external pressure:
  - inside (partially) empty, outside mud;
  - inside mud, outside (partially) cement;
  - squeezing formations (salt, shales).
- Calculation of different pressure situations for internal pressure, for example:
  - influx (brine, gas, etc.);
  - stimulation (Hydrofrac).
- Calculation of different (load) stress situations, for example:
  - casing string in mud (inside and outside mud);
  - casing string while cementation (inside cement, outside mud);
  - casing string at end of cementation (inside mud, outside cement);
  - casing string under additional stress (e.g., additional tension to compensate thermal elongation during production).

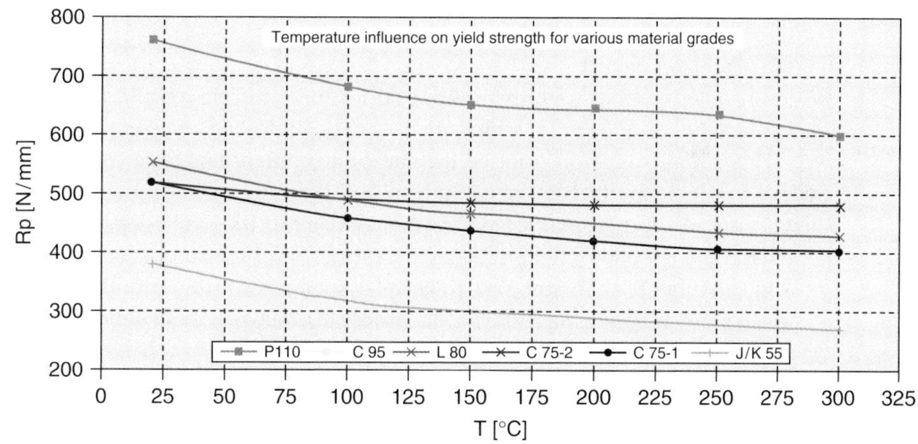

Figure 3.18 Yield strength of various casing material grades versus temperature. (By A. Sperber.)

The calculations of pressure capacities and load capacities are done normally according to the formulae given in API Bulletin 5C3. Here, API gives formulae for "burst" (= internal yield pressure) and for four different cases of the D/t relation (= nominal outside diameter of pipe divided by the nominal wall thickness) for collapse calculations. As mentioned above, the respective yield strength is a function of temperature. Typical yield strength reductions of different material grades are shown in (Figure 3.18).

Because – particularly on long casing strings – the external and internal pressures and loads change with depth very often, a combination of different wall thicknesses and/or grades can be selected for a single string in order to find an optimum solution of total string weights and cost. However, for all cases to be investigated the calculated values of all expected load situations must be less than the corresponding values of casing material selected.

3.6 Drilling a Well

After having finalized the planning process of the well the decision has to be made as to what type of drilling contract is planned because – depending on the contract type – the further steps may be different, particularly concerning the material supply and the contracting of subcontractors for special services. There will be also differences in the further project organization needed to proceed successfully.

3.6.1 Contract Types and Influence on Project Organization

Several contract types are common for the execution of deep drilling projects; four typical contract types are described below.

3.6.1.1 Turnkey Contract

With a turnkey contract a lump sum is paid for the complete well, including all materials, services, and energy which are necessary for drilling and completing the well. This contract type may not be the cheapest solution because a certain risk margin has to be granted to the contractor, but it is the easiest contract type for the operator. All responsibilities (except the "geological risk") are relegated to the contractor. On the other hand, the operator cannot order the contractor to do the work in a manner the operator would like to have it done.

The operator is also responsible for the proper organization of all subcontractor work.

The turnkey contract is mainly used when the operator is not experienced in drilling wells.

It has to be pointed out that a turnkey contract for only one or two wells is a significant risk for the contractor; the risk margin is therefore high and the target to reduce cost might be difficult to achieve.

3.6.1.2 Meter-contract

With a meter-contract normally, a fee for each meter drilled is to be paid to the contractor by the operator. Because the drilling progress is not constant for the total length of a well it is common to fix the meter-fee either for the different hole sections or to fix it for different formations to be drilled.

This contract type can be negotiated either with or without organization and delivery of material and services by the contractor; however, it should be clear that all services and materials which can affect the drilling progress (ROP) will be decided (and supplied) by the contractor.

It is common that all nondrilling related work will be done on an hourly based fee.

3.6.1.3 Time-based Contract

In this contract type all contractor work is paid for by the operator on a daily or hourly based fee which normally does not include any material, services, or energy. Typically, all responsibility – except drilling the contractor's responsibility for his equipment – is on the operator. The contractor works on operator's direct order. Therefore with this contract type the operator must have detailed knowledge of all necessary work related to drill and completion of a well; he also has to supply all materials, services, and energy and has to take care of proper organization of all work.

This contract type is standard in the oil and gas industry and may end with the lowest cost for a drilling project; however, it requires experienced personnel with the operator to plan and manage the project.

Because all risks are to be with the operator a certain "contingency" budget is mandatory.

3.6.1.4 Incentive Contract

An incentive contract is not a contract form on its own. Incentive elements can be part of any type of contract typically in time-based contracts. With an incentive contract, a kind of "bonus/malus" will be agreed on. Normally, this type of contract is used in a "team" approach to a drilling project, where the "team" consists of operator, drilling contractor and some service companies with high influence on progress (e.g., directional drilling service, drill bit manufacturer/trader, and drill mud service). Together, a "base case" for the project is decided in terms of time and cost, and a "key" in terms of how payment will change if performance is better (bonus) or worse (malus) compared to the base case.

The main problem may be the calculation and agreement of the base case, because the operator on one side is interested in lowest cost/highest performance while all contractors will try to be on the "safe side" and calculate more time/lower performance. So, a compromise has to be found.

This contract type offers a kind of risk sharing, but needs drilling experienced personnel with the operator.

3.6.2 Site Preparation and Infrastructure

3.6.2.1 General

Once a drill site has been selected and surveyed, the necessary lines for electric power supply, water supply (if necessary by drilling new groundwater wells), and waste disposal need to be organized. A drill site can only be adequately prepared, if the drill rig and its dimension are predefined since the rig fundament needs to achieve the statics. If necessary, the site will be cleared and leveled. For practical and safety reasons, a drill site should be separated in three areas: (i) internal area with drill rig, engines, and machinery; (ii) external area for necessary traffic of machines and transport; and (iii) peripheral area for parking, storage space for casings, tubings, and further equipment.

3.6.2.2 Excavating and Trenching

The scale and duration of excavating and trenching are minor and site-specific. On most drill sites, a below-ground-level cellar may be excavated. This pit provides additional space between the rig floor and the wellhead to accommodate the installation of BOPs, rat holes, and so on. It also collects drainage water and other fluids for disposal. A reserve pit lined with plastic to prevent soil contamination and settling pits may be excavated and used for water or mud discharges.

3.6.2.3 Environmental Impact (Noise, Pollution Prevention)

Attenuation of noise of the rig engines may be necessary in urbanized areas or in 24 hour operation in populated areas. Also, special care may have to be taken for noise protection against noises generated by handling the drillstring, for example, during roundtrips. Attenuation may be given by installation of sound insulating walls and covered drill rig engines as the rig assembly allows covering. Pollution of groundwater and soil needs to be avoided by appropriate sealed fundament and sealed excavations.

3.6.3 Drilling Operations

The normal drilling operation starts after the site construction is finished with the move and rig-up of the drill rig.

The next work steps are

- mix drill mud;
- run in hole drill bit and DCs;
- start drilling, add drillpipe as needed;
- pull out drill bit if it is worn out and/or if section depth is reached;
- perform geophysical and technical logging as needed;
- run in hole casing (or liner) to section depth;
- cement the annulus between casing/liner and open hole;
- nipple up and pressure test BOP stack;

- pick up and run in hole drill bit of fitting size for the casing;
- drill ahead (and so on until final well depth is reached).

3.6.4
Problems and Trouble Shooting

Many problems affect the success of a geothermal drilling operation. Since the main function of a drilling rig is to penetrate and to seal off formations, any single technical failure may halt this process thereby causing additional expenditures. Therefore, the success of a geothermal drilling operation is strongly dependent on avoiding problems causing downtime. Problems may occur mainly during drilling, but can also occur during other work steps, for example, during running casing strings. In all cases, problems will lead to delay and higher costs.

Bradely (1979) identified the human element as a key factor in avoiding borehole problems like stuck pipe. Therefore, in addition to sound engineering practices the operation culture may also strongly affect the outcome of a potential borehole problem. The mud density selection in context with formation geomechanics is essential for the success of a drilling operation, but good planning of related elements are important as well. Examples are torque and drag considerations in well path planning as discussed by Sheppard, Wick, and Burgess (1987).

Of course, proper hole cleaning is of importance for fast and safe drilling. Particularly in deviated well sections, care has to be taken for sufficient carrying capacity of the drill mud. Occasionally, some junk may be left in the hole (e.g., inserts of TCI bit) and has to be removed specifically when PDC bits are to be used

Following problems often occur during drilling operation:

- **Mud losses:** Trouble shooting by decrease of mud density and/or adding sealing additives to the mud.
- **Influx:** Trouble shooting by increase of mud density.
- **Borehole stability problems:** Borehole problems such as fracturing, collapse, lost circulation with severe drop of hydrostatic head of the mud column, and others may affect borehole stability and need to be considered in a rock mechanical context. It is shown in many wells that by maintaining the mud pressure close to the level of the *in situ* stresses, most borehole problems will be minimized (Aadony, 1999). On the other hand, a low mud density, near to formations pressure, will minimize formation damage through mud infiltration. Borehole stability and mitigation of target formation damage must be carefully weighted. In any case, a key issue in evaluating and avoiding borehole problems is understanding rock failure within the *in situ* stress field. Hence, stress distributions along the borehole wall and geomechanical parameters belong to the essential knowledge of determining the optimal mud density.

A case study of the deep geothermal well Gt GrSk 4/05 in Groß Schoenebeck, the key site in the Northeastern German Basin, exemplifies a combined approach

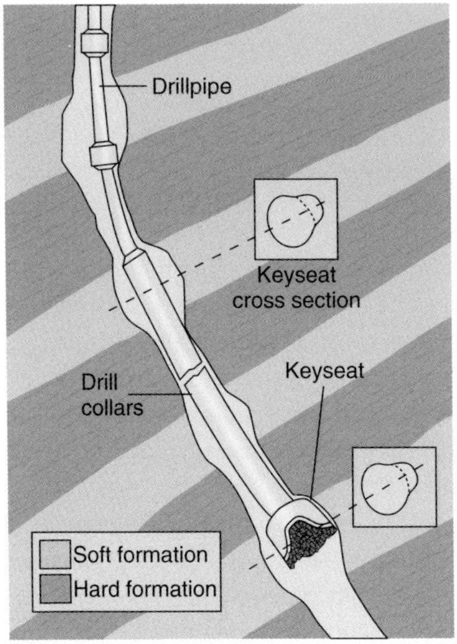

Figure 3.19 Keyseat. (From Schlumberger Oilfield Glossary, 2009.)

of structural geology and geomechanics to understand borehole breakout initiation and to specify a certain mud window (Moeck, Backers, and Schandelmeier, 2007; Backers, Stephansson, and Moeck, 2008).

- **Stuck pipes or tools:** Trouble shooting by down pumping a high viscose pill (for cleaning the borehole/remove filtercake), decrease of mud weight, and reducing friction coefficient of the mud by adding friction-reducing additives if differential sticking is assumed; increase of mud weight if cavings and borehole breakouts are assumed; eventually cleaning circulation run.
- **High dog legs (depending on diameter of casing):** Hole sections with high angle changes (inclination and azimuth) over a short distance (= "dog legs") may cause problems for both drilling and run casing. During drilling the rotation of the drillpipe body under a certain sideforce may cause to form a "keyseat" in the hanging side of the borehole (Figure 3.19). When pulling the string the next time the drillpipe may jam in the keyseat with the tool joint due to its larger diameter.

This may result in a fishing operation and – in worst case – even in the loss of the drillstring part (and of course the hole section) below the keyseat.

- **Fishing:** The activity of recovering parts (e.g., from a drill bit or a broken drillstring) which had been left in the borehole is called *fishing*; consequently, the lost part is called *fish*. Depending on the specific situation and shape and size of the "fish" special tools are available to increase the chance to catch the fish. There are tools to catch a drillstring part internally or externally, to catch small pieces of junk and a lot of other tools. However – like with real fishing – success is never guaranteed.

- **Freeing a stuck pipe:** Sometimes it is even necessary to part the drillstring in the hole intentionally in order to be able to recover at least the upper free part. This is done in several working steps.
- **Freepoint measurement:** The first step is normally to measure the "freepoint" by running a special measurement tool on an electric cable into the drillstring. Such a freepoint measuring tool consists mainly of three parts, an upper anchoring device, a lower anchoring device, and a strain gauge between both. Operation is as follows: The tool is anchored in a certain depth and then tension or torque is applied to the drillstring. If the anchoring depth is above the stuck point, a change in tension or torque will be measured because of the small elongation or twist of the drillstring between the two anchoring units. If no movement is measured, the freepoint is above the measuring depth. The measurement is repeated until the lowest free connection between drillpipe joints is found.
- **String shot:** In a second step, a "string shot" (= some strings of explosive cord) are lowered on wireline into the found lowermost free drillpipe connection. Then left-hand torque is applied to the drillstring and the shot is fired. If the amount of explosive was sufficient and the torque was properly applied the connection will be unscrewed due to the shock and the left-hand torque.

The working principle is similar to hammering onto a tight screw while applying torque to it.

- **Try to free the stuck part of drillstring:** After having pulled out the freed drillstring part normally, the recovery of the stuck part too is attempted. This is done by running in a heavy fishing jar which acts like a drilling jar described earlier but offers remarkably higher blow energy. The jar assembly then has to be screwed on to the top of the fish and made-up properly; the jarring can then start and will run for several hours.
  If the jarring attempt is not successful it can be tried to "wash-over" the stuck pipe. This is done with a kind of "coring operation" using a wash-over shoe on several casing joints which offer an outside diameter smaller than the hole diameter and an inside diameter large enough to pass over the drillstring or DC. When the string (or a part thereof) has been freed by the wash-over assembly that is left in the hole, the washpipe is pulled and again a fishing string with jar is lowered to be connected with the fish and to try to pull the fish out of the hole.

- **Sidetracking:** Sometimes a well has to be sidetracked, for example, if a fishing job had to be stopped without success or because the well did not find the reservoir in expected condition. Sidetracking is to leave the existing hole in a certain depth and start from there a new hole. This can be done either in open hole (mostly done from a cement plug) or inside a casing string (mainly done from a whipstock).
  - **Sidetrack from a cement plug:** In this case, a cement plug of a certain length (approximately 100 m) is set in the desired depth of the borehole. After a sufficient waiting time to assure to get hard cement – normally between

24 and 72 hours, a directional drilling assembly (with DHM and MWD, as described in Section 3.2) is used to sidetrack from the hard cement into the borehole wall, favorably into the proper direction from the early beginning. When the new hole is drilled out of the old one, the desired new well path can be followed up.

- **Sidetrack from a whipstock** (Figures 3.20 and 3.21): Inside the casing a "whipstock" (= a wedge) is set oriented, either on a cement plug or is fixed to a packer.
  Then a milling assembly is used to mill out the casing wall into the direction of the wedge which pushes the mill into the casing wall.
  Depending on the system another milling run may be necessary to cut the window to full size prior to run a directional drilling assembly as described above to drill the new well path.

3.7 Well Completion Techniques

The well completion starts with the casings which are planned to be used directly for injection or production and includes all tubulars and safety equipment need to handle the water flow in the wells.

3.7.1 Casing (Please Refer Also to "Casing String Design")

In a very early phase (planning the well) the decision has to be made whether the casing strings which will be used either for injection or for production will be allowed to move under temperature changes or not (Section 3.5).

3.7.1.1 Allowance of Vertical Movement of Casing

If a traveling of the upper part of a casing string due to thermal expansion or contraction should be allowed only the lower part is to be cemented. With this technique several meters of up- and downward movement of the casing head may occur and has to be taken into consideration! Critical with this technique may be that the uncemented casing may buckle under its own weight – particularly in caved hole sections – and that a sliding seal may be necessary to seal the annulus of the casing.

3.7.1.2 Pretensioning

If no casing movement should be allowed either the casing has to be cemented to surface or the noncemented upper section has to be pulled into sufficient tension to prevent later movement during production. To determine the "sufficient" pulling force for pretensioning the following work steps have to be done:

- Estimate the average expected temperature change over the noncemented string length.
- Calculate the elongation which will be generated by this temperature change.
- Calculate which load is necessary to generate the same elongation.

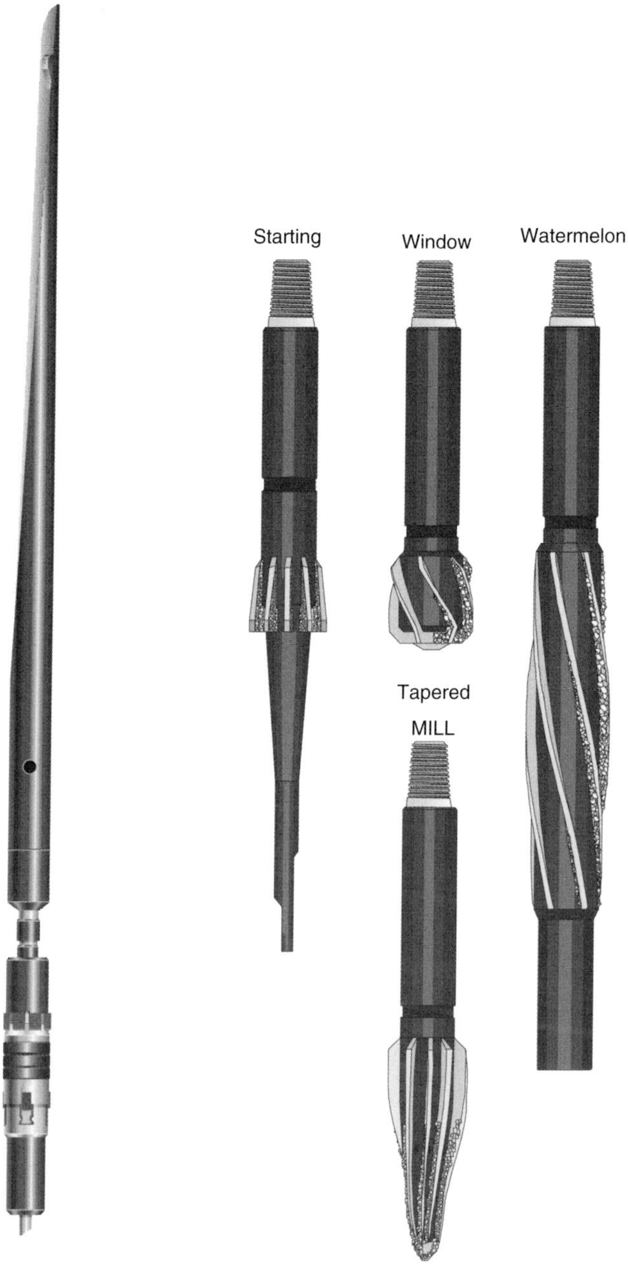

Figure 3.20 Assembly of a whipstock, which includes instrumentation to sense depth and orientation of the anchor groove on the packer (tools of Weatherford, as used in well GtGrSk4/05). (Courtesy of Weatherford, 2007. Whipstocks and Casing Exits.)

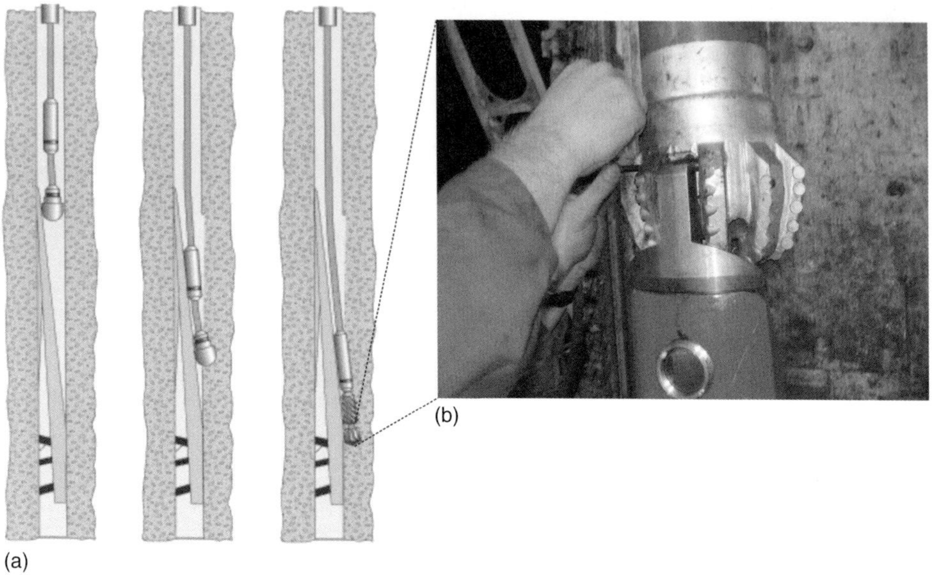

Figure 3.21 (a) Process of sidetrack drilling by setting a whipstock acting as a wedge (From *www.glossary.oilfield.slb.com*.). (b) Staged sidetrack mill as part of a whipstock. (From GtGrSk4/05.)

- Add a safety margin for the weight loss when hanging casing string into slips.
- Check whether the casing design allows application of additional tensile load.
- Check whether the casing design of the anchor string allows the application of the total load of all casing/tubing strings and the wellhead.

3.7.1.3 Liner in Pay Zone (Slotted/Predrilled) or Barefoot Completion

Depending on the borehole stability in the pay zone it has to be decided whether a slotted or predrilled liner should be run to prevent the borehole from collapsing during production/reinjection or whether the borehole can be used without a liner (= "barefoot"). It should be noticed that a barefoot completion offers the lowest flow resistance (pressure drop) and is therefore the best to lower total energy consumption during operational phase (production/injection).

3.7.2 Wellheads, Valves and so on

A wellhead is the top end of a borehole. Besides the purpose to allow flow into or out of the well – depending on whether the well is used for injection or production – the wellhead it is also equipped with valves which can be used to shut in the borehole, for example, for safety reasons. The wellhead normally

Figure 3.22 Example of a wellhead with valves resistant to a wellhead pressure of up to 10 000 psi (690 bar). (From GtGrSk4/05.)

starts with a bottom flange which is screwed-on or welded to the anchor casing. Depending on the number of surface casing strings additional flanged parts may be added on top of the each other; also a hanger may be needed for the pump string or an injection string. Normally each annulus between the casing strings can be accessed through valves which are mounted onto the side outlets. A typical wellhead for a production well with pump string is given in Figure 3.22.

3.7.3
Well Completion without Pumps with Naturally Flowing Wells

The general configuration of flowing well equipment includes from top to bottom:

- A production wellhead with the Christmas tree and the tubing head. The Christmas tree comprises a series of valves, a choke, and connections. It provides a means of controlling the effluents, ensuring the safety of the facilities, and giving measurement tools and instruments access to the well.
- The tubing, pipe to carry the effluents from bottom of the well to the surface.
- An annular seal or production packer, which is used first and foremost to isolate the casing from the pressure in the well and from physical contact with the effluents which are often highly corrosive.
- Downhole accessories such as sliding sleeve valves and landing nipples (tubing parts with specially designed inside profile). These components allow circulation

between the tubing and the annulus or are used to make it easier to use measurement and maintenance tools.

- An extra safety valve, called the *subsurface safety valve* for high risk wells under high pressure.

3.7.4 Well Completion with Pumps

There are two different pump types in use; it is the line shaft pump and the electric submersible pump.

- **Line shaft pump:** This type of pump is driven mechanically by a long hollow shaft which is rotated from surface while the pump is mounted at a certain depth of up to several 100 m inside the inner casing at this depth.
- **Electrical submersible pump (ESP):** This pump type is driven by an electric motor which is installed downhole below the pump; both hangs under a pump string (casing string) of sufficient diameter to handle the high flow rates. A cable for the electric power supply is clamped outside onto the pump string.
- **Production/pump string:** Pump strings normally consist of normal casing material; however, it is recommended to use a flush or near-flush connection instead of a standard coupled connection in order to reduce the risk of cable damage by the rectangular shoulder of the standard couplings.
  To reduce the pressure loss inside the pump string an internal plastic coating (IPC) can be used if the temperature of the hot water is not exceeding the temperature limits of the coating material.
  When using a pump string together with a high torque electrical submersible pump (ESP) a casing thread should be selected which is capable to withstand high torque without being unscrewed, particularly when starting the ESP.
- **Injection string:** For injection strings also standard casing material is used. Because normally the pump (if needed) is installed at surface no torque will be transmitted to the injection string during use standard coupled threads can be used. In order to reduce the pressure loss inside the injection string an IPC can be used.
- **Protection string (retrievable):** Sometimes it may be advised to install a retrievable protection string which protects the cemented casing against wear and/or corrosion (e.g., if highly corrosive fluids will be produced). However, such a string has to be taken into consideration from early planning stage because it reduces the available space for pumps and pump string.

3.8 Risks

Occurrence of risks leads always to delay due to additional measures and need for extra tools/services (fishing, additional casing/liner strings, additional cementations, and other services); always additional time and costs; often to additional costs due to damage or loss of downhole tools (e.g., DHM/MWD approximately €500 000); sometimes to partial or (worst case) total loss of the hole (sidetrack/new hole); seldom environmental impact; all risks are influencing costs.

Occurrence of risks cannot be totally avoided. Therefore it is highly recommended to calculate project budget on realistic assumptions and to take care for an additional part of funding (risk margin).

3.8.1 Evaluating Risks

3.8.1.1 Poor or Wrong Geological Profile Forecast

The predrill geological information can be poor if not all available geodata for a geothermal field are compiled together, for example, in a geological model. However, the data quantity and quality for an explorated field might be insufficient or limited. If preexisting data are used, a reprocessing and subsequent 3D structural modeling, integrating the knowledge of regional geology, petrophysics, and geophysics, will provide the best picture of subsurface structure. Well path trajectory can be directly planned within the geological models and can be updated on-site during operation supporting real-time surveillance of drilling progress. Finally, 3D geological models help also to specify uncertainties and evaluating risk. If geological interpretations are wrong or if a compilation of geological knowledge is poor, a lot of problems may arise as

- possibly wrong casing setting depth
  - too less strings, drilling problems;
  - target not or with insufficient diameter reachable;
- fault zones and/or targets not where expected
  - drilling problems;
  - "dry" hole.

3.8.1.2 Poor Well Design

Poor well design is a typical planning risk. It can happen due to a wrong geological profile forecast as described above, but also other reasons can lead to it as shown below.

- **Wrong casing setting depth:** Having chosen the wrong casing setting depth for a well (e.g., setting a casing string too high) may cause drilling problems and may ask for running the next casing size earlier than planned. Consequently too less casing strings may available for the well. This may cause that the target cannot be reached or only with insufficient diameter, causing problems (and higher costs) during production and/or injection.
- **Insufficient casing design:** An insufficient casing design (e.g., wall thickness too small or material strength (grade) not adequately chosen) can cause
  - **Casing collapse:** This hazard occurs specially in deep evaporitic environments if cementation is incomplete. Two mechanisms are recognized for casing collapse in salt rock: (i) sectorial pressurization and (ii) shear by slipping (Smart, Ford,

and Somerville, 1995). Choice of high strength steel and thick casing delays collapse but if the borehole and cement job are of poor quality, deterioration is still inevitable. Therefore the producer is strongly recommended to focus his efforts on drilling and cement job rather than on the choice of a casing (Charlez, 1982). The cement job is obviously all the more difficult when heavily weighted muds have to be displaced.
  - casing burst (particularly during stimulation).
  - parting of string (mainly in casing connection).
- **Casing diameters too small:** If the casing diameters which had been planned are too small
  - no contingency casing size may be available if needed;
  - so the borehole may not reach the planned target;
  - at least a high pressure drop through well during production/injection will have to be overcome leading to high energy costs for the pumps.

3.8.2 Technical Risks

A typical technical risk is the possible failure of drilling equipment, either at surface or downhole.

3.8.2.1 Failure of Surface Equipment

A failure of equipment at surface may cause a short interruption in the drilling operation only but can also lead to severe problems downhole. Mainly the following three components may be subject to failures.

- **Drawworks:** If the drawworks fail when the bit is on bottom the drillstring cannot be pulled; however, this may not become critical as long as circulation and rotation is possible until drawworks are repaired successfully. A critical situation occurs if the repair cannot be done at rig site.
- **Rotary table/topdrive:** If the drillstring cannot be rotated due to a failure of either rotary table or top drive this may not cause a critical situation as long as the drillstring can be pulled and mud can be circulated.
- **Mud pumps:** Normally a drill rig is equipped with more than one mud pump. So a failure of one mud pump may not be critical because of availability of at least a reduced pump rate from the remaining mud pumps. However, it is good practice always to pull the bit off bottom as long as the desired flow rate for proper hole cleaning cannot supplied.
- **General electric system:** Loss of power or failure in the electric system/Silicon-Controlled Rectifier (SCR)-Unit.

3.8.2.2 Failure of Subsurface Equipment

Generally, a failure of downhole tools may be more critical for the well than a failure of surface equipment. This is particularly valid for a breakage of drillstring components and resulting loss of parts in the borehole. Typical risks of downhole failures are

- **Breakage of drillstring components:** A breakage of drillstring components (such as DC, HWDP, DP, and mud motor) generally results in leaving all parts below the broken part in the hole. Depending on the specific situation a fishing operation with suitable special tools will be done in order to retrieve the lost part (fish) as described previously. If these operations are not successful within a certain time/certain attempts a plug-back and a sidetrack may be taken into consideration.
- **Failure or breaking of MWD system:** If an MWD system fails normally it can be retrieved and substituted by a backup tool so that there will be only some delay. Only in very few situations a breakage occurs leading to fishing operation.
- **Failure/loss of drill bit parts:** A typical drill bit failure is the loss of a cone from a tricone bit due to excessive wear in the bearing. In most of the cases the parts which remained in the borehole can either be caught by special fishing tools (e.g., junk basket) or by running a flat-bottom mill to the bottom of the hole and destroy the junk by milling.

3.8.3 Geological–Technical Risks

Geological–technical risks cover drilling technical problems during operation that are caused by reaction of the geological formation to applied drilling technology. Such problems mostly arise from unexpected geomechanical behavior of known or unknown formations or if drilling technology is insufficiently matched to known geological conditions (Figure 3.23). A modification of drilling parameters – especially mud technology – will commonly solve geological–technical problems. Basically, a rock mechanical understanding and geomechanical modeling will help to identify risk zones and to quantify rock failure under certain circumstances. Hence, drilling technical parameters like mud weight can precisely be specified by geomechanical models minimizing wellbore stability problems.

- **Borehole wall breakouts/cavings/washouts:** Breakouts are generated by high stress concentrations on the borehole wall, induced by too low mud pressures in weak brittle rock.
- **Squeezing or swelling formations:** Squeezing and creeping formations are salt rocks, creeping is increased by higher temperatures and thick overburden. From the Northeast German Basin it is known that particularly salt rock below 2000 m depth belong to risky zones in terms of casing failure. Swelling formations like clay or subordinately shale can be controlled by higher mud weight and additives like inhibitors which prevent clay minerals to assimilate water from drilling mud. If the mud weight is not adapted to the presence of swelling formations, a drillstring can be stuck within seconds.
- **Circulation losses (e.g., in fractured or karstic rocks):** Circulation losses are the reduced or total absence of fluid flow up to the annulus when fluid is pumped

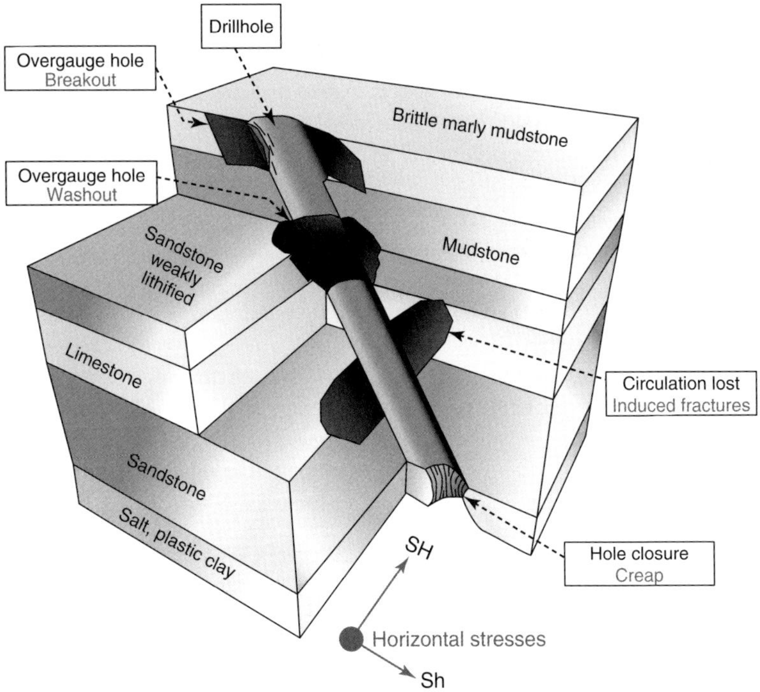

Figure 3.23 Possible geologically related drilling problems.

through the drillstring. Massive lost circulation can appear in highly fractured or karstic rocks like the Malmkarst of the Molasse Basin in South Germany. Karstic fault zones represent the geological target of high flow rates of geothermal fluids, but when unexpectedly entered the fault zone, a total circulation loss may cause a catastrophic loss of well control. If geothermal wells of large diameter are drilled in formations that are prone to lost circulation or if underbalanced drilling (UBD) is required foam drilling might the most suitable mud technique.

- **High pressure zones (e.g., shales, particularly dangerous if gas bearing):** Gas-bearing high pressure zones like sealed shales threaten the well by a sudden gas kick, especially with UBD.
- **Keyseats:** Keyseats often occur in doglegs and in alternating sequences of harder and softer layers. This phenomenon is already explained in Section 3.6.4. Eventually, larger tools of the bore hole assembly like tool joints, DCs, or stabilizers will not pass the small channels (keyseats) and drillstring could get stuck. Preventive measures especially in alternating strata include keeping any turns in the wellbore gradual and smooth otherwise the keyseating necessitate enlarging the worn channel by additional runs.
- **Differential pressure sticking:** This phenomenon belongs to the most critical drilling problems worldwide in terms of time and financial costs. It crops up especially in permeable sandstones. In general it occurs if drilling is done

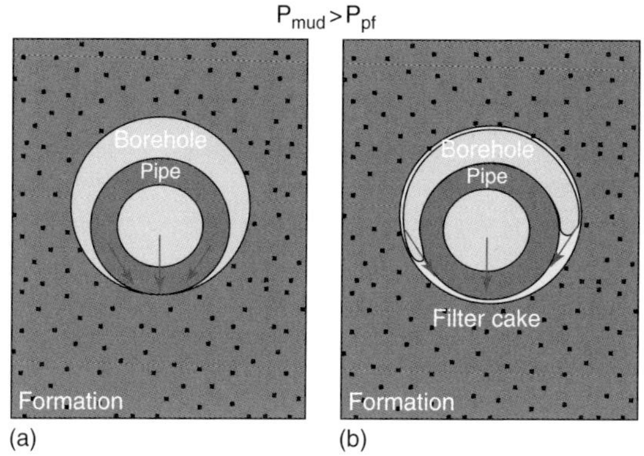

Figure 3.24 (a) Situation without filtercake and (b) stuck pipe along the borehole wall in a permeable formation with filter cake. (Modified after Schlumberger Oilfield Glossary 2009.)

overbalanced and the drilling mud forms a thick filtercake at the borehole wall. The differential pressure presses the DCs into the filtercake, and if the contact area is large enough the string gets stuck due to the high-contact forces. However, overbalanced drilling may be necessary in permeable sediments, for example, if gas influx needs to be kept out, but bears a high potential of differential sticking (Figure 3.24a and b). Differential sticking mainly occurs during stagnancy of rotary, for example, if a new pipe joint is added to the drillstring.

- **Thermally induced stress on borehole wall and/or casing/cement (due to drilling process):** A mud which is colder (hotter) than the formation, will reduce (increase) both pore pressure and hoop stresses (that are the tangential stresses on the borehole wall). Consequently, cooling the mud should have beneficial effects on borehole stability, whereas heating would have negative effects. As regards the temperature/pore pressure coupling, this becomes negligible for rocks having permeability greater than 10 μD (microDarcy).

3.8.4
Geological Risks

Geological risk arises from geological uncertainties or if the rock mechanical reaction of certain lithologies to the drilling process is insufficiently considered. In terms of well integrity as part of the well planning, the potential geologic risks should be specified for each geologic unit of the well profile before drilling. Some general risks as observed in different lithologies of boreholes are listed in Table 3.2.

Wells may encounter major unexpected structural geologic/stratigraphic changes during drilling. Therefore, real-time correlation must be systematically planed in advance, and must identify key marker beds and decision points. A systematic and careful cutting sampling belongs to the key issues in recognizing changes

Table 3.2 Potential geological risks of different lithologies.

Lithology	Risk							
	Washout	**Sand production**	**Caving/ breakouts**	**Collapse**	**Swelling**	**Creeping**	**Lost circulation**	**Gas kick**
Sandstone		xx	xxx	x			xxx	
Siltstone		X	xx	xx			x	
Mudstone			xx	xxx				
Claystone	x		xx		xx			x
Unconsolidated sand	xxx	xxx						
Clay/salt	xx				xxx	xxx		
Marl	xx		x					
Limestone			xxx	x			xxx	
Shale			xx	x	xx			xxx
Schist/phyllite			xx	x	x			
Gneiss			xx				x	
Granite			xxx				xx	
Basalt			xx				xx	

in geologic units during the drilling process. Generally, a sampling rate of 5 m steps will be sufficient unless critical strata like casing shoe formations, or entry points to the reservoir are expected. In critical strata and in the reservoir rock a sampling rate of 1–2 m is advisable in order to react prompt to changes in lithology or to recognize expected geologic boundaries. A careful and continuous analysis and description of the cuttings is an essential requirement to this special rock identification. Additionally, conventional LWD tools can provide the necessary information and insight for steering boreholes geologically and for petrophysical evaluation. However these costly tools require a detailed cost-benefit calculation for the geothermal project.

A sound geological well planning can considerably minimize geological risks and involves delineating the targets on the basis of (i) seismic maps, (ii) top structure maps, (iii) offset well logs, (iv) geographic target selection, and (v) 3D modeling and visualization. If one or more of these points are missing, following major complications may emerge:

- drilled geological profile differs from expected (poor seismic interpretation, deviated well path is insufficiently considered);
- fault/fracture zones found which had not been detected in seismic profiles (subseismic features);
- borehole stability/caving problems/sand production in weakly consolidated formations;
- borehole stability depending of trajectory in the current stress field (high stress concentration along well path);

- trouble shooting by geomechanical analysis (Moeck, Backers, and Schandelmeier, 2007; Backers, Stephansson, and Moeck, 2008);
- stress anisotropies/abnormal stresses in deep evaporites;
- temperature and/or temperature gradient lower than expected;
- low effective porosity/low matrix or fracture permeability in target formation, no natural productivity.

3.8.5
Geotectonical Risks

Geotectonical risks can arise in critically stressed regions like, for example, the Alpine foreland and the Oberrheingraben in Central Europe. Critically stressed regions are featured by natural seismicity. In particular, in these regions fault and fracture zones of certain orientation within the current stress field accommodate high shear stresses. If the shear stress exceeds the rock's frictional resistance, rock failure occurs by slip along the fault plane generating an earthquake or earthquake swarms. Drilling operations can be affected by earthquakes if the well is in the direct vicinity of an active fault or can even induce seismicity by increasing formation pressures. Possibly, fault activity can damage the well or the casing and results in the worst case to loss of the well. Especially hydraulic stimulations as part of work over operation in a well can induce seismicity. A detailed stress modeling of known or suspected faults and a slip tendency analysis for initial and changed fluid pressures contribute to the understanding of changing stresses and potential seismicity of man-made geothermal reservoirs (Moeck, Schandelmeier, and Holl, 2009).

3.9
Case Study Groß Schönebeck Well

A production well (GrSk 4/05) for completing the *in situ* geothermal laboratory at Groß Schönebeck has been drilled and stimulated 2006/2007.

The pay zones consisting of Rotliegend sandstones (about 80 m, average permeability 35 mD) and underlaying volcanics reach 150 °C at a depth of 4200 m. The investigations performed since 2000 in a reopened abandoned gas exploration well (GrSk 3/90) showed the feasibility of an EGS basing on the flow in the pore space between parallel situated hydrofrac planes (Huenges and Moeck, 2007).

After the stress field could be identified via the assessment of breakouts (Moeck and Backers, 2006; Moeck, Schandelmeier, and Holl, 2009) and the direction of frac ruptures in the first well (injection) the well path of the second well (production) was designed to allow for the planned frac treatments to be performed properly (Zimmermann *et al.*, 2007) (Figure 3.25).

The planned borehole design consisted of 26 in. surface casing, 18 5/8 in. anchor casing, a combined 16 in. × 13 3/8 in. production casing to the top of Zechstein, 9 5/8 in. liner to cover the 1500-m-thick rock salt layers, 7 in. liner for selective

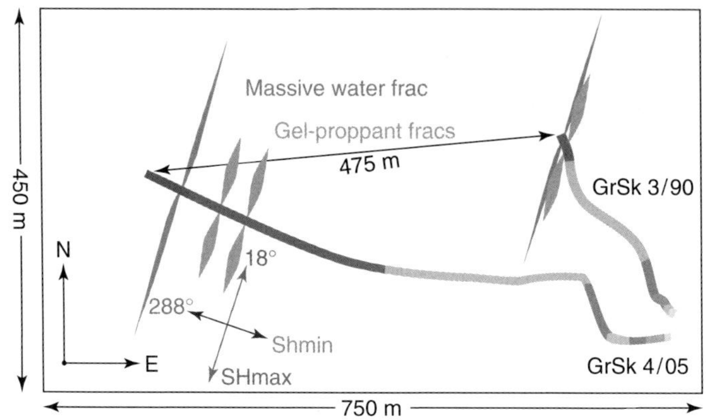

Figure 3.25 Well paths, stress field and direction of hydraulic fractures.

perforation of Rotliegend sandstones and a preperforated 5 in. liner for the open hole completion of the volcanics. The high diameter of the top hole section offers the option for installing a (corrosion-) protective casing and an ESP as well.

The super single drilling rig Streicher VDD.370.2 (*www.drilltec.de*) with automatic pipe handling system has been chosen in the tendering process and started drilling in April 2006.

Problems were encountered during the cementation of the combined production casing 16 in. × 13 3/8 in. The frac coefficient in the open borehole section from 741 to 2381 m was known from the adjacent well to be 1.57 bar/10 m. Therefore, light cement slurry with density of 1400 kg m^{-3} has been used. Already while pumping the lead cement a sudden pressure increase just up to the gradient mentioned above arose due to the increased flow restriction of the mud push solution (probably loaded with cuttings) at the entrance into the smaller annulus between the 16 in. casing and the preceding 18 5/8 in. casing. Total loss of circulation into carbonates of the Muschelkalk (Middle Triassic) formation could not be avoided. The top of cement detected at 1400 m was not acceptable for the production of 150 °C hot thermal brine due to the resulting thermal expansion. An unconventional solution was the top-down cementation of the annulus with a 1270 kg m^{-3} cement slurry for the displacement of the 1340 kg m^{-3} mud remained in the annulus. Prior to the top-down cementation the casing has been set in tension up to the hook load capacity of the drilling rig (370 t). So buckling of the production casing is avoided under all possible production temperature conditions. Temperature loggings proved the success of the "healing" cementation and the absorption of the displaced fluids by the formation at the intended depth. The total loss of circulation would had been avoided with a sufficient distance to the fracture gradient and an immediate reaction of the operator (i.e., by reducing the pump rate).

The thick series of creeping rock salt in the Zechstein (Upper Permian) formation have been drilled without any signs of tight hole problems due to the mud density as high as 2000 kg m^{-3}. Running and cementing of the 9 5/8 in. liner went smoothly within the 12 ¼ in. hole with the maximum inclination being 21° (kick-off point at

Figure 3.26 "Bit sample" of casing material. (Please find a color version of this figure on the color plates.)

2760 m). Immediately after reducing the mud density to 1030 kg m^{-3} as planned for drilling into the Rotliegend formation the rotary torque increased and finally the drillstring could not be retracted. It became evident, that the topmost stabilizer stuck inside the casing. Under high overloads and continuous rotating with the top drive the string became free and could be pulled out of hole. Logging operations and the borehole behavior while running various tools showed heavy deformation of the liner on its lower end of 900 m and an open window between 3851 and 3855 m. Remains of the casing material have been found in the bit after pulling out (Figure 3.26).

Within few days the creeping rock salt plugged the well totally (Figure 3.27). The affected well section had to be plugged with cement in order to prepare the bore hole for a side track.

There are no clear causes of the casing damage:

- The casing has been designed according to the rules considering the lithostatical pressure of the overburden and 1000 m empty casing.
- There is no evidence for lithological or geometrical (directional drilling) reasons.
- The appropriate number and kind of centralizers has been used to control the position of the casing within the smoothly calipered hole.
- The cement slurry had been weighted by hematite up to a density of 2100 kg m^{-3} with the mud density being 1980 kg m^{-3} so as to allow for the cementing model to result in good displacement results.

Geomechanical simulations showed the highest probability of incomplete mud displacement being the reason of the casing failure (Backers and Meier, 2008).

The remedy of choice consisted in modifying the borehole design in the lower section by installing an additional liner covering the rock salt bearing Zechstein after sidetracking at 3155 m. Within the "intact" 9 5/8 in. section already delivered 7 in. casings have been used while for the open rock salt section 7 5/8 in. casings with a large wall thickness had to be provided on the spot. Consequently the Rotliegend had to be drilled with 5 7/8 in. to the total depth at 4400 m with the maximum inclination in the pay zones reaching 48°. Thus the Rotliegend

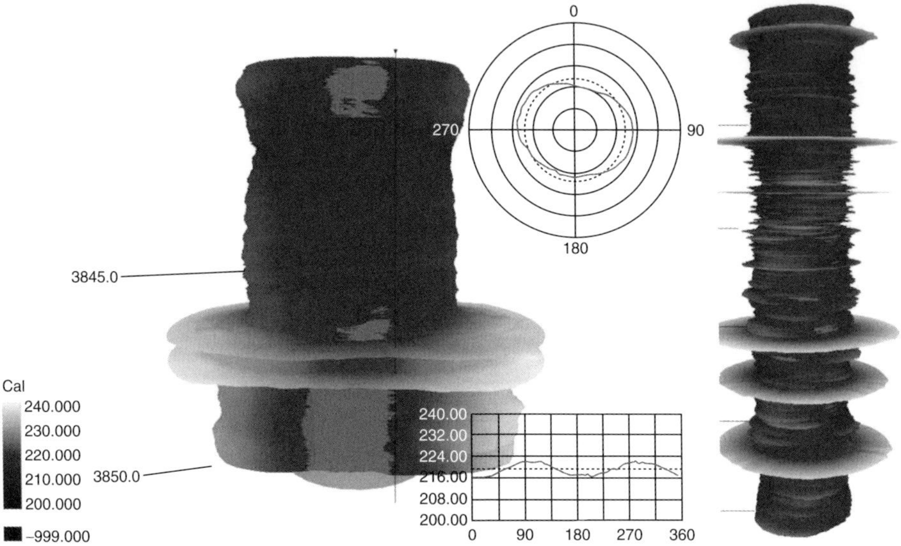

Figure 3.27 Example of multifinger caliper results in the damaged casing section. (Please find a color version of this figure on the color plates.)

sediments have been cased and cemented as planned. The obviously impermeable volcanics have been saved against cement intrusion instead of using an external casing packer by a highly viscous pill of mud weighted by chalk to 1600 kg m^{-3} which could be flushed out after drilling float collar and float shoe. Despite the loss of one casing dimension the target of the well could be reached finally (Figure 3.28).

With many other technical and drilling problems (e.g., four weeks top drive repair, fishing of boulders just beneath the shoe of the surface casing, successful side-tracking only after four attempts and modification of the anchor system, poor ROP and long round trip times, unexpected encountering of H_2S-gas traces in the lowest member of the Zechstein formation) drilling time and costs increased considerably in comparison with the planned ones based on the adjacent smaller offset well.

Nevertheless, the well reached the targets though real drilling time was beyond the time plan (Figure 3.29). After the multiple fracs carried out successfully in 2007 (Zimmermann *et al.*, 2008) it is now ready for thermal brine production within the intended communication experiment of the *in situ* geothermal laboratory.

3.10 Economics (Drilling Concepts)

Geothermal drilling projects normally are done to use the geothermal energy for, for example, heating and electric power generation. Very often (in most of the cases) the cost for the downhole part (drilling the wells) of a deep geothermal project is higher than the cost for the surface part (heat exchanger, power plant,

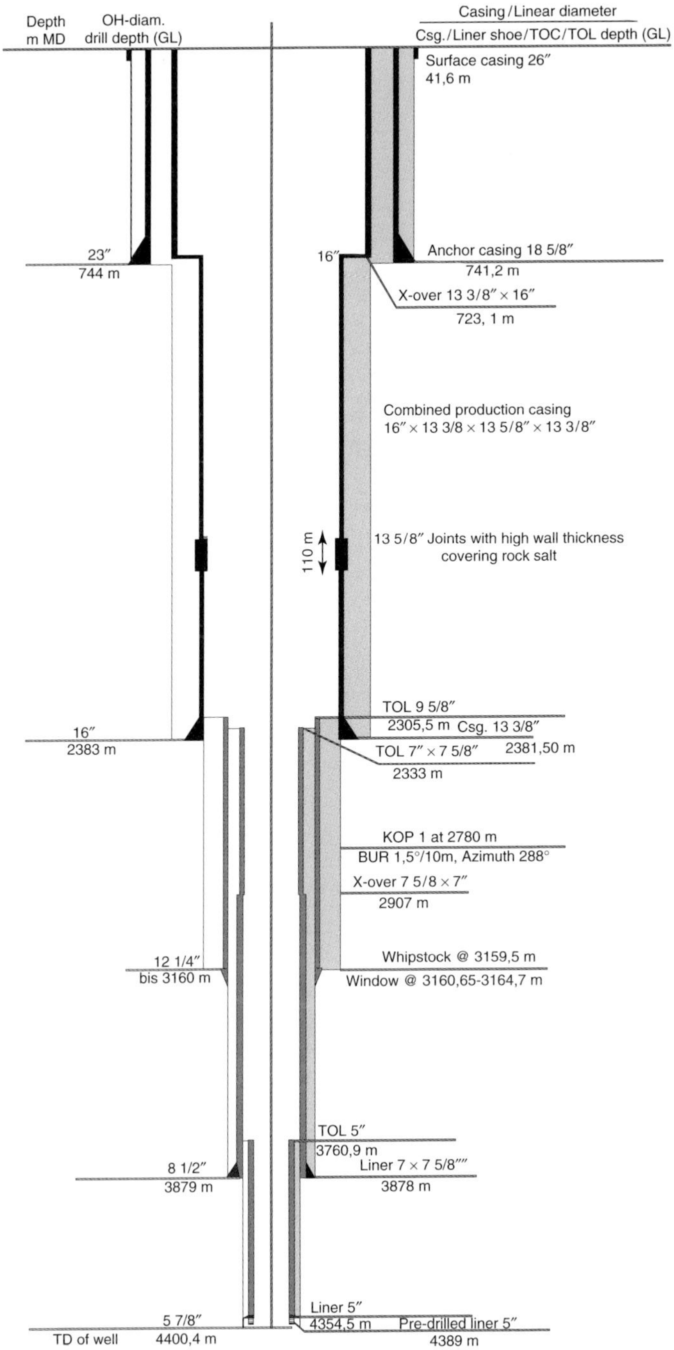

Figure 3.28 Final casing scheme.

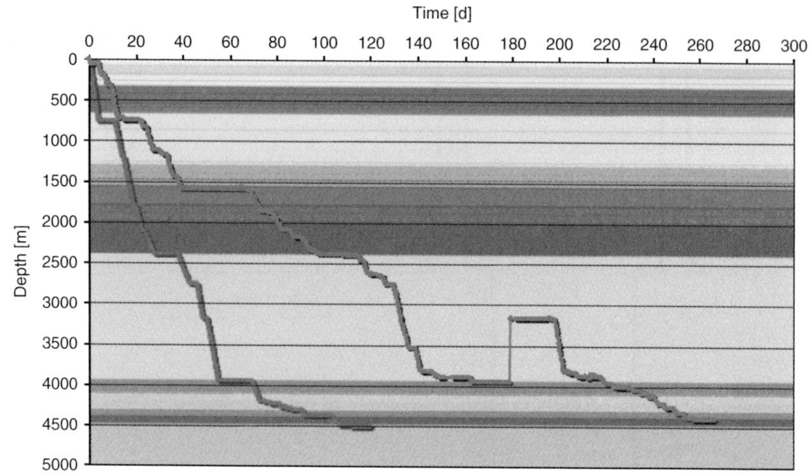

Figure 3.29 Depth versus time (red – planned, blue – real, colors indicate lithostratigrafic members). (Please find a color version of this figure on the color plates.)

cooling towers, etc.). It is obvious that the cost of the downhole part has a high impact on economics of a geothermal project; this is particularly valid for low-enthalpy projects. So, it is mandatory to have a close look on cost-influencing parameters in conjunction with the wells.

3.10.1 Influence of Well Design on Costs

Besides a proper drilling operation a good planning is of high importance, and inside the planning process the well design may have the highest impact on overall well cost.

3.10.1.1 Casing Scheme

Early investigations had shown that there is a certain relationship between drilled diameter and overall well cost. At this time it was found out that there is a direct relationship between total volume of destroyed rock and the cost, which was investigated for oil and gas wells in sedimentary rock. Whether this is still true or also valid for nonsedimentary rock and geothermal wells is not so important but the general influence of borehole size on cost. So, it should always be checked which casing scheme is to be favored for a specific drilling project as discussed in Section 3.5.

Generally the casing diameters of a deep geothermal well have to be larger than for oil and gas wells. The reason is that typical flow rates have to be higher for an economic hot water production than for normal production rates of oil wells. Additionally particularly the upper hole section needs a large diameter when an ESP has to be installed for production.

3.10.1.2 Vertical Wells versus Deviated Wells

Particularly in cases where special targets like fault systems have to be reached by the wells different approaches are possible. As an example a "hydrothermal doublet," consisting of a production and a reinjection well, will show the different possible alternatives.

When a doublet is planned it is necessary to keep a certain minimum distance between both wells at the pay zone if same formation is used for both wells (otherwise the cold water–front from the reinjection well will very quickly affect the temperature in the surrounding of the production well). Anticipating the minimum distance to be 2 km the following three alternatives are possible:

- Two vertical wells with 2 km distance between the slots and the targets of production and reinjection well. This gives in tendency the following advantages (+) and disadvantages (−):
  - two drill pads necessary (−);
  - long surface pipeline between the wellheads necessary (−);
  - short overall drill length (+);
  - no directional drilling costs (+).
- One vertical and one deviated well with 2 km departure
  - one drill pad (+);
  - short surface pipeline between the wellheads (+);
  - long overall drill length (−);
  - 1 × high directional drilling costs (−).
- Two deviated wells with 1 km departure
  - one drill pad (+);
  - short surface pipeline between the wellheads (+);
  - medium overall drill length (−);
  - 2 × moderate directional drilling costs (−).

Because of the complexity it is recommended to carry out a complete calculation for all alternatives which are applicable to a specific project in order to find out the optimum solution.

3.11 Recent Developments, Perspectives in R&D

3.11.1 Technical Trends

In most of the cases drilling-related technological developments start in offshore oil and gas applications due to the extremely high cost of such drilling operations which can reach up to a half million US dollars per day! Taking into account these costs a new technology which accelerates drilling operation can save money even if the additional costs for the new technology are high. If the new technology enters the drilling market successfully it normally lasts some years until it is common practice and the prices are coming down to a level where also onshore drilling can share the positive influence on overall drilling cost. There are numerous examples for such developments done in the past which are now common practice for the drilling business on oil and gas as well as on geothermal drilling.

Very often new technology not only improves the drilling progress but also reduces downhole risks. Some technical trends which are to become "standard" are described below.

3.11.1.1 Topdrive

As described earlier (Section 3.2) the topdrive is the modern type of equipment to rotate the drillstring; it has become already more or less standard since the last 10 years because of the advantage to be able to rotate the drillstring even while pulling. This allows back-reaming through "tight spots" and can avoid fishing under some circumstances.

3.11.1.2 Rotary Steerable Systems (RSS)

These systems are to substitute the conventional directional drilling technique with DHM and allow to rotate the drillstring all time, while with the conventional systems the drillstring cannot be rotated during steering phases. Particularly in deep and deviated hole sections a nonuniform drilling process is caused by the "stick-slip" effect when using conventional directional drilling technique; this can be avoided by the continuous rotation of the drillstring with RSSs. The reason is that due to continuous rotation the friction at the contact areas between drillstring and borehole wall is always in the lower "sliding" mode and avoids the "sticking" mode where the friction is higher. The result is normally a higher ROP and a safer drilling operation even under adverse conditions.

The first generation of those RSSs were developed for the ultradeep main hole of the "Continental Deep Drilling Project of the Federal Republic of Germany" (KTB) in the late 1980s/early 1990s. At this time the focus was to develop self-steering vertical drilling systems (VDSs) which were able to keep the hole trajectory as close as possible strictly vertical (Figure 3.30). Several different types of systems had been investigated, designed and built during this time, with and without integrated DHM, with and without expandable ribs, "point the bit" or "push the bit", and so on. One of the early systems is shown in Figure 3.30.

The different types and steering techniques have been improved over the years for various applications; however, the "roots" of these tools for all tools are leading back to the efforts which had been made during the KTB project.

Most of the systems use hydraulically actuated expandable ribs in the outer (stationary) sleeve which are pushed against the borehole wall on the opposite side to where the trajectory's direction is to be corrected. Other systems have a fixed outer diameter of the outer sleeve, and the inner rotating mandrel is pushed to the right direction inside the outer sleeve.

A recently used system is shown in Figure 3.31 (please note the similarity in design).

Figure 3.30 Vertical drilling system VDS-2 at KTB-HB1; preparation to run in hole. (Courtesy A. Sperber.)

The AutoTrak rotary closed-loop system delivers higher penetration rates. more pracise well placement and high-quality holes in directional and horizontal applications.

Figure 3.31 Modern RSS (type "Auto-Trac"). (Courtesy Baker Hughes Inteq, *http://www.bakerhughesdirect.com/*.)

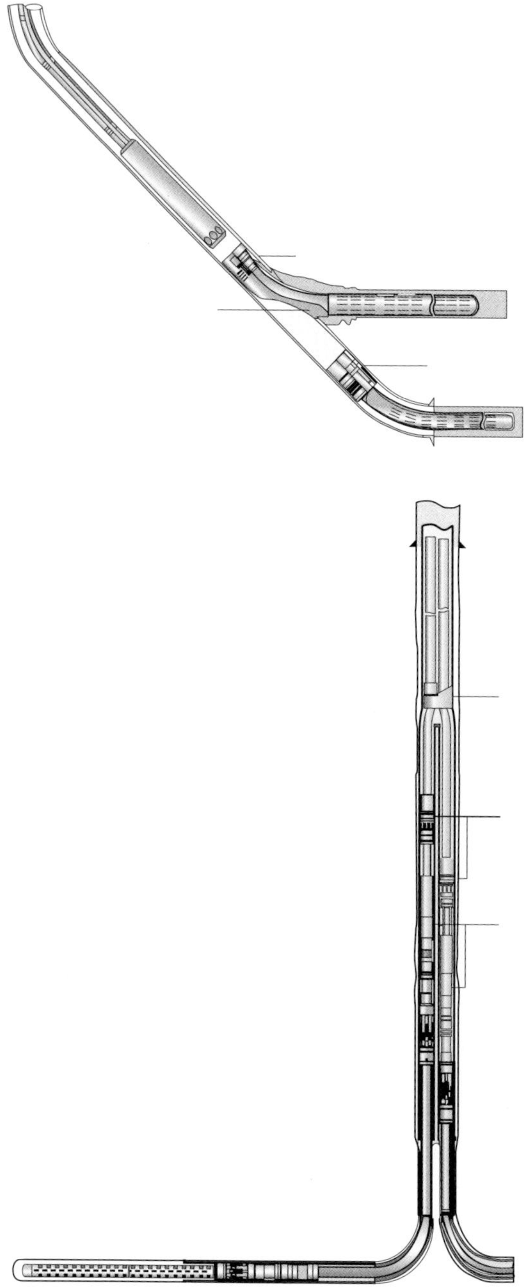

Figure 3.32 Examples for multilateral drilling. (Courtesy of Baker Hughes 2007; personal communication.)

3.11.1.3 Multilateral Wells

Multilateral wells are quite common in special oil and gas applications. The idea behind is to reduce overall drilling costs by using one top hole section for several "laterals" and to multiply production by these multiple laterals. Typical cases are shown in Figure 3.32.

The technique of drilling multilateral wells may be advantageous also for geothermal wells, but diameters of the laterals may be critical for the high flowrates needed in geothermal applications.

3.11.2
Other R & D-Themes of high Interest

Future research is needed in large-diameter drilling specially through plastic, creeping or swelling formations as salt or shale. Commonly, abnormal high fluid pressures in such formations cause abnormal stresses that can differ considerably from usual pressure gradients. To provide long-life completion systems in plastic formations, new cementing technologies accommodating the geomechanical behavior of plastic rock need to be defined, especially for deviated wells.

Another aim in accessing the reservoir is to minimize formation damage using low mud pressures by means of near-balanced drilling (NBD) or even UBD, particularly in reservoirs with depleted (underhydrostatic) formation pressure. Whenever drilling mud is introduced into the borehole in an overbalanced pressure situation, mud invasion, and formation damage occur. Hydrothermal wells are in particular sensitive to this invasion because of the long completion zones, complex chemistries, and high temperatures thus the aim is to keep the mud pressure in a low condition that is near or below the fluid pressure of the formation. The negative effect of NBD is, however, high stress concentrations on the borehole wall probably causing borehole failure. NBD and borehole stability under changing stress conditions must be well understood. In future, wellbore integrity needs to be investigated by fracture mechanical experiments and simulations. These important efforts require core testing and thus, an extensive core sampling campaign is essential to describe and compile rock mechanical parameters of geothermal fields. More importantly, the availability of drillcore sections are crucial for sophisticated formation evaluation at borehole scale that characterizes the subsurface conditions with less uncertainties than with logs, where disturbing factors, for example, from mud or tool itself can basically mislead log interpretations (Norden *et al.*, 2008).

The objective of new-generation geothermal drilling should be to promote ways and means to reduce the cost of geothermal drilling through an integrated effort which involves improved 3D well planning and geosteering, rock and fracture mechanical modeling in context of borehole stability consequently reducing borehole problems, and fostering an environment and mechanisms to share methods, experiences, and means to advance the state of the art.

References

Aadony, B.S. (1999) *Modern Well Design*, Balkema, Rotterdam, 240 p.

Backers, T., Stephansson, O., and Moeck, I. (2008) Fracture mechanics – examples of applications in rock engineering. Extended Abstracts Volume, EAGE 70th Conference and Exhibition, June 09-12, 2008, Rome, Italy, CD-ROM, 186 p.

Backers, T., and Meier, T. (2008). Analyse der möglichen Einfluesse auf den Linerkollaps GrSk 4/05. Unpubl. report of GeoFrames GmbH, Germany, pp. 52.

Barker, J.W. (1998) Reduced-clearance Casing Programs Offer Numerous Advantages. World Oil (May 1998).

Bourgoyne, A.T., Millheim, K.K., Chenevert, M.E., and Young, F.S. (1986) *Applied Drilling Engineering*, SPE Textbook Series,Society of Petroleum Engineers, Richardson, TX,USA, ISBN: 1-55563-001-4-2 Vol. 2, 502 p.

Bradely, W.B. (1979) Failure on inclined boreholes. *Journal of Energy Resources Technology*, **102**, 232–239.

(a) Charlez, P.A. (1997) *Rock Mechanics*, Petroleum Applications, Vol. 2, Editions Technip, Paris, p. 661; (b) (1982) *Drilling Mud and Cement Slurry Rheology Manual*, Editions Technip, Paris, 33 p.

Economides, M.J., Watters, L.T., and Dunn-Norman, S. (1998) *Petroleum Well Construction*, John Wiley & Sons, Ltd, Chichester, 640 p.

French, F.R. and McLean, M.R. (1993) Development drilling problems in high-pressure reservoirs. *Journal of Petroleum Technology*, **8**, 772–777.

Huenges, E. and Moeck, I., The Geothermal Project Group (2007) Directional drilling and stimulation of a deep sedimentary geothermal reservoir. *Scientific Drilling*, **5**, 47–49.

Krus, H. and Prieur, J.M. (1991) High-pressure well design. *SPE Drilling Engineering*, **6** (4), 240–244.

Moeck, I. and Backers, T. (2006) New ways in understanding borehole breakouts and wellbore stability by fracture mechanics based numerical modelling. EAGE 68th Conference & Exhibition, June 12–15 2006, Vienna, Austria.

Moeck, I., Backers, T., and Schandelmeier, H. (2007) Assessment of mechanical wellbore assessment by numerical analysis of fracture growth. *EAGE 69th Conference and Exhibition, 11–14 June 2007*, Extended Abstracts vol. D047, CD-ROM, London, and Co Productions, Houten/The Netherlands. ISBN: 978-90-73781-54-2.

Moeck, I., Schandelmeier, H., and Holl, H.G. (2009) The stress regime in Rotliegend reservoir reservoir of the Northeast German Basin. *International Journal of Earth Sciences (Geol Rundsch)*, **98** (7), 1643–1654. doi: 10.1007/s00531-008-0316-1.

Norden, B., Förster, A., Vu-Hoang, D., Marcelis, F., Springer, N., and Le Nir, I. (2008) Lithological and petrophysical core-log interpretation in CO2Sink, the European CO2 onshore research storage and verification project. SPE Asia Oil & Gas Conference and Exhibition, 20-22 October 2008, Perth, Australia, SPE 115247OMSCO Industries 1997. Catalogue "OMSCO's Major Products 8/97".

Perrin, D. (1999) *Oil and Gas Field Development Technique: Well Completion and Servicing*, Editions Technip, Paris, 325 p.

Prevedel, B. (2007) Drilling and completion challenges in fault zones: lessons learned from ICDP projects. *Scientific Drilling*, Special Edition No. 1, 2007, ISSN: 1818-8957. 98–99.

RockBit International (2005) Catalogue.

Schlumberger, D. (1984) *Cementing Technology*, Nova Communications, Ltd, London.

Schlumberger Oilfield Glossary (2009) *www.glossary.oilfield.slb.com*.

Seymour, K.P. and MacAndrew, R. (1994) Design, drilling and testing of a deviated HPHT well exploration well in the North Sea. *SPE Drilling Engineering*, **9** (4), 244–248.

Sheppard, M.C., Wick, C., and Burgess, T. (1987) Designing well path to reduce drag and torque. *SPE Drilling Engineering*, **2** (4), 344–350.

Smart, B.G.T., Ford, J.T., and Somerville, J.M. (1995) An overview of drilling of salt bodies – problems causes and solutions, *Salt Induced Casing Collapse Symposium*, British Gas Publication.

Tischner, T., Schindler, M., Jung, R., and Nami, P. (2007) HDR project Soulz: hydraulic and seismic observations during stimulation of the 3 deep wells by massive water injections. Proceedings of the 32nd Workshop on Geothermal Reservoir Engineering, Stanford University, Stanford, California, January 22–24.

Zimmermann, G., Reinicke, A., Blöcher, G., Milsch, H., Gehrke, D., Holl, H.-G., Moeck, I., Brandt, W., Saadat, A., and Huenges, E. (2007) Well path design and stimulation treatments at the geothermal research well GtGrSk4/05 in Groß Schönebeck. Proceedings Thirty-Second Workshop on Geothermal Reservoir Engineering, Stanford University, Stanford, California, January 22–24, 2007, SGP-TR-183, pp. 75–80.

Zimmermann, G., Reinicke, A., Brandt, W., Blöcher, G., Milsch, H., Holl, H.-G., Moeck, I., Schulte, T., Saadat, A., and Huenges, E. (2008) Results of stimulation treatments at the geothermal research wells in Groß Schönebeck /Germany. Proceedings, Thirty-Third Workshop on Geothermal Reservoir Engineering Stanford University, Stanford, California, January 28–30, 2008, SGP-TR-185.

4
Enhancing Geothermal Reservoirs

Thomas Schulte, Günter Zimmermann, Francois Vuataz, Sandrine Portier, Torsten Tischner, Ralf Junker, Reiner Jatho, and Ernst Huenges

4.1
Introduction

In many cases, drilling operations will not open up a geothermal reservoir under such conditions that an extraction of geothermal energy is economically viable without any further measures. Geothermal wells often have to be stimulated, in order to increase well productivity. Different stimulation concepts have been applied to enhance the productivity of geothermal wells. Formally, stimulation techniques can be subdivided with respect to their radius of influence. Techniques to improve the near-wellbore region up to a distance of few tens of meters are chemical treatments, and thermal fracturing. The only approved stimulation method with the potential to improve the far field, up to several hundreds of meters away from the borehole is hydraulic fracturing.

The foundations, on which today's enhanced geothermal systems (EGS) projects are built were laid out in the early 1970s, when a hot dry rock (HDR) development concept was worked out at Los Alamos National Lab, entailing drilling a well into hot crystalline rock, using water under high pressure to create a large vertical fracture by interaction with the *in situ* stress field, and finally to drill a second well to access that fracture at some distance above the first wellbore. The Rosemanowes experiment, performed by the Camborne School of mines, was set up in 1976, aiming at testing at lower temperatures and at lower depth, and some of the ideas that had emerged from earlier investigations, were experimented at Fenton Hill. One of the main findings from these two pioneering endeavors was that deep granites often are fractured and that potentially existing natural fracture systems will determine the orientation and the propagation of the stimulation. Both projects have been comprehensively presented by Armstead and Tester (1987), who also give a broad overview of the state of stimulation technologies at the end of the 1980s.

Geothermal Energy Systems. Edited by Ernst Huenges

ISBN: 978-3-527-40831-3

4.1.1 Hydraulic Stimulation

Hydraulic fracture stimulations are performed as waterfracs, gel-proppant fracs, or a combination of both called *hybrid fracs* (Sharma *et al.*, 2004) The procedures are well known in the hydrocarbon industry (Shaoul *et al.*, 2007a, b) as well as in the HDR technology (Hettkamp *et al.*, 2004; Baumgärtner *et al.*, 2004; Schindler *et al.*, 2008). However, the application for geothermal reservoirs requires a technique that is able to produce considerable higher amounts of fluids than the ones required for production of hydrocarbon reservoirs.

4.1.2 Thermal Stimulation

Thermal stimulation treatments are performed in order to increase the productivity or injectivity of a well by either, enhancing the near well permeability, which may have been reduced by drilling operations itself (drill cuttings or mud clogging feed zones), or by opening hydraulic connections to naturally permeable zones, which were not intersected by the well path. This can happen by either reopening of existing, possibly sealed fractures, or by creation of new fractures through thermal or additional hydraulic stresses.

4.1.3 Chemical Stimulation

Acid treatments were developed by the oil industry for improving the productivity of oil and gas wells (Smith and Hendrickson, 2005; Economides and Nolte, 1989; Schechter, 2006) Acid treatments were first applied to wells produced from limestone formations, about 100 years ago, and they became a technology widely used in the 1930s. The technique was partially adapted to the geothermal wells, most often to remove the mineral scaling deposited in the wells after several years of exploitation (Strawn, 1980; Epperson, 1983; Barrios *et al.*, 2002; Serpen and Türeyen, 2000), and also to enhance the fractures network in the reservoir.

4.2 Initial Situation at the Specific Location

4.2.1 Typical Geological Settings

EGS can be and are being engineered in a wide variety of geological settings: volcanic settings like Iceland (Axelsson, Thórhallson, and Björnsson, 2006), metamorphic environments like, for example, Larderello (Italy) (Bertini *et al.*, 2005) magmatic systems like deep granites in Soultz, France, (see Hettkamp *et al.*, 2004); or

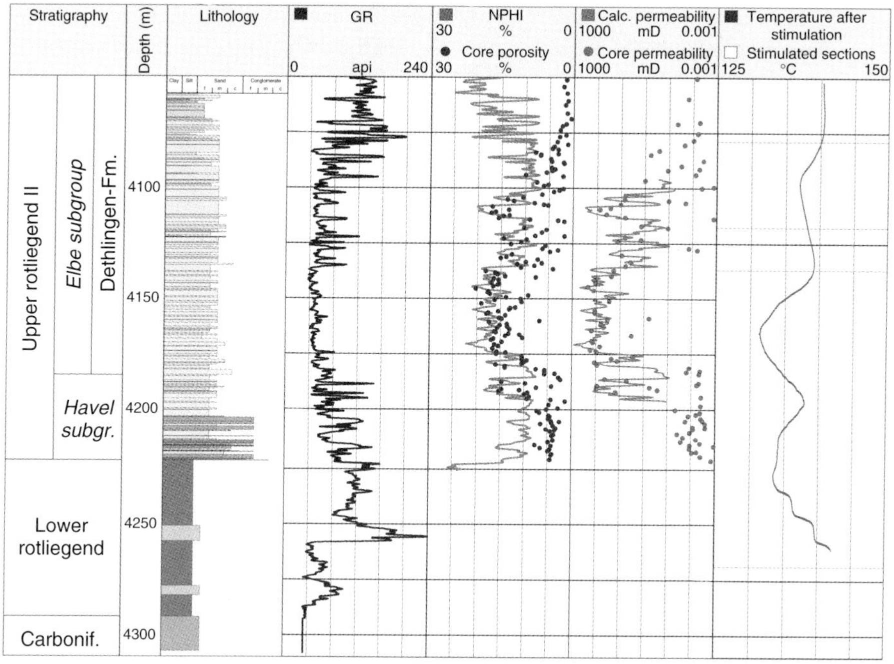

Figure 4.1 Example for well logging data recovered at Groß Schönebeck 3/90 showing stratigraphic and lithological units, which are determined mostly by analyzes of drill cuttings or derived from gamma ray (GR) measurements (3.column), porosity sensitive measurements (4.column), derived permeability values both compared with core measurements and a temperature profile, which is sensitive to cold water injection.

sedimentary environments like Groß Schönebeck, Germany, (see Huenges *et al.*, 2009); or Horstberg, Germany, (Orzol *et al.*, 2005).

Each of these typical and representative geological settings has its characteristics in terms of temperature regime, degree of reservoir complexity, influence and importance of stress field, structural features, lithological variability, type value and distribution of porosity and permeability, extent and form of natural fracture system, brine chemistry, and so on. Like with regard to exploration strategy or power plant layout, these characteristics also determine the most suitable stimulation method and hence the extent and details of an appropriate upfront investigation program. Reservoir characterization is crucial before the stimulation. After drilling a borehole well, logging provides important information about the target horizons such as the data given in Figure 4.1.

4.2.2
Appropriate Stimulation Method According to Geological System and Objective

In general there are three main potential reasons why geothermal wells are being stimulated.

First, stimulation measures may be necessary because the target formation have been clogged or damaged within the course of the drilling process. "Clogging" may even be caused deliberately by adding weighting material in order to protect the target formation throughout the drilling process. In these cases, acid stimulations may be applied to clean up the well. Natural fractures can become mechanically blocked by drill cuttings as well during the drilling process. This will especially happen, depending on the mechanical rock properties and the drill bit applied, if drill cuttings are very small. Thermally stimulating these wells can be an effective way of restoring their productivity (Axelsson, Thórhallson, and Björnsson, 2006; Tulinius *et al.*, 1996).

Secondly, and most importantly, geothermal wells are being stimulated in order to significantly increase their productivity such that they flow at economic levels. This can be achieved by generating new fractures or fracture systems, by connecting the well to an existing natural fracture network, or by expanding natural joints and fractures. In high temperature systems like in volcanic and in some metamorphic settings, the most cost-effective way of stimulation is a thermal treatment, provided reservoir temperatures are sufficiently high. It is difficult to give a lower temperature limit above which thermal stimulation will be possible, but successful thermal stimulations have been reported for reservoirs with temperatures of $T > 210\,^{\circ}C$ (see e.g., Tulinius, Correia, and Sigurdsson, 2000). For the vast majority of other cases, namely deep sedimentary and granitic settings, a hydraulic stimulation will be the method of choice, which can be followed by an acid treatment. On the contrary, in the particular case of the fractured or karstic carbonates, an acid stimulation or acid frac will be the preferred method, in order to sustainably increase hydraulic conductivity.

Thirdly, in case a well is damaged or clogged by mineral deposits or scaling during the production phase of a geothermal reservoir, an acid treatment may be indicated.

Although main stimulation methods will consecutively be dealt with separately, hybrid treatments are not uncommon. Thermal stimulations are being conducted applying a certain wellhead pressure, and in acid fracs, acid is injected into the reservoir above fracture pressure, thus creating hydraulic fractures. The faces of these fractures are being etched in a nonuniform way, which creates hydraulic conductivity.

4.3 Stimulation and Well path Design

There are several options to exploit a geothermal reservoir by obtaining the optimum design to extract the maximum heat with minimum costs. The simplest design consists of a single production well. Sustainability is only given if ground water recharge compensates production; otherwise one takes the risks of depletion of the geothermal reservoir. The basic types of a sustainable design are doublets or triplets consisting of one or two production wells and one injection well to complete the water cycle. These can be enhanced to a multiwell design with several patterns of vertical wells like hexagonal or five spots patterns to enlarge the effective heat transfer area (Armstead and Tester, 1987). The optimization of the number of wells includes several constraints which depend crucially on the target depth and hence the individual costs for drilling, the initial productivity of the reservoir rocks, and the required costs for stimulation treatments to enhance this productivity.

The arrangement of two wells follows two conflicting goals and can be generalized for a multiwell design. On the one hand, the wells should be located in such a way, that the pressure in the reservoir will not drop significantly during production resulting in a comparatively close distance of the wells. On the other hand, a short circuit between the wells, implying a temperature drop in the production well, should be avoided. In general, there are two options:

- Arrangement of stimulated fractures on the connection line of both wells, corresponding to the classical HDR approach, that is, the orientation of the doublet is in the direction of the maximum principle stress.
- Arrangement of stimulated fractures perpendicular to the connection line of both wells, that is, the orientation of the doublet is in the direction of the least principle stress and fluid flow propagation is through natural permeable rocks.

Both arrangements of the doublet (parallel and perpendicular to the minimum principal stress) in a reservoir with some matrix permeability result in a pressure decrease at the production well and a pressure increase at the injection well (Huenges *et al.*, 2006). The risk of a thermal short circuit of the system is most probable in the parallel case. Therefore the perpendicular case is the appropriate arrangement for a deep sedimentary reservoir with some matrix permeability and is valid as long as no extended natural fracture systems are connected. (Legarth *et al.*, 2005; Zimmermann, *et al.*, 2005).

A different concept includes the connection of an existing permeable fault zone in a reservoir, which can be on one side connected to the wells via stimulation treatments or on the other side through a well path intersecting the fault zone like in the European HDR project in Soultz-sous-Forêts (Baria *et al.*, 1999; Hettkamp *et al.*, 2004; Zimmermann *et al.*, 2009). In a natural fractured reservoir, a deviated well can intersect multiple fractures and connect them to the well. This design can be supported by multiple stimulation treatments to enhance the number of connected fractures and hence the productivity of the well.

Considering vertical or deviated (nonhorizontal) wells, the use of multiple layers for production and injection is another option to enhance the productivity of a geothermal system. This design depends on the particular geological setting of the reservoir. A potential option is a connection of two different layers via a hydraulic fracturing treatment (Zimmermann *et al.*, 2009). In this particular case, a vertical fracture propagates from the bottom of the well (a fractured rock layer) upward to the next layer (permeable sediments) where vertical fracture propagation may stop, if the fracture reaches a superimposing impervious (and "soft") layer, and is followed by a leak off into the sediments. If a doublet of wells is scheduled, the flow between the wells is then performed through the sediments and the respective connection to the wells is supported by the induced fractures.

4.4 Investigations Ahead of Stimulation

The choice of the most suitable stimulation measure for any particular EGS development project will depend on both, the specific objective of the intended stimulation measure, and the geological and thermal conditions at the particular location. The method chosen to be best suited on the other hand, will, besides the local geological setting, determine the scope as well as the necessary depth of any investigation ahead of stimulation measures. The prime purpose of any upfront investigation program is to provide a framework for the decision on geometrical and procedural frac layout, and key input parameters for a potential frac growth modeling to be performed ahead of stimulation.

Apart from all potential differences in the necessary investigation program, the decision on every stimulation measure will be based on a subsurface model: a detailed, structural, and stratigraphical interpretation of 2D or ideally 3D seismic data will deliver the main structural setting and features, qualitative information on the prevalent stress regime, (normal-, thrust-, or strike-slip faulting), the fault inventory, and information on the main stratigraphical and lithological units. If necessary and relevant, these investigations may be complemented by local field mapping or an outcrop analysis aiming at analyzing fault and fracture geometries and hierarchies, rock petrography, mineralogy, or any potential hydrothermal alteration.

Investigations may be extended on the well scale, by, for example, analyzing well-based seismic data (vertical seismic profiling, VSP), allowing a more detailed reservoir and fault delineation and identification of fractured and high permeable zones. Dipmeter, full waveform sonic, and electrical or acoustic image logs can provide information on fracture distribution patterns, orientations, magnitude, and aperture with an even higher resolution. They can also serve in discriminating between open and mineralized fractures. If applicable for the geological setting under investigation, a detailed wire line log analysis will yield further insight into lithological variations, porosity and permeability distribution, the mineralogical composition, and fluid saturation of the matrix rock. Key geomechanical parameters like Youngs modulus and poisson ration may be derived from monopole and dipole sonic logs.

In case a hydraulic stimulation is planned, a detailed evaluation of the local stress field will be performed aiming at ideally determining the full stress tensor for the candidate EGS reservoir. A comprehensive review on techniques to determine the orientation and magnitude of the in situ stress in deep wells was provided by Zoback *et al.* (2003) Within this context it is sensible also to predict the proximity to shear failure by performing a slip tendency analysis for the fault inventory (see e.g., Morris, Ferrill, and Henderson, 1996).

In order to understand the fabric of the EGS reservoir under investigation, and potentially to develop an empirical model relating rock type to failure resulting from hydraulic stimulation, a detailed mineralogical and petrographical characterization

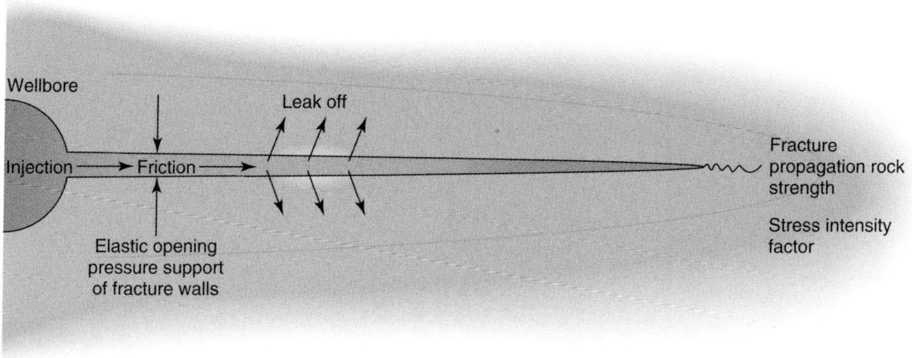

Figure 4.2 Top view of a fracture propagation due to hydraulic stimulation (Fokker, 2007).

of core or cutting material should be performed. This certainly is of particular importance in case the well is to be stimulated chemically.

A routine core analysis will provide key rock matrix properties like porosity and permeability to which a wireline analysis can be calibrated. A complementary special core analysis may be performed focussing on thermal properties (thermal conductivity and possibly radiogenic heat production) and fracture mechanical parameters (e.g., fracture toughness).

Within the process of planning a hydraulic stimulation treatment, frac modeling is a key step, for example, to understand the ongoing processes as shown in Figure 4.2. The geomechanical parameters most important for these models are the Young's modulus E, the Poissson ratio ν, the angle of internal friction ϕ, and the poroelastic constant η; From these the single, most important parameter certainly is Young's modulus E since it has a nearly direct relationship with the net pressure, fracture geometry, and fracture width. Young's modulus can easily be measured in the lab, or derived from sonic logs, and it is recommended in case hydraulic stimulation measures are being planned, to take core samples, and conduct triaxial tests in order to determine the elastic parameters Since these parameters may largely vary between the various lithological units over which the hydraulic fracture will grow, they should be determined for each of the main unit ahead of the stimulation, in order to enable the frac treatment to be planned with sufficient accuracy.

Every case-specific upfront investigation program will be compiled according to both the stimulation method of choice and the particular local geological conditions. However, in case a chemical stimulation is being envisaged, the program will certainly focus on mineralogical content and potential hydrothermal alterations, where in the case of hydraulic stimulation treatments, the focus will shift toward the lithological characterization and the rock mechanical properties of the matrix material, and a detailed description of the local stress field.

A joint interpretation of the structural geological setting, results of an image log analysis (Figure 4.3), wireline, and if applicable, lab-based stress field investigations

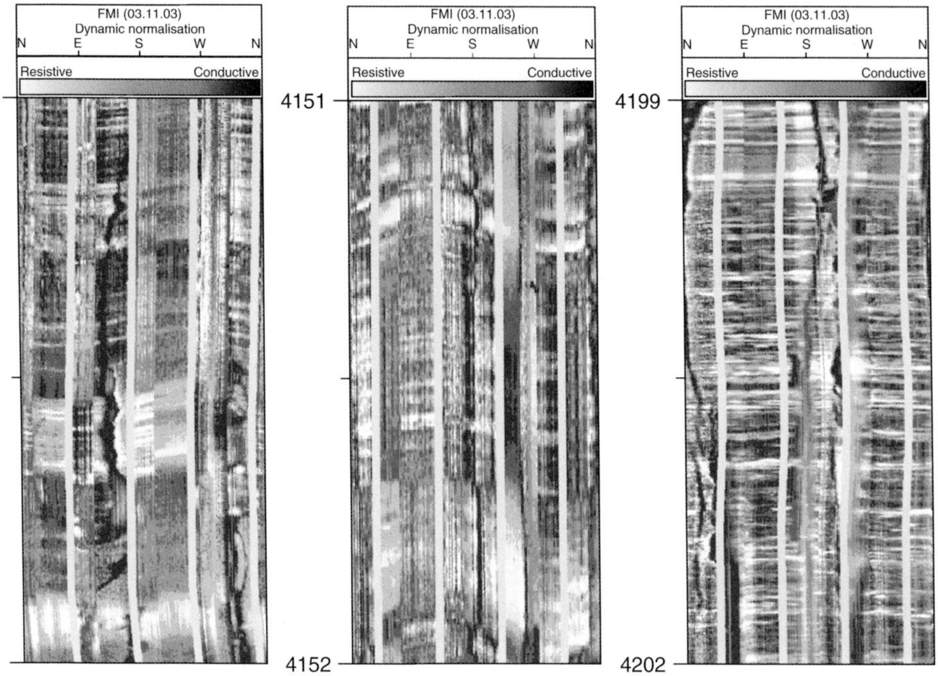

Figure 4.3 Formation MicroImaging measurement in the borehole Groß Schönebeck 3/90 exhibiting a "roll out" wall of the borehole. The dark color shows a vertical fissure in the rough direction south-north which was opened by a massive waterfrac treatment. The color scale runs from high electrical resistance (light) to low resistance (dark). (Please find a color version of this figure on the color plates.)

will help to clarify the structural controls on fold-related fracturing, which may also comprise performing a fracture network modeling. All relevant upfront investigations should be synoptically interpreted forming an integrated conceptual 3D subsurface model, which will ultimately build the base for the treatment design and layout.

4.5
Definition and Description of Methods (Theoretical)

4.5.1
Hydraulic Stimulation

4.5.1.1 General

Since the early 1980s, research at various sites confirmed that shearing rather than tensile fracturing is the dominant process (Pine and Batchelor, 1984; Baria *et al.*, 1999; Cornet, 1987). Natural joints, favorably aligned with the principal stress directions, fail in shear. As a consequence, formations with high stress anisotropy and hence, a high shear stress, should be best candidates for hydraulic fracturing in low permeable rock.

Knowledge about the stress regime is of great importance to understand or even to predict the hydraulic fracturing process (Cornet, Bérard, and Bourouis, 2007; Evans *et al.*, 2005b). Borehole breakouts, borehole fractures, microseismic events, and stimulation pressures have been evaluated to confine the orientation and amplitude of the principal stress components.

One method to reduce the risk of creating shortcuts is the isolation of intervals in the borehole and the successive stimulation of these intervals. Such a strategy is also favorable to reduce the risk of creating larger seismic events.

Cases of induced seismicity have been reported from hydraulic stimulation programs in geothermal wells, but not all geological formations are prone to these events. Induced seismic events, which could be felt at the surface, have been reported from hard rock environments. Since the permeability in these formations is a fracture- permeability, the pressures generated to frac the formation can only diffuse through the fracture and fault network, which will lead to a reduction in effective stress. In sedimentary environments, due to their matrix porosity and permeability, elevated pressures will not focus on fracture and fault pathways, but diffuse through the porous matrix. A potentially considerable sedimentary coverage of a hydraulically stimulated hard rock formation will also damp induced seismic events.

Controlling the reservoir growth while stimulating or circulating is an important issue for all projects in low permeable rock (Baria *et al.*, 2006). Microseismic monitoring gives 3-D time-resolved pictures of event location and magnitude from which the fractured rock volume can be inferred. This method has evolved to the key technique to map the reservoir in HDR projects (Niitsuma, 2004; Wallroth, Jupe, and Jones, 1996; Dorbath *et al.*, 2009). In current projects, (Soultz; Cooper Basin, Australia) the microseismic event distribution serves for the determination of the target area for new wells. More recently, microseismic monitoring has become important to detect and to control larger seismic events, which might occur during stimulation in geological active areas (Bommer *et al.*, 2006).

4.5.1.2 **Waterfrac Treatments**

Waterfrac treatments are applied in low permeable or impermeable rocks with high amounts of water to produce long fractures with low width compared to the following treatments. In general, waterfrac treatments produce long fractures in the range of a few 100 m with low apertures of approximately 1 mm and hence low conductivity. The success of the treatment depends on the self propping of the rock and on the potential of shear displacement (Figure 4.4).

The flow rate during waterfrac treatments can be constant during the whole treatment or vary in a cyclic manner with several high flow rates followed by low stages. Simulations have shown that the impact of high flow rates for the fracture performance is better, even if the intervals are limited in time, compared to a constant flow rate.

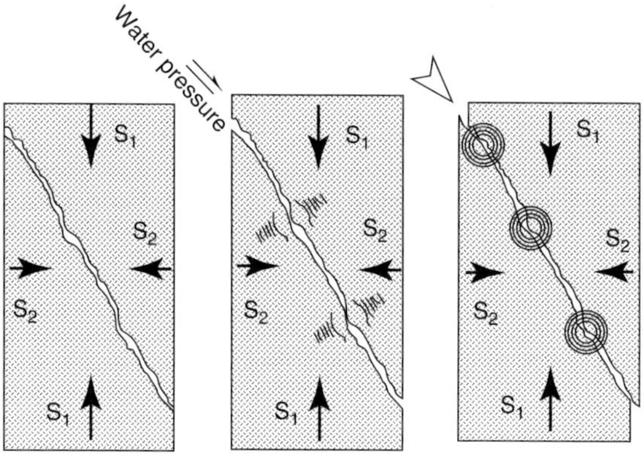

Figure 4.4 Potential self-propping mechanism after water frac treatment (from Jung, 1999).

Enhancing the treatment design comprises adding some abrasive agent such as sand or proppants in the fluid during the high flow rates. This will help etch conductivity into the fractures created, and using a proppant suspending agent which gives the proppant mechanical suspension while travelling through the frac, will increase the height that will be etched and allow the proppant to travel to the end of the fracture. It can be considered using a friction reducing agent in the fluid as opposed to using a guar-based gel in case pH values do not correspond between the fluids injected and the cross-linked gel.

4.5.1.3 Gel-Proppant Treatments

Gel-proppant treatments are used to stimulate reservoirs with cross-linked gels in conjunction with proppants of a certain mesh size (Figure 4.5). These treatments can be applied in a wide range of formations with varying permeabilities and a good control of stimulation parameters. The produced fractures have a short length of about 50–100 m, but a higher aperture of up to 10 mm compared to the waterfracs. It is especially used to bypass the wellbore skin in high permeable environments. In general, this kind of treatment is more expensive than a waterfrac treatment.

Typically, the gel-proppant treatments start with a Data FRAC to obtain information about friction and tortuosity of the perforated interval. In this DataFRAC, one would first pump an uncross-linked gel which would give an indication if any near-wellbore problems exist which could potentially adversely effect the placement of the frac treatment. This would then be followed by pumping a cross-linked fluid which would give an idea of leakoff as well as help to predict closure pressures, the frac geometry, and whether there is any indication of pressure dependent leakoff.

The MainFRAC treatment that followed after these pretesting measures is an injection of gel-proppants with a stepwise increase of proppant concentration

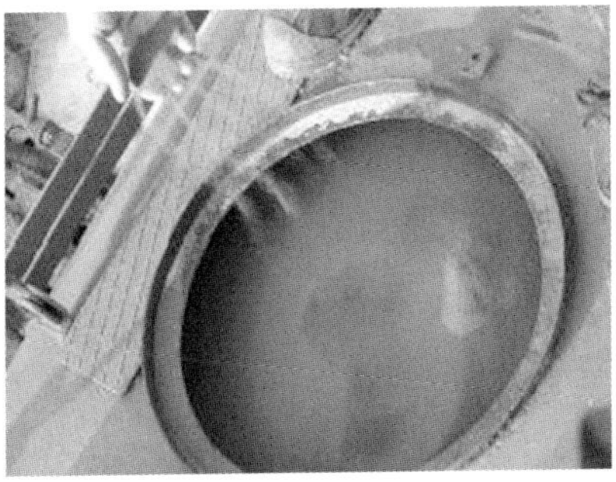

Figure 4.5 Proppants ready for gel-proppants treatments.

with a high viscous cross-linked gel into the fracture. The result of the treatment, that is, the propagation of the fracture, mainly depends on the slurry rate and the concentration of proppants added and their variation as a function of time.

An adjustment during the treatment is possible and often necessary to avoid a screen-out of the well. One can adjust the treatment by varying the flow rate and the proppant concentration in case the pressure progression suspects a failure of the treatment.

4.5.1.4 **Hybrid Frac Treatments**

In hybrid frac treatments, slickwater is pumped first to generate a fracture. Then, a gel-pad with cross-linked gel is injected, followed by proppants or sand of a certain mesh size with a cross-linked gel to fill the fracture. This method can be applied to low permeable reservoirs and provide sustainable production rates.

4.5.2 **Thermal Stimulation**

While thermal stimulation as a phenomenon is known to occur spontaneously at injection wells in hydrocarbon production (see e.g., Charlez *et al.*, 1996 or Gadde and Sharma, 2006) and although it has actively been used in high enthalpy fields in volcanic and metamorphic settings to increase the well productivity, the detailed process is not yet satisfyingly understood. The basic mechanism of thermally stimulating new fractures was delineated by, for example, Clifford, Berry, and Gu (1991). The injection of cold water into a hot well, leads to a cooling of the rock surrounding the wellbore, or adjacent to existing natural or induced fractures. The cooling rock matrix contracts and thus induces a tensile component of stress (thermo elastic stress) near the injection well or adjacent to the injection surface. The value of this thermally induced tensile stress depends on the shape of the cooled region, the thermal and elastic rock properties, the difference between the downhole and surface water temperatures, as well as the injection rate. It may exceed a value of $100\,\text{kPa}\,^{\circ}\text{C}^{-1}$. In case the water bottomhole pressure exceeds the minimum horizontal including the thermoelastic stress, a fracture can propagate from the wellbore.

Various numerical models have been developed to explain and predict thermally induced fracturing in sedimentary rocks. A numerical model to calculate the change in horizontal stress resulting from a change in temperature across a region of elliptical cross section and finite thickness has been developed by Perkins and Gonzales (1985) simulating the thermoelastic processes associated to the injection of cold fluids during waterflooding. Conditions, under which secondary fractures perpendicular to the primary main fracture may open, are also discussed. In fact, these considerations explain, how a line crack, representing a two wing vertical hydraulic fracture might gradually evolve into a fracture network of elliptical geometry, confocal with the line crack, by injection of cold fluid. Gadde and Sharma (2001) further refine the Perkins and Gonzales model also including fracture growth in injection wells due to particle plugging, thermal, and pore pressure effects.

4.5.3 Chemical Stimulation

Matrix acidizing treatments are designed to remove or bypass solid flow obstructions (damage material) from the wellbore or from the reservoir in the vicinity of the wellbore by injecting fluids of low pH into the wells. There are two main stimulation techniques: matrix stimulation and fracturing. Matrix stimulation is accomplished, in sandstones, by injecting a fluid (e.g., acid or solvent) to dissolve and/or disperse materials that impair well production. In carbonate formations, the goal of matrix stimulation is to create new, unimpaired flow channels from the formation to the wellbore. Matrix stimulation, typically called *matrix acidizing* when the stimulation fluid is an acid, generally is used to treat only the near-wellbore region. In a matrix acidizing treatment, the acid used is injected at a pressure low enough to prevent formation fracturing (Economides and Nolte, 1989). Very often carbonates show low matrix permeability and just creating wormholes in the near-wellbore area may not be sufficient to produce the reservoir economically. Fracture acidizing is the technique that is being used to achieve the task of providing a conductive path deeper into the formation (Burgos *et al.*, 2005).

Matrix acidizing is one of the oldest well-stimulation techniques used to remove damage near the wellbore. It was initially applied in carbonate reservoirs and over the years it has been extended to more complex mineralogies. So far, matrix acidizing is often considered by many people as risky to undertake, primarily due to the heterogeneous nature of formation minerals and an appreciable degree of unpredictability of their response to acid formulations. However, it is a relatively simple stimulation technique that has became one of the most cost-effective method to improve significantly the well productivity and/or injectivity.

In sandstone formations, matrix acidizing may enhance significantly the well performance by removing the near-wellbore damage, primarily associated with plugging of pores by siliceous particles as the consequence of drilling, completion,

Table 4.1 Original McLeod's sandstone acidizing use guidelines.

Formation	Main acid	Preflush
Whole rock solubility in HCl >20%	Use HCl only	–
High permeability (>100 mD)		
High quartz (>80%); low clay (<5%)	12% HCl–3% HF	15% HCl
High feldspar (>20%)	13.5% HCl–1.5% HF	15% HCl
High clay (>10%)	6.5% HCl–1% HF	Sequestered 5% HCl
High iron chlorite clay	3% HCl–0.5% HF	Sequestered 5% HCl
Low permeability (≤10 mD)		
Low clay (<5%)	6% HCl–1.5% HF	7.5% HCl or 10% Hac
High chlorite	3% HCl–0.5% HF	5% Hac

Hac: acetic acid.

stimulation, perforating, and production operations; therefore their natural permeability can be restored. During matrix acidizing treatments the acid reacts within a few meters form wellbore in sandstones.

The major issues that determine design of matrix acid treatments are the reservoir characterization which comprises the understanding of formation mineralogy, permeability, porosity, and reservoir fluid dynamics. Formation damage assessment in turn includes laboratory analysis, fluid compatibility, and core testing. As a general basis, the treatments normally follow this procedure:

- Preflush (HCl) stage
- Main acid (HCl–HF) stage
- Postflush stage.

In the late 1930s Dowell (Dow Well Service) introduced the now famous mixture of 12% HCl and 3% HF, called "*Regular Strength Mud Acid*," whose main objective was initially the removal of the drilling mud filter cake from the wellbore. Hydrochloric acid (HCl), hydrofluoric acid (HF) or both have been used since the 1980s in hydrothermal wells (Strawn, 2004) listed yet these two acids as the most effective ones. HCl was selected to treat limestone, dolomite, and calcareous zones whereas HF was used to dissolve clay minerals and silica. In 1984, McLeod presented the basic guidelines for proper designing of acid treatments based on formation mineralogy, an important issue which is often overlooked. These guidelines are presented in Table 4.1.

These guidelines have been modified since its introduction to fill certain gaps and must be considered as a starting point in treatment design. More recently, some authors have called for the necessity of using non HF-based systems because of the nature of damaging potential inherent in the reactions between sandstone minerals and HF (Crowe, Masmonteil, and Thomas, 1992; Malate *et al.*, 1998). They insist on the capability of these new systems to stimulate effectively, specially those formations with high content of HCl-soluble minerals.

HF is the only common acid that dissolves clay, feldspar, and quartz fines. For years mixtures of HF and HCl have been the standard acidizing treatment to dissolve the minerals that cause damage. These treatments are preceded by a preflush of either HCl varying between 7.5 and 15% or weaker acids such acetic acid to dissolve carbonates and avoid precipitation of calcium fluoride. The readily HCl-soluble minerals are mainly calcite, dolomite, and siderite which additionally do not generate precipitates. The reactions are as follows:

Calcite	$2HCl + CaCO_3 \longrightarrow CaCl_2 + H_2O + CO_2$
Dolomite	$4HCl + CaMg(CO_3)_2 \longrightarrow CaCl_2 + MgCl_2 + 2H_2O + 2CO_2$
Siderite	$2HCl + FeCO_3 \longrightarrow FeCl_2 + H_2O + CO_2$

Siliceous minerals are dissolved by hydrofluoric acid and its chemistry is much more complex than HCl when reacting with carbonates (Walsh, Lake, and Schechter, 1982; Pournik, 2004). Quartz, clay, and feldspars are the main siliceous particles involved in damage of sandstones. The primary chemical reactions in sandstone acidizing are as follows:

Quartz	$SiO_2 + 4HF \longrightarrow SiF_4$ (silicon tetraflourude) $+ 2H_2O$
	$SiF_4 + 2HF \longrightarrow H_2SiF_6$ (fluo silicicacid)
Clays (kaolinite)	$Al_4Si_4O_{10}(OH)_8 + 24HF + 4H^+ \longrightarrow 4AlF_2 + 4SiF_4 + 18H_2O$
(montmorillonite)	$Al_4Si_8O_{20}(OH)_4 + 40HF + 4H^+ \longrightarrow 4AlF_2 + 8SiF_4 + 24H_2O$
Feldspars (Mg, Na or K)	$KAlSi_3O_8 + 14HF + 2H^+ \longrightarrow K^+ + AlF_2 + 3SiF_4 + 8H_2O$

The main acid stage requires the greatest emphasis because of the damage mechanisms, directly associated to precipitation of products from the HF reactions (Pournik, 2004). Secondary reactions may occur between fluosilicic acid H_2SiF_6, a byproduct of primary reaction, and aluminum-silicates, clays, and feldspars. These reactions are considered to have adverse effects since silicon can be precipitated as hydrated silica, which contributes to damage if mobile inspite of the presence of HCl to reduce the pH to prevent silica and fluosilicate precipitation. Conversely, some authors believe these reactions are beneficial because they retard HF reactions allowing deeper penetration. Additionally, HF dissolves native clays and feldspars, and when reacting with quartz may also cause formation deconsolidation by weakening the matrix. Precipitation will always take place, associated with HF concentration among other things; however, it tends to be more severe if HF acid treatments are not properly displaced. Acid stimulation techniques have to account for both chemistry and treatment execution to accurately predict the effectiveness since the effect of these precipitates could be minimized if they are deposited far from the wellbore (Entingh *et al.*, 1999). Careful selection of mixtures, additives, acids formulations, and treatment volumes must be accounted to minimize these secondary adverse effects.

Reservoir geology and mineralogy are the relevant issues for successfully removing the acid-soluble particles present in reservoirs; removal mechanisms are strongly related to dissolution pattern of the matrix. Recent studies have shown that one of the most important factors that determine the etching pattern is the heterogeneity of the rock. In sandstones, the variations in permeability, porosity, and mineralogy may drive the acid to follow certain paths, the highly permeable channels called *wormholes*. There is experimental evidence that wormhole dissolution patterns can be achieved particularly using high HF concentrations and elevated temperatures; however, the risk of precipitation and rock deconsolidation might be significantly increased. Reaction rates are affected for kinetics; among the factors that strongly influence the mineral reactions are acid concentration and temperature. Dissolution reaction rates are proportional to the HF concentration for most sandstone minerals. The dissolution of minerals is a thermally activated phenomenon; thus, the rates increase greatly as a function of temperature, and the penetration depths of live acid diminish accordingly.

Sandstone matrix acidizing cannot be considered as an exact and predictable set of rules; thus, the appropriate design of treatments almost never has only one right answer. That is an inherent problem of the complex and heterogeneous nature of most sandstone matrices (Davies *et al.*, 1992). The interactions between different minerals and the injected acid depend on the chemistry, as well as temperature, pressure, pore-size distribution, surface morphology, and pore-fluid composition.

4.6 Application (Practical)

4.6.1 Hydraulic Stimulation

The equipment and materials as well as the modifications of existing equipment necessary to perform a hydraulic stimulation treatment are summarized in this chapter (Figure 4.6).

In most cases the well head (Figure 4.7) has to be modified to withstand increased wellhead pressures during the stimulation treatments. Furthermore, it is necessary

Figure 4.6 Main equipment necessary to perform a hydraulic stimulation treatment (left pump aggregats, middle tanks, left container for proppants) (from treatments in Groß Schönebeck 2007). (Please find a color version of this figure on the color plates.)

Figure 4.7 Wellhead Groß Schönebeck designed for 600 bar wellhead pressure.

to protect the tubings in the upper part of the well. This can be solved by installation of a frac string specially configured for the frac treatment. The installation of a blow out preventer is a highly recommended option if any gas indication or pressurized reservoir conditions are concerned.

A stimulation pump is a trailer-mounted fracturing service pumping unit designed for high-horsepower applications. The pump delivers fracturing fluids to the well at high pressure and rate. It is equipped with a remote panel allowing control of the fracturing job away. The panel allows gear selection, throttle control, and emergency shutdown to be performed.

A centrifugal pump will be required to feed water from the frac tanks and filtration system to the high pressure pumps.

Frac tanks are used to store water before stimulation treatment. If these tanks are internally coated or are manufactured from stainless steel, they tanks give the option to pre-filter the water and store it in advance of the treatment.

The blender is a fracturing service blender which is capable of blending and pumping fracturing slurry. It controls the solid/liquid ratio at design values throughout the entire treatment. A computer-controlled integral gate meters the amount of proppant added to the slurry.

Liquid additive systems are used for accurate metering of fracturing additives. Typically, each unit consists of stainless steel tanks, connected to positive displacement pumps and micromotion flow meters. This allows to accurately pump the required chemicals.

For storage and delivery of the proppant to the blenders, stimulation sand silos are used. Gravity silos are equipped with hydraulically controlled discharge outlets at the bottom of the tanks. These are connected to the sleeves which assure proppant delivery straight in the blender. The silo feeds sand to the blender during fracturing operations and serves as a sand storage facility at the wellsite.

The main concern in filtration is to prevent reservoir damage. Many contaminants can plug off the reservoir /production zones during completion and water injection. This filter system protects the reservoir from contaminants, including bacteria, scale, clay, rust, and so on.

A typical filtration system comprises a self cleaning filter system, which will pre-filter down to 50 μm. This is followed with cartridge filter units, which can filter down to 2 μm.

A cross-linked gel is a water-based system with a specially designed polymer loading composed of a refined guar gelling agent cross-linked by a borate solution. For geothermal purposes it should include a fluid's high-temperature stability and a time-delayed cross-linked reaction. Using a fluid with as low a polymer loading as possible is essential to ensure that both the effective fracture half-length and the polymer degradation are maximized.

A linear gel fluid may be used to carry proppant on slickwater treatments, as a breakdown fluid on conventional proppant frac treatments, or as a flush fluid.

In order to fine tune fluid, the following additives may also be included in this fluid, if required: temperature stabilizer, biocide, breaker to disintegrate the cross-linked gel, corrosion inhibitor to protect the casing, potassium chloride or acid to avoid scaling (iron etc.), and surfactant to reduce friction.

4.6.1.1 Induced Seismicity

Definition Generally speaking, induced seismic events can be described as "Earthquakes triggered by anthropogenic activity." Induced seismicity has been observed in association with the production of hydrocarbons from oil- and gas fields, at large water reservoir dams, during deep mining activities and in geothermal developments. In connection with geothermal projects, induced seismic events may occur either during the stimulation phase or in association with the production and injection of fluids during the operation of a geothermal field (Figure 4.8).

Source Mechanisms There are several potential source mechanisms which, according to Majer *et al.* (2007) may explain the occurrence of induced seismicity linked to the development or operation of EGS.

- **Reduction of effective stress:** increasing pore pressure can lead to a reduction of the effective stress on potential shearing planes like fractures and faults. In the presence of a deviatoric stress field this reduction of static friction may cause seismic slip.
- **Stress redistribution:** injection or production of fluid may lead to volumetric changes within the reservoir. These volume changes can result in large-scale stress redistributions, which may cause seismic slip at fractures and faults close to failure (critically stressed) within or close to the reservoir.
- **Thermoelastic strain:** A temperature drop due to the injection of cold fluid can cause a contraction of fracture surfaces which, like in the case of effective stress, reduces static friction, and triggers slip along a fracture that is already near failure in a regional stress field.
- **Chemical alteration of fracture surfaces:** Injection of fluids into the reservoir may cause geochemical alterations of fracture surfaces which might change the coefficient of friction on the fractures affected.

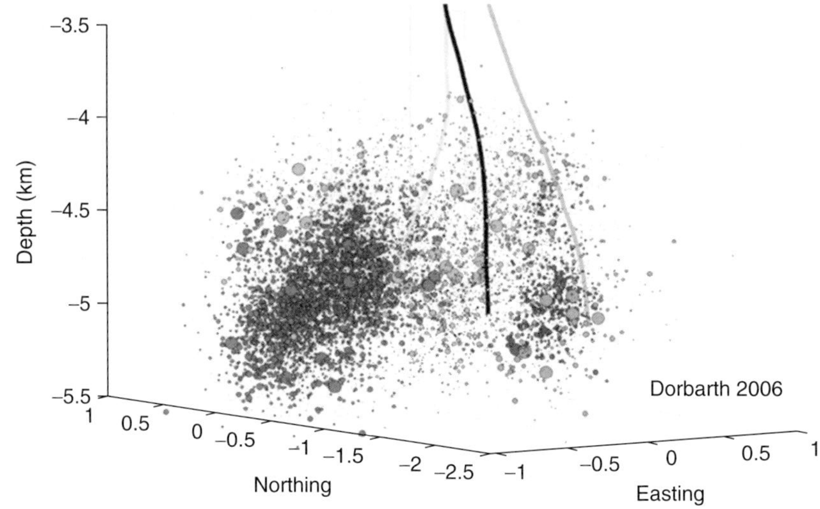

Figure 4.8 Relocation of induced seismicity event related to a stimulation treatment in Soultz sous Forets (Dorbath, 2006). (Please find a color version of this figure on the color plates.)

It has to be emphasized, that the aforementioned mechanisms give a qualitative description of potential source mechanisms of induced seismic events. But although a relationship between stimulation and induced seismicity appears to be obvious, no quantifiable correlations could be identified up to now, relating pressure, flow rate, volume of injected water, stress field and rock mechanical properties of the reservoir, and the number and magnitude of induced events (Häring, 2007).

Natural and Induced Seismicity After Brune and Thatcher (2002), the size of an earthquake generally depends on how much slip occurs along the fault, how much stress there is on the fault before slipping, how fast it fails, and over how large an area the failure occurs. This indicates, that the occurrence of induced seismic events to a great deal depends on local conditions, namely, and most importantly, the magnitude and orientation of the local stress field and the extent and orientation of local faults and fractures in relation to this local stress field. The actual ground motion, on the other hand, which might be induced by a seismic event of a certain magnitude, not only depends on the source distance, but also on the local soil conditions.

Historical Cases and Magnitudes Cases of induced seismicity have been reported from various EGS and conventional geothermal projects: Soultz-sous-Forêts, France (magnitudes $M < 2.9$), the Geysers, USA ($M < 4.6$), Cooper Basin, Australia, ($M < 3.0$), Berlin, El Salvador ($M < 4.4$), and most recently Basel, Switzerland ($M < 3.4$). It has been observed in various occasions that induced seismic events associated to hydraulic stimulation activities often occur also after shut-in, which means, after the actual injection of fluid has ceased. At the Berlin Geothermal field, a magnitude 4.4 seismic event has been observed about two weeks after the end of the injection, although there are discussions whether this particular event can undoubtedly be connected to the injection activity. Since the 4.6 earthquake at The Geysers field in Northern California, in the 1980's when its fluid production was at its peak, there have been only few magnitude 4 events. Almost all other induced seismicity at other geothermal fields has been in the range of magnitude 3 or less (e.g., Majer *et al.*, 2007; Rybach, 2006).

Although induced seismic events are inherent to EGS development, because hydraulic stimulation, by definition and design is a form of induced seismicity, there have been recent EGS development projects as well, at Groß Schönebeck (Kwiatek *et al.*, 2008), for example, where no microseismic activity strong enough to be perceptible at the surface was induced.

Structural Damage and Perceptibility Concerning any potential structural damage, it is important to notice, that within the frame of a probabilistic seismic hazard analysis for engineering purposes, it is common practice to specify a lower bound of magnitude 5.0, on the basis that smaller events are not likely to be of engineering significance (Bommer, Georgallides, and Tromans, 2001). Humans, nevertheless, can feel seismic events of lower magnitude already. According to the modified Mercalli intensity scale, most people indoors can feel the ground shaking caused by a seismic event of magnitude 3. A few people might even feel the movement caused by magnitude 2 events. These felt vibrations, though they may not be causing any structural damage, may give rise to disturbance or distress to the people living close to the source location. Complaints and protests may result, which can ultimately have the potential to jeopardize the entire project.

Traffic Light Approach Acknowledging the fact that a certain degree of induced seismicity is inevitably associated to hydraulic stimulation measures, and recognizing that hydraulic stimulations should not produce levels of ground shaking at the surface, which present a serious disturbance or threat to the local population, an induced seismicity warning system has been developed by Bommer *et al.* (2006). It is based on pre-defined thresholds of both intensity of induced ground movements and frequency of occurrence of any episodes of shaking, in the style of standards relating to acceptable levels of vibrations in buildings caused by construction-related activities. The parameter most indicative of any damage potential has turned out to be the peak ground Velocity (PGV). Using correlations between PGV and macroseismic intensity, and considering local attenuation relations, PGV thresholds are converted into equivalent magnitudes of seismic events at the depth in which the induced seismicity is most likely to occur. Incorporating also the frequency of occurrence, the microseismic events are finally classified into three classes (Figure 4.9) with associated operational directions: Green – Continue injection as planned; Amber – Continue injection with caution, be ready to act; Red – stop injection immediately.

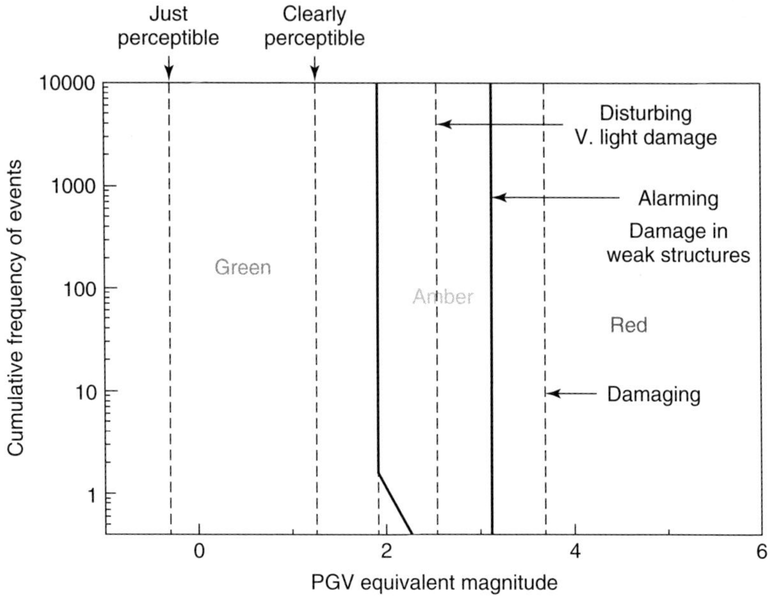

Figure 4.9 Pseudo Gutenberg Richter plot ("traffic light plot") with thresholds of PGV equivalent magnitudes (after Bommer *et al.*, 2006).

Although the "traffic light approach" has been successfully applied during three hydraulic injection experiments in the Berlin geothermal field, and in El Salvador, in 2003 and 2004, it has to be kept in mind, that it is a reactive measure. Induced seismic events have been reported by various authors (e.g., Bommer *et al.*, 2006; Hoover and Dietrich, 1969; Hsieh and Bredehoft, 1981; Baisch *et al.*, 2006; Charléty *et al.*, 2007; Cuenot, Dorbath, and Dorbath, 2008) well after shut-in. Also at Basel, the injection was halted after the first felt event, but this did not prevent subsequent perceptible events from happening (IEA-GIA, 2009).

Addressing Induced Seismicity In order for EGS to be successfully utilized, the issue of induced seismic events should be proactively addressed on various levels, as illustrated by IEA-GIA(2009) and Bromley and Mongillo (2008).

- **Estimate local potential for natural seismic hazard and induced seismicity:** When considering sites for EGS development, it is judicious to consult geological and seismological information in order to determine suitability in relation to background natural seismicity, the state of stress, and the existence of superficial deposits, with the potential of amplified ground shaking. Many of the necessary information might be publicly accessible via governmental institutions like geological surveys, for example
- **Technological innovation:** Pioneering concepts for stimulation, and field management strategies for operation of EGS ought to be used and developed, aiming

at mitigating induced vibrations to an acceptable level. Although there are that certain stimulation regimes (e.g., injecting at low pressures over longer periods, or avoiding large pressure gradients by slowly increasing and decreasing the wellhead pressure at the beginning and the end of injections (Bromley and Mongillo, 2008)) might be beneficial, these topics are subjects of ongoing research, and will still be so in the near future.

- **Information and education:** The benefits of an open information and education policy, targeting the local population, regional authorities, and the stakeholders cannot be overestimated. Such a proactive information policy should include regular information (meetings, newsletters, articles in local newspapers, etc.) about the general development objectives, recent project progress, and benefits for the community. This will help to reduce public concern and avoid complaints and unreasonable claims.
- **Monitoring concept:** It is generally desirable to have a kind of seismic monitoring system available in order to observe and keep a record on any microseismic events potentially induced during stimulation operations. This may either be done using an existing regional or national seismic grid, in case there are seismic stations of the grid in the vicinity of the EGS development site, or by installing a dedicated seismic monitoring system which can be tailored to the site-specific conditions and particular stimulation parameters.
- **Implement emergency action plan:** It is advisable, particularly if the upfront assessment has shown the potential for induced seismic events to be felt by the local population, to design and implement a clear action plan to monitor and assess induced seismicity and vibrations. The aforementioned "traffic-light approach" may serve as an example here, which can be modified to match local conditions.

4.6.2 Thermal Stimulation

Up to now, thermal stimulation was actively applied almost exclusively in hightemperature systems associated to volcanic activity (e.g., Iceland, Axelsson *et al.*, 2007, or Bouillante, Tulinius, Correia, and Sigurdsson, 2000) or in high temperature metamorphic environments like in Travale or Larderello (Italy). The stimulation processes normally starts with water circulation through the drillstring followed by pumping cold water into the well. Occasionally water is injected at increased wellhead pressures. The injection can be interrupted by intervals of non-activity, during which the well is allowed to heat up toward its natural temperature state. In this way, a combination of thermally induced cracking forces and pressure impulses can increase the permeability of existing fractures and possibly create new ones. Stimulations at low temperature fields (below 150 °C reservoir temperature) primarily involve pressure changes induced either directly at wellhead or downhole, where inflatable packers (seals) are placed to more effectively address deeper well sections. Air-lift pumping is commonly used in low-temperature stimulation operations. Thermal stimulation operations generally are applied for a period of a few hours to a few days. Only rarely are they performed over several months. They are generally carried out immediately after drilling operations are completed.

4.6.3 Chemical Stimulation

Matrix acidizing is suitable to both generate extra production capacity and to restore original productivity in damaged wells. Matrix acidizing of sandstones starts with the careful evaluation of the well and the accurate determination of the nature and severity of the problem. Then, a possible treatment fluid is selected. The first selection criterion is the nature and location of the damage. Then, the potential compatibility problems between the rock minerals and the fluids are examined. The composition of the fluid is further defined by performing flow tests and checking the absence of damaging reactions. Once the treating fluids and the sequence of fluids have been defined, treating parameters, such as volumes, rates, and pressure are estimated or calculated and simulated. If the extension and severity of the damage are known, economic evaluation (production prediction vs treatment cost) can be performed and the treatment results can be optimized. Various diverting techniques, including mechanical techniques (such as packers) and various chemical diverting agents allow better fluid placement. To enhance the production (or injection) capacity, most of the damage must be removed, and thus the treating fluid must be injected in the least permeable and most damaged zones. Finally, a comprehensive monitoring of the job effectiveness and a posttreatment evaluation are necessary.

Various types of chemical stimulation methods have been considered. High pH fluids seem to be a logical choice for some wellbore and/or reservoir stimulations. The solid silica, one of the major sources for injector plugging is highly soluble in many high pH fluids. But unfortunately, the native reservoir fluids as well as the injected brine are often highly sensitive to a high pH value. The precipitation of hydroxide and basic carbonate scales is a consequence of the chemical reactions between high pH stimulation fluid and reservoir or injection brine. These scales, particularly the hydroxides, are extremely voluminous in the pores even if their amounts are only very small.

Fluids having a neutral pH can be successfully used in chemical stimulation methods only in a very few and rare instances. There are some neutral pH fluids that could be excellent solvents for certain types of damaging materials. For example, ethylenediaminetetraacetic acid (EDTA) and nitrilotriacetic acid (NTA) salts are excellent chelating agents (Fredd and Fogler, 1998). Thus, scales could be removed by solutions of these materials in neutral or near neutral pH water without causing secondary precipitates if properly applied (Rose *et al.*, 2007). The major problem is cost: these materials are rather expensive and large amounts would have to be used for most stimulation jobs. Low pH fluids, that is, acids, have by far the best chance to be used for these chemical stimulation jobs. The standard acid treatments are HCl mixtures to dissolve carbonate minerals and HCl–HF formulations to attack those plugging minerals, mainly silicates (clays and feldspars).

The reactivity of the formation-dissolving fluid may be selected (for example, with the use of fracture and/or acidizing simulator computer programs) on the basis of the flow rate and formation and fluid parameters. The reactivity of the stimulation fluid can be controlled by varying the rate of reaction, the rate of mass transfer, or both. For example, the rate of reaction can be decreased by changing the type of stimulation fluid, by changing the form of the fluid from a solution to an emulsion, by adding appropriate salts (which change the equilibrium constant for the surface reaction), or by increasing the pH of the stimulation fluid. The rate of reaction can also be decreased by changing the physical processing conditions (e.g., by reducing the pump flow rate and/or pumping pressure, or by cooling the stimulation fluid). A matrix acidizing apparatus for conducting linear core flooding is capable of reproducing different conditions regarding flow rate, pressure, and temperature. The results obtained from the experiments carried out on core samples showed that the temperature activates the reaction rate of HF–HCl acid mixtures in sandstone acidizing. The use of higher concentrations of HF, particularly at high temperatures, may cause deconsolidation of the matrix adversely affecting the final stimulation results. It was also seen that the higher the flow rate the better the permeability response, until certain optimal flow rates are reached.

In general, in creating propped fractures having wormholes in the fracture faces far from the wellbore, simple mineral acids such as HCl, HF, or mixtures of HCl and HF, would be too reactive, and would spend too close to the wellbore. It would normally be necessary to use a less reactive formation-dissolving fluid (Crowe, Masmonteil, and Thomas, 1992). Acids are not the only reactive fluids that will dissolve formation minerals. Nonlimiting examples would be organic acids, retarded mineral acids (such as gelled or emulsified HCl), or chelating agents. The reactivities of organic acids such as acetic or formic acids, could be further adjusted by including varying amounts of sodium acetate or sodium formate respectively. The reactivities of chelating agents, such as EDTA or hydroxyethylethylenediamine-triacetic acid (HEDTA), could be further adjusted by converting them partially or completely into sodium, potassium, or calcium salts or by adjusting the pH with, for example, HCl. These chelant-based materials have low reactivity, low viscosity, but high dissolving capacity (Mella and Rose, 2006a,b).

At present, matrix acidizing treatments exhibit at least four serious limitations: inadequate radial penetration; incomplete axial distribution; corrosion of the pumping and wellbore tubing, and iron precipitation.

The first problem with acid treatment – inadequate radial penetration – is caused by the reaction between the acid introduced into the formation and the material in the wellbore and/or formation matrix, with which it first contacts. The material and/or formation first contacted by the acid is usually at or near the wellbore such that the formation near the wellbore is adequately treated and portions of the formation more far to the wellbore (as one moves radially outward from the wellbore) remain untouched by the acid, since all of the acid reacts before it can get there. In fact, dissolution of the material and/or formation encountered by the acid may be so effective that the injected acid is essentially spent by the time it reaches a few centimeters beyond the wellbore.

The second problem that limits the effectiveness of matrix acidizing technology is incomplete axial distribution. This problem relates to the proper placement of the acid- containing fluid, that is,, ensuring that the fluid is delivered to the desired zone or zones (i.e., the zone that needs stimulation) rather than another zone or zones. This behavior is exacerbated by intrinsic permeability heterogeneity (common in many formations), especially the presence of natural fractures and high permeability streaks in the formation. Again, these regions of heterogeneity attract large amounts of the injected acid, hence keeping the acid from reaching other parts of the formation along the wellbore where it is actually desired most. In response to this problem, numerous techniques have evolved to achieve more controlled placement of the fluid, diverting the acid away from naturally high permeability zones, and zones already treated, to the regions of interest.

Techniques to control acid placement (i.e., to ensure effective zonal coverage) can be roughly divided into either mechanical or chemical techniques. Mechanical techniques include packers and coiled tubing (flexible tubing through which the acid can be delivered with more precise location within the wellbore). Chemical techniques include foaming agents, emulsifying agents, and gelling agents to modify the acid-containing fluid itself. Coiled tubing plays a major role in matrix stimulation and is largely viewed as a tool to aid placement and diversion of acids (Pasikki and Gilmore, 2006). However, the challenge of zonal coverage becomes increasingly difficult with larger intervals and/or when there are large permeability contrasts within the formation to be stimulated. Conventional acid placement techniques are less effective for the long, open-hole, or liner-completed intervals typically encountered in geothermal wells. Hightemperature foam systems may improve zone coverage. Gelling agents for thickening acid have been shown to be ineffective in geothermal liner completions. The best way to maximize acid coverage in geothermal wells is by pumping at maximum injection rates.

The third problem with acid treatments is their susceptibility to the temperature of geothermal reservoir. The effects of high formation temperatures, for instance, vary widely according to the details of the particular fluid treatment. In some acid treatments, the high temperature has a tendency to accelerate corrosion of metal in the wellbore. High temperature reduces the efficiency of corrosion inhibitors and increases their cost. Protecting the tubulars against corrosion requires careful selection of acid fluids and inhibitors (Buijse *et al.*, 2000), while cooling the well by injecting a large volume of water preflush may reduce the severity of the problem.

Another limitation of known acid treatments is iron precipitation. The dissolved iron tends to precipitate, in the form of ferric hydroxide or ferric sulfide, as the acid in the treatment fluid becomes spent and the pH of the fluid increases. Precipitation of iron is highly undesirable because of damage to the permeability of the formation. Therefore, acid treatment fluids often contain additives to minimize iron precipitation, for example, by sequestering the iron ions in solution using chelating agents such as EDTA.

4.7 Verification of Treatment Success

4.7.1 General

The evaluations techniques that are commonly used to assess the success of a particular stimulation measure can be grouped into two main classes: those techniques whose significance is restricted to the borehole wall or its immediate vicinity, and those that are capable of gauging the extension of induced fractures into the reservoir and obtain estimates on reservoir volume and fracture properties. The former are the classical wire line measurements, the latter primarily the evaluation of standard well test, flow-back test, or inter-well tests, in case tracers were applied during the injection.

4.7.1.1 Wireline Based Evaluation

The most basic and cost-effective way of assessing stimulation success will be achieved, besides an evaluation of caliper logs, via a combined temperature and spinner survey. The temperature can be used to quantify the height of the induced fracture since the formation is cooled down, where it takes injected fluid during the course of the stimulation. Spinner surveys allow to quantitatively assess the actual flow into the well, coming from fractured intervals. In case radioactive tracers were admixed to the stimulation fluid or, if applicable, to the proppants, gamma ray logging can be used to identify those zones, which predominantly took fluid or proppants during the course of the stimulation. Wire line–based electric or acoustic imaging surveys have turned out, analog to a prestimulation fracture evaluation, to be a very meaningful way of evaluating the fracture inventory created during the stimulation.

4.7.1.2 Hydraulic Well Tests

Many different types of well tests can be performed, and the choice depends wholly on the information that is being sought. Being clear about the objectives of the test is paramount when deciding the type of test to carry out. The general objectives are as follows:

- Evaluate existing fracture systems
- Assess fractures induced by hydraulic stimulation
- Investigate matrix transport properties
- Determine reservoir boundaries
- Assess reservoir compartmentalization.

A Pressure drawdown survey, in which the flowing bottomhole pressure is measured while the well is flowing, is a primary method of measuring productivity index (PI). Establishing a stable rate over a long period can be difficult, creating some uncertainty in the analysis.

Pressure build-up surveys measure the bottomhole pressure response during the shut-in period which follows a pressure drawdown. This is useful for measuring reservoir properties and near well effects such as skin. In this test, the flow rate is known (zero).

Interference tests between two wells are used to estimate the transmissibility (k h) of the formation in the interval between the wells. A pressure change is created at the active well by shutting in or opening up the well, and a pressure gauge in the closed-in observation well awaits a pressure response, the arrival time of which can be used to estimate transmissibility.

A pulse test is a version of the interference test, but attempts to provide enough information to allow the interpreter to eliminate the effects of noise and gauge drift in pressures (to which the interference test is prone) as measured at the observation well.

A Stepped flow and stimulation test essentially is a small hydraulic stimulation, aimed at reducing near-wellbore inlet and outlet hydraulic impedances. It consists of injecting into a well at a constant flow rate, and recording the pressure rise within time.

In a multi-rate-pre-fracturing-hydraulic-test, the injection flow rate is increased in steps. In each step the flow is continued at a constant rate until the injection pressure attains an asymptotic value. The test delivers valuable information on transmissibility, the significance of turbulence, and details of fracture dilation (Murphy *et al.*, 1999).

A multi-rate-post-fracture-performance-test is performed to determine the hydraulic properties of the stimulated fracture system. It is performed like the corresponding prefracture tests except that the flow rates are usually much higher, and the test duration is longer.

Long-term injection and production tests in a single well are especially useful for determining the outer hydraulic boundary conditions of the stimulate fracture system. They can help to predict the long-term fluid loss from the reservoir during operation.

Generally, pressure differences during the test phase should be kept sufficiently small, in order to remain in the hydraulic mode instead of frac mode.

All of the aforementioned test procedures were developed in order to test wells which produce a single well fluid. In the more general case of multiphase flow, multiphase effects have to be taken into account in test interpretation. After Horne (1995), it is almost always better to design a well test ahead of time, to avoid multiphase conditions during the test.

4.7.1.3 **Tracer Testing**

Tracer testing is a technique which is frequently applied in geothermal development projects in order to detect and characterize hydraulic connections between deep geothermal wells, to understand the migration of injected and natural fluids, and to estimate their proportions in discharged fluids, their velocities, flow rates,and residence times. The method also enables to quantify fluid-rock contact surfaces, swept volumes, circulation paths, and reservoir volumes. It can provide very useful information on reservoir processes. Depending upon the tracer test methodology, that is, whether continuous tracer injection or slug injection is used and whether the test is backflow or inter-well, useful information on transport properties and hydraulic connections after hydraulic stimulation essential for heat exchange or for fluid re-injection in geothermal reservoirs can be obtained from the data collected during such tests after their interpretation and modeling (Sanjuan *et al.*, 2006; Ghergut *et al.*, 2007).

As the physicochemical behavior of the tracers under given reservoir conditions (high-salinity fluid, very low redox potential, low pH, etc.) is not always well known, the use of a minimum of two tracers (or comparison with a natural tracer or laboratory experiments) is recommended.

Following the required information, several types of modeling approaches can be used to analyze the tracer Return-Curve data from a slug injection: signal processing codes such TEMPO based on a model of dispersive transfer (Sanjuan *et al.*, 2006) or using the moment analysis method (Shook, 2005), hydraulic or hydrodynamic codes such as SHEMAT (Blumenthal *et al.*, 2007) or TOUGH2 (Pruess, O'Sullivan, and Kennedy, 2000), coupled hydro-mechanical codes, and so on.

Among the tracers recommended in the literature for use at high temperature conditions, we can distinguish the following compounds:

- Liquid phase tracers:
  - naphthalene (di, tri)sulfonates (nds, nts, ns) family: 1,5-, 1,6-, 2,6-, 2,7-nds, 1,3,5- and 1,3,6-nts, 1- and 2-ns (Rose, Benoit, and Kilbourn, 2003);
  - aromatic compounds: sodium benzoate (Adams *et al.*, 1992);
  - fluorescein ($T < 260\,^{\circ}C$; the other organic dyes are not recommended).

The gas tracer SF_6 is little used. The use of inorganic and radioactive tracers is limited because of the high natural background of the halides (Cl, Br, I . . .) and difficulty in obtaining permits for radioactive tracers.

- Vapor or two-phase tracers
  - Alcohols (isopropanol, butan-2-ol, etc.) and hydrofluorocarbons (volatile low molecular weight compounds R-134a (CF3CH2F) and R-23 (CHF3), Adams *et al.* (2001) as geothermal vapor-phase tracers;
  - homologous series of short-chain aliphatic alcohols as geothermal two-phase tracers: ethanol, *n*-propanol (Adams *et al.*, 2004; Mella *et al.*, 2006a, b).

The analytical methods usually used for each of these recommended compounds are reported below:

- naphthalene sulfonates and fluorescein: HPLC with Fluorescence Detection or spectrofluorimeter (for fluorescein only);
- sodium benzoate: extraction procedure coupled with HPLC with UV detection;
- SF_6: Fourier transformed infra red (FTIR)/gas chromatography with electron-capture detection (ECD);
- hydrofluorocarbons: enrichment procedure coupled with gas chromatography;
- ethanol, *n*-propanol: Gas chromatography with a flame ionization detection (FID). A new method of analysis for alcohol tracers using solid phase microextraction

(SPME) has reduced the limit of detection by a factor of about 30 (Mella *et al.*, 2006a).

4.7.1.4 **Monitoring Techniques**

In order to assess the success of the thermal stimulation measures, and to obtain information of the relevant reservoir properties, it is recommended to perform a monitoring program together with the well treatment. The most obvious and basic monitoring parameters will be the injection flow rate, the wellhead pressure during stimulation and the temperature of the injected fluid. Very useful information is provided by monitoring the injection pressure downhole, while profiles of temperature and pressure can serve in identifying relevant feed in zones.

Although not a standard measure up to now, it might be useful to perform a seismic monitoring program as well. Microseismic imaging of induced fractures is a technique originating from earthquake seismology. It significantly evolved over time, and has often been successfully applied in recent EGS development projects (e.g., Soma *et al.*, 2004; Charléty *et al.*, 2007, or Kwiatek *et al.*, 2008) Axelsson, Thórhallson, and Björnsson (2006) report about cases, in which the analysis of seismic monitoring data showed that the thermal resource might have a much wider extent, and therefore a much higher generation potential than previously assumed. In order to record compressional and shear waves emitted during fracture stimulation, three-component geophones (3C) are placed in a monitoring well, or, if possible, at the surface location, to determine the location of the seismic event.

Tiltmeter surveys of the borehole wall and/or the surface, measure the tilt of the earth from deformation caused by a displacement of a hydraulic fracture, with very high resolution (up to 1 nano radian). Depending on the registration geometry, they can be used to resolve for geometric parameters of the induced fracture like: azimuth, dip, height, and length.

Microseismic monitoring and tiltmeter surveys work most reliably when sensors are deployed at the target depth in an offset well. In case no offset monitoring well exists, the methodologies can be applied form surface locations, but in these instances, a successful application is limited through local parameters like treatment depth, volume of the fracture, overburden material, and strength of the source signal. Although techniques are available for application in the treatment well as well, these surveys are subject to technical restrictions in terms of applicability and of force of expression. Barree, Fisher, and Woodproof (2002) or Cipolla and Wright (2000) have given comprehensive and general outlines of fracture diagnostic technologies, frequently applied in the hydrocarbon industry.

A chemical monitoring program, which can be performed alongside conventional well tests, will provide useful information on any changes in the chemistry of the produced fluid or steam, and help to asses any changes in the hydraulic state of the reservoir (see e.g., Sanjuan, 2000), which may be caused by the stimulation program. Still the best evaluation of the treatment effect is achieved by a combination of pressure, flow, and temperature monitoring at the wellhead with temperature, pressures, and wellbore imaging downhole. This way, the overall effect of the thermal stimulation is monitored along with observations of local fracturing leading to productivity improvement.

4.7.2 Evaluation of Chemical Stimulations

As with any stimulation operation, it is important to evaluate the effectiveness of a matrix stimulation treatment. The effectiveness is gauged by apparent increases in the PI, without the benefits of posttreatment test. Wellhead pressures and injection rates are recorded during every matrix stimulation treatment. Ideally, these variables should be measured at bottomhole. However, the hostile nature of the stimulation fluids sometimes prevents the use of downhole pressure gauges and flowmeters. A calculation allows the derivation of bottomhole parameters from those measured at the surface in terms of progress of the remedial treatment. Each stage of injection or shut-in during the treatment is considered as a short individual well test. The transient reservoir pressure response to the injection fluids is analyzed and interpreted to determine the condition of the wellbore (skin) and the formation transmissibility (Economides and Nolte, 1989).

The common assumption is that dislodged and migrating fines should be forced into the formation after an acid stimulation job. However, fine particles may flow a considerable distance from the wellbore where they can form permanent damage (Nguyen, 2006). It may become impossible to remove these damages at a later time. It may be more advisable to backflow the well after the stimulation job and before routine injection operations. In addition, partial removal of damage with acid treatment may eventually result in complete damage removal when the treated well produces back. The high-rate and high-energy backflow from geothermal wells can blow out damage that was not dissolved by acid. Damage that was softened, broken up, or detached from downhole tubulars and fracture channels can be produced back through a large-diameter casing completion. Erosion of production lines may occur if drill cuttings are produced back during blow down of a well after stimulation and care must be taken in this regard. A temporary flow line may be required until solids production has stopped.

The risk seems to be too large to base the job evaluation solely on the increase of injectivity. It seems to be worthwhile to go to an additional job evaluation method which can be used if the stimulated well is backflowed after the stimulation job. The backflowed (i.e., produced) liquids contain the dissolved products of all chemical reactions taking place during and after the acid job. Using the chemical analyses, flow rates and produced volumes of the back-produced fluids will allow the operator to perform some rather accurate and reliable material balances. These material balances could be used as a valuable aid for the job evaluation. This method has been used successfully for the evaluation of the acid stimulation of geothermal production wells.

4.8
Outcome

4.8.1
Hydraulic Stimulation

4.8.1.1 Hydraulic Stimulation – Soultz

The three 5000 m deep geothermal wells in the Soultz granite were stimulated using the water frac technique where the 500 m long open hole sections were treated on the whole. Each hydraulic stimulation started with the injection of up to 800 m^3 of heavy brine with a density of up to 1.2 g cm^{-3} in order to reactivate natural fractures having low initial permeability, as deep as possible in the well. The main stimulation consisted of a fresh water injection as summarized in Table 4.2.

GPK2 and GPK4 were stimulated by a single-well-injection whereas in GPK3, the concept of a dual injection was tested. In order to achieve a pressure concentration and therefore fracture concentration between the two wells to be connected, the fresh water was injected simultaneously into GPK2 and GPK3.

The pressure curves of the GPK2 and GPK4 (2004) stimulation indicate an effective stimulation which could improve their productivities by a factor of 20 (Table 4.1). The wellhead pressure of GPK3 stimulation is more flow rate dependent and indicates a less effective stimulation in which the productivity could be improved by a factor of 1.5.

An important relationship was found between the injection rate during stimulation and the productivity of the well after stimulation (Jung and Weidler, 2000). The productivity of the wells appears to increase linearly with the injection rate during stimulation. Moreover, the series of stimulations at 5 km depth has shown that the productivity as determined during the stimulation persists completely after stimulation. Thus, a "perfect" self-propping was observed.

In a deep geothermal system with low initial permeability, the reservoir will be developed by stimulation. To control the reservoir development the operation of a seismic network is essential. The located events serve to monitor the reservoir

Table 4.2 Summary of the four stimulation operation on the Soultz wells.

Well	Duration (days)	Volume (m^3)	Flow rate ($l\ s^{-1}$)	Initial productivity ($l\ s^{-1}\ bar^{-1}$)	Improvement factor	Located microseismic events
GPK2	6	23 400	50	0.02	20	14 000
GPK3	11	34 000	50	0.2	1.5	21 600
GPK4	3.5	9 300	30	0.01	20	5 700
	4	12 300	45	–	–	3 000

development and to target the deviated wells into the rim of the stimulated zone. A downhole seismic network of six stations registers the microseismic events occurring during stimulation in Soultz, while a surface seismic network is also available and well adapted for the microseismic event magnitude determination.

The strength of those microseismic events was up to ML = 2.9 for the stimulation of GPK3. In order to prevent the development of strong microseismicity, the stimulation strategy was adopted to lower flow rates and shorter duration (compare Table 4.1).

4.8.1.2 **Hydraulic Stimulation Groß Schönebeck**

In the Groß Schönebeck wells, three stimulation treatments were performed in the wells EGrSk3/90 and GtGrSk4/05, respectively. Table 4.3 summarizes the productivity improvement of each stimulation. In the well EGrSk3/90, the first stimulation was carried out in the sandstone section, the other two stimulations included the underlying volvanic rocks as well.

The three hydraulic stimulations in the well GtGrSk4/05 were carried out separately in two sandstone sections and the volcanic rocks. To determine the improvement factor a production test was carried out for all sections together. Individual contribution of each section was determined by a profiling flowmeter.

Microseismic monitoring was performed during the last treatment in EGrSk3/90 from subsurface, but no events could be registered. During the first two treatments in GtGrSk4/05 microseismicity was monitored in the well EGrSk3/90 at 3800 m depth. Observation of the microseismic events reveals a very low seismicity during and after stimulation (80 events).

Table 4.3 Summary of the stimulation treatments in the wells EGrk3/90 and GtGrSk4/05 in Groß Schönebeck.

Well	**Duration (days)**	**Volume (m^3)**	**Flow rate ($l\ s^{-1}$)**	**Productivity ($l\ s^{-1}\ bar^{-1}$)**	**Improvement factor of productivity**	**Located microseismic events**
EGrSk3/90	0.51	167	3.75	0.027	Initial PI	Not measured
EGrSk3/90	0.58	307	6.22	0.059	2.2	Not measured
EGrSk3/90	0.24	338	16.4	0.112	1.9	Not measured
EGrSk3/90	1.0	859	4.3–14.7	0.207	1.9	No events
GtGrSk4/05 volcanics	0.06	4.4	0.83	0.004	Initial PI	–
GtGrSk4/05 volcanics	0.5	356	8.2	0.0849	22	78
GtGrSk4/05 sandstones1	0.3	250	9.5	0.028	Initial PI	–
GtGrSk4/05 sandstones1	0.5	356	8.2	0.142	5	2
GtGrSk4/05 sandstones2	0.2	170	10.3	0.034	Initial PI	–
GtGrSk4/05 sandstones2	0.5	356	8.2	0.057	1.67	Not measured

4.8.2 Thermal Stimulation

An evaluation of treatment success may be performed by means of simply comparing the ultimate productivity of a well to the productivity prior to the stimulation treatment. It is also possible to relate the final well productivity to the cumulative circulation losses during drilling. The results between both methods may vary significantly, since they will contain different effects. While in the first case, the observed improvements will primarily be attributed to reopening of feed zones, which were clogged during drilling operation, in the second case the observed improvement will most likely be attributed to the actual increase in feed zone permeability, for example, by raising the permeability of near well fractures. Unlike in the case of low-temperature wells, no clear correspondence between the injectivity at the end of the thermal stimulation operation and the productivity exists, for high temperature wells (Table 4.4).

4.8.3 Chemical Stimulation

Over the past years, great improvements in matrix acidizing have taken place for geothermal wells, paralleling the developments in hydraulic fracturing. Provided that the forecasted production/injection results make economic sense, matrix acidizing is still simpler, often less risky, and more economic to implement than

Table 4.4 Characteristics of production wells in the Reykjanes high temperature field, in SW-Iceland.

Well	Depth (m)	Temperature (°C)	Injectivity index $_1$ (ls^{-1} bar^{-1})	Injectivity index $_2$ ($l\ s^{-1}$ bar^{-1})	Productivity index ($l\ s^{-1}$ bar^{-1})
RN-10	2 050	310	–	6.6	2.3
RN-11	2 250	295	–	>10.0	10.0
RN-12	2 510	290	–	8.0–9.0	20.0–40.0
RN-13	2 460	290	–	4.0–5.0	1.0–2.0
RN-14	2 310	290	6.0	6.0–7.0	–
RN-15	2 510	280	3.5	4.0	1.0
RN-16	2 630	220	1.2	2.0	–
RN-18	1 820	>285	5.0	5.4	1.5
RN-19	2 250	250–260	5.0	5.0	–
RN-21	1 710	275	6.0	13.0	6.0
RN-22	1 680	305	10.0	10.0	15.0
RN-23	1 920	305	–	38.0–48.0	50.0
RN-24	2 110	>275	–	10.0–20.0	38.0

After Axelsson, Thórhallsson, and Björnsson (2006).
Injectivity index $_1$: end of drilling, Injectivity index $_2$: end of stimulation operations.

hydraulic fracturing. Preplanning and proper job design is essential for a successful matrix stimulation treatment. Different techniques are being used over a variety of reservoir and well conditions. Sophisticated laboratory equipment, expertise, and well testing software can help to identify production or injection damage effects and mechanisms – making it easier to select proper well candidates and optimize job design. Except as may be helpful in adjusting the pH of the fluid system, there are no restrictions on the order of addition of the components in the fluid mixture.

Alternatively, any combination of the components can be premixed on site or at a separate location (to enhance safety on location) and then another component or components may be added later. Standard mixing equipment and methods may be used; heating and special agitation are normally not needed but may be used. The only acid additives necessary in a geothermal acid job are corrosion inhibitor and inhibitor intensifier, as well as high-temperature iron-control (reducing) agent. Corrosion inhibitors of diverse description and composition have been proposed over the years for use with well treating acids. Treatment placement is better ensured through the use of chemical or mechanical diversion methods and technologies, and placement tools (coiled tubing, packers, etc.). On-site quality control is enabled by monitors and software, enabling to determine the evolution of skin with time, and radius of formation treated.

The application of chemical acids generally improves conditions in the reservoir, however, with largely varying success rates. During the 1990s, the acidification technique has been used more often, principally for the reservoir development or to treat formation damage caused by drilling mud and scaling (mineral deposits) in geothermal wells (Buning *et al.*, 1995; Buning *et al.*, 1997; Malate *et al.*, 1997; Yglopaz *et al.*, 1998; Malate *et al.*, 1999; Barrios *et al.*, 2002; Jaimes-Maldonado and Sánchez-Velasco, 2003). This protocol has not really evolved since these years. In each of the experiments proposed by the authors, the same technique is used. The acidification occurred in three main steps:

1) A preflush, usually with hydrochloric acid (10%). The objective of the preflush is to remove calcium carbonate materials in the formation. The preflush acid minimizes the possibility of insoluble precipitates.
2) A main flush with hydrochloric–hydrofluoric acid mixture. A mixture of 10% HCl – 5% HF (called *mud acid*) is generally prepared by dissolving ammonium bifluoride (NH_4HF_2) in HCl. A mixture of 1% of HCl and 56 kilos of NH_4HF_2 will generate 1% HF solution. Regular mud acid (RMA) (12% HCl–3% HF) is made from 15% HCl, where 3% HCl is used to hydrolyze the fluoride salts.
3) A postflush/overflush usually by either HCl, KCl, NH_4Cl, or freshwater.

Concerning the injected amounts for the cleaning out of the geothermal wells, the preflush volume was based on a dosing rate of $600\,l\,m^{-1}$ of target zone. The mainflush volume was based on a dosing rate of $900\,l\,m^{-1}$ of target payzone (Malate *et al.*, 1997; Barrios *et al.*, 2002).

In geothermal well acidizing, more acid often is better. A very successful method of acidizing geothermal wells has been a basic, high-rate, brute-force method. High

Figure 4.10 Installation of a pumping unit for injection of chemical compounds at Coso geothermal field (photo P. Rose, EGI, Univ. of Utah).

acid concentrations have been shown to be effective in geothermal wells producing from natural fractures not containing separate, large carbonate zones. Naturally fractured volcanic formations can withstand high HF concentration. The HCl–HF stage can be 10% HCl–5% HF, or 10% HCl–7% HF, for example. These acid mixtures have been used successfully in stimulating geothermal wells in South and North America (Figure 4.10), as well as in Southeast Asia (the Philippines), where a large number of acid treatments have taken place. Acid volumes can vary quite a bit. A summary of the main chemical stimulation experiments carried out in geothermal fields is given in Table 4.5, showing variable results.

Only few chemical stimulation experiments and laboratory tests have been attempted until now in EGS wells and reservoirs. Limited reported data were found at the projects of Fenton Hill in USA (Sarda, 1977) and Fjällbacka in Sweden (Sundquist, Wallroth, and Eliasson, 1988; Wallroth, Eliasson, and Sundquist, 1999). At the EGS reservoir of Soultz-sous-Forêts however, several consistent and documented chemical stimulation tests have been carried out since 2003. Different techniques were consecutively used in the three 5 km-deep wells: soft acidizing, RMA, chelating agents (NTA), and organic clay acid (OCA). Although they were not executed with the same comparable protocol, various but encouraging results were observed after this first series of tests using chemical stimulation methods in a fractured granitic EGS reservoir.

4.9 Sustainability of Treatment

4.9.1 Hydraulic Stimulation

4.9.1.1 Proppant Selection

Experiences from fracture stimulations (Legarth, Huenges, and Zimmermann, 2005; Zimmermann *et al.*, 2009) highlight the importance of making the right

Table 4.5 Results of chemical treatments for scaling removal and connectivity development in selected geothermal fields.

Geothermal field	Methods used	Number of treated wells	Variation of the injectivity index before and after acid treatment ($kg\ s^{-1} bar^{-1}$)	Improvement factor	References
Bacman (Philippines)	HCl–HF	2	0.68–3.01 0.99–1.40	4.4 1.4	Buning *et al.* (1995)
Leyte (Philippines)	HCl–HF	3	3.01–5.84 0.68–1.77 1.52–10.80	1.9 2.6 7.1	Malate *et al.* (1997) and Yglopaz *et al.* (1998)
Tiwi (Philippines)	HCl–HF	1	2.52–11.34	2.6	Buning *et al.* (1995)
Mindanao (Philippines)	HCl–HF	1	Successful	2.8	Buning *et al.* (1997)
Salak (Indonesia)	HCl–HF	1	4.70–12.10	2.6	Pasikki and Gilmore (2006)
Berlin (El Salvador)	HCl–HF	5	1.60–7.60 1.40–8.60 0.20–1.98 0.90–3.40 1.65–4.67	4.8 6.1 9.9 3.8 2.8	Barrios *et al.* (2002)
Las Tres Virgenes (Mexico)	10% HCl–5% HF	2	0.8–2.0 1.2–3.7	2.5 3.1	Jaimes-Maldonado and Sánchez-Velasco (2003)
Los Azufres (Mexico)	HCl–HF	1	3.3–9.1	2.8	Flores Barajas, and Rodriguez (2006)
Beowawe (USA)	12% HCl–3% HF	1	Successful	2.2	Epperson (1983)
The Geysers (USA)	5% HCl–10% HF	1	No effect	–	Entingh (1999)
Coso (USA)	HCl and NTA	30	24 Wells successful	–	Evanoff Yeager, and Spielman (1995) and Rose *et al.* (2007)
Larderello (Italy)	HCl–HF	5	11–54 4–25 1.5–18 Successful 11–54	4.9 6.3 12 4 4.9	Capetti (2006)

choice of proppants and proppant concentrations to guarantee sufficient fracture conductivity in sedimentary environments. Experiments have shown a crushing of proppant pack in addition to no self-propping. To maintain long-term productivity of the reservoir, in advance to the field experiments different proppant types should be tested in the laboratory for long-term conductivity under simulated insitu reservoir conditions as well as for mechanical effects that would lower the permeability of the proppant pack and the reservoir.

Proppants are used to keep the fracture open after pumping has stopped and pressure drops below the fracture opening pressure. The proppant pack in the fracture provides a conductive path from the reservoir rock to the wellbore. Placing the appropriate concentration and type of proppant in the fracture are critical parameters for the success of a hydraulic fracturing treatment (Economides and Nolte, 2000).

Proppant selection must consider hydraulic conductivity at *in situ* stress conditions. Hydraulic conductivity is influenced by stress on proppant pack, leading to proppant crushing and embedment as well as to a reduction of fracture width and fines production. Proppant size and proppant concentration has to be taken into account. In general, large-diameter proppants yield a better hydraulic conductivity but they are more sensitive for stress. Small diameter proppants offer less initial hydraulic conductivity, but the average hydraulic conductivity over the life cycle of the well is higher. Proppant concentration affects the hydraulic width and is important for long-term hydraulic conductivity under production conditions (Wen *et al.*, 2007).

Reinicke *et al.* (2006) have shown that during fracture closure the majority of destruction is located at the fracture face, at the rock proppant contact. The destruction at the fracture face leads to fines production and pore blocking resulting in a reduced permeability at the fracture face (Legarth, Huenges, and Zimmermann, 2005) (Figure 4.11).

This reduced permeability can be expressed as fracture face skin (FFS). The FFS is referred to as an impairment affecting flow normal to the fracture face (Cinco-Ley and Samaniego, 1977). The FFS (denoted as s_{ff} according to the original publication) can be described in terms of fracture half-length x_f, damage penetration w_s and the ratio of unaffected reservoir permeability to damaged permeability k_i/k_s:

$$s_{ff} = \frac{\pi w_s}{x_f}\left(\frac{k_i}{k_s} - 1\right) \tag{4.1}$$

The FFS has a direct influence on the productivity of a reservoir; it can be used to calculate the dimensionless PI PI_D within the pseudo steady state flow regime:

$$PI_D = \frac{1}{\frac{1}{PI_{D,s=0}} + s_{ff}} \tag{4.2}$$

with $PI_{D,s=0}$ representing the dimensionless PI of the well with zero fracture face skin (Romero, Valkó, and Economides, 2003).

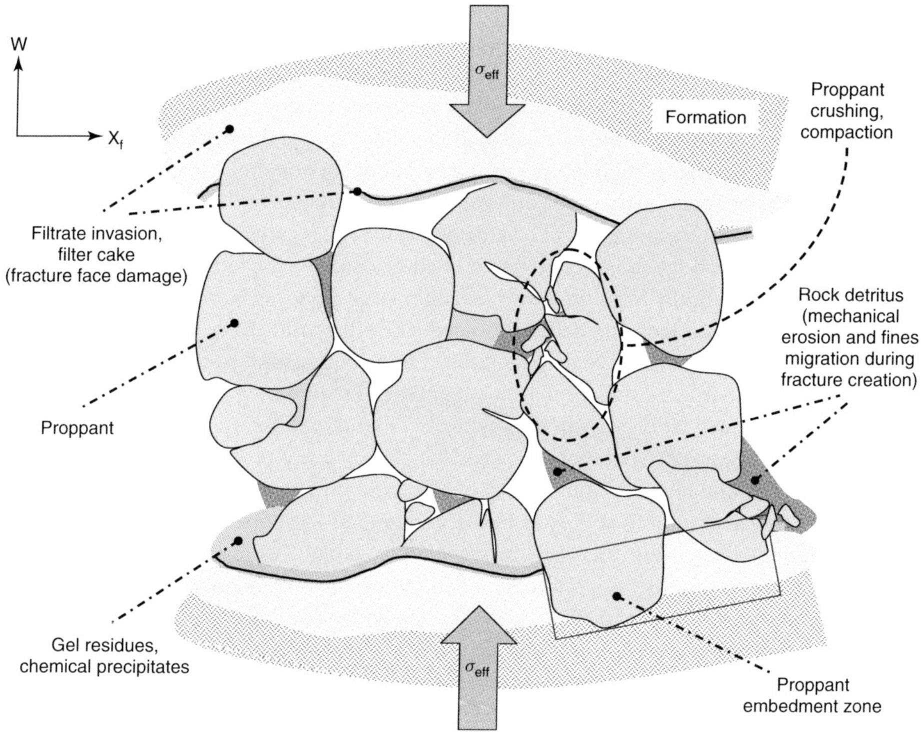

Figure 4.11 Conceptual model of fracture face damage and fines production and pore blocking of a system fracture with not well designed proppants (Legarth, Huenges, and Zimmermann, 2005).

4.9.1.2 **Coated Proppants**

Adding coated proppants (a kind of resin on the surface of the proppants) at the end of the treatment (typically 20% of the total amount) leads to sustainable and long-term conductivity of the fracture especially if a production well is concerned. These proppants are interconnected due to the special coating and act as a barrier in the near-wellbore region and avoid a flow-back of proppants into the well during production.

4.9.2
Thermal Stimulation

Little is generally known, about the sustainability of thermal stimulation measures, and to the authors' knowledge, no systematic investigations into this topic are publicly available to this day. However, it has been recognized at thermal stimulation measures on Icelandic wells that thermally induced fractures seem to be self-propping, in that no additives have to be admixed during the stimulation treatment.

4.9.3 Chemical Stimulation

Geothermal injection wells are prone to having a naturally low injectivity and/or exhibiting serious injectivity losses at various stages of their life. This is especially true in the case of high temperature geothermal operations. The main reasons for sometime rapid injectivity losses are the need to inject very large amounts of brine per well and the plugging of originally good injectors due to specific conditions of a geothermal operation. The new drilled or completed geothermal injection well differs drastically from the same well after large amounts of heat-depleted (cooled) brine are injected. The damage caused by drilling and completions operations is quite different from the damage caused by a prolonged injection of heat-depleted brine and also require special considerations (O'Sullivan and McKibbin, 1993).

Some field studies involving suspended particles measurement and monitoring in a geothermal operation showed that the majority of the suspended particles entering the injection well consist of silica and iron compounds. These suspended particles can eventually result in the plugging of the pore spaces of the reservoir, thereby decreasing the well injectivity. In addition, various chemical inhibitors are added during the production of the brine to combat scale (and/or corrosion) problems in the producing of brine. Excess amounts of these chemicals can remain in the brine and enter the injection wells. This can damage the well by blocking the pores. The heat-depleted brine has a composition somewhat different from the formation brine and can create incompatibility problems. The stimulation of injection wells generally consists of repairing the near-wellbore damage described above. Chemical methods (acidizing or use of chemicals other than acids) are commonly used for such stimulation.

4.10 Case Studies

4.10.1 Groß Schönebeck

4.10.1.1 Introduction

The aim of stimulation treatments in the geothermal research wells in Groß Schönebeck is the enhancement of productivity of the reservoir targets as a prerequisite for geothermal power generation of the Rotliegend Formation as an EGS (Gérard *et al.*, 2006; Huenges *et al.*, 2007). Optimum economic utilization of reservoirs is mainly due to adequate planning including reservoir modeling and understanding of the processes and interaction of the system "borehole – reservoir."

Hydraulic stimulation treatments were carried out in the two wells at the Groß Schönebeck drill site. In the first well GrSk3/90, several treatments were performed (in 2002 and 2003). The second well GrSk4/05 was drilled in 2006 and stimulation treatments were performed in August 2007.

4.10.1.2 **Hydraulic Fracturing Treatments in GrSk3/90**

A series of stimulation experiments were carried out at the geothermal research well in Groß Schönebeck (EGrSk 3/90) located in the north-eastern part of Germany. The aim was the development of concepts for the productivity enhancement of geothermal wells in that region. In a first attempt, hydraulic gel-proppant fracturing treatments were conducted in two sedimentary reservoir zones with high permeability at about 4 km depth. These treatments were performed under challenging conditions in the open hole section at a temperature of about 150 °C. On the one hand, they proved to be technically demanding and on the other less successful than expected due to a suboptimal design. Most likely, the small injection volumes combined with a low proppant density did limit the success of these operations. Nevertheless, the productivity of the well could be increased by a factor of two. The characterization of the inflow zones after the proppant fracturing treatments and derived values for the minimal horizontal stress led to a completely different fracturing concept. Massive waterfrac treatments were now applied over the entire open section of the well below 3874 m to the final depth at 4294 m. Again, a significant increase of productivity was achieved, demonstrating that waterfracs can be a successful and effective stimulation concept for this geological situation. Evidence of the creation and properties of a very long vertical fracture were retrieved from pressure response analyses demonstrating a bilinear flow regime. The stimulation effect in terms of a productivity increase was determined for the described concepts and improvements are derived for similar field experiments.

4.10.1.3 **Hydraulic Fracturing in Sandstones (Gel-Proppant Stimulation)**

The first stimulation experiments were of the conventional kind, that is, on the basis of expertise of the hydrocarbon industry. Two experiments were performed in January 2002 using proppant-gel-frac techniques in two intervals of the Rotliegend sandstones (Zimmermann *et al.*, 2003; Legarth, Huenges, and Zimmermann, 2005). Experimental design comprised the isolation of the bottom boundary of the interval of interest by filling the bottom of the well with sand. The top of the interval was sealed with a mechanical packer. High viscosity fluid (gel) with proppant was employed for stimulation. The flowmeter-log indicated a significant increase of inflow due to this frac operation. Visualization by borehole televiewer (BHTV) and formation micro imager (FMI) confirmed the creation of the new open vertical fracture in the stimulated intervals with a height of more than 100 m and in the direction of the maximum horizontal stress ($S_H = 18.5 \pm 3.7^{\circ}$) (Holl *et al.*, 2003, 2004). Nevertheless, the observed flow rates were not sufficient for economic power production (Zimmermann *et al.*, 2003), but the PI could be enhanced to 2.2 m^3 (h MPa)$^{-1}$ due to the stimulation treatments.

The mean flow rate obtained during the casing lift test (CLT) was 22.4 m^3 h^{-1} at a differential pressure of 10.5 MPa at the end of the test. A total volume of 307 m^3 was produced at the duration of the test of about 14 hours, which is a similar time of production as during the previous CLT. Hence this result can be compared to the previous test indicating a doubling of the PI of the well. Legarth *et al.* (2003) conclude that the limited achievement was strongly influenced by the proppant properties during the treatment and prevented a better result of the stimulation treatments.

To determine the hydraulic parameters of the stimulated reservoir in more detail and to obtain stable conditions over a longer period, a long-term pumping test was performed in summer 2002 (Zimmermann, 2004; Reinicke *et al.*, 2005). A downhole pump was installed at 330 m depth (the water level is in equilibrium at 250 m). The flow rate was set to approximately 1 $m^3\ h^{-1}$ over a period of 37 days. In total, 580 m^3 formation fluids were extracted during this test, which is equivalent to approximately five borehole volumes. The draw down reached a constant level after 10 days, but steady state conditions were not reached until the end of the test. The productivity-index was estimated at pseudo steady state conditions to 0.6 m^3 $(h\ MPa)^{-1}$.

Transmissibility of the productive formations was estimated from pressure build up toward the end of the shut-in to assure pseudo radial flow conditions and was calculated to $3.1 \times 10^{-14}\ m^3$. The minimum extension of the reservoir was calculated from maximum radius of investigation to $R = 617\,m$ (assuming total height of reservoir = 100 m; porosity = 0.05; fluid viscosity = 4×10^{-4} Pa s; total compressibility = $5 \times 10^{-10}\ 1\ Pa^{-1}$) (e.g., Carslaw and Jaeger, 1959; Lee, 1981).

Another parameter to improve the results of reservoir stimulation is the total volume of injected frac fluid in a forthcoming experiment. Therefore, it was intended to continue the stimulation experiments with a procedure injecting at least two orders of magnitude higher volume into the reservoir.

4.10.1.4 **Hydraulic fracturing in Volcanics (Waterfrac Stimulation)**

The first waterfrac treatment started in January 2003 with a moderate injection test with a flow rate of 3.6 $m^3\ h^{-1}$ over a period of 200 hours. The aim of this pretest was to obtain initial injection properties of the reservoir and to compare these results with the former short-term and long-term production tests carried out in spring and summer 2002. After 48 hours the injectivity index (ratio of injection flow rate and differential pressure) was 1.15 $m^3\ (h\ MPa)^{-1}$ and decreased to 0.83 $m^3\ (h\ MPa)^{-1}$ at the end of the test after 200 hours (Tischner, 2004). The observed injectivity corresponds to the productivity derived in former production tests at similar low differential pressure. For this reason it can be assumed that the hydraulic response of the reservoir is similar for production and injection for a pressure change up to 10 MPa (decreasing for injection as well as increasing for production).

Thereafter, the first waterfrac treatment was performed in whose progression a total amount of 4284 m^3 fluid was injected under high pressure into the reservoir. In the first part, a pressure step test with gradually increasing injection rates up to 24 $l\ s^{-1}$ was performed. The results show that starting with an injection rate of 8 $l\ s^{-1}$ the pressure increase is reduced due to an enhanced injectivity of the formation. This effect can be interpreted as a mechanical reaction of the rock due to an opening of the generated artificial fractures as well as the extension of pre-existing fracture in the conglomerates and volcanic rocks at the bottom of the well (Huenges *et al.*, 2006).

In a subsequent flow-back test, 250 m^3 of water was produced within a time of 5 hours and a mean flow rate of 50 $m^3\ h^{-1}$. This indicates in comparison with tests after the sandstone fracturing treatment, a significant increase of productivity (Zimmermann *et al.*, 2005). PI is above 4 m^3 (h MPa)$^{-1}$ during the whole test. This is an indication that the massive water injection produced additional fractures, so that the experiment was rated successful and represented roughly another doubling of the PI. However, borehole breakouts occurred resulting in an obstruction just at the upper part of the tested section at about 3900 m depth. Therefore, further technical borehole operations were necessary. In October 2003, the obstruction in the well was removed and the well was deepened to 4309 m and an additional liner from 3850 m down to the final depth was installed. Prior to the liner installation, an extensive logging program was performed in order to get information about the geological structure and the lithology of the borehole section of interest. The liner was installed with perforated tubes in the lower part beneath 4135 m installation depth (diameter of holes 15 mm; 93 holes per meter circumferential) to ensure the hydraulic contact to the formation. In the stabilized well, the massive water frac experiment was continued in fall of 2003.

The injection treatment was continued with a pressure step-rate test to obtain the fracture opening pressure (Huenges *et al.*, 2006). Thereafter, a massive stimulation test of 108–144 $m^3\ h^{-1}$ over several days and one short peak of up to 288 $m^3\ h^{-1}$ for approximately 2 minutes were performed (Figure 4.5). The total injection volume accumulated to 7291 m^3. Initially, it was intended to perform the last high rate flow injection over 8 hours, but due to a cable break during the first few minutes the test had to be abandoned. The pressure step-rate test indicates multiple fracture opening events. Fracture closure pressure was determined by pressure decline analyses during shut-in at 6.4 MPa above formation pressure.

The stimulation treatments were accompanied by a passive microseismic monitoring in the five wells (50 m deep) in the vicinity of the drill site. Unfortunately, micro-seismicity could not be observed during the various stimulations due to attenuation between surface and reservoir in the Zechstein salt (Weber *et al.*, 2005).

4.10.1.5 Hydraulic Fracturing Treatments in GrSk4/05

Three stimulation treatments were performed in the completed well GtGrSk4/05 in August 2007 (Figure 4.12) starting from the bottom of the well within the volcanic section. The whole well is cased and cemented with the exception of the last 40 m (perforated liner and open hole section).

A first leakoff test in the volcanics was applied to obtain the fracture gradient. Then a massive waterfrac treatment in the volcanics followed. After isolation of this part of the well with a bridge plug, the lower sandstone horizon was perforated and tested (injection test). The particular interval was selected due to the interpretation of borehole measurements carried out before completion. Subsequently, the gel-proppant stimulation followed. After isolating this interval as well, the upper sandstone layer was perforated and stimulated with another gel-proppant treatment. Finally, the bridge plugs were removed and a production test was carried out to test all stimulated intervals.

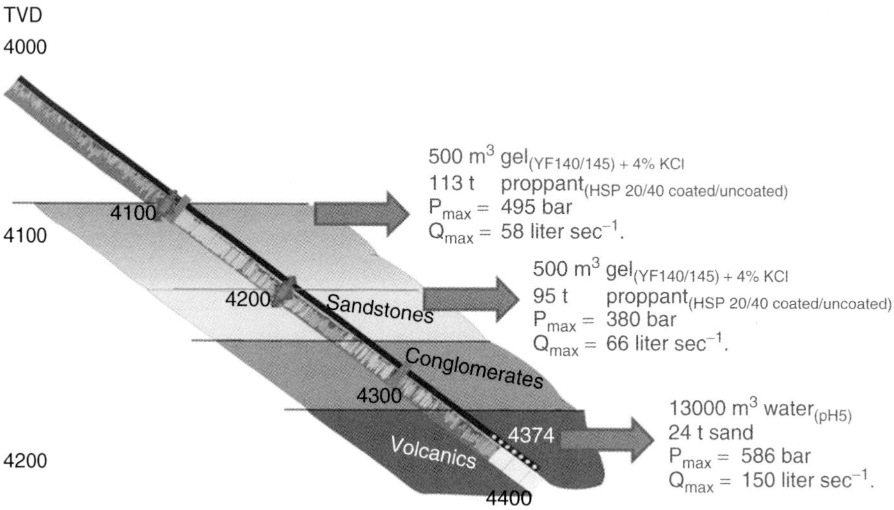

Figure 4.12 Summary of three stimulation treatments in the deviated well GtGrSk4/05 in Groß Schönebeck (see text). (Please find a color version of this figure on the color plates.)

4.10.1.6 Hydraulic Fracturing Treatment in Volcanics (Waterfrac Stimulation)

The design comprises several high flow rate intervals due to the limitation of water availability from the existing water wells (approx. $50\,l\,s^{-1}$) and storage capacity in containers of approximately 1000 m^3. We expect that the impact of high flow rates ($150\,l\,s^{-1}$) for the fracture performance is better, even if the intervals of high flow rates are limited in time compared to a constant flow rate of $50\,l\,s^{-1}$. To fill the containers, it is provided to reduce the flow rate to $20\,l\,s^{-1}$. This should be far above the transition from the fracture mode to the hydraulic mode and should keep the fractures open. The total duration of the treatment is limited for budget reasons to approximately five days.

The design comprises adding sand during the high flow rates to support the long-term opening of the fractures created.

The waterfrac stimulation treatment was carried out between 9 August and 14 August in 2007. Figure 4.13 shows the data. During the high flow rates a friction reducing agent was used to reduce the friction in the well and limit the maximum wellhead pressure to 580 bars. To avoid iron scaling of the injected water, acetic acid was added to set the pH to 5. During the high flow rates of $150\,l\,s^{-1}$ low concentrations of quartz sand (20/40 mesh size) were added to support a sustainable fracture width. Transport of the sand in the fracture and the well was realized solely by the high flow velocity, because a gel to support the transport was not an option due to the pH value.

In total, 13 170 m^3 of fluids and 24.4 t of quartz sand were injected into the volcanic rocks. Maximum wellhead pressure achieved 586 bars at the maximum flow rate of 9 $m^3\,min^{-1}$ ($150\,l\,s^{-1}$).

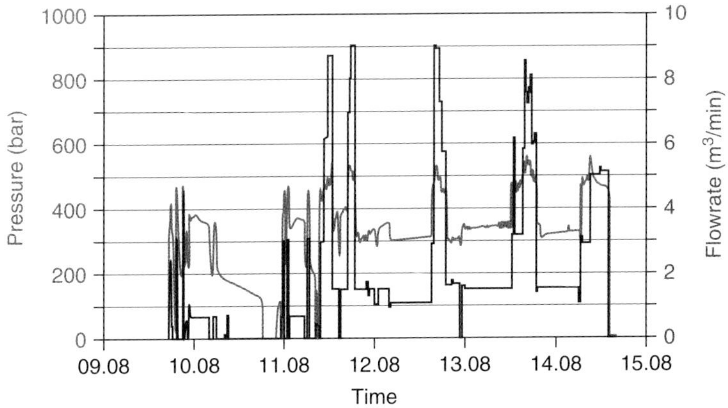

Figure 4.13 Pressure and flowrate data of the massive waterfrac treatment in Groß Schönebeck in August 2007.

4.10.1.7 **Hydraulic Fracturing in Sandstones (Gel-Proppant Stimulation)**

The stimulation treatment in the sandstones of the lower Dethlingen was carried out from 18 August to 19 August 2007 (see Figure 4.14). The intended interval from 4204 to 4208 m was isolated with a bridge plug in 4300 m and then perforated with big hole perforations (20 per m, circumferential). Transport of the proppants was provided with a cross linked gel with high viscosity. Two kind of proppants were applied: high strength proppants coated and uncoated. Both consist of a diameter of 0.4–0.8 mm (20/40 mesh size); the coated proppants were used at the end of the treatment to support the sustainable fracture opening in the near-wellbore vicinity.

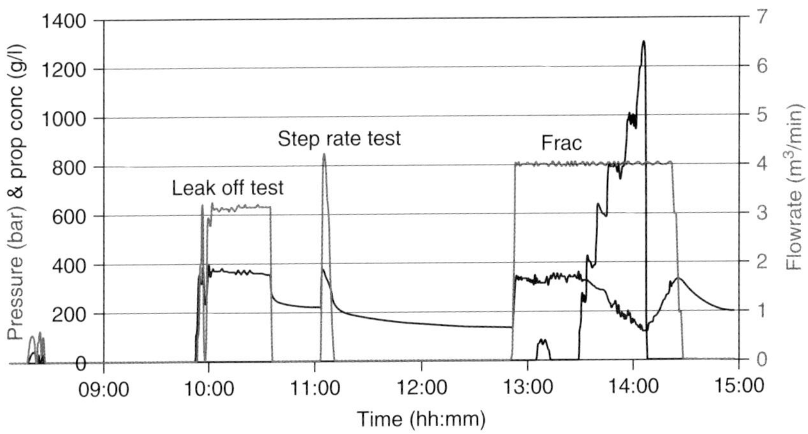

Figure 4.14 Pressure and flowrate data of the 1. gel-proppant treatment in Groß Schönebeck in August 2007.

The treatment started with an injection test with flow rates between 0.3 and 0.57 m^3 min^{-1}. In total 250 m^3 were injected into the reservoir at a maximum wellhead pressure of 416 bars.

Subsequently, a leakoff test was carried out to obtain the frac gradient (0.16 bar m^{-1}). Then, a step-rate test followed to calculate the friction and tortuosity at the perforation. Finally, the gel-proppant treatment was performed where 95 tons of proppants and 280 m^3 of cross-linked gel were injected into the Lower Dethlingen formation with a flow rate of 4 m^3 min^{-1}.

The second gel-proppant treatment was carried out from 23 August to 24 August 2007 in the sandstones of the Upper Dethlingen. The treatment had a similar design as the previous one. The bridge plug was set in 4123 m depth and the interval above from 4118 to 4122 m was perforated. The treatment started with an injection test with rates between 0.3 and 0.62 m^3 min^{-1} and a total volume of 170 m^3. The leakoff test analysis yielded a frac gradient of 0.15 bar m^{-1}. In the following stimulation treatment, 113 tons of proppants and 310 m^3 of cross-linked gel were injected at flow rates between 3 and 3.5 m^3 min^{-1}.

After stimulation of the reservoir sections a CLT with a nitrogen lift in conjunction with a flowmeter profiling was performed to test the stimulated intervals. In advance, additional perforations (deep penetration charges) were carried out in the sandstone sections. Over a period of 12 hours approximately 300 m^3 were produced. During production two runs (up and down respectively) were performed by the flowmeter to obtain the inflow profile. The results of the flow profile show that 30% of flow originates from the volcanic rocks. Nearly 50% of flow can be attributed to the first gel-proppant treatment and 15% is due to the second gel-proppant treatment. Only 5% can be assigned to the post perforations. A possible reason might be the drilling fluid, which was used to build a filter cake at the borehole wall to protect the reservoir. To enhance the performance of the post perforations it is intended to acidize these intervals.

4.10.1.8 **Conclusions**

Development of a technology to stimulate deep geothermal reservoirs in sedimentary basins was the purpose of installing the downhole geothermal laboratory in Groß Schönebeck. The results reflect the learning curve from several reservoir hydraulic fracturing treatments in both wells. These experiments are major steps toward developing a procedure to increase the thermal water productivity from a prior low permeable sedimentary reservoir. The obtained values of productivity seem to show the feasibility of geothermal power production from a sedimentary geothermal environment.

From these experiences we give the following recommendations for other potential drill sites within the same environment:

1) In general, the stimulation treatments should be specially designed in relation to the different characteristics and properties of the reservoir rocks. This implies that several treatments in a well, which have to be isolated in advance are to be performed.

2) The volcanic rocks should be stimulated by a waterfrac treatment in conjunction with low proppant concentrations to support the sustainability of the produced fractures. At the end of the treatment one can complete the same with a short gel-proppant treatment to support the access to the fracture in the near wellbore, where the highest pressure changes are expected during production. The design of the treatment should be laid out in a way that the fracture propagates upwards as well to connect the sandstone layers above the volcanic rocks and offers the possibility to drain the high permeable sandstones from below and build up a hydraulic connection to the other well. This can be achieved by a special adjustment of the flow rate during the fracture treatment. The recommendation is to carry out cyclic changes of flow rate with high and low stages to control the fracture propagation in horizontal as well as vertical direction.
3) In the Rotliegend sandstone sections gel-proppant treatments should be performed to give access to the high permeable layers. These treatments should comprise high proppant concentration to obtain a multilayer proppant pack and hence a high fracture conductivity.
4) To accomplish several treatments independently from each other, it is recommended to drill a deviated well with a deviation in the reservoir rock. In this case, no hydraulic interferences must be suspected.
5) Deviation in the reservoir rock should be in the direction of minimum horizontal stress as far as hydrothermal systems are concerned. This assumes the existence of a high permeable layer between both wells to establish a hydraulic connection. The distance between both wells should be far enough to avoid a temperature short cut over the live cycle of the geothermal power production, which is in the time frame of tens of years.

4.10.2 Soultz

4.10.2.1 Hydraulic Stimulation

The test site of the EGS project Soultz is located in France on the western edge of the Rhine Graben, some 50 km north of Strasbourg near the German border (Figure 4.15). This area is characterized by a thin continental crust resulting in a high geothermal heat flow. The granitic basement at Soultz-sous-Forêts lies beneath approximately 1400 m of sedimentary rock (Hoijkaas, Genter, and Dezayes, 2006). Soultz is located in the heart of an abandoned oil production area which was intensively exploited at the beginning of the last century. Important reasons for the selection of Soultz as location for a HDR project were

- high geothermal gradient (about 110 °C at 1000 m depth);
- good geological data basis from oil exploration.

The geothermal research program started at Soultz-sous-Forêts in 1987 as a joint German and French project. In three main consecutive phases, subsurface reservoirs in granite were developed and tested (Hettkamp *et al.*, 2004). First, a reservoir at about 2 km depth was created and later at about 3 km. During the years

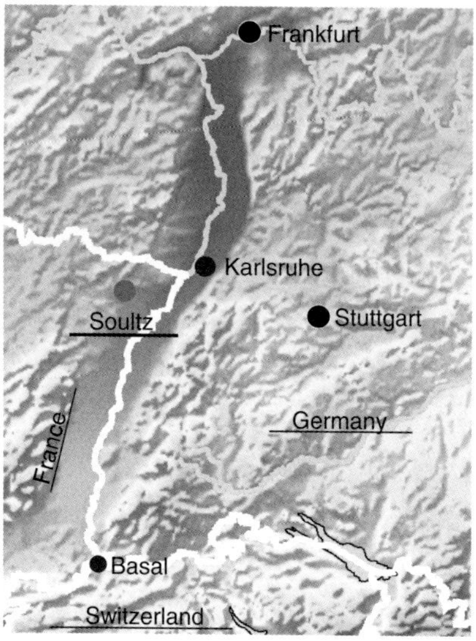

Figure 4.15 Location of the Soultz site at the border between France and Germany.

1999–2004, three wells were drilled in granitic rock down to 5 km. Two wells (GPK2 and GPK4) will be used for production whereas the central well (GPK3) will be the injection well. All the three wells were drilled from the same platform, whereas the horizontal distance between the re-injection well and each of the production wells is approximately 700 m at target depth (Figure 4.16).

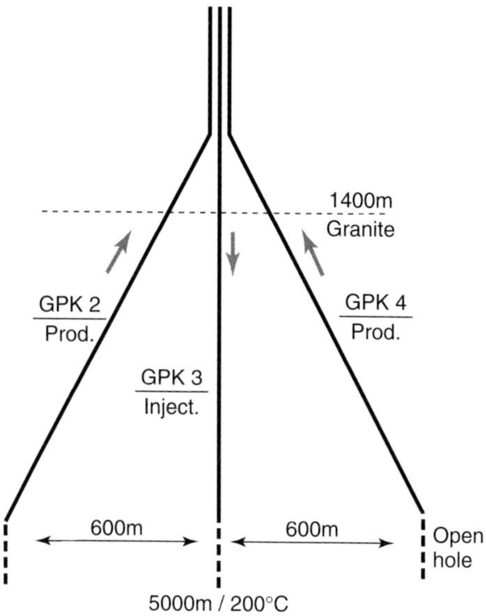

Figure 4.16 Scheme of the borehole triplet (5 km depth) in Soultz. Inject.: Injection well; Prod.: Production well.

Table 4.6 Overview upon the massive hydraulic fracturing operations performed in Soultz.

Date	Well	Depth-TVD (m)	Volume (m^3)	Maximum rate ($l\ s^{-1}$)	Comment
July 1991	GPK1	1 966–2 000	1 250/2 400	7/15	2 Fracs in a Packer isolated zone
September 1993	GPK1	2 850–3 400	25 300	36	Open hole frac
October 1993	GPK1	2 850–3 590	19 300	50	Open hole frac
June 1995/September 1996	GPK2	3 211–3 876	28 000/27 000	50/78	Open hole fracs
June 2000	GPK2	4 403–5 026	23 400	50	Open hole frac
May 2003	GPK3	4 488–5 021	34 000	50	Open hole frac
September 2004/February 2005	GPK4	4 479–4 972	9 300/12 300	30/45	Open hole fracs

Only frac operations are listed with an injected water volume of more than 1000 m^3. Maximum rate means the maximum injection rate which was maintained for a period of 6 hours at least.

A characteristic feature of the Soultz wells are their long open hole sections. The lower 400–600 m of the wells are uncased. In principle, the advantage of a long open hole section is the access to several permeable structures.

The aim of the project is the development of a subsurface heat exchanger, in a hydrothermally altered and fractured granite by means of hydraulic and chemical stimulation, to be used for commercial geothermal electricity production. A hydraulic link between the wells over a distance of several hundred meters has to be created, to allow water circulation between the deep wells.

In low permeable crystalline rock, massive frac operations are the only means of forming this link, to create a connection between the wells and a natural fracture system, and to reactivate and expand a hydrothermally altered and fractured zone, which acts as geothermal reservoir at Soultz.

Investigations at Soultz and other sites have shown that shearing is the dominant fracturing process induced by massive water injections (Cuenot *et al.*, 2006; Pine and Batchelor, 1984). The dislocation of the fracture surfaces leads to a self-propped fracture – the mechanism of which is assumed to be responsible for permeability enhancement due to waterfracs.[1)]

In the early project, some waterfracs were performed in packer isolated intervals of the open hole section of GPK1 (Table 4.6). Since 1991, successful fracs were conducted over the complete open hole sections.

Waterfrac operations in Soultz often commenced with the injection of brine (density up to 1.2 kg l^{-1}) to initiate the stimulation as deep as possible in the borehole. Subsequently, large volumes of fresh water were injected at a constant

1) The terms *waterfrac*, *frac* and *stimulation* are used synonymously.

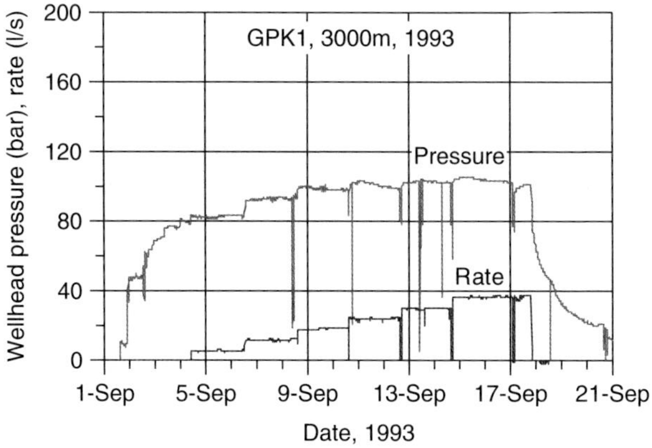

Figure 4.17 Wellhead pressure and injection rate during stimulation of GPK1 in 1993 at about 3 km.

or an increasing rate. Especially in the reservoir at about 3 km depth, stepwise increasing rates were applied during hydraulic stimulation (Figure 4.17). An increasing injection rate is assumed to prefer shearing, first on fractures that are most critically stressed, followed by the activation of fractures less exposed to shear stress. In the deep reservoir, the concept of stepwise hydraulic stimulation was only partly pursued, mainly for technical and economical reasons.

Compared to other sites the fracturing pressure in Soultz is low. At about 2 km depth the overpressure during fracturing was approximately 60 bar and in 5 km depth an overpressure of 130–180 bar was observed. The latter numbers lead to a fracture gradient of approximately 26–34 bar km^{-1} at 5 km depths. In most cases, only a weak dependence of the pressure on flow rate and time was obtained. A slightly increasing fracturing pressure versus time may be an indication for shearing as observed during the last two days of stimulation of GPK2 (Figure 4.18) for instance. A continuously decreasing pressure however, may

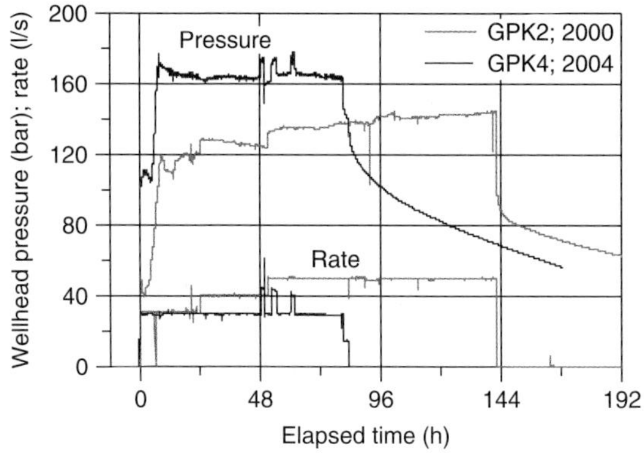

Figure 4.18 Wellhead pressure and injection rate during the stimulation of GPK2 and GPK4 in the deep reservoir (5 km).

suggest tensile fracturing as the dominant fracturing process – see stimulation of GPK4 (Figure 4.18).

Indications about the flow outlets in the wells were derived from flowlogs (Evans, Genter, and Sausse, 2005a; Evans *et al.*, 2005b). A general feature of all Soultz wells is that the flow is controlled by a few outlets only. In GPK3 (5 km) one dominant outlet, which corresponds to a hydrothermally altered and fractured zone, takes about 70% of the total flow. The flowlogs further demonstrate that the fractures which were stimulated most show also the best performance after the treatment. The flow distribution in the wells is stable after the stimulation.

Injection tests were carried out before and after the waterfrac operations in order to determine the initial productivity and the productivity after fracturing. In these injection tests, the rate was much lower than during stimulation to avoid any further fracturing. Usually, the productivity of the wells could significantly be enhanced. A 20-fold increase in productivity of the wells GPK2 and GPK4 (5 km) was achieved up to 0.35 l $s^{-1}bar^{-1}$ (GPK2) and 0.20 l $s^{-1}bar^{-1}$ (GPK4). The productivity of the well GPK3 could only slightly be improved. But here the well was very productive already before fracturing.

In these injection tests at 5 km depths the pressure continuously increased at a constant rate but with decreasing slope. Steady state conditions have never been observed. To allow the comparison between different injection tests, the productivity has to be evaluated at a similar time period. The productivities were usually determined after two to three days of injection.

In Soultz, the empirical rule was established that the injection period during fracturing should last several days. Only after such long extended stimulations a sufficient and persisting productivity enhancement was observed.

An important relationship was found between the injection rate during stimulation and the productivity of the well after stimulation (Jung, 1999; Jung and Weidler, 2000). The productivity of the wells appears to increase with the injection rate during stimulation. Results from all the three reservoirs developed in Soultz confirm this observation. In the deep reservoir (5 km), it could be shown that the productivity of the well after stimulation is essentially the same as during stimulation (Tischner *et al.*, 2007). There is obviously no closing or relaxation of the fractures after the long extended waterfracs and the productivity of the well during stimulation persists after stimulation. This observation may be explained by a self-propping effect of fractures, failing in shear under the local stress regime.

As a consequence, the productivity enhancement due to the waterfrac operation becomes predictable, simply by adjusting the injection rate during stimulation. The predictability of the productivity enhancement is a great advantage of those operations and must be emphasized compared to other stimulation methods.

Microseismic monitoring has evolved to the key technique to map the reservoir in HDR projects (Niitsuma, 2004). In Soultz, six wells in the depth range of 1500–3500 m have been used as seismic observation wells (Figure 4.19). The recorded and localized seismic events during hydraulic stimulation allow tracing the development of the reservoir and serve as indication for the hydraulic connection

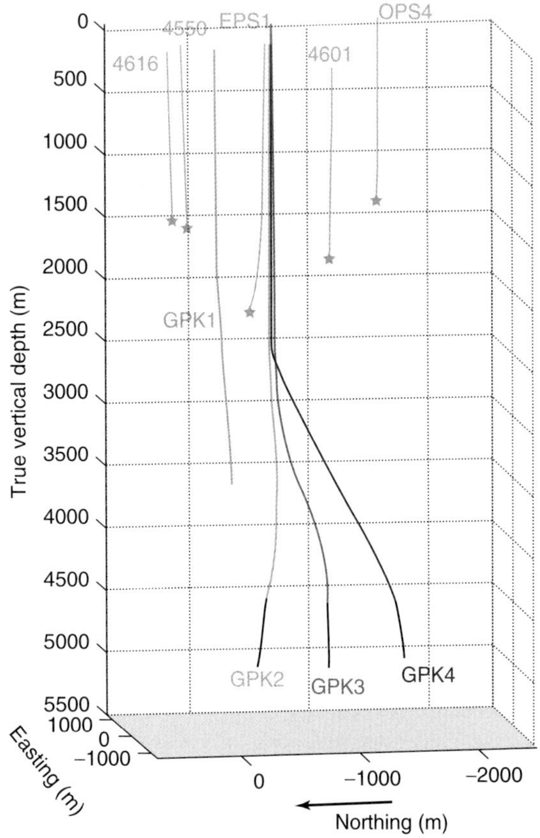

Figure 4.19 Sketch of the three deep wells and the surrounding seismic observation wells. The observation wells are orange, the stars indicate the depth level of the installed downhole seismic sensors. (Please find a color version of this figure on the color plates.)

between the wells. Further, the target area of the wells were determined based on the orientation of the seismic cloud.

Usually several thousands of microseismic events were recorded during each of the frac operations (Baria *et al.*, 2006). There is a clear tendency toward more seismicity with increasing depths. During the hydraulic stimulation of GPK3 (5 km), more than 90 000 events were recorded, much more than in the upper reservoir (3 km).

As an example, the density distribution of seismic events observed during the hydraulic stimulations 2000–2005 is shown in a lateral view (Figure 4.20).

The localized seismic events are essentially located along single structures implying that predominantly a few single structures have been stimulated (Jung, 1999). This finding has been confirmed by the flow distribution in the borehole

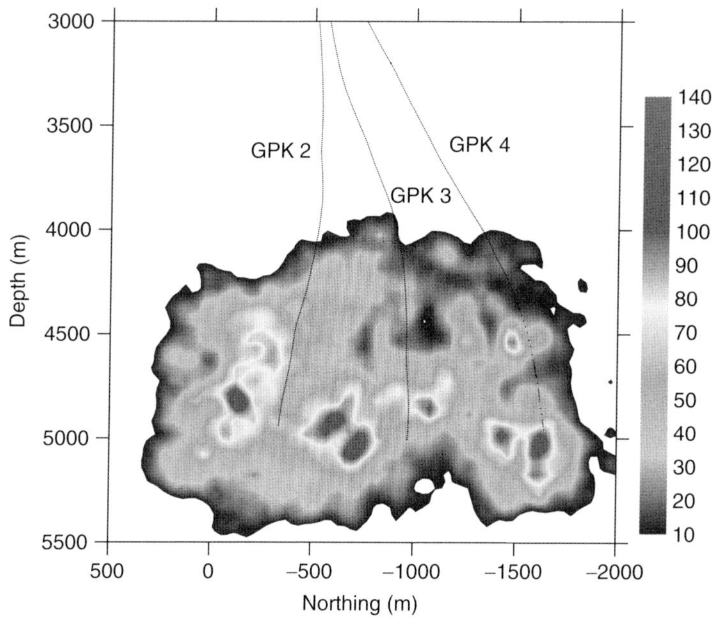

Figure 4.20 Color contour plot of the event density in 50 × 50 m cells in the plane of the graph. Perpendicular to this plane the cells are unlimited. All events localized during the stimulations 2000, 2003, 2004, and 2005 were included. (Please find a color version of this figure on the color plates.)

derived from flowlogs. In terms of the life duration of an EGS-reservoir, the stimulation of single structures may be critical. In further projects the application of a multifrac concept should therefore be an issue to achieve a more volumetric stimulation.

The concept of massive water injections may have the disadvantage of triggering larger seismic events (Baisch *et al.*, 2006). A few events above the threshold for human detection have occurred during the massive stimulation of GPK2 and GPK3 (5 km). The biggest induced seismic event ($M \approx 2.9$) in Soultz was recorded in 2003 during the shut-in period of the GPK3 stimulation. Although this event caused no structural damage or any casualties, it triggered serious concerns of the population, so that the succeeding stimulation of GPK4 was performed with significant smaller water volume.

4.10.2.2 **Chemical Stimulation**

The operations of chemical stimulation in the geothermal wells of Soultz-sous-Forêts are relatively recent because they have started in 2003 in the GPK-2 well. Moreover, few operations were carried out and the chemical treatments were limited in terms of time, volume, and concentration. The main operations occurred in GPK-4 and GPK-3 wells between 2006 and 2007.

Although the public concern about induced seismic events during hydraulic stimulation was one important reason for undertaking chemical treatments as an additional or even an alternative method to hydraulic stimulations, the main argument for chemical stimulation was the evidence of fracture filling carbonates and other soluble minerals, based on drill cuttings and cores analysis, as well as on geophysical logs.

Chemical stimulation and especially, the use of strong acids are rather known to attack the rocks, minerals, or fractures located at the vicinity of a well. It was the reason for which other chemical treatments such as NTA, chelating agent dissolved in caustic soda solution or OCA, mixing of weak organic acids, that slower dissolve rocks and minerals, were tested in GPK-4 well after the injection of HCl and RMA in order to extend the chemical stimulation as far as possible.

Chemical stimulations were performed by injecting acid from the wellhead through the casing string (9.5/8" for GPK-3 and GPK-4 and 7" for GPK-2). The stimulation zone was therefore, the whole open hole section of the wells (500–650 m length). Corrosion inhibitors were used to protect the inner casing string. With exception to chemical treatments with HCl, the other operations were conducted by specialized service companies. Hydraulic tests were performed before and after the chemical stimulations, to evaluate the progress in productivity or injectivity. A geochemical monitoring of the discharged fluid was carried out after most of the chemical stimulation experiments (Sanjuan *et al.*, 2007).

Although they were not executed with the same comparable protocol, different but encouraging results were observed after these series of tests using several chemical stimulation methods in a fractured granitic EGS reservoir. If GPK-3 well has shown weak variations of its injectivity, GPK-4 well presented a real increase of injectivity and productivity after the treatments (GEIE, 2006; Nami *et al.*, 2008), and GPK-2 well also presented a very sensible improvement despite the fact that the treatments were limited in terms of time, volume, and concentration (Gérard, Fritz, and Vuataz, 2005). As a summary of the chemical stimulation tested in the three deep wells of the Soultz EGS project, the Table 4.7 gives a synthesis of the principal results.

HCl and OCA, the only compounds injected into GPK-3, were not efficient to improve the injectivity and productivity indices of this well whereas the treatments used in GPK-4 increased the PI of GPK-4 from 0.2 to 0.5 $l\ s^{-1}\ bar^{-1}$. However, as the treatments used in each well were different and their combination in the case of GPK-4 could bring more efficiency, it is difficult to compare their effects between each well.

The injection of a caustic soda solution has been particularly efficient to clean up the well GPK-4 and neighboring fractures by removing significant amounts of drilling grease and drilling fragments such rock debris and cuttings. Consequently, the injection of a caustic soda solution accompanied by a fluid production test seems to be an efficient operation of cleaning and is recommended to remove drilling wastes and residues from the geothermal wells and their vicinity just after their drilling. Given the results obtained on GPK-4, it would have been preferable to use NTA after a prior injection of caustic soda solution, which would have

Table 4.7 Summary of the chemical stimulation operations carried out in the three 5-km deep wells. at Soultz-sous-Forêts EGS project.

Well	Date	Concentration of chemical agents	Results: injectivity/productivity increase
GPK2 (production well)	February 2003	HCl 0.09 + 0.18%	Wellhead pressure drop and productivity increase (0.5 $l\,s^{-1}$ bar^{-1}).
GPK3 (re-injection well)	June 2003	HCl 0.45%	No increase: 0.35 $l\,s^{-1}$ bar^{-1}
	February 2007	Organic clay acid OCA HT	Weak impact: 0.4 $l\,s^{-1}$ bar^{-1}
GPK4 (production well)	February 2005	HCl 0.2%	Productivity: 0.2–0.3 $l\,s^{-1}$ bar^{-1}
	May 2006	HCl 15% (3 tons) – HCl 12% + HF 3%	Productivity: 0.3–0.4 $l\,s^{-1}$ bar^{-1}
	October 2006	Chelant: NTA 19%	The formation of a plug increased wellhead pressure.
	March 2007	Organic clay acid OCA HT	Productivity: 0.4–0.5 $l\,s^{-1}$ bar^{-1}

allowed cleaning the well and neighboring fractures, and injecting NTA farer in the fractures.

Although few data are available, the efficiency of a chemical treatment to improve the injectivity or productivity of a well seems to be dependent on the distribution, size, and hydrothermal deposits of the fractures located at the vicinity of this well. So, the injection of HCl into GPK-2 and GPK-4 has increased the productivity of these wells (Gérard, Fritz, and Vuataz, 2005) whereas HCl and OCA injected into GPK-3 had no effect in this well crossed by a major fault governing more than 75% of its communication with the reservoir at 4750 m true vertical depth (TVD).

The chemical stimulation program performed at Soultz generated an improvement factor of 1.1–2.5 of the injectivity/productivity (Nami *et al.*, 2008). The integration of results from seismic monitoring, temperature, and flow logging helps in detecting the productive zones of the wells and their changes through chemical stimulations. The effectiveness of chemical stimulation could be further improved by using techniques to divert the treatment fluid toward selected zones in the reservoir (drillpipe, coiled tubing, packers). Particularly in fractured crystalline formations, where the reservoir permeability is strongly controlled by the pre-existing natural fracture network, a "focused" acidizing of these highpermeable joints and fracture zones is essential.

Due to the recent and limited operations, the experience in chemical stimulation in the geothermal wells acquired from the site of Soultz is not very important. Regrettably, the same chemical treatments have not been used in GPK-3 and GPK-4 and hence, it is not possible to compare the efficiency of these treatments.

For example, a prior cleaning operation with a caustic soda solution to remove significant amounts of drilling grease and drilling fragments from GPK-3 (injection well), and from neighboring fractures, could have improved the injectivity index of this well and allowed a best efficiency of the OCA treatment. A low seismic activity has been observed during the chemical stimulation.

It is obvious that further on-site and laboratory experiments of chemical stimulation must be carried out in the future and probably combined with moderate hydraulic stimulation operations in order to reduce the injection pressures and the possible resulting microseismic events, and improve the stimulation effects. Such combined operations were not carried out on at Soultz.

4.10.3 Horstberg

4.10.3.1 Introduction

In Germany, hydrothermal reservoirs in which geothermal energy can be extracted without hydraulic stimulation occur only in a few regions. Low permeable rock dominates in the deep underground and hence, the utilization of geothermal resources can be considerably extended, if concepts are available for heat extraction from relative tight formations.

The investigations in Horstberg address the geothermal utilization of low permeable sedimentary rocks in the North German Basin.

For this objective, the experiences made in the EGS project "Soultz" serve as an important basis. The concept of massive waterfracs as applied in Soultz is the basic stimulation concept for the investigations in Horstberg too.

In Horstberg, single well schemes are investigated, where the well is simultaneously used for production and re-injection, in difference to the common dublet scheme. Via large artificially created fractures a hydraulic link should be formed between two layers at some vertical distance. This hydraulic link allows then the circulation of water within one well. A second method of heat energy extraction from one well is a Huff-Puff-scheme: Cold water is injected in a large fracture and extracted as hot water after some recovery time. However, the basis for both single well concepts is the creation of large fractures in a low permeable formation by applying the waterfrac concept.

A one-well-concept can be operated economically even for a relatively low power output in the order of few MW_{th} and, is suitable for providing heat to large buildings, or districts.

The scientific investigations at the test site "Horstberg" are an integral part of the GeneSys-project. The GeneSys project finally aims at the realization of a one well concept in Hannover to geothermally heat the buildings of the GEOZENTRUM based on the experiences of "Horstberg."

The well Horstberg Z1 is located some 80 km north-east of Hannover (Figure 4.21). It has been drilled into an inversion structure, striking NEE-SWW, that is bound by saltdomes on either sides. The encountered stratigraphy is typical for the Northern German Basin (Figure 4.22).

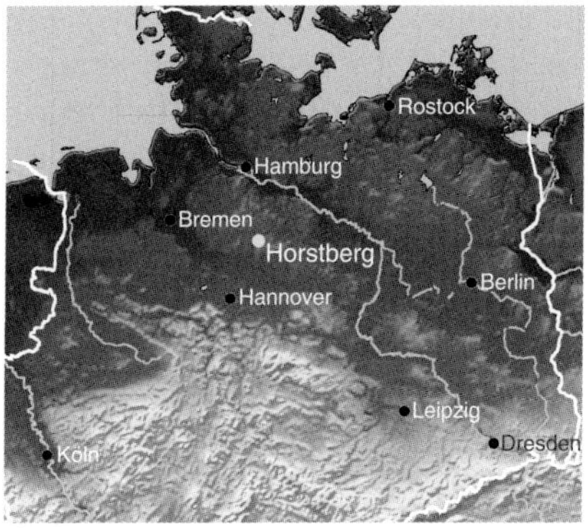

Figure 4.21 Location of the test site "Horstberg" in the Northern German Basin.

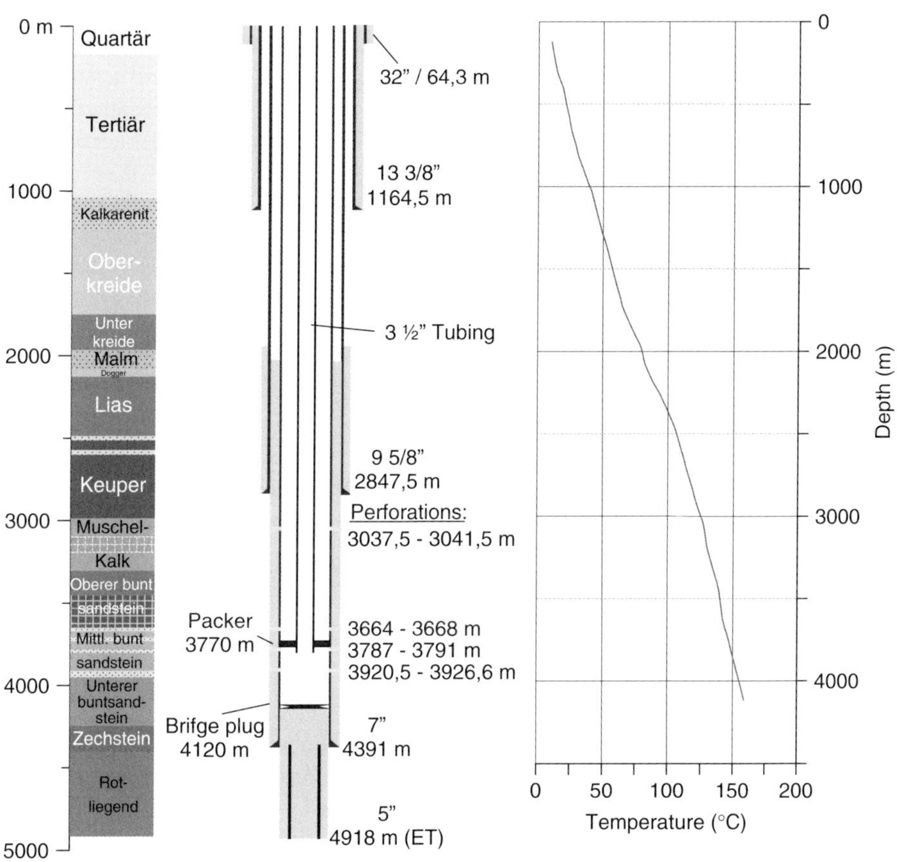

Figure 4.22 Stratigraphy, schematic of well completion and equilibrium temperature profile of the Horstberg well.

The well was originally intended for the production of gas but no economic production was achieved. The well was abandoned as gas well and transferred to Federal Institute for Geosciences and Natural Resources (BGR) as a geothermal test well. Before takeover, the gas bearing formation "Rotliegend" was sealed by plugging back the well from final depth of 4918–4120 m.

For geothermal investigations the Middle Bunter (3636–3926 m) was selected because of its widespread occurrence in the Northern German Basin and because of the high temperature at that depth level. The Middle Bunter is dominated by clay and siltstone and only a few sandstone subformations (Volpriehausen-, Detfurth-, and Solling) are embedded (see Figure 4.23). These sandstone layers are in the focus of investigations. The porosities of these layers vary from 3 to 11%. Due to the low porosities and permeability, an economic production rate over a relevant time period cannot be maintained by simply producing these layers.

"The well is equipped with a 7" casing that is cemented from final depth to 2035 m. These specifications allow injection operations at high pressures and flow rates, and separate perforation of single horizons. Waterfracs were performed in two formations, Volpriehausen at about 3920 m depth and Dettfurth at about 3790 m after individual perforation.

In 2004, a packer was installed at 3770 m depth in order to separate the two subformations Solling and Detfurth as a precondition for circulating water between both layers.

Information on the stress field was not available before fracturing. No indicators for stress (borehole breakouts, drilling induced fractures) were detected during drilling. Besides, it is known that on a regional scale the stress field is very inhomogeneous because of the salt tectonics. Hence, the evaluation of stress indicators from other wells around is likely not representative for the test site Horstberg.

4.10.3.2 **Fracturing Experiments**

Waterfrac-operations were performed through the perforated intervals in the Volpriehausen-sandstone and later, in the Detfurth-sandstone. Pure freshwater without any additive was applied at these frac operations.

Volpriehausen The waterfrac tests in the Volpriehausen comprised several injections with increasing flow rate.

In a first injection, a breakdown pressure of about 430 bar was recorded at a flow rate of only $0.17\,l\,s^{-1}$. These values indicate a low initial productivity and an almost impermeable formation.

During the most extensive fracturing test the volume of $600\,m^3$ was pumped at a rate of $7\,l\,s^{-1}$ and at a wellhead pressure of 460 bar. As the technical equipment was not dimensioned for this high pressure a flow rate larger than $7\,l\,s^{-1}$ could not be injected. Therefore, the stimulation of the Volpriehausen ended ahead of schedule.

Production tests after fracturing revealed still a poor productivity of the Volpriehausen. A low rate of about $0.5\,l\,s^{-1}$ led to a pressure drop of more than 300 bar after a few hours of production.

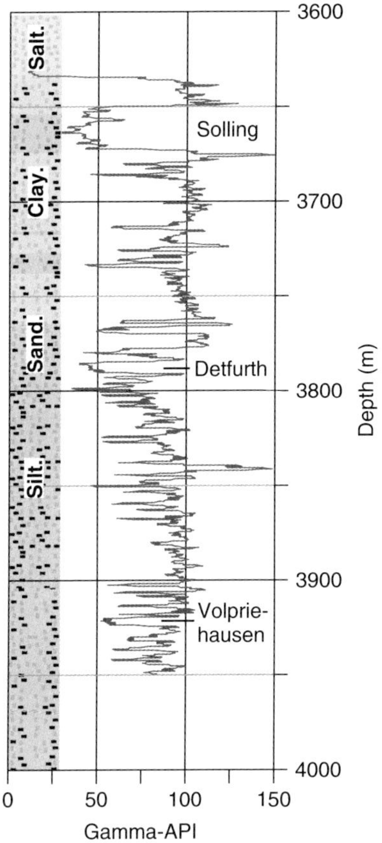

Figure 4.23 Lithology and Gamma-Ray-Log in the depth range of relevance (3600–4000 m).

Detfurth After perforating the Detfurth formation, a breakdown pressure of about 340 bar was recorded at the beginning of injection (Figure 4.24). A small injection rate of $1.7\,\mathrm{l\,s^{-1}}$, pumped for only 2 hours, was already sufficient to reach this fracturing pressure. By including the reservoir pressure of about 230 bar (wellhead), as determined later, the initial productivity of the Detfurth-sandstone could be estimated as lower than $0.015\,\mathrm{l^{-1}s\,bar^{-1}}$. This number is an upper limit because the pressure was cut by fracturing and not by water diffusion. A further reason for overestimating is its derivation after a short time period.

The fracture was enlarged and propagated by injecting 20 000 m^3 fresh water during three frac tests (Figure 4.24). The fracturing pressure was approximately 330 bar at a typical flow rate of $50\,\mathrm{l\,s^{-1}}$.

It should be noted that the fracturing pressure in the Detfurth formation is about 100 bar lower than in the Volpriehausen, only 120 m below. Obviously, the minimum stress varies significantly over short vertical distances – an issue for further investigations.

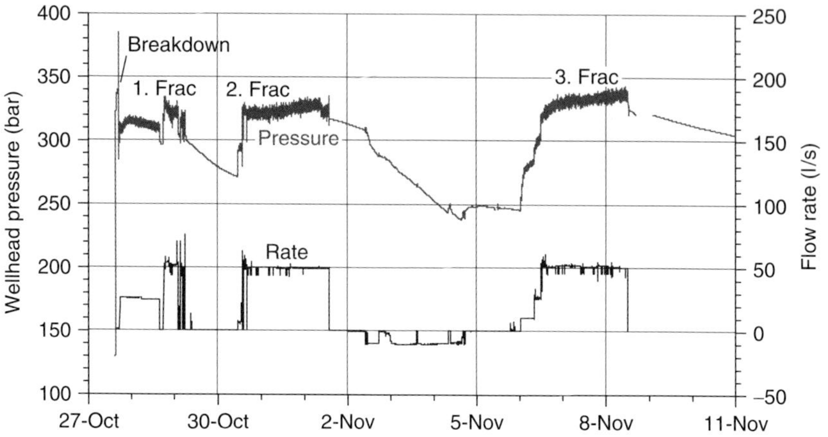

Figure 4.24 Wellhead pressure and flow rate during stimulation (waterfrac) of the Detfurth. Positive flow means injection and negative means production.

To estimate the fracture dimensions, the pressure decline after fracturing is of special importance. The pressure decreases very slowly but almost linearly. Even when water was produced at a rate of $10\,\mathrm{l\,s^{-1}}$, after the second frac test, the linear pressure decay continued. Three important conclusions about the frac propagation can be derived from the slow and linear pressure decline:

- The frac behaves like a huge ballon with a storage capacity of about $100\ \mathrm{m^3\,bar^{-1}}$. The huge storage capacity is indicative for tensile fracturing rather than shearing.
- The fracture area is larger than $100\,000\ \mathrm{m^2}$ as derived from the storage capacity and assuming typical mechanical rock properties.
- The hydraulic connection to permeable structures in the far field is poor.

The temperature profile, measured half a year after the frac operations confirms the assumption of a large fracture (Figure 4.25). The cooled region extended over a vertical distance of more than 150 m. Adjacent clay and siltstone layers were obviously penetrated by the fracture showing that the frac propagation is not confined by these layers as is commonly expected. In particular, a hydraulic link was formed between the Detfurth-sandstone and the upper Solling-sandstone. This link forms the basis for testing the concept of a closed circulation between both layers, separated by a packer in the well.

The hydraulic fracture characteristics at near reservoir pressure have been investigated in several tests. Figure 4.26 displays the pressure and flow rate for a cyclic test performed in 2004.

The test consisted of an injection period with cooled water and the succeeding production of warm water out of the frac after a period of heating-up. As shown, the pressure close and below reservoir pressure can be matched with an unique fit. The pressure varies linearly versus the square root of time over a wide range

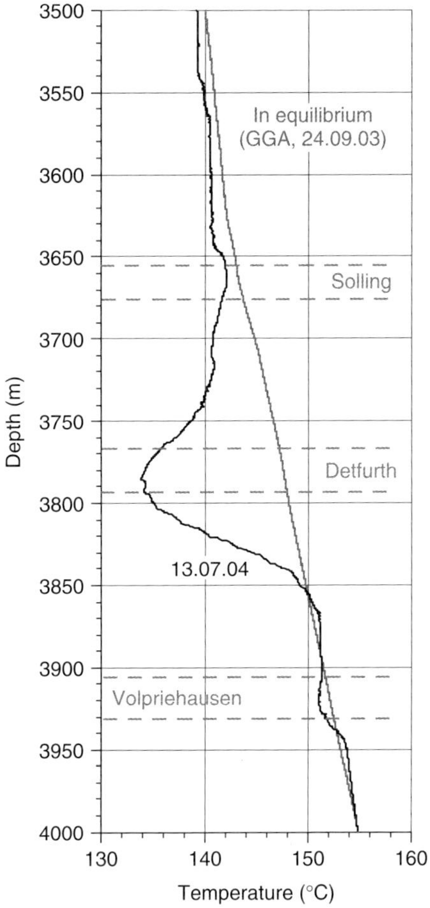

Figure 4.25 Temperature profile recorded half a year after fracturing the Detfurth-sandstone (black) in comparison to the equilibrium profile (red).

of pressure. Formation linear flow from a high conductive fracture into the matrix can unambiguously be derived from these characteristics.

Because of the long-lasting formation linear flow, a radial flow regime has not been observed and the independent determination of the matrix permeability is not possible. Assuming an effective matrix permeability of 1 mD, a fracture area of 10 000 m^2 can be derived. For an assumed permeability of 0.1 mD the fracture area would be approximately threefold higher (30 000 m^2). The persisting fracture area is significantly smaller than the fracture area during stimulation.

The most important result is, however, that a large part of the fracture remains open (infinite conductive) even at low pressure. It demonstrates that the concept of massive waterfracturing without proppants is a successful method for stimulating low permeable siliciclastic sedimentary rock.

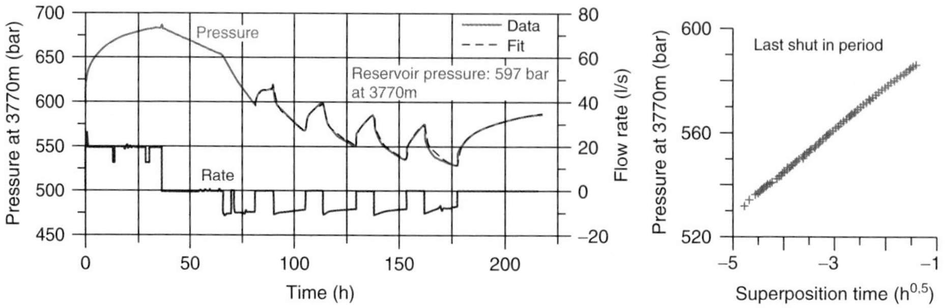

Figure 4.26 Cyclic injection/production test performed in 2004. The right picture displays the pressure build up during the last shut-in period of the test versus the square root of time-superposition time.

The productivity of the well after stimulation was determined as $0.15\,\mathrm{l\,s^{-1}\,bar^{-1}}$ after 6 hours production and at a pressure level below the reservoir pressure. Accordingly, the productivity of the well was improved due to fracturing by a factor of 10, at least. The still limited productivity after stimulation reflects the impact of a low permeable matrix. For a concept that is based mainly on the properties of the fracture the productivity however, is less meaningful. In particular, the circulation of water between two layers via this high conductive fracture, as it is envisaged in Horstberg, will mainly depend on the fracture property and less on those of the matrix.

It has to be emphasized that the good hydraulic properties of the created fracture persist already over a long time. A low rate injection test performed in 2007 provided essentially the same hydraulic parameters as derived shortly after stimulation, in 2004.

4.10.3.3 **Summary and Conclusion**

Waterfrac operations in two different sandstone formations of the Horstberg well were performed after perforating the individual layers. In the Volpriehausen-sandstone, only a small-scale operation was carried out with a water volume of less than 1000 m^3. Here, no significant productivity enhancement was achieved. A second formation (Detfurth-sandstone) was perforated and stimulated by a much higher volume. About 20 000 m^3 of fresh water at a typical rate of $50\,\mathrm{l\,s^{-1}}$ was injected here. A huge fracture with an area of more than 100 000 m^2 was created whereof a significant part of more than 10 000 m^2 retained a high (infinite) hydraulic conductivity at reservoir pressure. Although, only a part of the original created fracture remains high conductive after fracturing, the dimensions of this retaining fracture are still large compared to fractures that are typically created in more permeable rock by applying the proppant frac concept. Based on the successfully created fracture new concepts are tested for the heat extraction from one well.

The question why the waterfrac operation was successful only in one formation is still an issue for investigation. One reason might be cooling and inelastic shrinking of the rock. In the Detfurth formation, a much higher volume and a higher rate was applied and hence the rock mass was much more cooled than in the Volpriehausen. The usually assumed mechanism of shearing as reason for self-propping is implausible in the Detfurth. The observed ballooning of the fracture while injecting is not typical for a shear fracture.

References

Adams, M.C., Beall, J.J., Enedy, S.L., Hirtz, P.N., Kilbourn, P., Koenig, B.A., Kunzman, R., and Bill Smith, J.L. (2001) Hydrofluorocarbons as geothermal vapor-phase tracers. *Geothermics*, **30**, 747–775.

Adams, M.C., Moore, J.N., Lazlo, G.F., and Jong-Hong, A. (1992) Thermal stabilities of aromatic acids as geothermal tracers. *Geothermics*, **21** (3), 323–339.

Adams, M.C., Yamada, Y., Yagi, M., Kasteler, C., Kilbourn, P., and Dahdah, N. (2004) Alcohols as two-phase tracers. Proceedings of the 29th Workshop on Geothermal Reservoir Engineering, Stanford University, January 26-28, 2004, SGP-TR-175, Stanford.

Armstead, H.C.H. and Tester, J.W. (1987) *Heat Minning – A New source of energy*, E & F.N. Spon, Ltd, London and New York.

Axelsson, G., Thórhallson, S., and Björnsson, G. (2006) Stimulation of geothermal wells in basaltic rock in Iceland. ENGINE – Workshop 3, Stimulation of Reservoir and Microseismicity, Kartause Ittingen, June 29–July 1, Zürich.

Baisch, S., Weidler, R., Vorös, R., and Jung, R. (2006) A conceptual model for post-injection seismicity at Soultz-sous-Forêts. *Geothermal Resources Council Transactions*, **30**, 601–605.

Baria, R., Baumgärtner, J., Rummel, F., Pine, R.J., and Sato, Y. (1999) HDR/HWR reservoirs: concepts, understanding and creation. *Geothermics*, **28**, 533–552.

Baria, R., Jung, R., Tischner, T., Nicholls, J., Michelet, S., Sanjuan, B., Soma, N., Asanuma, H., Dyer, B., and Garnish, J. (2006) Creation of an HDR reservoir at 5000 m depth at the European HDR project. Proceedings of the Thirty-First Workshop on Geothermal Reservoir Engineering, Stanford University, Stanford, January 30-February 1, 2006, SGP-TR-179.

Barree, R.D., Fisher M.K., and Woodproof, R.A. (2002) A Practical Guide to Hydraulic Fracture Diagnostic Technologies, Society of Petroloem Engineers, Paper No. 77442.

Barrios, L.A., Quijano, J.E., Romero, R.E., Mayorga, H., Castro, M., and Caldera, J. (2002) Enhanced permeability by chemical stimulation at the Berlin Geothermal Field, El Salvador. *Geothermal Resources Council Transactions*, **26**, 73–78.

Baumgärtner, J., Jung, R., Hettkamp, T., and Teza, D. (2004) The status of the hot dry rock scientific power plant at Soultz-sous-Forêts. *Zeitschrift für Angewandte Geologie*, **2**, 12–16.

Bertini, G., Casini, M., Ciulli, B., Ciuffi, S., and Fiordelesi, A. (2005) Data revision and upgrading of the structural model of the travale geothermal field (Italy). World Geothermal Congress, Antalya, Turkey.

Blumenthal, M., Kühn, M., Pape, H., Rath, V., and Clauser, Ch. (2007) Hydraulic model of the deep reservoir quantifying the multi-well tracer test. EHDRA Scientific Conference, June 28-29 2007, Soultz-sous-Forêts, 5 p.

Bommer, J.J., Georgallides, G., and Tromans, I.J. (2001) Is there a near-field for small to moderate magnitude earthquakes? *Journal of Earthquake Engineering*, **5** (3), 395–423.

Bommer, J.J., Oates, S., Cepeda, J.M., Lindholm, C., Bird, J., Torres, R., Marroquín, G., and Rivas, J. (2006) Control of hazard due to seismicity induced by a hot fractured rock geothermal project. *Engineering Geology*, **83** (4), 287–306.

Bromley, C.J. and Mongillo, M.A. (2008) Geothermal energy from fractured reservoirs – Dealing with induced seismicity. *IEA Open Energy Technology Bulletin*, **48**, 1–7.

Brune, J. and Thatcher, W. (2002) *International Handbook of Earthquake and Engineering Seismology*, Vol. 81A, International Association of Seismology and Physics of Earth's Interior, Committee on Education, pp. 569–588.

Buijse, M., Maier, R., and Casero, A., (2000) Successful high pressure, high temperature acidising with in situ crosslinked acid diversion. Proceedings of the SPE InternationalSymposium on Formation Damage Control, Lafayette, LA, USA, February 23-24, p. 8.

Buning, B.C., Malate, R.C.M., Austria, J.J.C., Noriega, M.T., and Sarmiento, Z.F. (1997) Casing perforation and acid treatment of well SK-2D Mindanao 1 geothermal project, Philippines. Proceedings of the 22nd Workshop on Geothermal Reservoir Engineering, January 27-29, 1997, Stanford.

Buning, B.C., Malate, R.C.M., Lacanilao, A.M., Sta Ana, F.X.M., and Sarmiento, Z.F. (1995) Recent experiments in acid stimulation technology by PNOC-Energy development corporation, Philippines. *Proceedings of World Geothermal Congress*, **3**, 1807–1812.

Burgos, B., Buijse, M., Fonseca, E., Milne, A., Brady, M., and Olvera, R. (2005) Acid fracturing in Lake Maracaibo: how continuous improvements kept on raising the expectation bar. Schlumberger and Shell Venezuela S.A, 2005 SPE Annual Technical Conference and Exhibition, October 9-12, Dallas, SPE 96531.

Capetti, G. (2006) How EGS is investigated in the case of the larderello geothermal field? Engine Launching Conference, February 12-15, 2006, Orleans.unpublished presentation.

Carslaw, H.S., and Jaeger, J.C., (1959) *Conduction of Heat in Solids*, 2nd edn, Clarendon Press, Oxford.

Charléty, J., Cuenot, N., Dorbath, L., Dorbath, C., Haessler, H., and Frogneux, M. (2007) Large earthquakes during hydraulic stimulations at the geothermal site of Soultz-sous-Forêts. *International Journal of Rock Mechanics and Mining Sciences*, **44**, 1091–1105.

Charlez, P., Lemonnier, P., Ruffet, C., Boutéca, M.J., and Tan, C. (1996) Thermally Induced Fracturing: Analysis of a Field Case in North Sea, Society of Petroleum Engineers, SPE 36916.

Cinco-Ley, H. and Samaniego, V.F. (1977) Effect of Wellbore Storage and Damage on the Transient Pressure Behaviour of Vertically Fractured Wells.SPE, Paper No. 6752.

Cipolla, C.L. and Wright, C.A. (2000) State-of-the – art in Hydraulic Fracture Diagnostics, Society of Petroleum Engineers, Paper No. 64434.

Clifford, P.J., Berry, P.J., and Gu, H. (1991) Modeling the vertical confinement of injection-well thermal fractures, SPE (Society of Petroleum Engineers). *SPE Production Engineering*, **6** (4), 377–383.

Cornet, F.H. (1987) Results from Le Mayet de Montagne project. *Geothermics*, **16** (4), 355–374.

Cornet, F.H., Bérard, T., and Bourouis, S. (2007) How close to failure is a granite rock mass at a 5 km depth? *International Journal of Rock Mechanics and Minning Science*, **44** (1), 47–66.

Crowe, C., Masmonteil, J., and Thomas, R. (1992) Trends in matrix acidizing. *Oilfield Review*, **4**, 24–40.

Cuenot, N., Charléty, J., Dorbath, L., and Haessler, H. (2006) Faulting mechanisms and stress regime at the European HDR site of

Soultz-sous-Forêts, France. *Geothermics*, **35**, 561–575.

Cuenot, N., Dorbath, C., and Dorbath, L. (2008) Analysis of the microseismicity induced by fluid injections at the Hot Dry Rock site of Soultz-sous-Forêts (Alsace, France): implications for the characterization of the geothermal reservoir properties. *Pure and Applied Geophysics*, **165**, 797–828.

Davies, D.R., Faber, R., Nitters, G., and Ruessink, B.H. (1992) A novel procedure to increase well response to matrix acidizing treatments. 2nd SPE Latin American Petroleum Engineering Conference, March 8-11, Caracas, SPE 23621.

Dorbath, L., Cuenot, N., Genter, A., and Frogneux, M. (2009) Seismic response of the fractured and faulted granite to massive water injection at 5 km depth at Soultz-sous-Forêts (France). *Geophysical International Journal* **177** (2), 653–675.

Economides, M.J. and Nolte, K.G. (1989) *Reservoir Stimulation*, 2nd edn, Schlumberger Educational Services.

Economides, M.J. and Nolte, K.G. (2000) *Reservoir Stimulation*, 3rd edn, John Wiley & Sons, Ltd.

Entingh, D.J. (1999) A review of geothermal well stimulation experiments in the United States. *Geothermal Resources Council Transactions*, **23**, 175–180.

Epperson, I.J. (1983) Beowawe acid stimulation. *Geothermal Resources Council Transactions*, **7**, 409–411.

Evanoff, J., Yeager, V., and Spielman, P. (1995) Stimulation and damage removal of calcium carbonate scaling in geothermal wells: a case study. World Geothermal Congress, Florence, pp. 2481–2485.

Evans, K.F., Genter, A., and Sausse, J. (2005a) Permeability creation and damage due to massive fluid injections into granite at 3.5 km at Soultz: Part 1 – Borehole observations. *Journal of Geophysical Research*, **110**, B04203, 19.

Evans, K.F., Moriya, H., Niitsuma, H., Jones, R.H., Phillips, W.S., Genter, A., Sausse, J., Jung, R., and Baria, R. (2005b) Microseismicity and permeability enhancement of hydrogeologic structures during massive fluid injections into granite at 3 km depth at the Soultz HDR site. *Geophysical Journal International*, **160**, 388–412.

Flores, M., Barajas, E.N., and Rodriguez, M.A. (2006) Productivity analysis and acid treatment of well AZ-9AD at the Los Azufres geothermal field, Mexico. *Geothermal Resources Council Transactions*, **30**, 791–796.

Fredd, C.N. and Fogler, H.S. (1998) The influence of chelating agents on the kinetics of calcite dissolution. *Journal of Colloid and Interface Science*, **204**, 187–197.

Gadde, P.B. and Sharma, M.M. (2001) Growing Injection Well Fractures and their Impact on Waterflood Performance, Society of Petroleum Engineers, SPE 71614.

GEIE (2006) Results of GPK4 RMA Stimulation of May 2006, unpublished data.

Gérard, A., Fritz, B., and Vuataz, F.-D. (2005) Results of soft acid injection tests performed at Soultz in wells GPK2, GPK3 and GPK4 – extended summary: revised status on 14 March 2005. Proceedings of EHDRA Scientific Conference, March 17-18, 2005, Soultz-sous-Forêts, France.

Gérard, A., Genter, A., Kohl, T., Lutz, P., Rose, P., and Rummel, F. (2006) The deep EGS (Enhanced Geothermal System) project at Soultz-sous-Forêts (Alsace, France). *Geothermics*, **35** (5-6), 473–483.

Ghergut, I., Sauter, M., Behrens, H., Licha, T., McDermott, C.I., Herfort, M., Rose, P., Zimmermann, G., Orzol, J., Jung, R., Huenges, E., Kolditz, O., Lodemann, M., Fisher, S., Wittig, U., Güthoff, F., and Kühr, M. (2007) Tracer tests evaluating hydraulic stimulation at deep geothermal reservoirs in Germany. Proceedings of the 32nd Workshop on Geothermal

Reservoir Engineering, Stanford University, January 22-24, 2007, SGP-TR-183, Stanford.

Häring, O. (2007) *Geothermische Stromproduktion aus Enhanced Geothermal Systems (EGS); Stand der Technik*, Geothermal Explorers, Ltd, Pratteln.

Hettkamp, T., Baumgärtner, J., Baria, R., Gérard, A., Gandy, T., Michelet, S., and Teza, D. (2004) Electricity production from hot rocks. Proceedings of 29th Workshop on Geothermal Reservoir Engineering, Stanford University, January 26-28, 2004, SGP-TR-175, Stanford.

Hoijkaas, G.R., Genter, A., and Dezayes, C. (2006) Deep seated geology of the granite intrusions at the Soultz EGS site based on data from 5 km-deep boreholes. *Geothermics*, **35**, 484–506.

Holl, H.-G., Hurter, S., Saadat, A, Köhler, S., Wolfgramm, M., Zimmermann, G., Trautwein, U., Winter, H., Legarth, B., and Huenges, E. (2003) First hand experience in a second hand borehole: Hydraulic experiments and scaling in the geothermal well Groß Schönebeck after reopening. Proceedings of the International Geothermal Conference, IGC-2003 Reykjavik, Multiple Integrated Uses of Geothermal Resources, S01, paper 060, 8–13.

Holl, H.-G., Moeck, I., and Schandelmeier, H. (2004) Geothermal well Groß Schönebeck 3/90. A low enthalpy reservoir (Rotliegend, NE Germany), Proceedings 66th EAGE Conference & Exhibition, F032, Paris.

Hoover, D.B. and Dietrich, J.A. (1969) Seismic Activity during the 1968 Test Pumping at the Rocky Mountain Arsenal Disposal Well, US Geological Survey Circular 61.

Horne, R.N. (1995) *Modern Well Testing Analysis*, 2nd edn, Petroway, Inc., Palo Alto, p. 257

Hsieh, P.A. and Bredehoft, J.D. (1981) A reservoir analysis of the Denver earthquake: a case of induced seismicity. *Journal of Geophysical Research*, **86** (B2), 903–920.

Huenges, E., Holl, H.-G., Bruhn, D., Brandt, W., Saadat, A., Moeck, I., and Zimmermann, G. (2007) Current state of the EGS project Groß Schönebeck - Drilling into deep sedimentary geothermal reservoirs. European Geothermal Congress - EGC (Unterhaching, Germany 2007).

Huenges, E., Saadat, A., Brandt, W., Moeck, I., Holl, H.-G., Zimmermann, G., Blöcher, G., Köhler, S., Legarth, B., and Tischner, T. (2006) Current status of the EGS Gross Schönebeck project: on the way to demonstrate sustainable brine production from deep sediments of the North German basin. *Geothermal Resources Council Trans*, **30**, 341–346.

IEA-GIA (2009) Protocol for Induced Seismicity Associated with Enhanced Geothermal Systems, IEA-GIA Report Produced in Annex I Task D, *http://www.iea-gia.org/publications.asp* (accessed 10 Juli 2009).

Jaimes-Maldonado, J.G. and Sànchez-Velasco, R. (2003) Acid stimulation of production wells in Las Tres Vírgenes geothermal field, BCS, México. *Geothermal Resources Council Transactions*, **27**, 699–705.

Jung, R. (1999) Erschließung Permeabler Risszonen für die Gewinnung Geothermischer Energie aus Heißen Tiefengesteinen. Final report, Project 0326690A funded by the German Ministry for Education and Research (BMBF), p. 82.

Jung, R. and Weidler, R. (2000) A conceptual model for the stimulation process of the HDR-system at Soultz. *Geothermal Resources Council Transactions*, **24**, 143–147.

Kwiatek, G., Bohnhoff, M., Dresen, G., Schulze, A., Schulte, T., Zimmermann, G., and Huenges, E. (2008) Microseismic event analysis in conjunction with stimulation treatments at the geothermal research well GtGrSk 4/05 in Groß Schönebeck/Germany. Proceedings, Thirty third Workshop on Geothermal Reservoir Engineering, Stanford University, January 28-30, 2008, Stanford.

Lee, W.J. (1981) *Well Testing, SPE Textbook Series*, Vol. 1, Society of Petroleum Engineers, Houston, Texas.

Legarth, B., Huenges, E., and Zimmermann, G. (2005) Hydraulic fracturing in sedimentary geothermal reservoirs. *International Journal of Rock Mechanics and Minninig Sciences*, **42** (7–8), 1028–1041.

Legarth, B., Tischner, T., and Huenges, E. (2003) Stimulation experiments in sedimentary, low-enthalpy reservoirs for geothermal power generation Germany. *Geothermics*, **32** (4-6), 487–495.

Majer, E.L., Baria, R., Stark, M., Oates, S., Bommer, J., Smith, B., and Asanuama, H. (2007) Induced seismicity associated with enhanced geothermal systems. *Geothermics*, **36**, 185–222.

Malate, R.C.M., Austria, J.J.C., Sarmiento, Z.F., DiLullo, G., Sookprasong, A., and Francia, E.S. (1998) Matrix stimulation treatment of geothermal wells using sandstone acid. Proceedings of the 23rd Workshop on Geothermal Reservoir Engineering, January 26-28, 1998, Stanford.

Malate, R.C.M., Sookprasong, P.A., Austria, J.J.C., Sarmiento, Z.F., and Francia, E.S. (1999) Wellbore soaking: a novel acid treatment of geothermal injection wells. Proceedings of the 24th Workshop on Geothermal Reservoir Engineering, January 25-27, 1999, Stanford.

Malate, R.C.M., Yglopaz, D.M., Austria, J.J.C., Lacanilao, A.M., and Sarmiento, Z.F. (1997) Acid stimulation of injection wells in the Leyte geothermal power project, Philippines. Proceedings of the 22nd Workshop on Geothermal Reservoir Engineering, January 27-29, 1997, Stanford.

Mella, M.J., Rose, P.E., Adams, M.C., Dahdah, N.F., McCulloch, J., and Buck, C. (2006a) The Use of n-propanol at the site of the coso engineered geothermal system. Proceedings of the 31st Workshop on Geothermal Reservoir Engineering, Stanford University, January 30-February 1, 2006, SGP-TR-179, Stanford.

Mella, M.J., Rose, P.E., McCulloch, J., and Buck, C. (2006b) A tracer test using Ethanol as a two-phase tracer and 2-Naphthalene Sulfonate as a liquid-phase tracer at the Coso geothermal field. *GRC Transactions*, **30**, 919–921.

Morris, A., Ferrill, D.A., and Henderson, D.B. (1996) Slikp tendency analysis and fault reactivation. *Geology*, **24** (3), 275–278.

Nami, P., Schellschmidt, R., Schindler, M., and Tischner, R. (2008) Chemical stimulation operations for reservoir development of the deep crystalline HDR/EGS system at Soultz-sous-Forêts (France). Proceedings of the 32nd Workshop on Geothermal Reservoir Engineering, January 28-30, 2008, Stanford.

Niitsuma, H. (2004) *Detection of Hydraulically Created Permeable Structures in HDR/HWR Reservoir by High Resolution Seismic Mapping Techniques*, Geo-Engineering Book Series, Vol. 2, Elsevier, pp. 73–79.

Nguyen, P.D. (2006) Controlling Formation Fines at their Sources to Maintain Well Productivity, Halliburton, SPE 97659.

Orzol, J., Jung, R., Jatho, R., Tischner, T., and Kehrer, P. (2005) The genesys project: extraction of geothermal heat from tight sediments. Proceedings World Geothermal Congress, April 24-29, 2005, Antalya.

O'Sullivan, M. and McKibbin, R. (1993) Geothermal Reservoir Engineering, Course notes, Institute of Geothermal Engineering, University of Auckland, New Zealand.

Pasikki, R.G. and Gilmore, T.G. (2006) Coiled tubing acid stimulation: the case of AWI 8-7 production well in salak geothermal field, Indonesia. Proceedings of the 31st Workshop on Geothermal Reservoir Engineering, January 30-February 1, 2006, Stanford.

Perkins, T.K. and Gonzales, J.A. (1985) The effect of thermoelastic stresses on injection well fracturing. *SPE Journal*, **25**, 77–88.

Pine, R.J. and Batchelor, A.S. (1984) Downward migration of shearing in jointed rock during hydraulic injections. *International Journal of Rock Mechanics and Minning Science and Geomechanics Abstracts*, **21**, 249–263.

Pournik, M. (2004) Evaluation of sandstone acidizing with high strength HF solutions.Master of Science in Engineering, University of Texas, Austin.

Pruess, K., O'Sullivan, M.J., and Kennedy, B.M. (2000) Modelling of phase-partitioning tracers in fractured reservoirs. Proceedings of the 25th Workshop on Geothermal Reservoir Engineering, Stanford University, January 24-26, 2000, SGP-TR-165, Stanford.

Reinicke, A., Zimmermann, G., Huenges, E., and Burkhardt, H. (2005) Estimation of hydraulic parameters after stimulation experiments in the geothermal reservoir Groß Schönebeck 3/90 (North-German Basin). *International Journal of Rock Mechanics and Mining Sciences*, **42** (7-8), 1082–1087.

Romero, D.J., Valkó, P.P., and Economides, M.J. (2003) Optimization of the Productivity Index and the Fracture Geometry of a Stimulated Well With Fracture Face and Choke Skin.SPE, Paper No. 81908.

Rose, P.E., Benoit, W.R., and Kilbourn, P.M. (2001) The application of the polyaromatic sulfonates as tracers in geothermal reservoirs. *Geothermics*, **30**, 617–640.

Rose, P., Xu, T., Kovac, K., Mella, M., and Pruess, K. (2007) Chemical stimulation in near-wellbore geothermal formations: silica dissolution in the presence of calcite at high temperature and high pH. Proceedings of the 32st Workshop on Geothermal Reservoir Engineering, January 22-24, 2007, Stanford.

Rybach, L. (2006) Induced seismicity during EGS operation? From Beaujard, C., Kohl, T., and Signorelli, S., Actes, Proceedings of the ENGINE Special Workshop No. 3, "Stimulation of Reservoir and Induced Microseismicity", June 29-July 01, Ittingen.

Sanjuan, B., Pinault, J.-L., Rose, P., Gérard, A., Brach, M., Braibant, G., Crouzet, C., Foucher, J.-C., Gautier, A., and Touzelet, S. (2006) Tracer testing of the geothermal heat exchanger at Soultz-sous-Forêts (France) between 2000 and 2005. *Geothermics*, **35** (5-6), 622–653.

Sanjuan, B., Pinault, J.-L., Rose, P., Gérard, A., Crouzet, C., Touzelet, S., Gautier, A., and Charlot, A. (2007) Geochemical monitoring at Soultz-sous-Forêts (France) between October 2006 and March 2007, after the chemical stimulations (RMA, NTA and OCA) Carried out in the wells GPK-4 and GPK-3. Proceedings of EHDRA Scientific Conference, June 28-29, Soultz-sous-Forêts.

Sarda, J.P. (1977) Chemical leaching. Proceedings of the 2nd NATA-CCMS Information Meeting on Hot Dry Rock Geotheral Energy, June 28-30, 1977, Los Alamos.

Schechter, R.S. (1992) *Oil Well Stimulation*, Prentice Hall, Englewood Cliffs.

Schindler, M., Nami, P., Schellschmidt, R., Teza, D., and Tischner, T. (2008) Summary of hydraulic stimulation operations in the 5 km deep crystalline HDR/EGS reservoir at Soultz-sous-Forêts. Proceedings of Thirty-Third Workshop on Geothermal Reservoir Engineering, Stanford University, January 28-30, SGP-TR-185, Stanford.

Serpen, U. and Türeyen, O.I. (2000) Acidizing geothermal wells. *Geothermal Resources Council Transactions*, **24**, 683–688.

Shaoul, J., Ross, M., Spitzer, W., Wheaton, S., Mayland, P., and Singh, A.P. (2007a) Massive Hydraulic Fracturing Unlocks Deep Tight Gas Reserves in India, SPE, Paper No. 107337.

Shaoul, J., Ross, M., Spitzer, W., Wheaton, S., Mayland, P., and Singh, A.P. (2007b) Hydraulic Fracturing Using Heated Fluids Brings Success in High-Pour-Point Waxy-Oil, SPE, Paper No. 107338.

Sharma, M.M., Gadde, P.B., Sullivan, R., Sigal, R., Fielder, R., Copeland, D., Griffin, L., and Weijers, L. (2004) Slick Water and Hybrid Fracs in the Bossier: Some Lessons Learnt, SPE, Paper No. 89976.

Shook, G.M. (2005) A systematic method for tracer test analysis: an example using Beowawe tracer data. Proceedings of the 30th Workshop on Geothermal Reservoir Engineering, Stanford University, January 31-February 2, 2005, SGP-TR-176, Stanford.

Smith, C.F. and Hendrickson, A.R. (1965) Hydrofluoric acid stimulation of sandstone reservoirs. *Journal of Petroleum Technology*, **17**, 215–222.

Soma, N., Asanuma, H., Kaieda, H., Tezuka, K., Wyborn, D., and Niitsuma, H. (2004) On site mapping of microseismicity at Cooper Basin, Australia HDR project by the Japanese team. Proceedings of the Twenty ninth Workshop on Geothermal Reservoir Engineering, Stanford University, January 26-28, 2004, Stanford.

Strawn, J.A. (1980) Results of acid treatment in hydrothermal direct heat experiment wells. *Geothermal Resources Council Transactions*, **4**, 427–430.

Sundquist, U., Wallroth, T., and Eliasson, T. (1988) The Fjällbacka HDR Geothermal Energy Project: Reservoir Characterisation and Injection Well Stimulation. Report Number Fj-9, Chalmers University of Technology.

Tischner, T. (2004) *Stimulations experimente und hydraulische Untersuchungen in den Vulkaniten der Bohrung Groß Schönebeck, Abschlussbericht.* Bundesanstalt für Geowissenschaften und Rohstoffe, Hannover.

Tischner, T., Schindler, M., Jung, R., and Nami, P. (2007) HDR project Soultz: Hydraulic and seismic observations during stimulation of the 3 deep wells by massive water injections. Proceedings of the 32th Workshop on Geothermal Reservoir Engineering, Stanford University, January 22-24, Stanford.

Tulinius, H., Axelsson, G., Tómason, J., Kristmannsdóttir, H., and Gudmundsson, A. (1996) Stimulation of well SN-12 in the Seltjarfnes low-temperature filed in SW-Iceland. Proceedings of the Twenty-First Workshop on Geothermal Reservoir Engineering, Stanford University, January 22-24, Stanford.

Tulinius, H., Correia, H., and Sigurdsson, O. (2000) Stimulating a high enthalpy well by thermal cracking. Proceedings World Geothermal Congress, Japan, 1883-1887.

Wallroth, T., Eliasson, T., and Sundquist, U. (1999) Hot Dry Rock research experiments at Fjällbacka, Sweden. *Geothermics*, **28** (4), 617–625.

Wallroth, T., Jupe, A.J., and Jones, R.H. (1996) Characterisation of a fractured reservoir using microearthquakes induced by hydraulic injections. *Marine Petroleum Geology*, **13** (4), 447–455.

Walsh, M.P., Lake, L.W., and Schechter, R.S. (1982) A description of chemical precipitation mechanisms and their role in formation damage during stimulation by hydrofluoric acid. *Journal of Petroleum Technology*, **34**, 2097–2112.

Weber, M., Zetsche, F., Ryberg, T., Schulze, A., Spangenberg, E., Huenges, E. (2005) Seismic Detection Limits of Small, Deep, Man-Made Reflectors: A Test at a Geothermal Site in

Nothern Germany. *Bulletin of the Seismological Society of America*, **95** (4), 1567–1573.

Wen, Q., Zhang, S., Wang, L., Liu, Y., and Li, X. (2007) The effect of proppant embedment upon the long-term conductivity of fractures. *Journal of Petroleum Science and Engineering*, **55**, 221–227.

Yglopaz, D.M., Buning, B.C., Malate, R.C.M., Sta Ana, F.X.M., Austria, J.J.C., Salera, J.R.M., Lacanilao, A.M., and Sarmiento, Z.F. (1998) Proving the Mahanagdong B Resource: A Case of a large-scale well stimulation strategy, leyte geothermal power project, Philippines. Proceedings of the 23rd Workshop on Geothermal Reservoir Engineering, January 26-28, 1998, Stanford.

Zimmermann, G. (2004) Results of moderate pumping tests in the deep well Groß Schönebeck 3/90 in summer 2002. *Scientific Technical Report*, GeoForschungsZentrum. Potsdam, STR04/03, 123–135.

Zimmermann, G., Hurter, S., Saadat, A., Köhler, S., Trautwein, U., Holl, H.-G., Wolfgramm, M., Winter, H., Legarth, B., and Huenges, E. (2003) The in-situ geothermal laboratory Groß Schönebeck – stimulation experiments of sandstones in 4200 m depth. *Proc. Twenty- Eight Workshop on Geothermal Reservoir Engineering, Stanford University, Stanford, California*, SGP-TR-173.

Zimmermann, G., Reinicke, A., Holl, H., Legarth, B., Saadat, A., and Huenges, E. (2005) Well Test Analysis after Massive Waterfrac Treatments in a Sedimentary Geothermal Reservoir. Proceedings of the 2005 World Geothermal Congress, April, Antalya, Turkey, paper 1129, 6 p..

Zimmermann, G., Tischner, T., Legarth, B., and Huenges, E. (2009) Pressure dependent production efficiency of an Enhanced Geothermal System (EGS) – Stimulation results and implications for hydraulic fracture treatments. *Pure and Applied Geophysics*, **166** (5-7), 1089–1106.

Zoback, M.D., Barton, C.A., Brudy, M., Castillo, D.A., Finkbeiner, D., Grollimund, B.R., Moos, D.B., Peska, P., Ward, C.D., and Wipruth, D.J. (2003) Determination of stress orientation and magnitude in deep wells. *International Journal of Rock Mechanics and Minning Sciences*, **40**, 1049–1076.

Further Reading

Kalfayan, L. (2001) *Production Enhancement with Acid Stimulation*, Pennwell Books.

Chemical Stimulation

Amistoso, A.E., Aqui, A.R., Yglopaz, D.M., and Malate, R.C.M. (2005) Sustaining steam supply in Palinpinon 1 production field, Southern Negros geothermal project, Philippines. World Geothermal Congress, April 24-29, 2005, Antalya.

Barrelli, A., Cappetti, G., Manetti, G., and Peano, A. (1985) Well stimulation in Latera field. *Geothermal Resources Council Transactions*, **9** (2), 213–219.

Crowe, C. (1986) Precipitation of hydrated silica from spent hydrofluoric acid: how much of a problem is it? *Journal of Petroleum Technology*, **38**, 1234–1240.

Economides, M.J. and Frick, T.P. (1992) Optimization of horizontal well matrix stimulation treatments. SPE International Meeting on Petroleum Engineering, March 24-27, SPE 22334, Beijing.

Erga, F. (2000) Esperimenti di Acidificazione in Flusso Continuo di Soluzione Acida.Erga, Gruppo Enel, 20 p.

Flores, M., Davies, D., Couples, G., and Palsson, B. (2005) Stimulation of geothermal wells, can we afford it? World Geothermal Congress, April 24-29, 2005, Antalya.

McLeod, H.O. (1984) Matrix acidizing. *Journal of Petroleum Technology*, **36**, 2055–2069.

Mella, M., Rose, P., Kovac, K., Xu, T., and Pruess, K. (2006) *Calcite Dissolution in Geothermal Reservoirs Using Chelants*, Geothermal Resources Council Transactions, San Diego.

Mendez, A., Neumann, L.F.de, Almeida Pinto, E., Torres, R., Farias, R., and Acosta, M. (2005) Achieving true sandstone-reservoir stimulation in deepwater horizontal wells. BJ Services Company and Petrobras, 2005 SPE Annual Technical Conference and Exhibition, October 9-12, SPE 95826, Dallas.

Molina, P.O., Malate, R.C.M., Buning, B.C., Yglopaz, D.M., Austria, J.J.C., and Lacanilao, A.M. (1998) Productivity analysis and optimization of well SK-2D, Mindanao I Geothermal Project Philippines. Proceedings of the 23rd Workshop on Geothermal Reservoir Engineering, January 26-28, 1998, Stanford.

Morris, C.W., Verity, R.V., and Dasie, W. (1984) Chemical stimulation treatment of a well in the Beowawe Geothermal Field. *Geothermal Resources Council Transactions*, **8**, 269–274.

Mukherjee, H. and Cudney, G. (1993) Extension of acid fracture penetration by drastic fluid loss control. *Journal of Petroleum Technology*, **45** (2), 102–105.

Nami, P., Schindler, M., Tischner, R., Jung T., and Teza, D. (2007) Evaluation of stimulation operations and current status of the deep soultz wells prior to power production. Proceedings of EHDRA Scientific Conference, June 28-29, Soultz-sous-Forêts.

Paccaloni, G. and Tambini, M. (1990) Advances in matrix stimulation technology. 65th SPE Annual Technical Conference and Exhibition, September 23-26, New Orleans, SPE 20623.

Portier, S., André, L., and Vuataz, F.-D. (2007) Review on chemical stimulation techniques in oil Industry and applications to geothermal systems, a technical report prepared for the EC-financed co-ordination project ENGINE (Enhanced Geothermal Innovative Network for Europe), Work Package 4: Drilling, stimulation and reservoir assessment. CREGE-Centre for Geothermal Research, Neuchâtel, Switzerland, November 2007.

PTTC (2000) Well stimulation advances. Workshop Co-sponsored by PTTC's North Midcontinent Region and the Wichita Chapter of Society of Petroleum Engineers, February 9, 2000, Wichita.

Templeton, C.C., Richardson, E.A., Karnes, G.T., and Lybarger, J.H. (1975) Self-generating mud acid. *Journal of Petroleum Technology*, **27**, 1199–1203.

Thomas, R.L. and Crowe, C.W. (1978) Matrix treatment employs new acid system for stimulation and control of fines migration in sandstone formations. 53 rd SPE Annual Technical Conference and Exhibition, October 1-3, SPE 7566, Houston.

Tuedor, F.E. (2006) A Breakthrough Fluid Technology in Stimulation of Sandstone Reservoirs, Schlumberger, SPE 98314.

Williams, B.B. (1979) Acidizing fundamentals. New York and Dallas Society of Petroleum Engineers, European Formation Damage Control Conference, SPE Monograph No. 6, May 15-16, The Hague, the Netherlands.

Xie, T. (2004) A parametric study of sandstone acidizing using a fine-scale simulator.Master of Science in Engineering, University of Texas, Austin.

Xu, T., Ontoy, Y., Molling, P., Spycher, N., Parini, M., and Pruess, K. (2004) Reactive transport modeling of injection well scaling and acidizing at Tiwi field, Philippines. *Geothermics*, **33** (4), 477–491.

Post Stimulation Evaluation

Charlety, J., Cuenot, N., Dorbath, L., Dorbath, C., Haessler, H., and Frogneux, M. (2007) Large earthquakes during hydraulic stimulations at the geothermal site of Soutz-sous-Forêts. *International Journal of Rock Mechanics and Minning Sciences*, **44**, 1091–1105.

Tracer Testing

Adams, M.C., Yamada, Y., Yagi, M., Kondo, T., and Wada, T. (2000) Stability of methanol, propanol, and SF6 as high-temperature tracers. *World Geothermal Congress*, 3015–3019.

Fukuda, D., Asanuma M., Hishi Y., and Kotanaka K. (2005) Alcohol tracer testing at the Matsukawa vapour-dominated geothermal field, Northeast Japan. Proceedings of the 30th Workshop on

Geothermal Reservoir Engineering, Stanford University, January 31-February 2, 2005, SGP-TR-176, Stanford.

Lovelock, B.G. (2001) Steam flow measurement using alcohol tracers. *Geothermics*, **30** (6), 641–654.

Geothermics, (2001) Special Issue on tracers, **30** (6), 571–781.

Stimulation Thermal

Charlez, P., Lemonnier, P., Ruffet, C., Boutéca, M.J., and Tan, C. (1996) Thermally Induced Fracturing: Analysis of a Field Case in North Sea. Society of Petroleum Engineers, SPE 36916.

Flores, M., Davies, D., Couples, G., and Palsson, B. (2005) Stimulation of geothermal wells. Can We Afford It? Proceedings World Geothermal Congress April 24-29, Antalya.

Sanjuan, B., Lasne, E., and Brach, M. (2000) Bouillante geothermal field (Guadeloupe, West Indies): geochemical monitoring during thermal stimulation operation. Proceedings of the Twenty-fifth Workshop on Geothermal Reservoir Engineering, Stanford University, January 24-26, 2000, Stanford.

Tester, J.W., Murphy, H.D., Grigsby, C.O., Potter, R.M., and Robinson, B.A. (1989) Fractured geothermal reservoir growth induced by heat extraction. *SPE Reservoir Engineering*, **4**, 97–104.

Review Articles Representative Sites

Bertini, G., Casini, M., Gianelli, G., and Pandelli, E. (2006) Geological structure of a long-living geothermal system, Larderello, Italy. *Terra Nova*, **18** (3), 163–169.

Case Study Soultz

Weidler, R., Gerard, A., Baria, R., and Jung, R. (2002) Hydraulic and microseismic results of a massive stimulation test at 5 km depth at the European Hot-Dry-Rock test site Soultz, France. Proceedings of the 27th workshop on Geothermal Reservoir Engineering, January 28-30, Stanford.

Case Study Horstberg

Jung, R., Orzol, J., Kehrer, P., and Jatho, R. (2005) Verbundprojekt: GeneSys, Vorstudie – Erprobung der Wasserfractechnik und des Einsonden-Zweischichtverfahrens für die Direktwärmenutzung aus gering permeablen Sedimentgesteinen. BMU-Abschlussbericht, Förderkennzeichen: 0327116, 70 p.

Jung, R., Orzol, J., Tischner, T., Jatho, R., and Kehrer, P. (2005) The geothermal project GeneSys – Results of massive waterfrac-tests in the Bunter sandstone formation in the Northern German Basin. Proceedings of the 30th Stanford University workshop Geothermal Reservoir, January 31–February 2, 2005, Stanford.

Wessling, S., Junker, R., Rutqvist, J., Silin, D., Sulzbacher, H., Tischner, T., and Tsang, C.-F. (2009) Pressure analysis of hydromechanical effects in a fractured tight sedimentary geothermal reservoir. *Geothermics*, **38** (2), June, 211–226.

Case Study Gross Schönebeck

Bommer, J.J., Oates, S., Cepeda, J.M., Lindholm, C., Bird, J., Torres, R., Marroquín, G., and Rivas, J. (2005) Control of hazard due to seismicity induced by a hot fractured rock geothermal project. *Engineering Geology*, **83** (4), 287–306.

Calcagno, P. and Sliaupa, S. (eds) (2008) *Enhanced Geothermal Innovative Network for Europe: Proceedings of the Engine Final Conference, February 12-15, 2008, Vilnius*, BRGM Editions, Orleans.ISBN 978-2-7159-2993-7, Collection Actes/Proceedings, ISSN 1773-6161.

Charlez, P., Lemonnier, P., Ruffet, C., and Boutéca, M.J. (1996) Thermally Induced Fracturing: Analysis of a Field Case in North Sea. SPE, Paper No. 36916.

Evans, K.F. (2005) Permeability creation and damage due to massive fluid injections into granite at 3,5 km at Soultz: 2. Critical stress and fracture strength. *Journal of Geophysical Research*, **110**, B04204.

Hettkamp, T., Baumgärtner, J., Baria, R., Gerard, A., Gandy, T., Michelet, S., and Teza, D. (2004) Electricity production from hot rocks, Proceedings of 29th Workshop on Geothermal Reservoir Engineering, Stanford University, January 26-28, 2004, SGP-TR-175, Stanford.

Huenges, E., Erbas, K., Moeck, I., Blöcher, G., Brandt, W., Schulte, T., Saadat, A., Kwiatek, G., and Zimmermann, G. (2009) The EGS project Groß Schönebeck – Current status of the large scale research project in Germany, 33rd Annual Meeting Geothermal Resources Council, Reno, 403–404.

Nami, P., Schellschmidt, R., Schindler, M., and Tischner, T. (2008) Chemical stimulation operations for reservoir development of the deep crystalline HDR/EGS system at Soultz-sous- Forêts (France). Proceedings of the Thirty-Third Workshop on Geothermal Reservoir Engineering, Stanford University, SGP-TR-185, Stanford.

5
Geothermal Reservoir Simulation

Olaf Kolditz, Mando Guido Blöcher, Christoph Clauser, Hans-Jörg G. Diersch, Thomas Kohl, Michael Kühn, Christopher I. McDermott, Wenqing Wang, Norihiro Watanabe, Günter Zimmermann, and Dominique Bruel

Since 1971, the hot dry rock (HDR) concept has been developed from a notion to the point where system feasibility is under demonstration. HDR technology can be applied in most countries and is not dependent on specific geological conditions. During the last decade, significant progress has been made in drilling and in borehole measurements, as well as in the understanding of the processes involved in the creation and operation of stimulated geothermal reservoirs. The design and creation of enhanced geothermal systems (EGSs) requires the development of simulation models to predict the growth behavior of hydraulically induced and active reservoirs. Geothermal reservoir simulation has been conducted at several European research sites in the past. Figure 5.1 shows a map of site studies that are presented in this work.

In the first part of this chapter we briefly describe the theoretical background of geothermal reservoir simulation with emphasis on thermohydromechanical (THM) processes. A short discussion of results from the Groß Schönebeck, Bad Urach, Soultz, Rosemanowes, the German Continental Deep Drilling Program (KTB) in Windischeschenbach and Stralsund case studies is given in the second part of this chapter.

5.1
Introduction

The numerical analysis of multifield problems in porous media is an important subject for many geoengineering tasks such as the management of georesources (e.g., engineering of geothermal, oil, and gas reservoirs) as well as waste management (e.g., chemotoxic and nuclear waste, CO_2 sequestration), see for example, (Doughty and Pruess, 2004; Stephansson *et al.*, 2004; Alonso *et al.*, 2005). For geothermal reservoir management, simulation tools are required that take into account both coupled THM processes and the uncertainty of geological data, that is, the model parametrization.

Geothermal Energy Systems. Edited by Ernst Huenges

ISBN: 978-3-527-40831-3

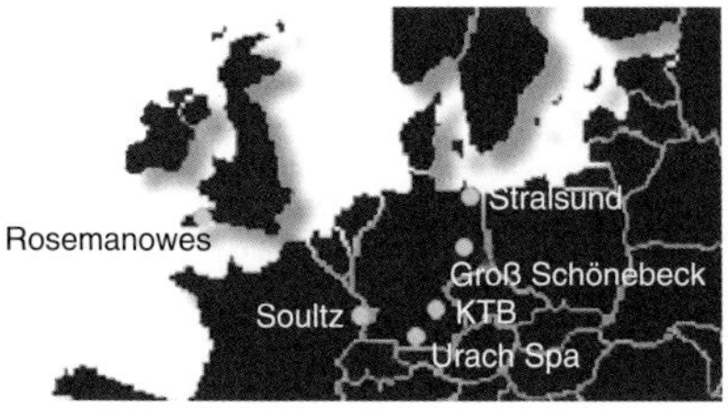

Figure 5.1 Selected geothermal research sites in Europe.

Geothermal reservoir simulation requires an adequate mathematical representation of the physical and chemical phenomena during the long-term heat extraction process. Mainly deterministic methods are used by solving the underlying partial differential equations (Section 5.1.1). Owing to data availability for deep geo-reservoirs, the parameterization of numerical models is complicated. Therefore, quantifying the parameter uncertainty using stochastic methods such as Monte Carlo simulation is an important part of the reservoir analysis (Section 5.1.2).

5.1.1 Geothermal Modeling

The use of computer modeling in the planning and management of the development of geothermal fields has become standard practice during the last 20 years. During that time models have been developed for more than 100 geothermal fields worldwide (Willis-Richards and Wallroth, 1995; O'Sullivan *et al.*, 2001). Owing to geological complexity and the number of processes involved, such as geometry, hydraulics, thermal effects, geochemical reaction, and stress changes, numerical methods have been widely used for geothermal reservoir simulation (Zyvoloski *et al.*, 1988).

The analysis of coupled processes, in particular, feedbacks of mechanical, thermal, and geochemical effects to the flow system, is important for both hydrothermal (Clauser, 2003) and HDR systems (Tsang, 1991; Bower and Zyvoloski, 1997; McDermott and Kolditz, 2006). Numerical THM models have been developed and applied to several HDR sites such as Soultz-sous-Forêts in the Rhine Valley (Kohl *et al.*, 1995)(Hicks *et al.*, 1996) and Urach Spa in the Swabian Alb by McDermott *et al.* (2006); Watanabe *et al.* (2009). More recently, chemical effects have been included into the coupled analysis (Kühn, 2004; Kiryukhin *et al.*, 2004; Bachler and Kohl, 2005). One of the key questions hereby is how dissolution and precipitation processes can change the pore structure and therefore the reservoir permeability. Discrete fracture network (DFN) models are available for the simulation of fluid, mass, and heat transport, even for realistic geological structures, for example, (Bruel, 1995b; Kolditz, 2001; Bruel, 2002) for the Soultz HDR reservoir. Their applicability in the context of fully coupled THM analysis, however, is still restricted to simplified problems (Walsh *et al.*, 2008). Equivalent porous media approaches are used for THM analysis of fractured rock instead (Birkholzer *et al.*, 2008).

From the mathematical point of view, THM processes lead to a nonlinear coupled initial boundary value problem (IBVP), which needs to be solved numerically. Among the available numerical methods, finite differences, volumes and elements (FEM) are mainly used (Noorishad *et al.*, 1984; Kohl, 1992; Lewis and Schrefler, 1998; de Boer, 2005; Wang and Kolditz, 2007; Rutqvist *et al.*, 2008). Irrespective of the specific numerical method, the calculation of coupled THM problems is very expensive. This is mainly due to two reasons: degree of freedom (i.e., number of field variables) and strong coupling among nonlinear processes.

There are several ways to improve the computational efficiency, for example, more efficient numerical algorithms, optimization of memory management in the code, and parallelization techniques. Among them, parallel computing provides the most powerful speed-up. Thanks to the decreasing hardware cost in the past years, parallel computation is becoming very attractive for applied research (Topping and Khan, 1996; Schrefler *et al.*, 2000; Shioya and Yagawa, 2005; Wang *et al.*, 2009).

5.1.2 Uncertainty Analysis

Data uncertainty is one of the major problems in subsurface reservoir analysis. Direct borehole measurements are very limited due to technical issues and costs. Normally, data are available from core samples and wellbore logging for the local scale, and geophysical measurements (e.g., microseismic monitoring) for a larger scale (e.g., Tenzer *et al.* (2000) for Urach Spa site and (Weidler *et al.*, 2002) for Soultz site). Thus, subsurface models are derived from limited information and include uncertainties. Dealing with uncertainty analysis is a common tool, for example, for safety assessment of nuclear waste repositories (Rautman and Treadway, 1991).

For HDR geothermal systems, aspects of uncertainty have been investigated in the framework of sensitivity analysis and parameters identification so far. Fractal and statistical DFN models have been developed, for example, by Watanabe and Takahashi (1995); Willis-Richards, (1995); Tezuka and Watanabe (2000). DFN models can represent the structural reservoir geology, but they are still restricted to simplified processes. Inversion methods have been used to identify physical rock parameters in order to reproduce the observed reservoir behavior (Finsterle and Pruess, 1997; Lehmann *et al.*, 1998).

Monte Carlo analysis is one common method to quantify parameter uncertainty and the corresponding system evolution. Using geostatistical techniques, for example, sequential Gaussian simulation and indicator simulation, enables the generation of multiple stochastically equivalent realizations, which take into account the status of the incomplete knowledge (Goovaerts, 1997; Pebesma and Wesseling, 1998; Chilés and Delfiner, 1999; Deutsch, 2002). Statistical properties of sequential Gaussian processes are described, for example, by Rouhani *et al.* (1996). The conditional distribution of an unknown statistical variable at a particular location, given a set of known values at nearby locations, is the Gaussian (normal). Therefore, the distribution of sample data has to be transformed to the Gaussian distribution beforehand. The mean and variance of the conditional distribution can be given by the simple Kriging estimate of the unknown value and its associated error variance. Before starting stochastic simulations, assumptions concerning parameter distribution, for example, histogram, spatial correlation, correlation with other parameters, have to be decided. Usually these assumptions are determined from site observation data. However, little information about histogram and variogram analysis for deep crystalline rocks is available in the literature.

5.2
Theory

In this section, we briefly introduce the conceptual models for geothermal reservoir simulation as well as summarize the governing equations of THM processes in fractured porous media. The constitutive equations for the reservoir characterization including geothermal fluid properties are presented in the subsequent Section 5.3. The list of symbols can be found at the end of this chapter.

5.2.1
Conceptual Approaches

Figure 5.2 depicts an overview of existing modeling concepts for geothermal reservoir simulation of both fractured and porous media. Before fracture network models have been introduced into geothermal reservoir simulation (Bruel and Cacas, 1992; Bruel *et al.*, 1994), more simple geometric concepts such as single and parallel fracture systems have been investigated, for example, (Cornet, 1985). The alternative for modeling fractured rock is the use of equivalent porous media, for example, (Stober, 1986). The behavior of rock masses is significantly influenced by fractures; therefore, many efforts have been undertaken to develop appropriate numerical methods for the representation of fractures (Cacas *et al.*, 1990). Fracture network models are a closer representation of geological reality as real rock masses are penetrated by different fracture populations. Figure 5.2 shows two different concepts for fracture network modeling, deterministic and stochastic approaches. Stochastic models are based on the statistic properties of the fracture network observed in borehole loggings. Monte Carlo methods with a representative number of different realizations are used in order to investigate, for example, the hydraulic behavior of fracture networks (Bruel *et al.*, 1994). Deterministic models deal with a few number of major fractures in order to study the interaction between fractures and rock matrix, for example, for heat extraction processes (Kolditz, 1995).

5.2.2
THM Mechanics

The physical processes involved are nonisothermal saturated flow, heat transport, and thermoporoelastic deformation. The corresponding field variables of the multifield problem are liquid phase pressure, temperature, and displacement vector.

Hydrogeological system	Unconsolidated and fractured aquifers		
Conceptual model	Fracture model		Continuum model
Model geometry	Single and multiple fractures	Fracture network (deterministic,stochastic)	Multiple porous media
	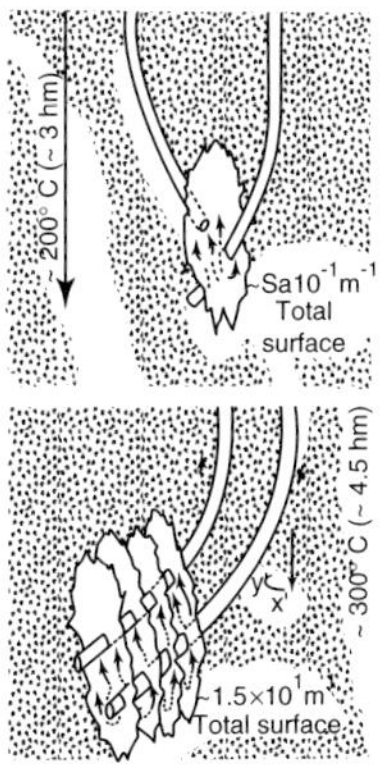	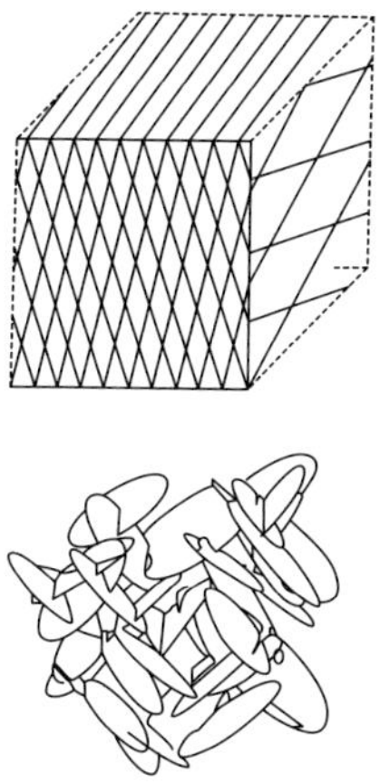	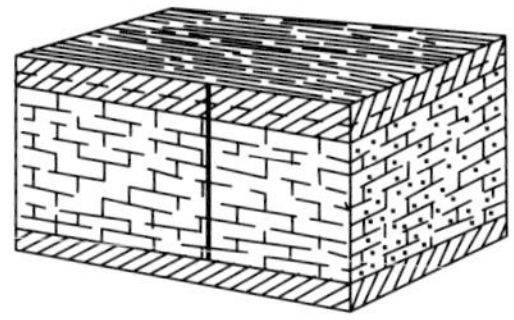
Sources	Cornet (1985)	Kolditz (1994a) Bruel & Cacas (1992)	Stober (1986)

Figure 5.2 Conceptual models for fractured rock (Kolditz, 1997).

Material properties of geothermal fluids are nonlinear functions of salinity, temperature, and pressure (McDermott *et al.*, 2006). More details on THM mechanics can be found, for example, in Lewis and Schrefler (1998); Ehlers and Bluhm (2002); Wang and Kolditz (2007).

5.2.2.1 **Heat Transport**

For the heat transport problem, we consider advective and diffusive fluxes in saturated porous and fractured media. The difference between porous and fractured media is the flow calculation according to Darcy's (Equation (5.3)) and Forchheimer's law (Equation 5.4), respectively. The governing equation of heat transport is

$$c\rho \frac{\partial T}{\partial t} + c^{\mathrm{l}}\rho^{\mathrm{l}}\mathbf{v} \cdot \nabla T - \nabla(\lambda \nabla T) = Q_{\mathrm{T}} \tag{5.1}$$

with following porous media properties

$$\begin{aligned} c\rho &= nc^{\mathrm{l}}\rho^{\mathrm{l}} + (1-n)c^{\mathrm{s}}\rho^{\mathrm{s}} \\ \lambda &= n\lambda^{\mathrm{l}} + (1-n)\lambda^{\mathrm{s}} \end{aligned} \tag{5.2}$$

5.2.2.2 Liquid Flow in Deformable Porous Media

Incompressible fluid flow in deformable porous media is described by the following fluid mass balance equation,

$$S_s \frac{\partial p}{\partial t} - \nabla \cdot \left(\frac{\mathbf{k}}{\mu} \left(\nabla p + \rho^l \mathbf{g} \nabla z \right) \right) + \nabla \cdot \left(\frac{\partial \mathbf{u}}{\partial t} \right) = q_f \tag{5.3}$$

For describing nonlinear flow behavior we use the Forchheimer equation (Forchheimer, 1914)

$$\Delta p = a_1 Q + a_2 Q^2 \tag{5.4}$$

which states that the pressure drop can be described by a quadratic function of the flow rate. If the quadratic term dominates as the case for large flow rates, nonlinear flow behavior can be expressed as a pressure-dependent permeability.

$$\Delta k(p) = \frac{\mu}{\rho^l} \frac{\sqrt{\rho^l D_h}}{\sqrt{\lambda}} \left(\frac{\Delta p}{L} \right)^{-1/2} \tag{5.5}$$

5.2.2.3 Thermoporoelastic Deformation

Thermoporoelastic deformation is described by the momentum balance equation in the terms of stress tensor as

$$\nabla \cdot (\sigma - \alpha_b p \mathbf{I} - \beta_T E \mathbf{I} \Delta T) + \rho \mathbf{g} = 0 \tag{5.6}$$

The density of the porous medium composed of two phases, liquid and solids is

$$\rho = n\rho^l + (1-n)\rho^s$$

Displacement is the primary variable to be solved by substituting the constitutive law for stress–strain behavior

$$\sigma = \mathbb{C}\,\varepsilon \tag{5.7}$$

$$\varepsilon = \frac{1}{2}(\nabla \mathbf{u} + (\nabla \mathbf{u})^T) \tag{5.8}$$

The fourth-order elasticity tensor $\mathbb{C}$ is

$$\mathbb{C} := \lambda \delta_{ij} \delta_{kl} + 2G \delta_{ik} \delta_{jl} \tag{5.9}$$

All symbols are listed at the end of the chapter.

Different numerical methods are applied to solve the above system of partial differential equations. FEFLOW (TH) (Kolditz and Diersch, 1993; Diersch, 2002), FRACTure (THM) (Kohl, 1992), and GeoSys/RockFlow (THMC) (Wang and Kolditz, 2007) use the FEM. SHEMAT (THC) is based on the finite difference method (Clauser, 2003).

5.3 Reservoir Characterization

A characterization of geothermal reservoirs is complicated due to several facts:

- Heterogeneity: Deep geothermal reservoirs are extremely heterogeneous. Fractures determine the flow, transport, and geomechanical properties to a large extent (Section 5.3.1).
- Nonlinearity: Owing to the large imposed changes of the thermodynamical state variables (e.g., pressure, temperature, stress), fluid and rock properties behave nonlinearly (Section 5.3.2).
- Uncertainty: Data on material properties rely on a few measurements; therefore, the information about geothermal reservoirs is to a large degree uncertain (Section 5.3.4).

5.3.1 Reservoir Properties

5.3.1.1 Reservoir Permeability

Reservoir permeability is one of the most important hydraulic parameters governing advective transport processes. At the same time, it is most difficult to determine as pumping tests give information only about the near field area. In most sedimentary rocks the porosity is interconnected, which makes the rock permeable for flow. A rough estimate of reservoir permeability can be calculated by the steady-state method directly using Darcy's Law and thus assuming laminar flow conditions

$$k = -\frac{Q^{\mathrm{l}}\mu}{A\nabla p^{\mathrm{l}}} \tag{5.10}$$

with the flow rate, dynamic viscosity, cross-section of the flow path, and pressure gradient, respectively (see list of symbols for parameter definitions and units).

5.3.1.2 Poroperm Relationships

The suitability of existing empirical relationships in order to correlate porosity and permeability changes such as the well-known Kozeny–Carman equation (Kozeny, 1927; Carman, 1937) is a matter of question for fractured rocks. During geothermal power production using a borehole doublet consisting of a production and injection well, the reservoir conditions will change. Besides, a temperature decrease at the injection well results in a thermoelastic response, and the pore pressure will also vary. This leads to a poroelastic response of the reservoir rocks depending on effective stress (difference between confining stress and pore pressure), resulting in a change in permeability and porosity. Various previous studies continued to investigate the effective pressure dependency of these rock properties. (Carroll and Katsube, 1983) developed a theory of hydrostatic poroelasticity in terms of porosity and bulk volume. By means of this theory changes in effective pressure can be related to changes in porosity (Zimmerman, 1991) (Figure 5.3)

$$\Delta n = -\left[(1-n)\beta_p - \beta_p^{\mathrm{s}}\right]\Delta(\sigma^{\mathrm{s}} - p^{\mathrm{l}}) \tag{5.11}$$

where the parameters are bulk compressibility and compressibility of the solid.

The permeability can always be expressed as a function of confining and pore pressure (Al-Wardy and Zimmerman, 2003). If the permeability follows the effective stress definition

$$\sigma = \sigma^{\mathrm{s}} - \alpha_{\mathrm{b}} p^{\mathrm{l}}$$

with the Biot coefficient, then it can be expressed as a function of this effective pressure,

$$k = f(\sigma)$$

Several experiments determined the effective pressure coefficient. (Morrow *et al.*, 1986) and (Bernabe, 1987) found that a value of 1 was the limiting value for crystalline rocks; (Zoback and Byerlee, 1975) and (Al-Wardy and Zimmerman, 2003) found that the effective stress coefficient depends on the clay content. Often the effective pressure coefficient is taken to be constant, yielding a linear expression for effective pressure, but (Kwon *et al.*, 2001) mentioned that it may itself be a function of effective stress or microstructural changes in pore structure. It has to be taken into account that the pore geometry parameters cannot be directly measured under pressure. Therefore, the permeability has to be correlated to other rock properties such as porosity or formation factor (Bernabe, 1988), which can be measured under pressure.

The dependency of permeability on pore structure was investigated by various studies: A general overview can be found in Bear (1972); hydraulic radius models (Kozeny, 1927; Carman, 1937) (for soils and porous rocks); geometric parameters determined by a fractal approach (Pape *et al.*, 2000); and, mercury injection (Katz and Thompson, 1987). For the hydraulic radius model the equation

$$k = \frac{nD_{\mathrm{h}}^2}{6}$$

is used with porosity and hydraulic radius as parameters. For the second permeability calculation the equation

$$k = 155n + 37315n^2 + 630(10n)^{10}$$

can be used for Rotliegend sandstones of northeast Germany and was calibrated at several core sample measurements. By means of measured porosity changes due to effective pressure changes, and also a more general definition of permeability (Bear, 1972) can be used (Figure 5.3):

$$k = f(\sigma) f(n) L^2 \tag{5.12}$$

(see "Nomenclature" for parameter definitions). The reference length can be the hydraulic radius or particle size as well. The often used porosity function

$$f(n) = n^3/(1-n)^2 \tag{5.13}$$

(Bear, 1972) can be directly calculated by means of the laboratory results. The geometry term depends on effective pressure itself, but the influence of the geometry term on permeability is small in comparison to the porosity function.

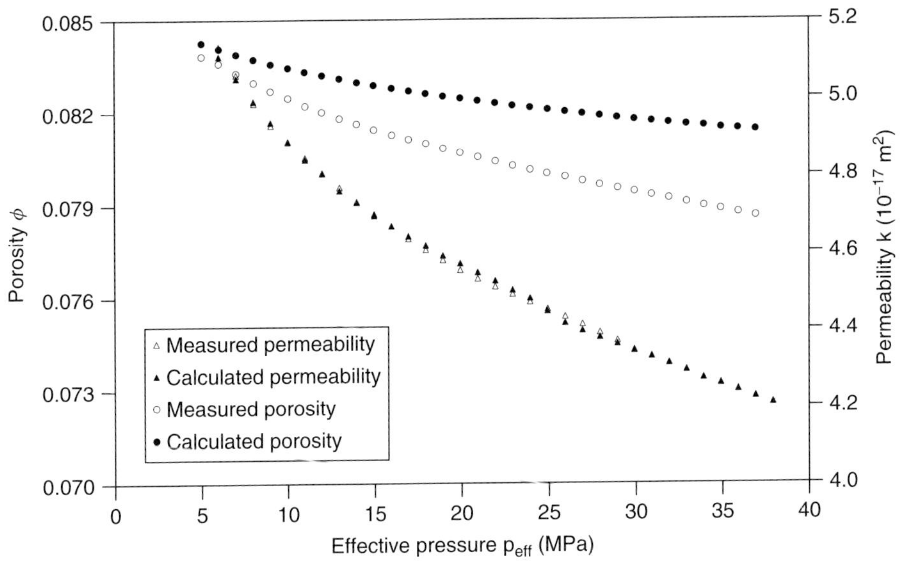

Figure 5.3 Measured and calculated porosity and permeability changes of a Rotliegend sandstone (Flechtinger sandstone) due to effective pressure change (Blöcher *et al.*, 2009). The porosity is calculated by means of bulk and solid compressibility, which depend on effective pressure as well. The permeability is calculated by means of porosity function (Equation 5.13) and measured geometry term $f(\sigma)L^2 = 0.1627\sigma + 98.348$. There the unit of $f(\sigma)L^2$ is $[10^{-15}\text{m}^2]$ and thus of σ is [MPa].

For shallow fractured crystalline rocks, (Pape, *et al.*, 1999) derived a functional relationship based on fractal theory. The corresponding poroperm function for microfissured granite at the Falkenberg site

$$k = \left(2.34 \times n^{1.25} + 20.94 \times (10n)^{3.88}\right) \times 10^{-16} \tag{5.14}$$

is illustrated in Figure 5.4. For porosities above 1% there is an enlarged increase of permeability. Both branches below and above the 1% porosity can be represented by power laws. In order to represent the typical behavior of crystalline rock the above relationship is used for the uncertainty analysis in the following.

Hydraulic conductivity (meters per second) is a combined fluid and solid phase reservoir parameter

$$K = \frac{k\rho^{l}g}{\mu^{l}} \tag{5.15}$$

(see list of symbols for parameter definitions).

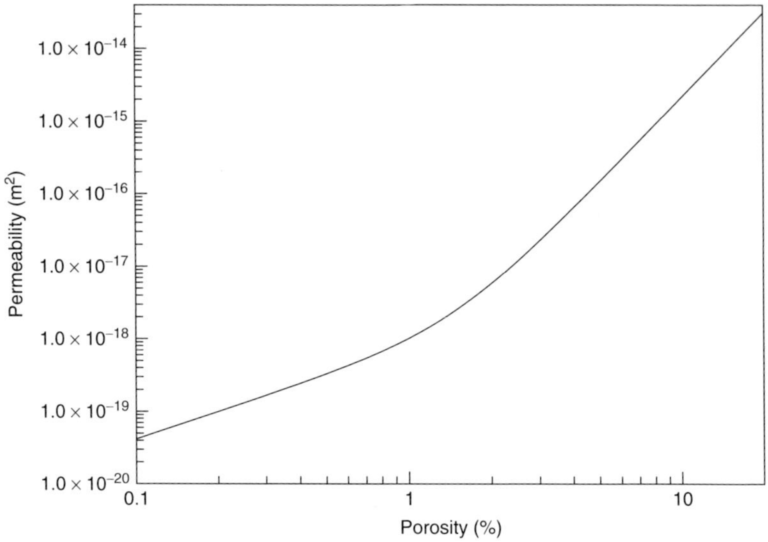

Figure 5.4 Porosity–permeability relationship observed at the Falkenberg site (Pape *et al.*, 1999).

5.3.2
Fluid Properties

The properties of geothermal fluids strongly depend on temperature and salinity, see, for example, (Wagner and Kurse, 1998), who use a very general thermodynamic description of fluid properties based on the free Helmholtz energy.

5.3.2.1 **Density and Viscosity**

In terms of the influence of the state variables, the effects of changes in pressure are smaller than those of temperature and salt concentration (Figure 5.5). Examples of the reservoir pressures and temperatures typical of the geothermal systems have been analyzed by McDermott *et al.* (2006) in water vapor phase diagrams. All the reservoirs plot in the subcritical region of the diagram. Although supercritical conditions are not reached in potential EGS reservoirs, significant changes are seen in the fluid properties of the hot, highly compressed reservoir fluids.

In addition to general thermodynamic description of fluid properties, frequently phenomenological equations of state are used. The dynamic viscosity of the fluid phase is typically regarded as a function of concentration and temperature (Diersch, 2002).

$$\frac{\mu}{\mu_0} = \frac{1 + 1.85\omega - 4.1\omega^2 + 44.5\omega^3}{1 + 0.7063\varsigma - 0.04832\varsigma^3} \tag{5.16}$$

with mass fraction and relative temperature coefficients

$$\omega = C/\rho, \quad \varsigma = (T - 150)/100$$

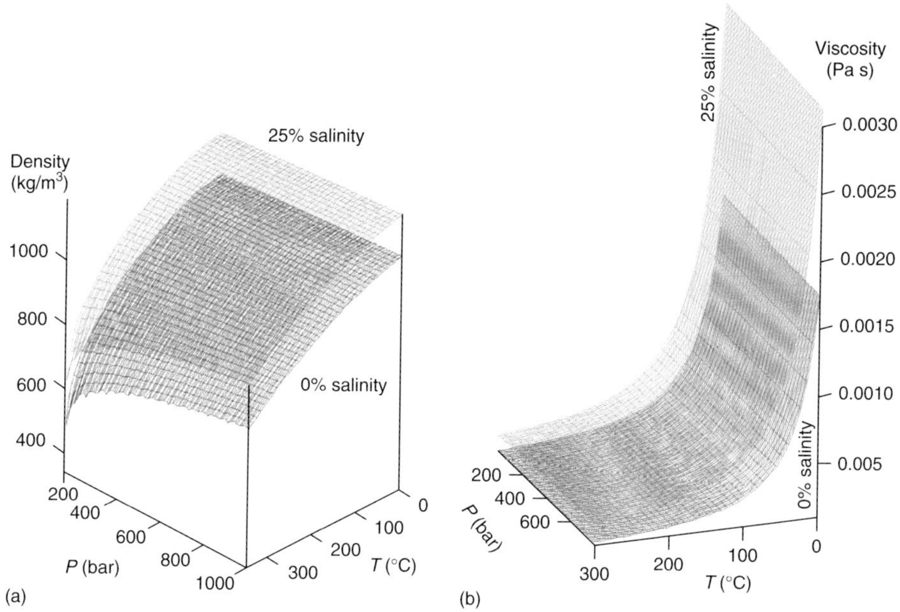

Figure 5.5 Variable hydromechanical fluid properties, density (a), viscosity (b) (McDermott *et al.*, 2006).

The equation of state for the fluid density (Equation 5.17) is related to reference temperature, reference pressure, and reference concentration.

$$\rho^{\mathrm{l}} = \rho_0^{\mathrm{l}} \left(1 - \beta_T (T - T_0) + \beta_p (p - p_0) + \frac{\beta_C}{C_s - C_0}(C - C_0)\right) \tag{5.17}$$

Therefore, the flow and transport equations for thermohaline convection are nonlinear and strongly coupled since temperature, pressure, and salinity control the fluid density.

5.3.2.2 Heat Capacity and Thermal Conductivity

Not only viscosity and density, also thermal fluid properties depend on temperature and salinity changes (Figure 5.6).

The functional relation for rock thermal conductivity is given by Somerton (1992).

$$\begin{aligned}\lambda(T) = {} & \lambda_{20} - 10^{-3}(T - 293)(\lambda_{20} - 1.38) \\ & \times \left(\lambda_{20}(1.8 \times 10^{-3} T)^{-0.25\lambda_{20}} + 1.28\right) \lambda_{20}^{-0.64}\end{aligned} \tag{5.18}$$

where the suffix 20 denotes the heat conductivity measured at 20 °C by laboratory experiments. By means of Equation (5.18) it is possible to calculate the heat conductivity of the solid at any point of the reservoir with defined temperature.

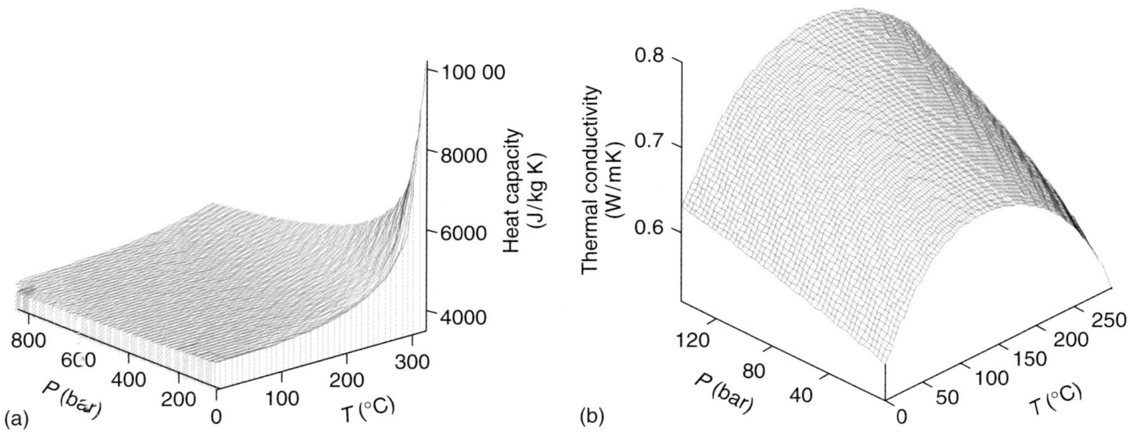

Figure 5.6 Variability of thermal fluid properties, heat capacity (a), thermal conductivity (b), (McDermott *et al.*, 2006).

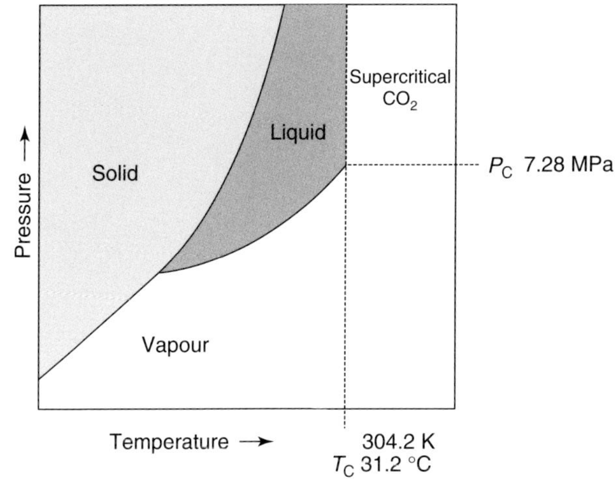

Figure 5.7 Phase diagram for CO_2.

5.3.3 Supercritical Fluids

Using supercritical CO_2 as a fluid for extracting heat from the basement rocks has often been suggested as a more efficient means of getting the heat energy to the surface (Pruess and Spycher, 2007). Figure 5.7 presents the phase diagram for CO_2. The critical point is illustrated as being 7.28 MPa and 31.2 °C. At higher pressures and temperatures than these, there is no phase boundary between the liquid and the gas phase, and so a gas can be considered to have liquid properties and vice versa. The consequence is supercritical CO_2 has a high density and a low viscosity, making it an attractive fluid for heat extraction. Under normal hydraulic conditions, a fluid pressure of over 7.28 MPa can be expected at depths deeper than about 800 m.

Considering only the use of supercritical CO_2 as a fluid in the reservoir to collect heat energy, and examining the energy balance equation (5.1). The advective term describes the transport of heat by fluid movement. Considering this, and assuming that the fluid attains the reservoir temperature as soon as it is in contact with the reservoir, it can be shown that a comparison between the efficiency of the heat removal between two fluids is dependent on the heat capacity of that fluid, the density of that fluid and the velocity of that fluid under the same pressure gradient. The advective velocity of the fluid in a heterogeneous media is given by Darcy's law. The intrinsic permeability and the effective porosity are material parameters, the only fluid-dependent parameter in the consideration of the velocity is the dynamic viscosity. Therefore, to consider the relative efficiency of the extraction of heat by two different fluids, consideration of the ratio of

$$c^{f1}\rho^{f1}\mathbf{v}^{f1} : c^{f2}\rho^{f2}\mathbf{v}^{f2}$$

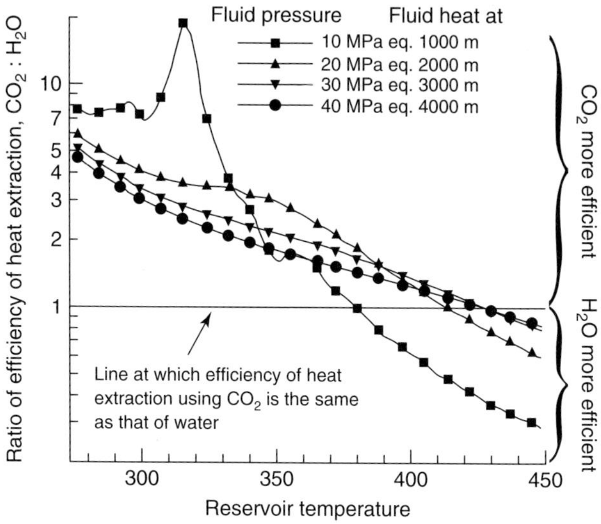

Figure 5.8 Consideration of the efficiency of CO_2 as a circulating fluid against H_2O as a circulating fluid.

will give an indication of which fluid is better applied under which conditions, as presented in Figure 5.8. The CO_2 properties were taken from (Span and Wagner, 1994) and compiled in Sultanov (2006).

The peak seen in the CO_2 under 10 MPa and at a temperature of around 310–320 K is related to the proximity of the critical point. Looking at the ratio of efficiencies it becomes apparent that for reservoir conditions found under a typical geothermal gradient of 30 °C km^{-1} that at from a depth of 800 m to approximately 4 km CO_2 will be more efficient that H_2O. At depths of more than 4 km (under normal geothermal gradients) H_2O will be more efficient at transporting heat energy out of the reservoir than CO_2. It is also important to note that for the higher reservoir temperature required for base load power production and better efficiencies (about >130 °C) that H_2O turns out to be the more efficient fluid. For closer surface reservoir systems and smaller heat requirements, for example, 100 °C or less, CO_2 proves to be the more efficient fluid.

5.3.4 Uncertainty Assessment

Data uncertainty is one of the major problems in subsurface reservoir analysis (Section 5.1.2). (Watanabe *et al.*, 2009) used spatially correlated random fields to generate parameter distributions. The stochastic properties of the random field are probability distribution (frequency) and spatial correlation. The probability distribution represents the quantitative trend of parameter values. For example, the distribution of hydraulic conductivity in aquifers often appears as a lognormal shape. As explained above, we assume normal distributions for all THM

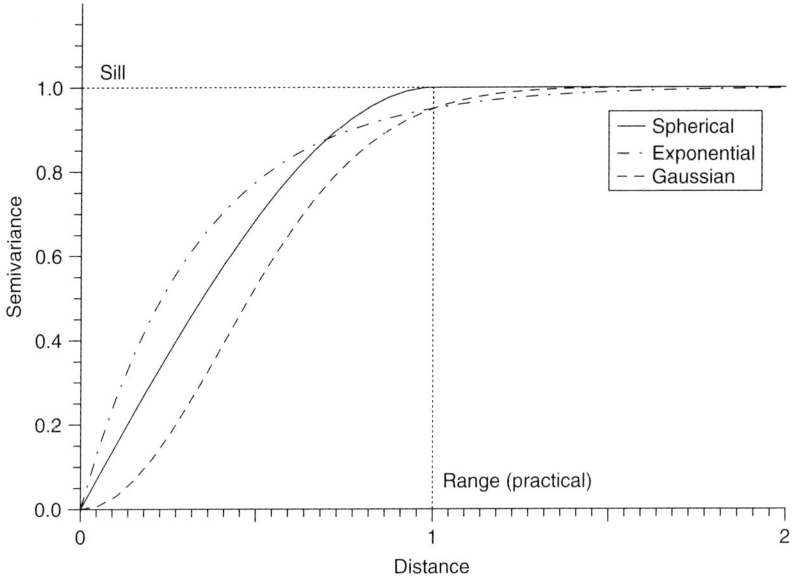

Figure 5.9 Typical variogram models with spherical, exponential, and Gaussian shapes (sill = 1.0, practical range = 1.0).

parameters in this work. Spatial correlation, which represents how parameter values at certain positions can affect or constrain values at a different position, can be defined with a variogram model. A variogram is a function of a separation lag distance and it describes spatial variability. A variogram model consists of fitting an empirical variogram, which is determined from measured data. There are different shapes of variogram models such as spherical, exponential, and Gaussian distributions (Figure 5.9).

The spatial dependency of a parameter reduces as the distance between locations becomes larger and it will converge to an asymptotical value (sill) from a certain lag (range) as illustrated on Figure 5.9. These models are different in how spatial dependency reduces. Spherical models decrease in a linear fashion. The decrease of an exponential is steeper than the spherical model; the local variability is stronger. Gaussian models decrease gently near the origin and linearly further out so that it has stronger continuity at short distances. In the uncertainty analysis, the spherical model is used for all parameters because of the simple linearity and to demonstrate our methodology.

As an example of a stochastic model, we consider the permeability of an undisturbed (i.e., before hydraulic stimulation) geothermal reservoir in crystalline rock based on data from the Urach Spa location (Haenel, 1982). Figure 5.10 shows one realization result of a permeability distribution for an undisturbed reservoir. The stochastic generation is based on a conditional Gaussian simulation with measured data at borehole locations. The parameter values are mapped to the finite element (FE) mesh for the numerical THM simulation (Section 5.6)

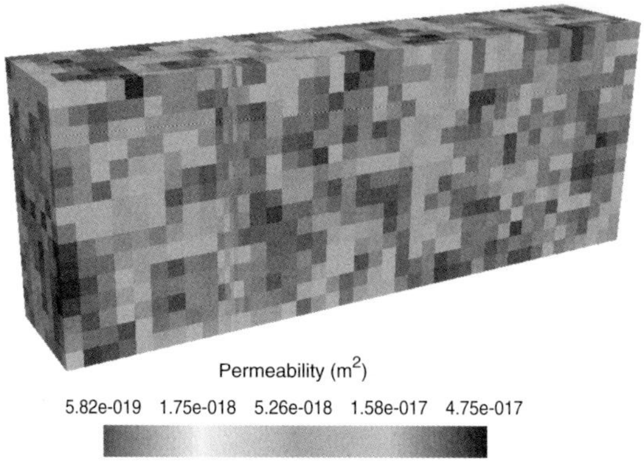

Figure 5.10 Realization of a permeability distribution for an undisturbed reservoir.

The problem with stochastic analysis of deep reservoirs is data scarcity. Normally only data from a few boreholes are available to derive statistical properties. Some oil reservoirs can afford 3D seismics in order to obtain a better picture of the subsurface structures.

5.4 Site Studies

In the following text, we present several site studies (Figure 5.1) using geothermal reservoir simulation based on different numerical models. These geothermal reservoirs are located in different geological formations such as sandstone and crystalline rock requiring the application of various conceptual models (Figure 5.2). New and ongoing investigations that is, at Groß Schönebeck (Section 5.5) and Bad Urach (Section 5.6) are explained in more detail. Former case studies at Rosemanowes (Section 5.7), Soultz (Section 5.8), KTB (Section 5.9), and Stralsund (Section 5.10) are briefly summarized, and literature references are given.

5.5 Groß Schönebeck

In this section, the hydrothermal simulation of the geothermal research doublet E GrSk03/90 and Gt GrSk04/05 at the drill site Groß Schönebeck is described.

5.5.1 Introduction

The technical feasibility of geothermal power production will be demonstrated by means of the geothermal research wells Groß Schönebeck (40 km north of Berlin/Germany) using a borehole doublet. The first well Groß Schönebeck E

GrSk3/90 was tested to investigate scenarios of enhancing productivity of thermal fluid recovery from the underground (Legarth *et al.*, 2005; Reinicke *et al.*, 2005; Zimmermann *et al.*, 2005). In order to complete the doublet system, a second well Gt GrSk4/05 with a total depth of −4198 m had been drilled in 2006, followed by three stimulation treatments to enhance productivity. In order to increase the apparent thickness of the reservoir horizon, the new well is inclined in the reservoir section by 48° and was drilled in the direction of the minimum horizontal stress ($S_h = 288°$ azimuth) for optimum hydraulic fracture alignment in relation to the stimulated preexisting well E GrSk3/90. Hence, the orientation of the fractures will be 18° azimuth in the direction of the maximum horizontal stress (Holl *et al.*, 2005). During the elapsed time of the geothermal project, it became obvious that an appropriate numerical model becomes increasingly important for planning the well path and fracture design, for interpretation of hydraulic tests and stimulations as well as for prediction of reservoir behavior during the time of geothermal power production.

To satisfy the requirements of the simulations, the model should implement all the acquired knowledge of the reservoir. This includes the reservoir geology and structure; the geometry of wells and fractures; the hydraulic, thermal, and mechanical conditions of the reservoir; and generated fractures due to changes of the reservoir conditions.

5.5.2 Model Description

5.5.2.1 Geology

The reservoir is located at a depth of −3850 to −4258 m within the Lower Permian of the northeast German Basin. The reservoir rocks are classified into two rock units from base to top: volcanic rocks (Lower Rotliegend of the Lower Permian) and siliciclastics (Upper Rotliegend of the Lower Permian) ranging from conglomerates to fine grained sand-, silt- and mudstones. These two main units can be subclassified depending on their lithological properties, which are, in particular, of importance for the hydraulic-thermal-mechanical (HTM) modeling:

- I: Hannover formation – silt- and mudstone (−3850 to −3996 m true vertical depth subsea (TVDSS))
- IIA: Elbe alternating sequence: siltstone to fine grained sandstone (−3996 to −4026 m TVDSS)
- IIB: Elbe base sandstone II: fine grained sandstone (−4026 to −4086 m TVDSS)
- IIC: Elbe base sandstone I – fine to medium-grained sandstone (−4086 to −4133 m TVDSS)
- III: Havel formation – conglomerates from fine sandstone to fine grained gravel (−4133 to −4178 m TVDSS)
- IV: Volcanic rocks – ryolite and andesite (−4178 to −4258 TVDSS)

Owing to the high hydraulic conductivity and porosity, the Elbe base sandstone IIB and IIC are the most prominent horizons for geothermal power production.

5.5.2.2 **Structure**

To model the geothermal reservoir, it is important to define the model area depending on the reservoir structure. As mentioned above, the geothermal reservoir is built up by six subhorizontal layers. The low permeable overlying and underlying horizons can be taken as no-flow boundaries. The mean and total thickness of the layers can be calculated as shown above. The horizontal extension was chosen depending on the maximum hydro-thermal-mechanical influence of stimulation treatments and scheduled geothermal power production, and geological boundary conditions. Therefore, we defined a model area of 6 km in the east–west direction and of 5 km in north–south direction around the research wells. Two northwest striking fault systems are included in the model at the north and south border as no-flow boundaries.

Beside these geological boundary conditions of the reservoir, the geometries of the hydraulic fractures and the research wells (Figure 5.11) have to be implemented into the model. Actually, four hydraulic fractures exist (Table 5.2). The fractures are orientated perpendicular to the minimum horizontal stress. This means that

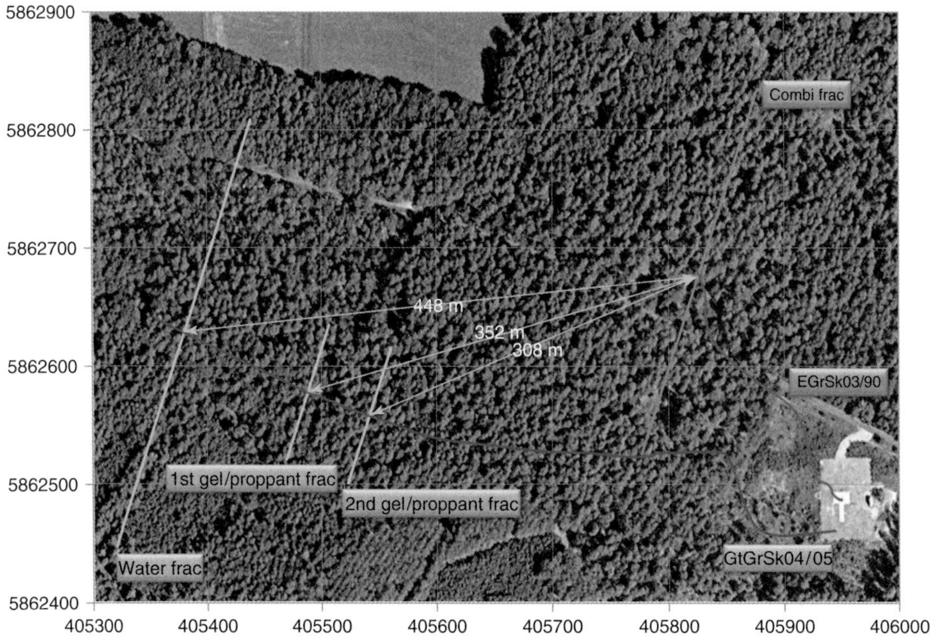

Figure 5.11 Projected well paths and hydraulic fractures of the geothermal research doublet E GrSk03/90 and Gt GrSk04/05 at the drill site Groß Schönebeck.

Table 5.1 Hydraulic and thermal properties of the reservoir rocks under in situ conditions. The hydraulic conductivity was estimated by means of a reference dynamic viscosity of 0.3 mPa·s (at $T = 150\,^\circ\text{C}$ and $C = 265\ \text{g l}^{-1}$).

Layer	k (mD)	k (m^2)	K ($m\ s^{-1}$)	n (%)	T (°)	λ [W $(mK)^{-1}$]	c [J $(kg\ K)^{-1}$]
I	0.05	4.93E-17	1.34E-09	1	138.2	1.9	920
IIA	2	1.97E-15	5.37E-08	3	141.7	1.9	920
IIB	4	3.95E-15	1.07E-07	8	143.2	2.9	920
IIC	8	7.90E-15	2.15E-07	15	145.2	2.8	920
III	0.1	9.87E-17	2.68E-09	0.1	146.5	3.0	1000
IV	0.1	9.87E-17	2.68E-09	0.5	147.4	2.3	1380

for the recent *in situ* stress field an azimuth of 18° is required. The arrangement of the two wells has to fulfill two important conditions. First, the wells should be located in such a way that the pressure in the reservoir would not drop significantly during production, and second, a temperature drop in the production well should be avoided. At the surface the two wells have a distance of 28 m. Owing to the fact that the well E GrSK3/90 is vertically orientated and to ensure a distance of 500 m within the reservoir, the second well Gt GrSk4/05 is drilled as a deviated well. At the top of the reservoir (−3850 m) the inclination is 18° and increases progressively to 48° at −4200 m. Therefore, the distance between the two wells increases from 254 to 473 m from the top to the bottom of the reservoir (Figure 5.11). Besides the realization of the required distance within the reservoir, the deviation leads to an increase of the well–reservoir intersection as well. Thus, the thickness increases from 350 m for the vertical well E GrSK3/90 to an apparent thickness of 442 m for the deviated well Gt GrSk4/05.

5.5.2.3 **Thermal Conditions**

According to the continental geothermal gradient, the lowest temperature of the reservoir can be found at the Hannover formation with 138 °C and increases continuously to 147 °C for the volcanic rocks. Further parameters are the heat conductivity λ and the heat capacity c of the reservoir fluid and solid rock. A detailed overview of thermal parameters of the North German basin is given by Lotz (2004) and Gehrke (2007); the results of the latter work are summarized in Table 5.1.

5.5.2.4 **Hydraulic Conditions**

In our reservoir model, the fluid flow is characterized by two different processes: matrix flow and drainage by induced hydraulic fractures.

Matrix Flow In most sedimentary rocks, the porosity is interconnected, which makes the rock permeable for flow. The permeability of a rock can be measured by

Table 5.2 Dimensions and hydraulic properties of the artificial hydraulic fractures under in situ conditions.

Well	Type	Layer	Depth (m)	Height (m)	Length (m)	K_f (m s^{-1})
E GrSk3/90	2 × gel/proppant 2 × water	IIB, IIC, III	−4004 to −4147	143	160	0.106
Gt GrSk4/05	Water	III, IV	−4098 to −4243	145	190	0.142
Gt GrSk4/05	Gel/proppant	IIB, IIC	−3996 to −4099	103	60	0.142
Gt GrSk4/05	Gel/proppant	IIA, IIB	−3968 to −4063	95	60	0.142

the steady-state method directly using Darcy's Law and thus assuming laminar flow conditions according to Equation (5.10). The highest *in situ* reservoir permeability of 4–8 mD (compare Table 5.1) (1 D = 1 Darcy = $9.869E - 13$ m^2) was determined by laboratory experiments on core samples from of the Elbe basis sandstone (Trautwein and Huenges, 2005). These values are confirmed by the results of log interpretation performed by Holl *et al.* (2004).

Drainage by Induced Hydraulic Fractures To drain a geothermal reservoir efficiently, hydraulic fracture stimulations are performed by pressurizing the wellbore above the minimum horizontal stress at defined intervals. For the research well E GrSk3/90, an *in situ* fracture closing pressure of 49.8 MPa was determined (Huenges *et al.*, 2006). To assure a sufficient fracture opening and hydraulic conductivity during production, proppants are placed into the fracture. The volcanic rock was propped with a low concentration of quartz sand (diameter 0.4–0.8 mm, 20/40 mesh), the sandstones were propped with sintered bauxite spheres (20/40 mesh) (Zimmermann *et al.*, 2008). The dimensions and the hydraulic properties determined by numerical simulations are listed in Table 5.2.

For all hydraulic fractures an aperture equal to 2.28×10^{-4} m was calculated by means of fracture transmissibility (permeability multiplied by aperture) equal to 1 Dm. The hydraulic fracture conductivity was estimated by means of a reference dynamic viscosity of 0.3 mPa s for the production and 0.4 mPa s for the injection well. The thermal properties vary with depth according to the thermal properties of the associated geological layers.

5.5.3 Modeling Approach

The governing equations of thermohaline convection in a saturated porous media are derived from the conservation principles for linear momentum, mass, and energy (e.g., Bear (1991); Kolditz *et al.* (1998); Nield and Bejan (1999)). The resulting system is fully implemented in a FE simulator FEFLOW (Diersch, 2002; Magri *et al.*, 2005, and references therein).

Beside the fluid and solid properties the geometries and properties of the hydraulic fractures and the boreholes have to be integrated into the model as well. The structure of the hydraulic fractures can be easily represented by vertical 2D quadrilateral fracture elements and those of the injection well E GrSk03/90 by vertical 1D channel fracture elements. Owing to the deviation of the production well Gt GrSk04/05, we used arbitrary 1D channel fracture elements to connect the three involved hydraulic fractures. The properties of the hydraulic fractures are summarized in Table 5.2. There, the hydraulic conductivity is corrected as mentioned above. For both wells a cross-sectional area of 126.7 cm^2 (5 in. diameter) was used. The hydraulic conductivity of 1236 m s^{-1} for the injection well and 1648 m s^{-1} for the production well were estimated according to Hagen–Poiseuille equation. For all discrete feature elements, the Equation of state (EOS) for fluid density was applied.

5.5.4 Results

By means of the above described model, we simulated the 30-year life cycle of geothermal power production. The first step included the calculation of the initial pressure and temperature field by means of a stationary model after a simulation period of 100.000 years. On the basis of the results of the stationary model, the 30-year life cycle was simulated. For this purpose, we assumed a production and injection rate of 75 m^3 h^{-1} and an injection temperature of 70 °C. After 30 years of simulation time it becomes obvious that the injected cold water has reached the production well, as shown in Figure 5.12. Further, the influence of the hydraulic fractures is shown. At the injection well, the water accesses first the induced hydraulic fracture and afterwards the connected matrix. At the production side, the hydraulic fractures drain the matrix. If the cold water front reaches one of the hydraulic fractures, then it will be directly forwarded to the production well.

For a detailed observation of pressure and temperature changes during the total time of simulation, four observation points were set up along the wells. Observation point 1 is located at the top of the hydraulic fracture at the injection well. The observation points 2, 3, and 4 are located at the top of each of the hydraulic fractures of the production well. The lowest observation point, 4, only shows the pressure and temperature behind the waterfrac in the volcanic rocks, observation point 3 shows the sum of waterfrac and the first gel/proppant frac. Observation point 2 gives a cumulative value of all three fractures. The results are shown in Figure 5.13.

The hydraulic head increases approximately 400 m due to injection and decreases approximately 500 m due to production. Taking into account, that the wells, the hydraulic fractures, and the reservoir matrix are in full hydraulic contact and no skin effects are present, the real hydraulic head change should be higher than simulated. By means of the simulation, a quasistationary state was achieved after one year of production and injection. In contrast, the temperature does not reach a stationary state during the time of simulation. After five years, the cold water front reaches the nearest production fracture (second gel/proppant frac). Starting from this time,

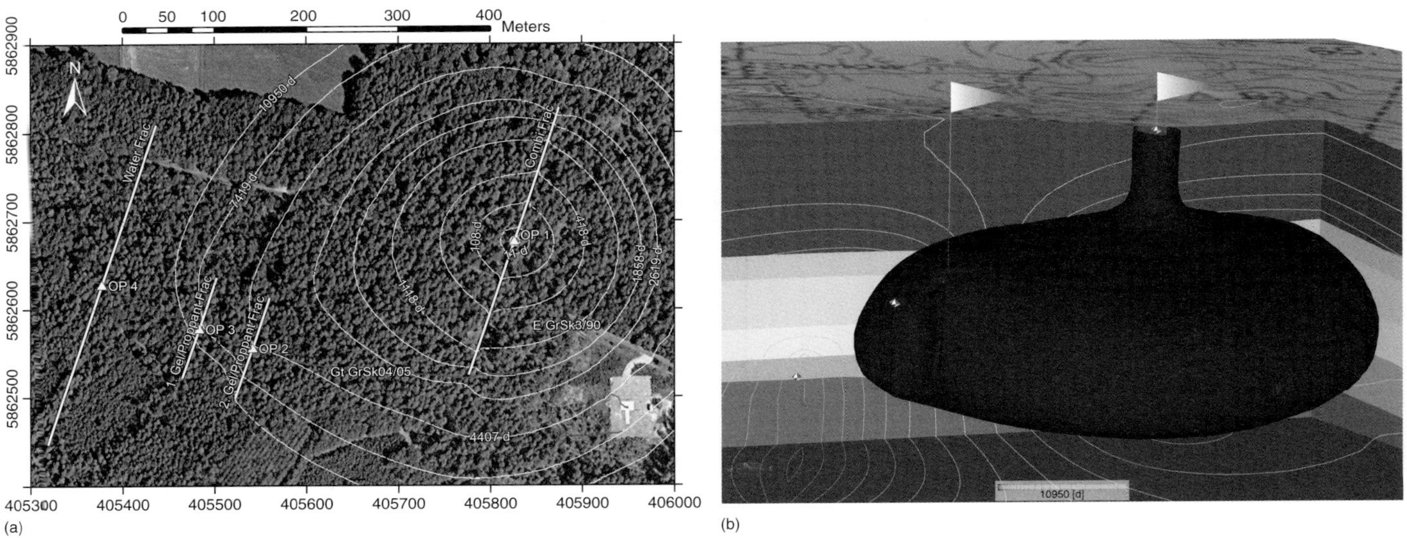

Figure 5.12 Propagation (a) and final stage (b) of simulated 130 °C isosurface around the injection well during 30 years of production. The propagation is illustrated by a horizontal cut through the Elbe base sandstone II unit (b). (Please find a color version of this figure on the color plates.)

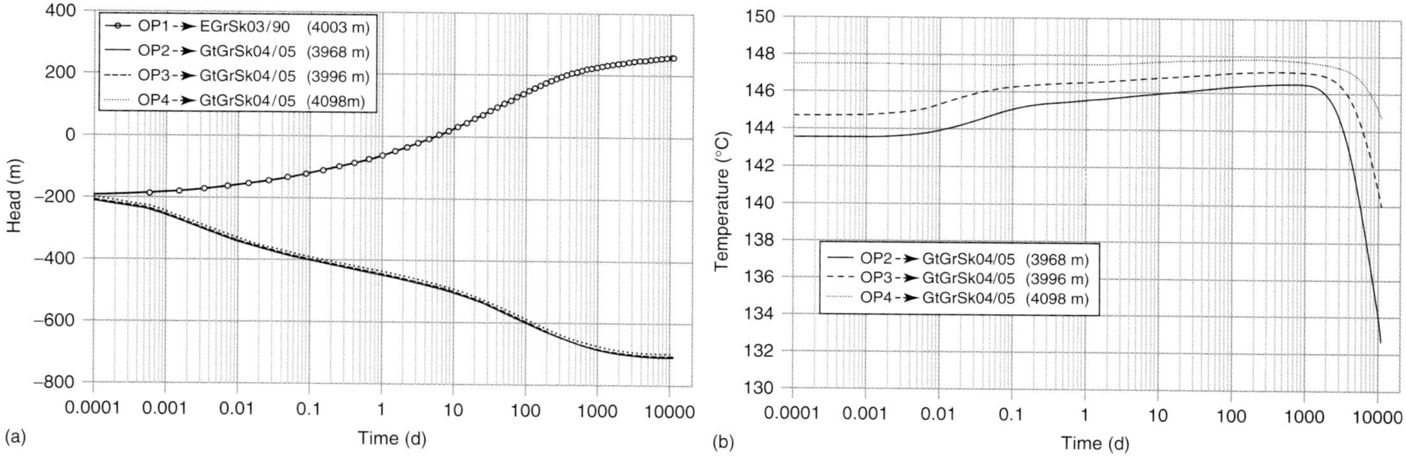

Figure 5.13 Chronological sequence of hydraulic heads (a) and temperature (b) at the four observation points over a period of 30 years.

a continuous decrease in temperature can be observed at observation point 2. After seven years of production, the cold water reaches the second production fracture (first gel/proppant frac) and after an additional five years, the waterfrac is also affected by the cold water front, leading to a further reduction of temperature. Consequently, a production temperature of 132 °C was simulated after 30 years production.

5.5.5 Conclusions

By means of the well-known reservoir geometry, structure geology, hydrothermal conditions, and the occurring coupled processes, we simulated the change of the geothermal reservoir conditions at the Groß Schönebeck site. This delivers an improved understanding of the reservoir behavior and leads to an interpretation of the long-term reservoir characteristics during geothermal power production. For a better prediction of the long-term reservoir behavior it becomes essential to integrate mechanical properties and chemical interaction. These requirements cannot be captured by FEFLOW simulation software. Therefore, the presented results are a basis for further investigations by numerical simulations with a coupled THMC simulator.

5.6 Bad Urach

The case study for the application of the stochastic THM model is based on a large data set for the Urach Spa geothermal site compiled in several research projects (Haenel, 1982; Tenzer *et al.*, 2000). The idea of the application case study is to demonstrate the methodology for an uncertainty analysis of THM-coupled processes in a typical geothermal reservoir in crystalline rock. A basic assumption of the conceptual model for the Urach Spa site is that due to dense fracturing, the geothermal reservoir can be represented as a heterogeneous porous medium. There is no database available in order to construct DFN models.

5.6.1 The Influence of Parameter Uncertainty on Reservoir Evolution

5.6.1.1 Conceptual Model

Urach Spa location was originally designed as a scientific geothermal pilot project. The proposed boreholes (U3 and U4) dipole flow circulation system (i.e., a "doublet") are located 400 m apart. On the basis of the large amount of scientific data available on the Urach Spa reservoir, we developed a three-dimensional model of the reservoir system. Parameters relevant to reservoir fluid flow and heat transport that were used in the model were based on the results of previous studies. The hydraulically active areas allowing the reservoir to be represented geometrically as a

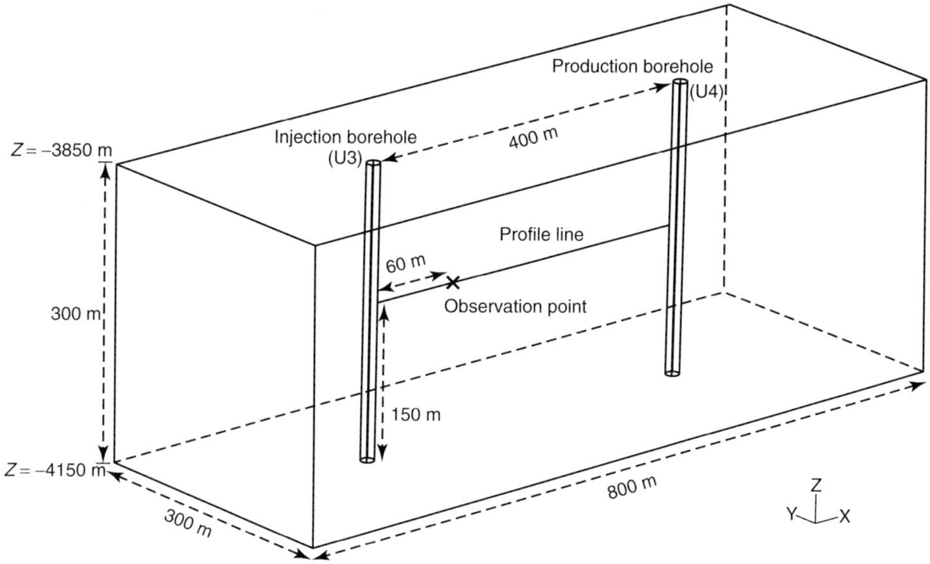

Figure 5.14 Cuboid reservoir model with a borehole doublet (U3 and U4).

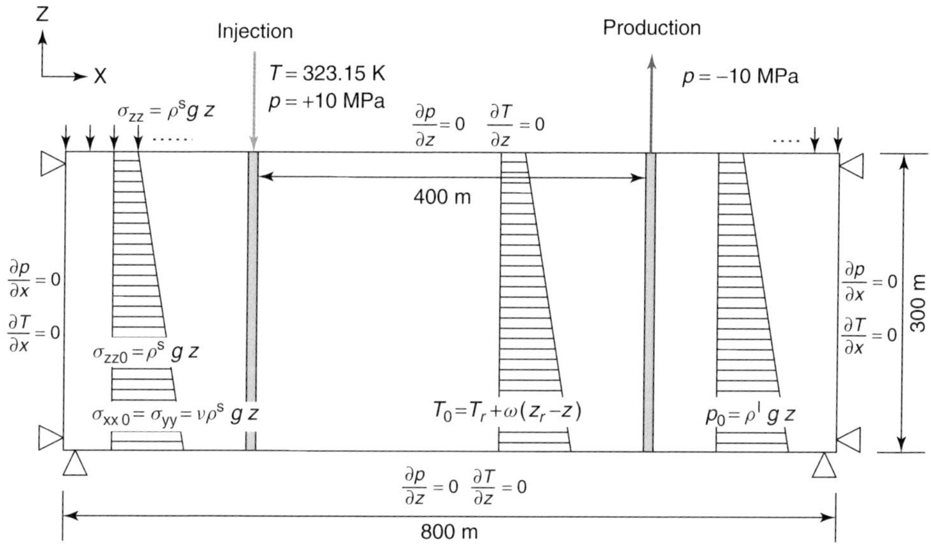

Figure 5.15 Thermodynamic system conditions for the geothermal reservoir.

cuboid are 300 m high, 300 m wide, and 800 m long (Figure 5.14). The observation point is selected to be rather close to the injection borehole as the significance of THM coupling during heat extraction is more evident in this area.

The reservoir depth is between 3850–4150 m. The corresponding THM system conditions are depicted in Figure 5.15. Concerning the initial conditions ($t = 0$), we assume linear depth-dependent hydrostatic pressure, lithostatic stress, and temperature distribution. The geothermal gradient according to the temperature logs in the reservoir depth range of U3 is $\omega = 0.3\text{K m}^{-1}$:

- $T(t=0) = 435.15 + \omega\,(-4445.0 - z)$ (K)
- $p^{l}(t=0) = \rho^{l}\,g\,z$ (Pa)
- $\sigma_{zz}(t=0) = \rho^{s}\,g\,z$ (Pa)
- $\sigma_{xx}(t=0) = \sigma_{yy}(t=0) = \nu\,\rho^{s}\,g\,z$ (Pa)

The injection well is considered to have an overpressure of 10 MPa and the production well an underpressure of 10 MPa. Fluid injection temperature is assumed to be 50 °C (McDermott *et al.*, 2006).

- $T^{in} = 323.15$ (K)
- $p^{in} = p_0 + 10 \times 10^6$ (Pa)
- $p^{out} = p_0 - 10 \times 10^6$ (Pa)
- $\mathbf{u} \cdot \mathbf{n} = 0$ at the lateral and bottom surface
- $\frac{\partial \sigma_{zz}}{dz} = \rho^{s}\,g\,z$ at the top surface

We use a FE model for the stochastic analysis, which takes into account fully coupled THM processes according to the governing equations given in Section 5.2. The present code provides an object-oriented FE concept; different element types can be easily used for THM analysis. The characteristic element length is about 20 m resulting in 6600 elements and 7920 nodes for the hexahedra mesh and 59 599 elements and 11 856 nodes for the tetrahedra mesh. Grid adaptation is necessary in order to resolve the boreholes geometrically and to obtain a smooth change of element sizes.

5.6.1.2 **Simulation Results**

A parameter sensitivity analysis for an undisturbed reservoir using the fully coupled THM model confirmed that the most important parameters are permeability and fluid viscosity for reservoir hydraulics, as well as heat capacity for reservoir thermodynamics.

5.6.1.3 **Stimulated Reservoir Model**

(Watanabe *et al.*, 2009) analyzed the statistical sensitivity of THM parameters as well as variogram properties in detail. Besides parameter uncertainty, a realistic reservoir model should address the effect of reservoir stimulation as well. As a result of massive hydraulic stimulation, the reservoir permeability could be increased by factor of 100, at least in the vicinity of the boreholes (Baisch *et al.*, 2004). The stimulation length is in the order of 100–200 m. In this section, we conduct a Monte Carlo analysis to assess the thermal reservoir evolution by a superposition of stochastic parameter heterogeneity and hydraulic reservoir stimulation. We represent the heterogeneity of the undisturbed reservoir using a spherical variogram model with a correlation length of 50 m range. Hydraulic stimulation is mimicked by a scaling factor between 1 (undisturbed) and 100 (fully stimulated), which depends on the borehole distance.

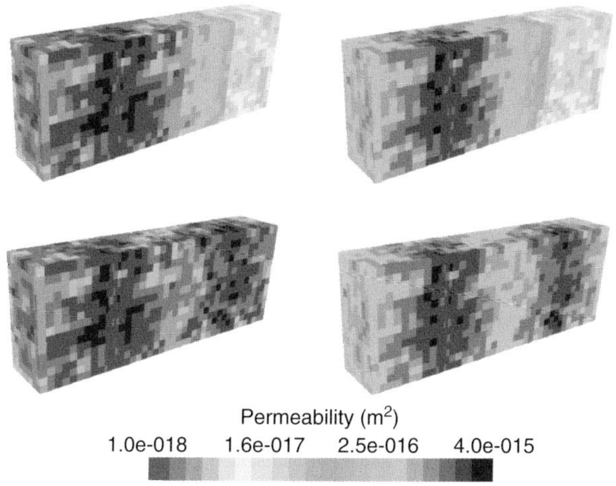

Figure 5.16 Permeability distribution for a stimulated reservoir by single borehole (top) and two boreholes (bottom): linear (left) and quadratic enhancement functions (right).

Figure 5.16 illustrates the reservoir permeability distribution if both boreholes are used for hydraulic stimulation. From experimental observation it is known that the stimulation radius is in the range of the borehole distance (Weidler *et al.*, 2002; Baisch *et al.*, 2004). The effect of hydraulic stimulation on the permeability (enhancement factor) is strongly dependent on the injection borehole distance. The functional behavior of the decay of the permeability enhancement factor is difficult to characterize. Therefore, we investigate two models in order to scale the permeability enhancement depending on borehole distance: linear and quadratic decline of enhancement factor (Figure 5.16).

Figure 5.17 depicts the THM results for both cases, single and double hydraulic borehole stimulation using linear and quadratic permeability enhancement functions. Hydraulic stimulation of the borehole doublet leads to an overall increase of reservoir permeability; therefore we obtain a much larger temperature drawdown in the observation point. The numerical study shows very impressively the consequences of single and multiple borehole stimulations to the long-term thermal reservoir evolution. These scenario analyses provide useful information for optimization of reservoir management.

5.6.1.4 Monte Carlo Analysis

We consider a reservoir type, where hydraulic stimulation is conducted in two boreholes with a quadratic permeability enhancement factor and the porosity–permeability relationship corresponds to that from the Falkenberg site (Figure 5.4). To perform a representative Monte Carlo simulation, we conduct 100 stochastic simulations. In order to be able to run this number of fully coupled

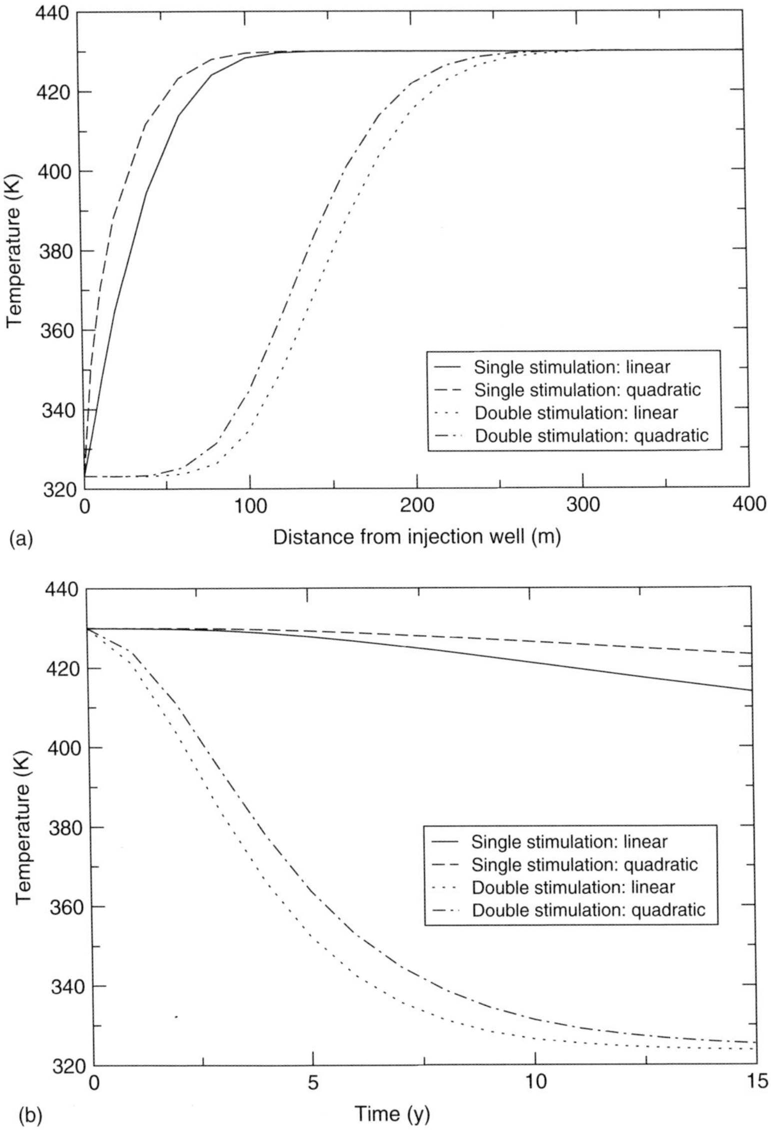

Figure 5.17 Analysis of thermal reservoir evolution for single and double borehole stimulations, profile (a) and thermal drawdown (b).

THM simulations, a parallelized version of the numerical code was used (Wang *et al.*, 2009).

In order to present and discuss the results we use different illustrations: variances, frequencies, and envelope curves. Figure 5.18 (b) shows the average temperature as well as the standard deviation in a horizontal cross section through the reservoir

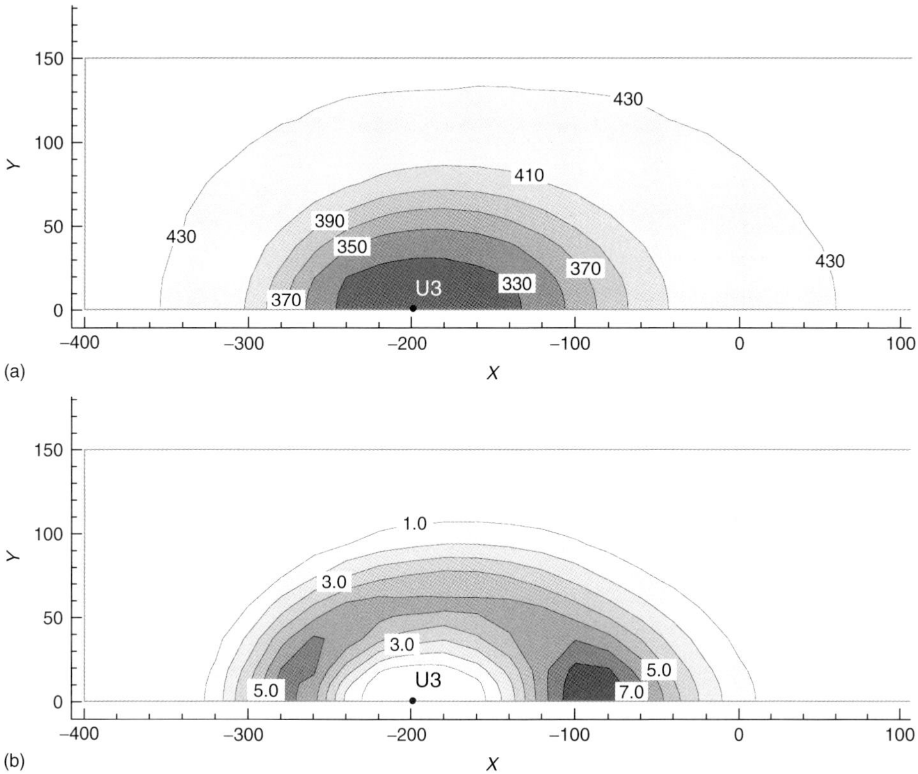

Figure 5.18 Monte Carlo analysis: average [K] (a) and standard deviation (b) of temperature in a horizontal cross section after 15 years.

center after 15 years of operation. The largest variances appear at places where temperature gradients are highest, that is, around the cooling fronts. Maximum standard deviation is large, that is, about 8. Figure 5.19 depicts temperature "frequencies" in the observation point at different operation times (1, 2, 5, 10, and 15 years). The frequency plot represents the number of calculated temperature values. The frequency range is time dependent: it is narrow in the early (one year) and late (15 years) stages (i.e., almost undisturbed or cooled reservoir), and it is widely distributed in the middle of the reservoir "lifetime" (5–10 years). During this time, a prediction of reservoir temperature is most uncertain. This finding corresponds to Figure 5.18 that the uncertainty is largest around the propagating cooling front due to the dominating advective heat transport. Figure 5.20 illustrates the 20, 80, and 100% uncertainty zones of the 100 temperature profiles between the boreholes after 15 years. The 20% zone covers 20% of 100 obtained temperatures around the median. The 100% zone provides an envelope to all 100 realizations with a maximum temperature difference of about 40 K.

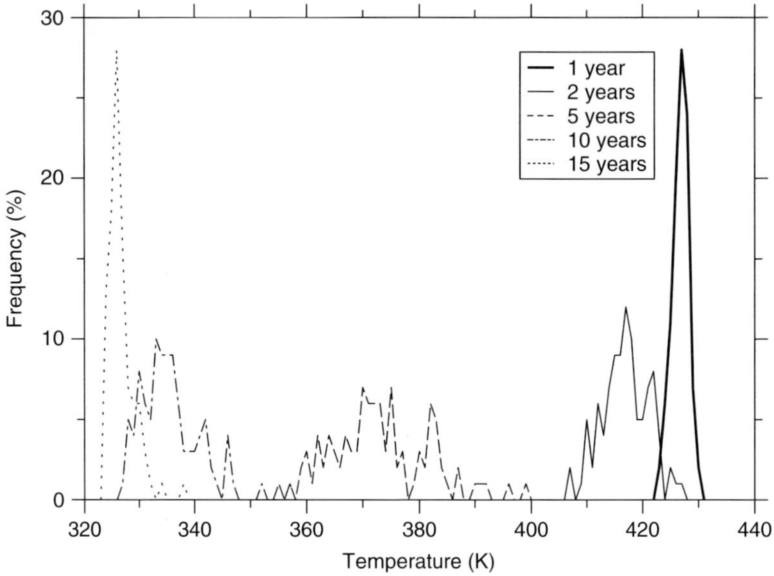

Figure 5.19 Monte Carlo analysis: temperature frequencies at the observation point after different operation periods, 1, 2, 5, 10, and 15 years.

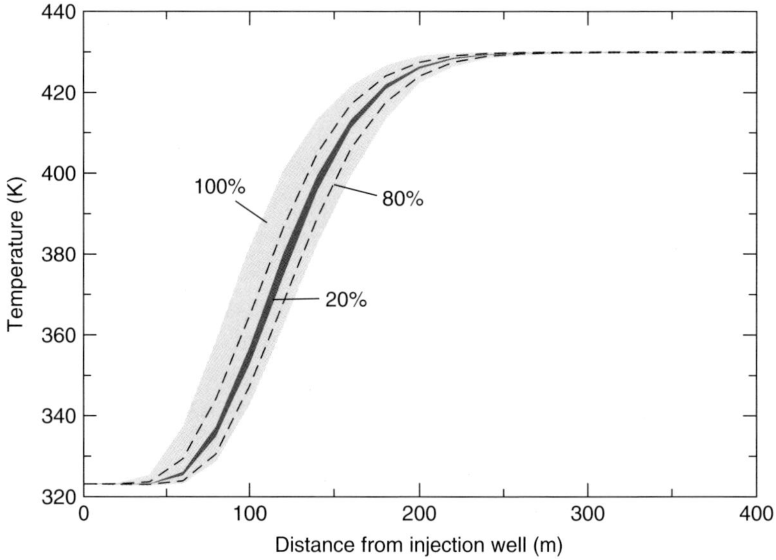

Figure 5.20 Monte Carlo analysis: uncertainty ranges of temperature profiles after 15 years heat extraction, 20% (dark gray area), 80% (dotted line), 100% (light gray area).

5.6.1.5 Conclusions

A fully coupled THM model is developed based on the general balance equations for fluid mass, momentum, and thermal energy as well as constitutive equations for variable fluid properties, thermoporoelastic deformation, and a phenomenological porosity–permeability relationship for crystalline rock, taking into account hydraulic stimulation effects. The stochastic concept is a combination of the fully coupled numerical THM model and the Monte Carlo method. On the basis of the stochastic THM model, we present an uncertainty analysis of thermal, hydraulic, and mechanical parameters on long-term geothermal reservoir evolution.

The most important findings using the stochastic THM model are:

- Accounting variable fluid properties is very important within THM analysis: the most sensitive parameter is fluid viscosity.
- Statistical heterogeneity of geothermal reservoirs is considered: sensitivity analysis shows that permeability and rock heat capacity are most important reservoir parameters; less relevant is thermal conductivity. The variability of the mechanical parameters in the site-specific range, porosity, Young's modulus, and Poisson ratio is negligible.
- Hydraulic stimulation effects: as the Urach Spa site has been hydraulically stimulated several times, we included those stimulation effects by a permeability enhancement factor depending on the borehole distance (Figure 5.16).
- Combination of parameter heterogeneity: we considered interrelated spatial permeability–porosity distribution using a constitutive model by Pape *et al.* (1999) for crystalline rock (Falkenberg site).
- As a result of the stochastic THM analysis we found a maximum temperature uncertainty range of about 40 K after 15 years of reservoir exploitation (Figure 5.19).
- Computational efficiency: parallel computing is an important technical prerequisite for THM Monte Carlo analysis.

5.6.2 The Influence of Coupled Processes on Differential Reservoir Cooling

5.6.2.1 Conceptual Model

Processes operating during the extraction of heat in fractured rock dynamically influence the fluid flow and heat transport characteristics. The incorporation of pressure and temperature-dependent parameters of the rock mass coupled with geomechanical deformation is particularly important for predictive modeling of hard rock geothermal reservoirs as discussed above. Utilizing an experimentally validated geomechanical model (McDermott and Kolditz, 2006), the changes in the flow and transport parameters within crystalline fractures due to changes in local effective stress were simulated by McDermott *et al.* (2006). The changes in local effective stress are linked to the dynamic reservoir fluid pressure, *in situ* stress conditions and the build up of thermal stresses during rock mass cooling. These processes are simulated in a case study of the Spa Urach (South West Germany) potential geothermal reservoir, using an FE model comprising tetrahedral elements.

The model couples mechanical deformation and alteration of fracture parameters with state of the art fluid parameter functions dependent on pressure, temperature, and salinity for heat capacity, conductivity, viscosity, and density. The effects of the coupling on the reservoir productivity characteristics and potential reservoir damage were investigated to assist in the identification of optimal heat recovery schemes for the long-term economical operation. Preferential flow paths and hydraulic short cuts dependent on the geomechanical and thermal stress release behavior were predicted for overexploitation.

5.6.2.2 **Development of Preferential Flow Paths due to Positive Feedback Loops in Coupled Processes and Potential Reservoir Damage**

The effect of the coupling of the processes investigated, led to systematic and localized changes in the flow and transport characteristics, systematic in terms of the progressive development of the phenomena and localized in terms of the impact at specific locations. This led to the development of preferential flow paths (Hicks *et al.*, 1996; Su *et al.*, 2001; Zheng and Gorelick, 2002) and geometrical

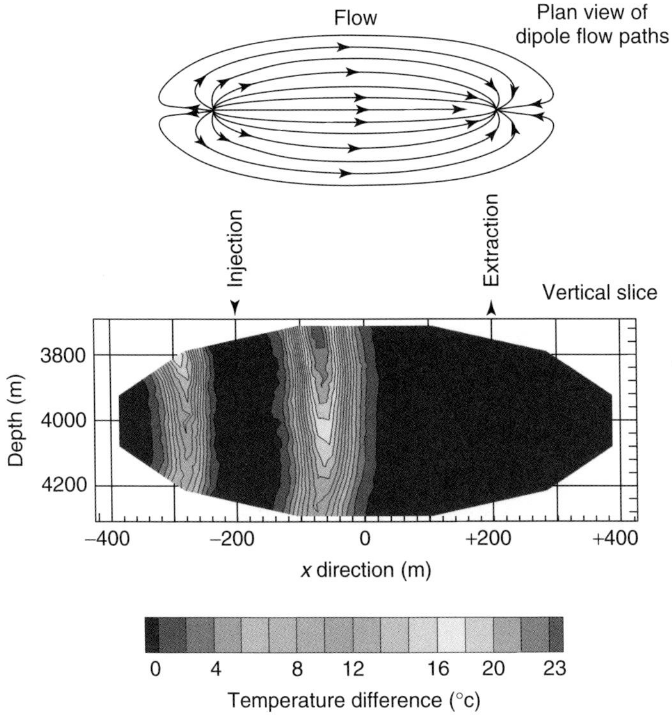

Figure 5.21 Hydromechanical coupling, the reservoir is more permeable under the lower stress conditions at the top of the reservoir than the higher stress conditions at the base of the reservoir. (Please find a color version of this figure on the color plates.)

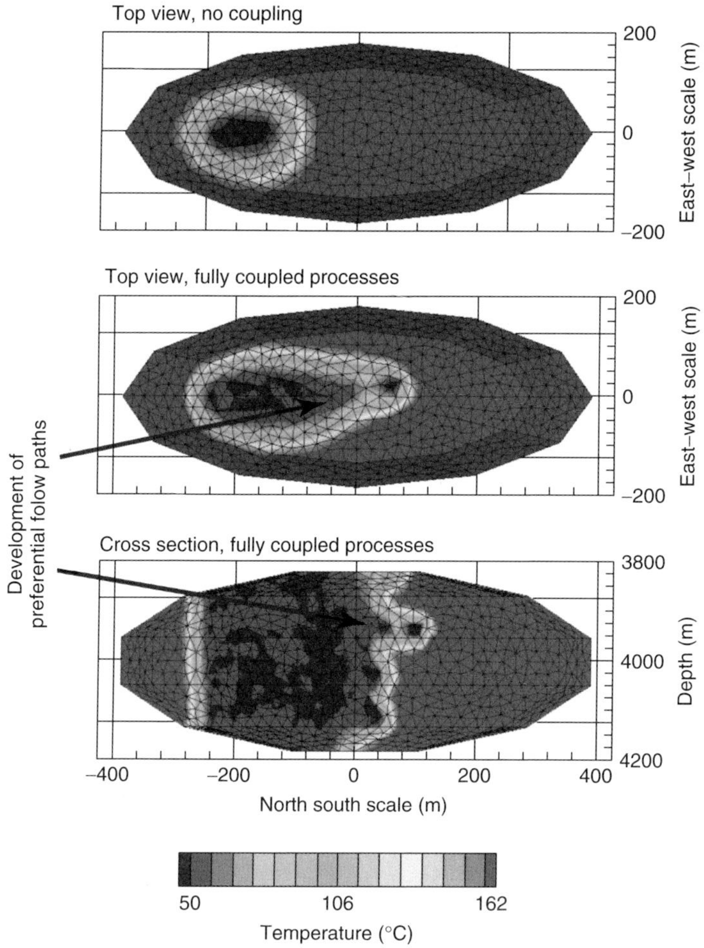

Figure 5.22 Development of preferential flow paths as a result of positive feedback between the coupled processes.

heterogeneities during the heat extraction. Figure 5.21 illustrates the difference in the reservoir temperature between a noncoupled constant parameter scenario HT and the coupled hydraulic mechanical HT (MHC) scenario after about 11 years. The larger differences are shown in dark. Here flow and therefore cooling is being concentrated in the upper part of the reservoir, and preferential flow paths start to develop. Figure 5.22 illustrates the same scenario with coupling to the thermal stress, and this leads to the development of preferential flow paths becoming much more significant. In this figure after seven years, the thermal coupling has not only led to a significantly higher flow rate in the system, but also to the development of preferential flow paths. These preferential flow paths represent potential short cuts in the operation of the reservoir.

The trigger mechanism to start the development of a preferential flow path was numerical heterogeneities in the mesh describing the reservoir. Slight differences in the size and orientation of the elements led eventually to the start of the preferential flow paths, which due to a positive feedback mechanism then continued to develop. The location of the preferential flow paths is not a coordinate description of the location in reality, rather an indication that under certain coupling of processes, such a development can be expected to occur. In nature, heterogeneities will also significantly effect the development of potential preferential flow paths and suggests the presence of ample trigger mechanisms.

Returning to the development of these preferential flow paths, there is a significant application in terms of the management of HDR geothermal systems. The development of such flow paths is due to flow concentration caused by an increase in hydraulic conductivity as a result of the build up of thermal or hydraulic stress. As flow is concentrated in a certain area, the amount of cooling in these areas is increased. This in turn leads to a further increase in the thermal stress, which turns the system into a positive feedback of everincreasing hydraulic conductivity. Significantly, the cooling of the fluids in these overcooled areas, and the consequent increase in viscosity, results in a brake to the feedback. The increase in viscosity of the fluids reduces the rate of fluid flow in the overcooled areas, and therefore helps to divert flow to the warmer areas where the viscosity is still low. Such considerations have a significant impact on understanding the operation of the reservoir, and deriving effective heat extraction schemes. In such systems, the inclusion of a viscosity reducer to increase productivity may have short-term benefits, but may be disastrous in the long term, encouraging the development of preferential flow paths. The use of lower viscosity fluids, such as supercritical CO_2 would be particularly prone to this type of reservoir damage scenario (Pruess, 2008b; Pruess, 2008a). Reservoir damage as a consequence of the development of preferential flow paths would prevent an efficient recovery of the heat energy present.

5.6.3
The Importance of Thermal Stress in the Rock Mass

One key issue resulting from the study by McDermott *et al.* (2006) was the consideration and quantification of the effect of thermal stress in the rock mass. The thermal stress induced in a bar due to thermal cooling can be approximated as

$$\sigma_T = K_r \beta_T E I \Delta T$$

assuming no viscous flow in the rock, after (Nilsson, 2001). Applying this to the rock mass environment, the coupling between the temperature change and the thermal stress is influenced by the coefficient of restraint. In the rock mass this can be understood as being related to the degree of fracturing in a rock, or the typical size of volume of blocks of rock between fractures. Figure 5.23 illustrates this concept. During cooling the thermal stress is assumed to work to open the fractures, opposite to the tectonic normal stress keeping the fractures closed but in alignment with the fluid pressure. For a typical quartz-rich rock, the elastic

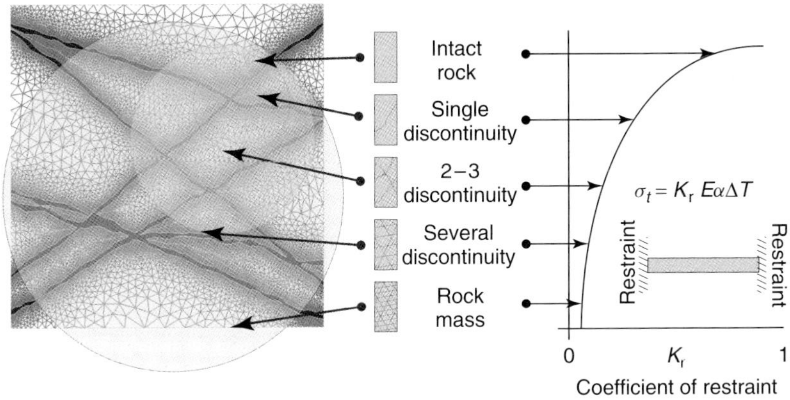

Figure 5.23 Coefficient of restraint and relationship to rock mass.

modulus may be of the order of 50 GPa, and the thermal coefficient of expansion of the order of 10^{-5}. For a fully restrained system, given a 10 K cooling, this would lead to a stress of 5 MPa, equivalent to a fluid head change of about 500 m. It can be seen that for relatively small changes in temperature, relatively large changes in the thermal stress can be expected comparative to the fluid pressure and in some cases the tectonic stress. Additionally understanding the importance of the relationship between the degree of fracturing and the coefficient of restraint becomes clear.

5.7 Rosemanowes (United Kingdom)

The Rosemanowes HDR site is located in the Carmenellis granite structure at Cornwall, southwest England (Figure 5.1). Stimulation and circulation experiments from three wells (RH11, RH12, RH15) extended over a period of more than 10 years (Richards *et al.*, 1994). The long-term circulation test was carried out within the RH12/RH15 reservoir (Figure 5.24a). The size and the shape of the accessible reservoirs were estimated from monitoring the microseismicity during the stimulation of the wells. From statistical analysis of the data, two subvertical sets of fractures were identified, which scatter around the two major strike directions of N165 and N250. Therefore, a deterministic fracture network was established according to these average orientations. Further, the number of fractures is restricted to those that actually absorb water as detected by the well logs (Figure 5.24b).

The observed anisotropy in the tectonic stress field implies anisotropy in the hydraulic behavior of the fracture system as well. For each case of hydraulic anisotropy, fracture apertures were identified, which correspond to the measured hydraulic reservoir impedance of 0.6 MPa l^{-1} s^{-1}. The thermal model was calibrated on the basis of the production (bottom hole) temperatures in the borehole RH15. The influence of several factors on the simulation results was studied in particular, such as the initial temperature distribution, hydraulic anisotropy, and matrix

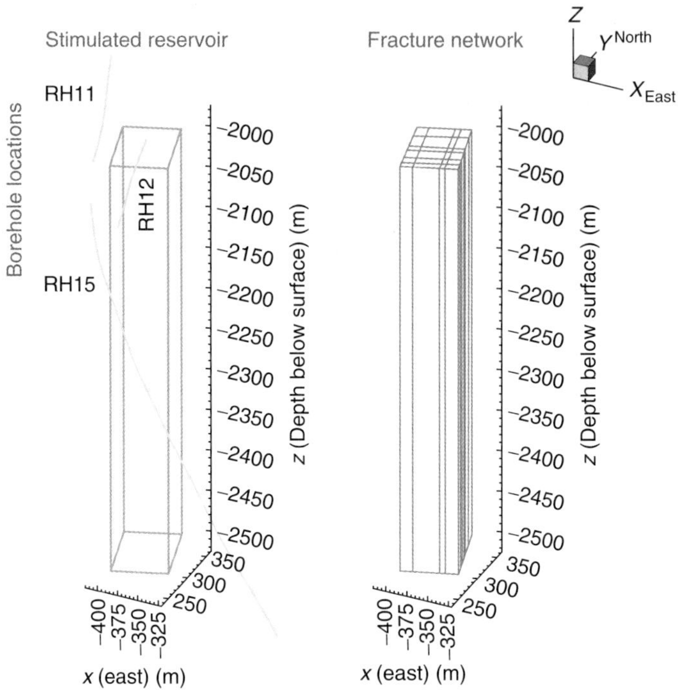

Figure 5.24 Rosemanowes geothermal reservoir model (Kolditz and Clauser, 1998).

porosity. Since normal and shear forces are acting on the fracture surfaces, the hydraulic characteristics of the fracture network are strongly affected by the *in situ* tectonic stress field, as well as in the tectonic stress field; therefore, anisotropy is observed in the hydraulic behavior of the fractured reservoir. A model with an anisotropy factor of $b1/b2 = 5$ (ratio of apertures of the two fracture sets) and with matrix porosity of about 1% provides the best fit to the data (Figure 5.25). The results of this hybrid fracture-matrix model are compared with earlier findings based on a parallel fracture array model by Nicol and Robinson (1990) and a stochastic fracture network model by Bruel (1995a). More details of the Rosemanowes geothermal reservoir model can be found in Kolditz and Clauser (1998). Hydromechanical coupling effects have been not yet investigated in these studies.

5.8 Soultz-sous-Forets (France)

In order to simulate the hydraulic behavior of HDR reservoirs, basic flow processes in fractured rock needs to be understood. Fractured rocks are strongly heterogeneous media. They consists of different structural components such as matrix blocks and fractures with varying orientations as well as different length scales. Owing to their geometric complexity several conceptual models were developed in

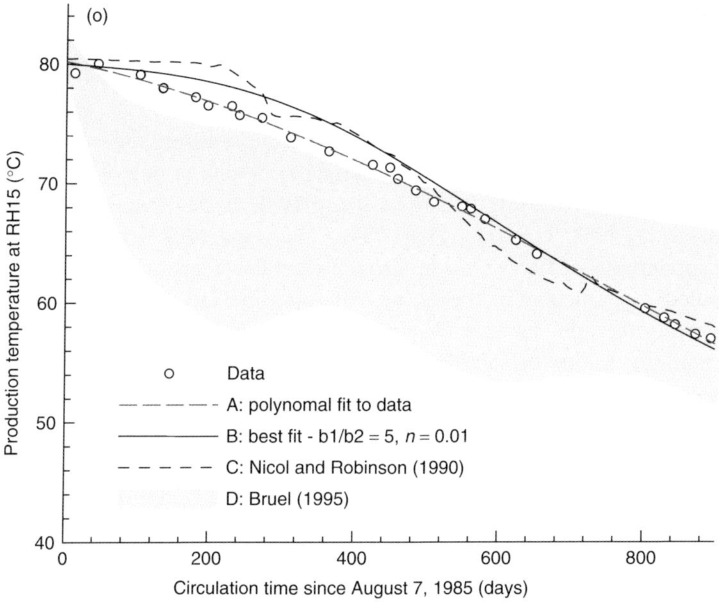

Figure 5.25 Production temperature of the long-term circulation test at Rosemanowes: measurements and simulations.

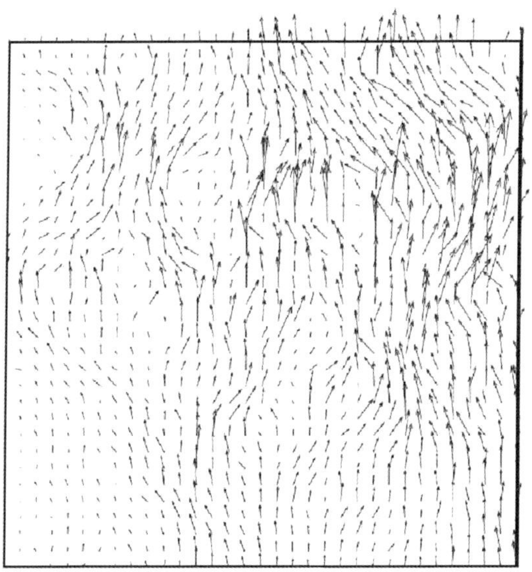

Figure 5.26 Flow channeling in fractures.

the past, such as continua approaches (e.g., Barenblatt *et al.* (1990)) and discrete fracture approaches (e.g., Witherspoon *et al.* (1980); Brown (1987)) as well as combinations of both. The Soultz geothermal reservoir is located in crystalline rocks, where water flows mainly through fractures. Therefore, we are dealing with deterministic discrete fracture models in the following.

The Soultz-sous-Forets HDR site is located in the upper Rhine valley between the Black Forest and Vosges mountains (Figure 5.1). Soultz is one of the most investigated geothermal research locations (Jung 1991; Baria 1992; Baria *et al.*, 1999; Weidler *et al.*, 2002; Baria *et al.*, 2006; Valley and Evans, 2007; Schindler *et al.*, 2008). The applicability of Darcy's law for groundwater flow (equation) to fractured rock is limited in particular in the vicinity of injection and production wells. Hydraulic testing at Soultz, for example, the pumping tests 94JUN16 and 94JUL04 showed that nonlinear flow behavior must be taken into account (Kohl *et al.*, 1997). For describing nonlinear flow behavior we use the Forchheimer equation (5.4). In addition to possible nonlinear flow behavior we have to take into account that rock fractures are not plane. Fracture roughness can cause flow channeling effects (Figure 5.26). A detailed study of nonlinear flow in fractured rock can be found in Kolditz (2001)

On the basis of the pumping test 95JUL01 (Jung *et al.*, 1995) investigated the nonlinear flow behavior in a fracture network system (Kolditz, 2001). Figure 5.27 shows the numerical analysis of the pumping test. Circles illustrate measured data and solid lines mark simulated pressure for both cases of linear flow (Darcy) and nonlinear flow (Forchheimer). Linear flow behavior shows a linear relationship between pumping rate increase and corresponding pressure increase to force the fluid volume through the system (lower solid curve in Figure 5.27). The pumping test clearly indicates nonlinear flow behavior. The flow rates of the four steps increase nearly in a linear stepwise way: 6, 13, 19, and 26 l/s. As can be seen from Figure 5.27, the pressure increase is nearly quadratical. The calibrated permeability values are close to that found by Kohl *et al.* (1997). The storativity values however differ. Note, storativity values by Kohl *et al.* (1997) correspond to the rock matrix, whereas storativity in this study corresponds to the fracture system. In fact, this is the conceptual difference between both models. (Kohl *et al.*, 1997) assumed that fluid can be stored in the rock matrix. In this study we found that in the short timescale of the experiment fluid loss into the rock matrix must be very small and, therefore, it was assumed that the fluid is stored in the fracture system. If relating both storativity values by the rock porosity of about $n = 10^{-3}$, it can be seen that the volume of storable fluid is comparable for both models.

From Figure 5.27, it can be seen that the pumping test data are well matched by using the nonlinear flow model except the shut-off period. During this period, fluid pressure is decreasing to the hydrostatic level. The overestimated pressure drawdown means that the storativity of the reservoir is underestimated in the shut-off period. This indicates storativity changes during the hydraulic tests. This can be explained with the increased volume of the stimulated fracture system. Because of small relative displacements of rough fracture surfaces during pressure increase they will not close perfectly after reducing the reservoir pressure again.

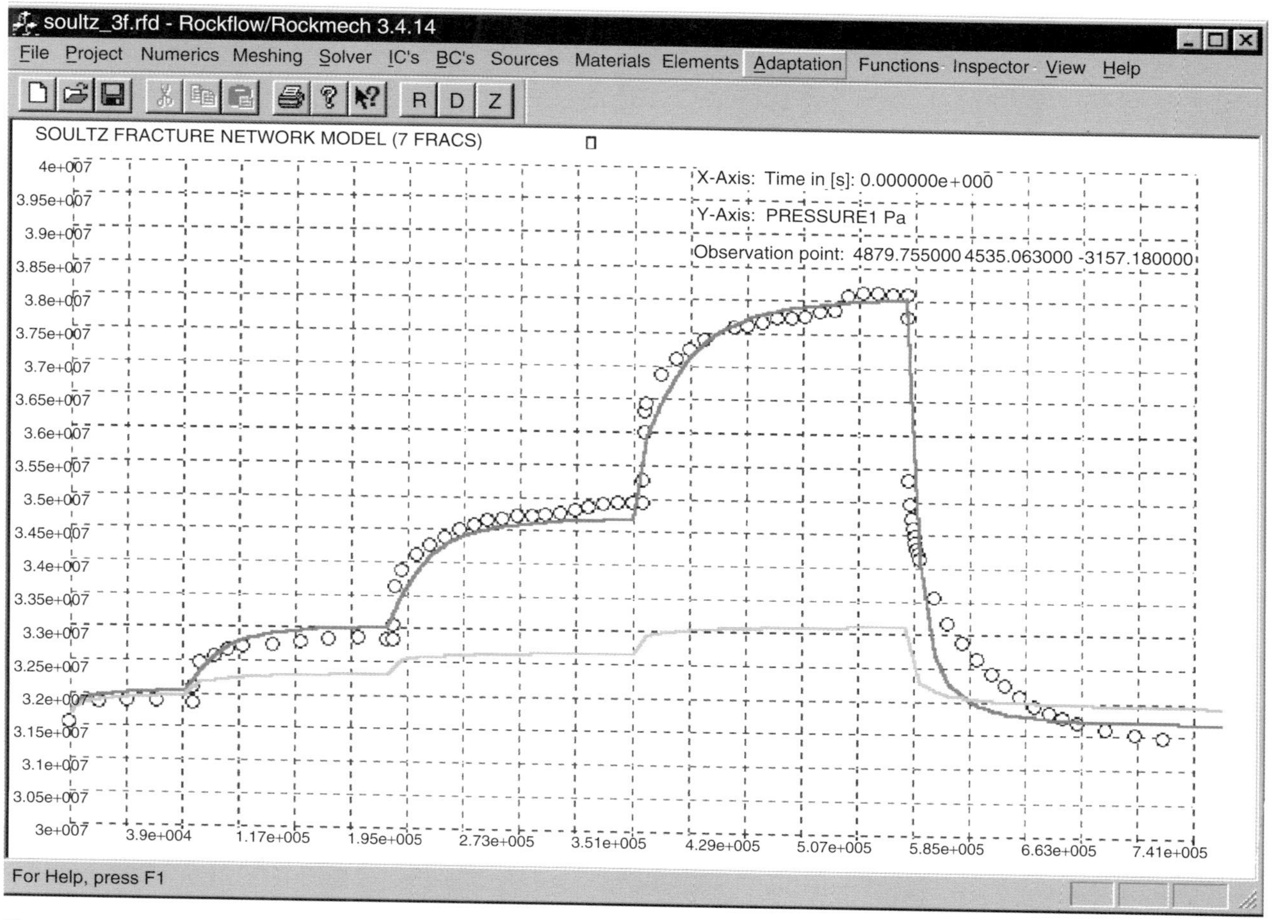

Figure 5.27 Pumping test analysis using a deterministic discrete fracture network model (up to 10 major fractures have been considered).

Those effects of residual apertures after loading and unloading of rough fractures was observed, for example, by Brown (1987). We repeated the simulations with increased storativity coefficient in the shut-off period by factors of 2 and 3. A better fit of the shut-off period required an adaptation of the storativity coefficient, that is, hydromechanical coupling effects become important.

Results of recent studies can be found, for example, in Kohl and Megel (2007); Megel *et al.* (2006). Bruel (2002) investigated the impact of induced thermal stresses during circulation tests in the Soultz site at the regional scale. More recently (Baujard and Bruel, 2006) studied the influence of fluid density differences on the pressure distribution in the reservoir during stimulation and fluid circulation tests.

5.9 KTB (Germany)

5.9.1 Introduction

The KTB, located at Windischeschenbach, Germany (Figure 5.1), comprising a pilot borehole down to 4000 m and a main borehole down to 9101 m in southeast Germany continues to provide a unique opportunity for the identification of important factors and processes in deep seated fluid and energy transfer directly relevant to the exploitation of geothermal energy (http://www.geozentrum-ktb.de/). *In situ* stress conditions significantly impact flow, transport, and exchange characteristics of fracture networks, which dominate the permeability of crystalline reservoir rocks. To model such systems several scales of information need to be combined to present a fully three-dimensional model of the principle KTB fault zones, and linked to a geomechanical model describing the alteration of the hydraulic parameters with stress changes caused by fluid extraction. The concept of geomechanical facies was introduced to define and characterize architectural elements in the subsurface system. The evaluation of a long-term pump test in the KTB pilot hole, June 2002 to July 2003, coupled with a geomechanical model gave an insight into some of the elastic and nonelastic processes controlling hydraulic transport in the basement rocks.

The geometrical basis for a three-dimensional fracture network model was provided by the interpretation of several sets of data (Table 5.3). This included large-scale geophysical surveys (Harjes, 1997) indicating the presence of reflectors considered to be pathways for geofluids, geological investigations, particularly, structural and tectonic interpretations (Hirschmann, 1996; 1997), and the interpretation of fluid bearing fracture zones (Lodemann, 1998). Geologically individual fault planes are seldom encountered, and in reality one is dealing with a whole swarm of similarly orientated fractures extending sometimes over considerable distances, within which the movement responsible for the formation of the fractures is accommodated as is flow within the fractures, for example, (Hoehn *et al.*, 1998; Talwani *et al.*, 1999). As such the concept of shear zones to represent the major geological tectonic features observed in the pilot and main boreholes was

Table 5.3 Fault zones of the KTB site.

	Fault Zones	Orientation		Depth of intersection with	
		Azimuth	Dip	Main hole	Pilot hole
	Altenparkstein Fault Zone (SEI)	45	52	7000 m	Does not intersect
Fraconian line	0.5SE1a	45	41	4750 m	Does not intersect
Fraconian line	Waldeck–Klobenreuth fault zone (SE2)	45	41	3500 m	3326 m
Fraconian line	SE2a	45	55	4225 m	3997 m
	Erbendorf line	120	54	6000 m	Does not intersect
	Nottersdorf fault zone	60	65	1100 m	721 m
	Fichtelnaab fault zone (SE4)	15	52	Does not intersect[a]	Does not intersect

[a]Closest surface trace +900 m east, +900 m north of main hole.

applied to understand the subsurface fluid flow and geometry. A shear zone in this context is a plane with a certain thickness comprising a number of discrete interacting fractures. Several analyses of the tectonic features at the KTB site have been undertaken, particularly (Hirschmann, 1996) and the KTB work group under Hirschmann provided much detailed fracture picking information. (Winter *et al.*, 2002) analyzed drill cuttings from the KTB main hole from 1700 to 2400 m and from 4500 to 5000 m. Their results indicate that cataclastic shear zones can be found in the cuttings extending well over 100 m thickness. Table 5.3 presents a summary of information gained from the literature on the geological fault zones recognized at the KTB site, along with their orientation and dip.

Figure 5.28 presents the density of fractures and tectonic features found in the KTB main hole down to a depth of 8000 m picked by the KTB work group. The arrows in Figure 5.28 refer to the calculated crossing points of the geological fault zones (Table 5.3) represented as a single plane with the main borehole. From Figure 5.28 it can be seen that the identification of a single fracture corresponding to the geological fault zones of Table 5.3 is impossible, rather areas of increased fracture density, with a thickness of a few hundreds of meters can be defined, as per (Winter *et al.*, 2002). This interpretation is naturally dependent on a number of factors including uncalibrated measurements of the effect of changing drilling techniques; however, a general pattern can be observed.

The hydraulic model developed is based on a single well test matching rather than a hydraulic characterization of the entire KTB HB (Hauptbohrung, main borehole)-VB (Vorbohrung, preparing borehole) reservoir.

5.9.2
Geomechanical Facies and Modeling the HM Behavior of the KTB Pump Test

Geological deposits are not random groups of deposits but rather there is both depositional and structural process control on the *in situ* properties and parameters.

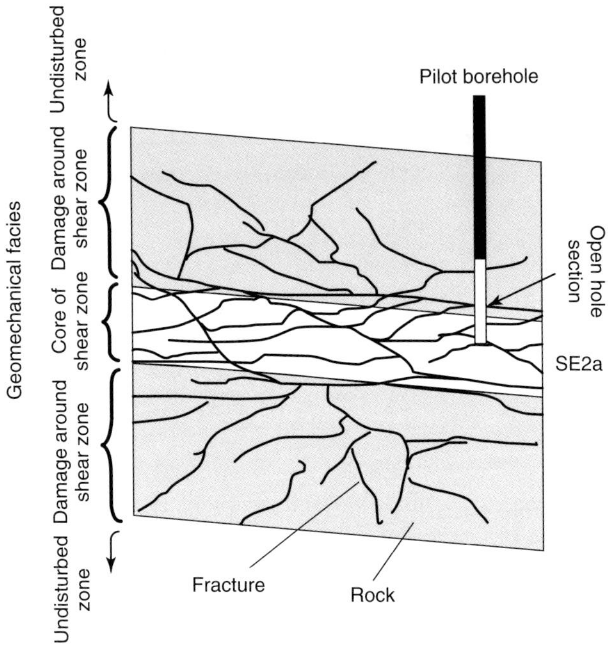

Figure 5.28 Fractures and tectonic features picked out by the KTB working group, interpreted according to a shear zone model. Also shown are the geomechanical facies interpretation.

This consideration has led a number of authors to consider the concept of architectural elements within geological deposits, particularly sedimentary deposits, for example, (Hornung and Aigner, 1999; 2002; Klingbeil *et al.*, 1999; Rea and Knight, 1998; Stephens, 1994; Genter *et al.*, 2002) identified fracture zones in Soultz Forets. An architectural element in this context defines a principal building block of the geological deposit being considered to which specific parameters are assigned. The whole system being assessed can be considered to be construct of the architectural elements. Adapting this to a THM hydrogeological and geomechanical situation, that is, the coupling of hydraulic, mechanical, and thermal properties, allows the definition of geomechanical facies. The division between the facies is defined by the parameters for the processes to be investigated. In the context of the KTB site, the geomechanical facies approach allowed the description of separate architectural elements of the shear zone with definite flow, transport, and mechanical characteristics. In the model presented, the shear zone is divided into a core zone where flow and transport is more prevalent and a damage zone where there is an increase in microcracking. Between the shear zones there is an undisturbed zone where the action of shearing has not influenced the material. This conceptual structure is illustrated in Figure 5.29, and the three-dimensional hydromechanical (HM) model is presented in Figure 5.30.

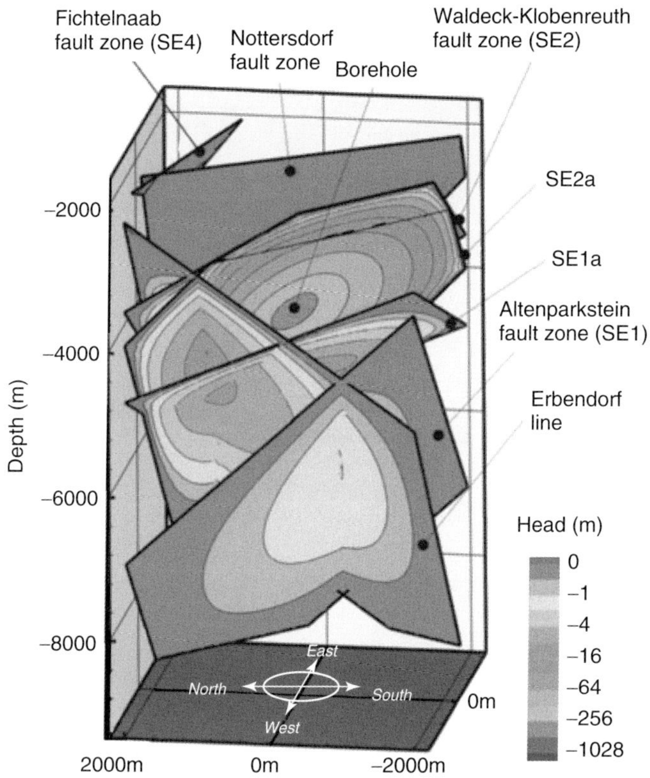

Figure 5.29 Conceptual deterministic fracture network model of the shear zones, illustration of the geomechanical facies concept (McDermott *et al.*, 2006). (Please find a color version of this figure on the color plates.)

The HM behavior of the geomechanical facies was defined using a geomechanical model for fracture deformation (McDermott and Kolditz, 2006); a characteristic curve relating effective stress and deformation is given in Figure 5.31. Note that the confining pressure is equal to the normal effective stress. Figure 5.32 presents the fit of the long-term pump test where fluid pressure differences of a few megapascals were induced on the shear zones. Recognized fault zones in the vicinity of the KTB site are given in Table 8.1 .

5.10 Stralsund (Germany)

The variation of reservoir properties in a sandstone reservoir due to heat mining with reinjection is simulated over the entire assumed time of operation of 80 years. The subsurface flow and hydrogeothermal simulation code SHEMAT (Clauser,

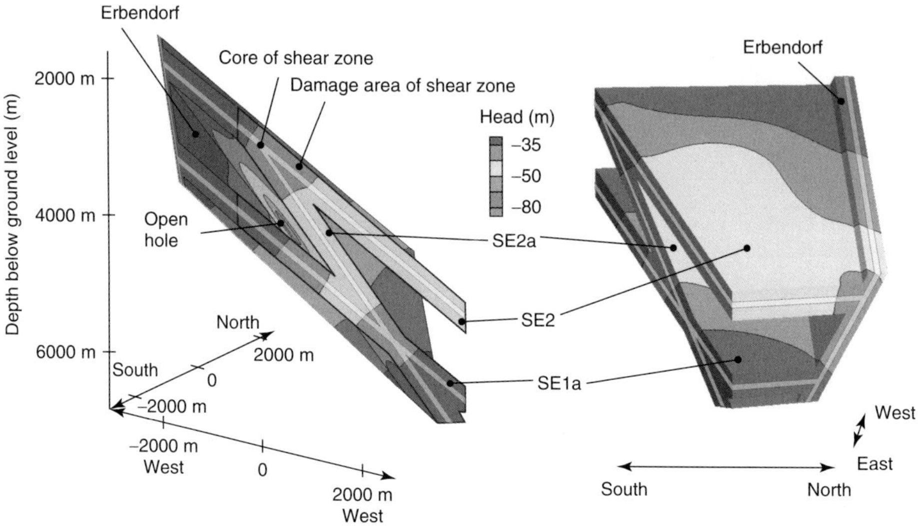

Figure 5.30 Three-dimensional geomechanical facies model of the shear zones for steady-state fluid extraction conditions. (McDermott *et al.*, 2006).

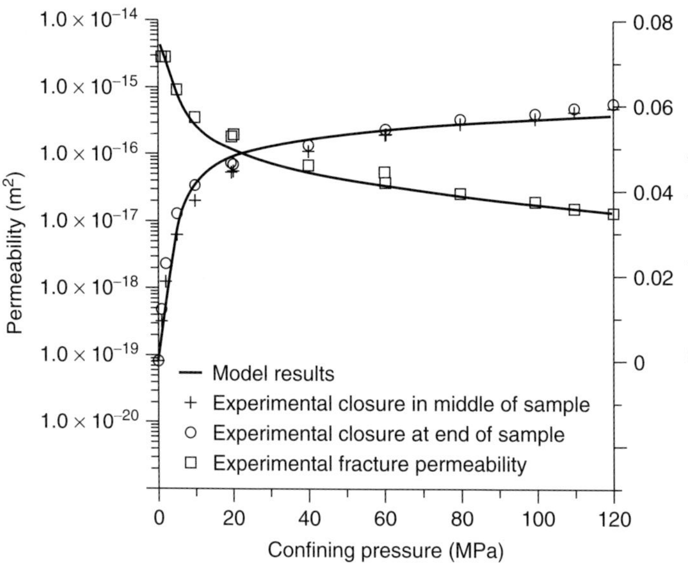

Figure 5.31 Comparison of measured closure of a fracture taken from the KTB site (Durham, 1997), and numerically predicted values after (McDermott and Kolditz, 2006), HM coupling to effective stress.

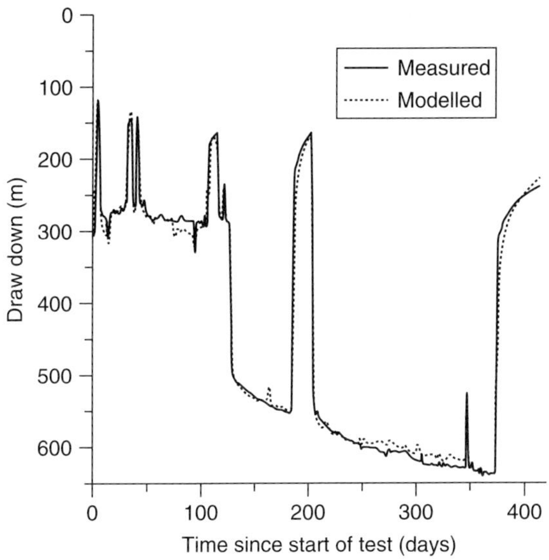

Figure 5.32 Hydromechanical coupled model used to fit measured data. (McDermott *et al.*, 2006.)

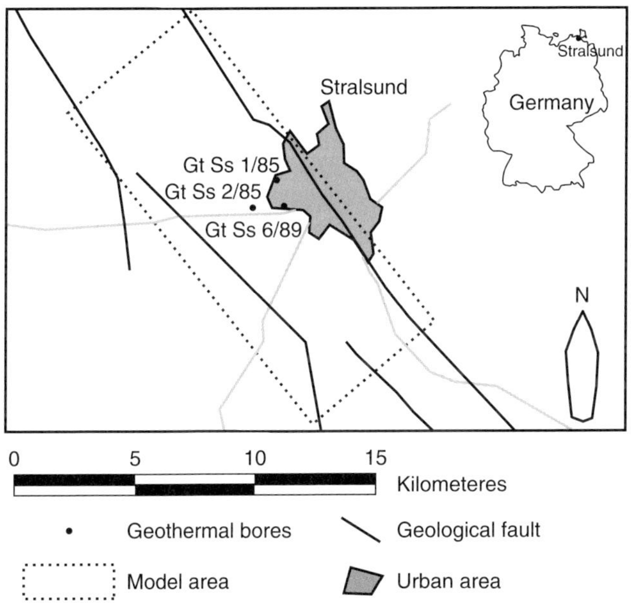

Figure 5.33 Stralsund geothermal site with wells Gt Ss 1/85, Gt Ss 2/85, and Gt Ss 6/89 (black dots). The reservoir is partly delineated by impervious faults (black lines). The model area (dotted rectangle) measures about 12 km × 6 km.

2003) is used to study the location Stralsund situated in the North German Basin (Figure 5.33). This study focuses on the prediction of long-term behavior of reservoir properties (Kühn *et al.*, 2002). The Stralsund location in northeast Germany and the Detfurth sandstone horizon (Buntsandstein) are chosen due to the combination of their already confirmed geothermal potential and the availability of a complete data set. An installation of two production wells and one well for reinjection implements heat exploitation. In order to quantify injectivity changes and allow the separation of thermal from chemical effects, changes in the hydraulic parameters of the reservoir are at first studied without chemical reactions. This modeling study of the long-term behavior of the reservoir focuses on the simultaneous temporal and spatial evolution of hydraulic, thermal, and chemical parameters and their contribution to injectivity trends.

5.10.1
Site Description

The city of Stralsund is situated on the Baltic Sea in northeast Germany at the northern edge of the North German Basin. Three wells have already been drilled and, within the depth range of 1500–1600 m, they reached the Detfurth sandstone whose thickness ranges between 33 and 36 m. This aquifer, suitable for geothermal exploitation, belongs to the Buntsandstein formation (Lower Triassic sandstones). From borehole profiles and core samples it appears that the Detfurth sandstone is a well sorted, weakly consolidated fine-to-medium sandstone interlayed by silt and coarse sandstone within an alternated stratification. It is feldspathic-quartz sandstone, low-graded with clay (<2%) and cement minerals (4–5%). The cementation mainly consists of calcite and a minor amount of anhydrite (Bartels and Iffland, 2000). The highly saline formation water is of the Na-(Ca-Mg)-Cl type with a solute content of total dissolved solids (TDS) 280 g l^{-1} and a formation temperature of about 58 °C. The composition of the water reveals that it is in equilibrium relative to the mineral phases, anhydrite and calcite, for the formation temperature.

5.10.2
Model Setup

The modeled horizontal area of the Stralsund location measures 12 km × 6 km and is partly delineated by the existing geological faults (Figure 5.33). The boundaries in all directions have been considered to be impermeable, because of the impervious faults to the northeast and southwest, the negligible regional flow, and the fact that the boundaries to the northwest and southeast are set at a sufficient distance from the central part of the model. The two drillings nearest to the town are used for production and the third one for reinjection to minimize transport distances for the hot water. Production rate is 50 m^3 h^{-1} for each production well. The produced water is reinjected at a temperature of 20 °C. For diagnostic reasons a conservative tracer is also injected to visualize transport of dissolved ions in the model area. The assumed period of operation of the geothermal heating plant is 80 years. Changes in porosity can be calculated without difficulty from the reaction rates of the mineral components knowing their molar volumes. However, it is not as straightforward to estimate the associated changes in permeability because permeability depends not only on the bulk porosity but also strongly on the structure of the pore space. We applied the so-called "pigeon hole" model especially suited for sedimentary rocks of (Pape *et al.*, 1999). This model yields petrophysically justified relations between the various geometric, storage, and transport parameters of these reservoir rocks. Like other fractal models, it is based on the observation that the shape of the internal surface of rock pores follows a self-similar rule.

5.10.3 Long-Term Development of Reservoir Properties

The study of the long-term behavior of the reservoir properties of the Stralsund deep aquifer requires, as far as possible, a quantitative separation of the single contribution of the various processes involved from the hydraulic changes in the aquifer that occur as a result of reinjection. Reinjection of cooled water of higher viscosity than the natural reservoir fluid leads to a continuous reduction of the injectivity (Figure 5.34). This effect is partially balanced by thermally induced mineral reactions. Dissolution of anhydrite in the vicinity of the injection well dominates the effect of anhydrite precipitation at the propagating thermal front leading to a net increase of injectivity (Kühn *et al.*, 2002). Observed calcite precipitation around the injection well and dissolution at the thermal front are too small to alter reservoir properties significantly. Coupled numerical simulation indicates that the

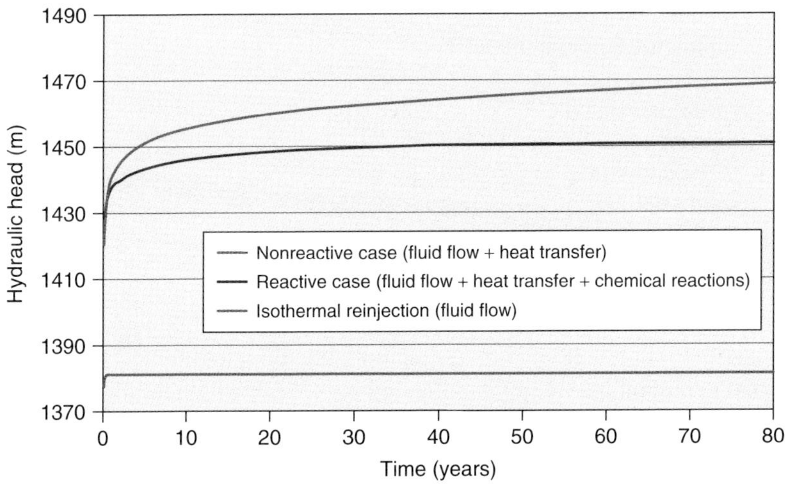

Figure 5.34 Temporal evolution of the pressure head at the injection well shown as a hydraulic head during 80 years of reservoir exploitation (Kuhn *et al.*, 2002).

injectivity of the reservoir is influenced primarily by the viscosity effect, but that mineral reactions weaken this negative trend. Operation of a geothermal heating plant at the Stralsund location would not be restricted by a long-term reduction in the injectivity of the reinjection well.

Nomenclature

A	cross-section area (m^2)
a_1, a_2	Forchheimer coefficients (m)
b	fracture aperture (m)
$\mathbb{C}$	a forth-order elasticity tensor (–)
c	specific heat capacity of porous medium ($J\ kg^{-1}\ K^{-1}$)
D_h	hydraulic radius (m)
E	Young's modulus (Pa)
$f(n)$	porosity function (–)
G	shear modulus (Pa)
$\mathbf{g}$	gravity acceleration vector ($m\ s^{-2}$)
h	lag distance in Variogram model (m)
$\mathbf{I}$	identity tensor
$k, \mathbf{k}$	intrinsic permeability (tensor) (m^2)
K	hydraulic conductivity ($m\ s^{-1}$)
K_r	coefficient of restraint (–)
L	characteristic length scale (m)
n	porosity
$\mathbf{n}$	normal vector
p	fluid pressure (Pa)
q_f	specific flow rate (s^{-1})
Q_f	volumetric flow rate ($m^3\ s^{-1}$)
Q_T	heat source ($J\ m^{-3}\ K$)
S_s	specific storage (Pa^{-1})
t	time (s)
T	temperature (K)
$\mathbf{u}$	solid displacement vector (m)
$\mathbf{v}$	fluid phase velocity ($m\ s^{-1}$)
x, y, z	coordinates (m)

Greek Symbols

α_b	Biot's constant (-)
β_p	compressibility coefficient (Pa^{-1})
β_T	thermal expansion coefficient (K^{-1})
β_C	coefficient of expansivity due to solute concentration ($m^3\ kg^{-1}$)
δ_{ij}	Kronecker delta
Δ	increment()
ε	strain tensor

λ	1st Lamé constant (Pa)
λ	heat conductivity of porous media (W m^{-1} K^{-1})
λ_{20}	heat conductivity at 20 °C (W m^{-1} K^{-1})
λ	hydraulic drag or friction coefficient, Equation 5 ( )
μ	dynamic fluid viscosity (Pa s)
ν	Poisson ratio
ω	thermal gradient (K m^{-1})
ω	mass fraction, Equation 16 (−)
ρ	density of porous medium (kg m^{-3})
σ	effective stress tensor (Pa)
ς	relative temperature coefficient (−)

Superscripts

in	value at the injection borehole
l	liquid phase
out	value at the production borehole
s	solid phase
T	transpose of matrix

Subscripts

f	fracture
r	reference value
0	value at initial condition ($t = 0$) or reference value

Special Symbols

∇A	gradient of a scalar
$\nabla \cdot \mathbf{A}$	divergence of a vector
$\nabla \mathbf{A}$	gradient of a vector

References

Al-Wardy, W. and Zimmerman, R.W. (2003) Effective Stress Law for the Permeability of Clay-rich Sandstones. *16th ASCE Engineering Mechanics Conference, July16–18, University of Washington, Seattle.*

Alonso, E.E., Alcoverro, J., Coste, F., Malinsky, L., Merrien-Soukatchoff, V., Kadiri, I., Nowak, T., Shao, H., Nguyen, T.S., Selvadurai, A.P.S., Armand, G., Sobolik, S.R., Itamura, M., Stone, C.M., Webb, S.W., Rejeb, A., Tijani, M., Maouche, Z., Kobayashi, A., Kurikami, H., Ito, A., Sugita, Y., Chijimatsu, M., Borgesson, L., Hernelind, J., Rutqvist, J., Tsang, C.F., and Jussila, P. (2005) The FEBEX benchmark test: case definition and comparison of modelling approaches. *International Journal of Rock Mechanics and Mining Sciences*, **42**(5–6), 611–638.

Bachler, D. and Kohl, T. (2005) Coupled thermal-hydraulic-chemical modelling of enhanced geothermal systems. *Geophysical Journal International*, **161**(2), 533–548.

Baisch, S., Weidler, R., Voeroes, R., Tenzer, H., and Teza, D. (2004) Improving Hydraulic Simulation Efficiency by Means of Real Time Monitoring. *Proceedings of the 29th Workshop on Geothermal Reservoir Engineering*, Stanford University, pp. 209–215.

Barenblatt, G., Entov, V., and Ryzhik, V. (1990) *Theory of Fluid Flows through Natural Rocks*, Kluwer Academic Publishers.

Baria, R., Baumgärtner, J., Rummel, F Pine, R.J., and Sato, Y. (1999) HDR/HWR reservoirs: concepts, understanding and creation. *Geothermics* **28**, pp. 533–552.

Baria, R., Jung, R., Tischner, T., Nicholls, J., Michelet, S., Sanjuan, B., Soma, N., Asanuma, H., Dyer, B., and Garnish, J. (2006) Creation of an HDR/EGS reservoir at 5000 m depth at the European HDR project, *Proceedings of the 31st Workshop on Geothermal Reservoir Engineering*, Stanford University, Stanford, CA, USA, **8** p.

Bartels, J. and Iffland, J. (2000) Hydraulic, Thermal, and Mechanical Behavior of Geothermal Aquifers under Exploitation: Analysis and Assessment of Petrographical / Petrophysical and Hydrochemical Data and Investigations as Well as Simulations. Project Final Report BMBF-0326995D, Landesamt für Umwelt, Naturschutz und Geologie Mecklenburg-Vorpommern, German.

Baujard, C. and Bruel, D. (2006) Numerical study of the impact of fluid density on the pressure distribution and stimulated volume in the Soultz HDR reservoir. *Geothermics*, **35**(5–6), 607–621.

Bear, J. (1972) *Dynamics of Fluids in Porous Media*, Elsevier, 764 p.

Bear, J. (1991) Modelling transport phenomena in porous media. *Convect Heat Mass Transfer Porous Media*, **7**, 69.

Bernabe, Y. (1987) The effective pressure law for permeability during pore pressure and confining pressure cycling of several crystalline rocks. *Journal of Geophysical Research*, **92**, 649–657.

Bernabe, Y. (1988) Comparison of the effective pressure law for permeability and resistivity formation factor in chelmsford granite. *Pure and Applied Geophysics*, **127**(4), 607–625.

Birkholzer, J., Rutqvist, J., Sonnenthal, E., Barr, D. (2008) Decovalex - Task D: long-term permeability changes due to THC and THM processes. Technical Report 2008:43, Swedish Nuclear Power Inspectorate, Stockholm.

Blöcher, M.G., Zimmermann, G., and Milsch, H. (2009) Impact of poroelastic response of sandstones on geothermal power production. *Pure and Applied Geophysics*, **166**(5–7), 1107–1123. Special Volume.

Bower, K. and Zyvoloski, G. (1997) A numerical model for thermo-hydro-mechanical coupling in fractured rock. *International Journal of Rock Mechanics and Mining Sciences*, **34**(8), 1201–1211.

Brown, S. (1987) Fluid flow through rock joints: the effect of surface roughness. *Journal of Geophysical Research*, **92**(B2), 1337–1347.

Bruel, D. (1995a) Heat extraction modelling from forced fluid flow through stimulated fractured rock masses: application to the Rosemanowes Hot Dry Rock reservoir. *Geothermics*, **24**(3), 361–374.

Bruel, D. (1995b) Modeling heat extraction from forced fluid-flow through stimulated fractured rock masses - evaluation of the Soultz-sous-Forets site potential. *Geothermics*, **24**(3), 439–450.

Bruel, D. (2002) Impact of induced thermal stresses during circulation tests in an engineered fractured geothermal reservoir: example of the Soultz-sous-Forets European hot fractured rock geothermal project, RhineGraben, France. *Geothermics*, **57**(5), 459–470.

Bruel, D. and Cacas, M. (1992) *Numerical Modeling Technique: Contribution to the Soultz Project*. Gordon and Breach Science Publisher, pp. 267–279.

Bruel, D., Cacas, M., Ledoux, E., and de Marsily, G. (1994) Modelling storage behaviour in a fractured rock mass. *Journal of Hydrology*, **162**, 267–278.

Cacas, M., Ledoux, E., de Marsily, G., Barbreau, A., Tillie, B., Durand, E., Feuga, A., and Peaudecerf, P. (1990) Modeling fracture flow with a discrete fracture network: 1. The flow model. *Water Resources Research*, **26**(3), 479–489.

Carman, P.C. (1937) Fluid flow through granular beds. *Transactions of the Institution of Chemical Engineers*, **15**, 150.

Carroll, M. and Katsube, N. (1983) The role of terzaghi effective stress in linearelastic deformation. *Journal of Energy Resources Technology*, **105**, 509–511.

Chilés, J.-P. and Delfiner, P. (1999) *Geostatistics: Modeling Spatial Uncertainty*, Wiley, New York.

Clauser, C. (2003) *Numerical Simulation of Reactive Flows in Hot Aquifers*, Springer, Berlin / Heidelberg.

Cornet, F. (1985) An Overview on Research on Energy Production From Hot-dry-rocks. Technical Report, Statusreport 1985, Geotechnik und Lagerstätten, Jülich.

de Boer, R. (2005) *Trends in Continuum Mechanics of Porous Media: Theory and Applications of Transport in Porous Media*, Springer-Verlag, Heidelberg.

Deutsch, C.V. (2002) *Geostatistical Reservoir Modeling*, Oxford University Press, New York.

Diersch, H.J.G. (2002) *Feflow Finite Element Subsurface Flow and Transport Simulation System - User's Manual / Reference Manual / White Papers*. WASY Ltd, Berlin, Release 5.0.

Diersch, H.-J. and Kolditz, O. (2002) Variable-density flow and transport in porous media: approaches and challenges. *Advances in Water Resources*, **25**(8–12), 899–944.

Doughty, C. and Pruess, K. (2004) Modeling supercritical carbon dioxide injection in heterogeneous porous media. *Vadose Zone Journal*, **3**, 837–847.

Durham, W. (1997) Laboratory observations of the hydraulic behavior of a permeable fracture from 3800 m depth in the KTB pilot hole. *Journal of Geophysical Research - Solid Earth*, **102**, 8405–18416.

Ehlers, W. and Bluhm, J. (2002) *Porous Media: Theory, Experiments and Numerical Applications*, Springer.

Finsterle, S. and Pruess, K. (1997) Development of Inverse Modeling Techniques for Geothermal Applications. *Proceedings, Geothermal Program Review XV, San Francisco, CA* , March 24–26, 1997, pp. 2–47–2–54.

Forchheimer, P. (1914) *Hydraulik*, Teubner Verlag.

Gehrke, D. (2007) *Thermisch-hydraulische 3d Modellierung des Geothermischen Reservoirs um die Dublette (e grsk 3/90 / gt grsk 4/05) von großschönebeck unter Berücksichtigung von Stimulationsmaßnahmen*. Institutfür Geographie und

Geologie, Ernst-Moritz-Arndt Universität, Greifswald Diploma Thesis.

Genter, A., Dezayes, C., Gentier, S., LedØsert, B., and SaussØ, J. (2002) Conceptual fracture model at Soultz based on geological data. *Geologisches Jahrbuch*, **SE1**, 93–102.

Goovaerts, P. (1997) *Geostatistics for Natural Resources Evaluation*, Oxford University Press, New York.

Haenel, R. (Ed) (1982) *The Urach Geothermal Project, Swabian Alb, Germany*. Schweitzerbartsche Verlagsbuchhandlung.

Harjes, H. (1997) Origin and nature of crustal reflections: results from integrated seismic measurements at the KTB superdeep drilling site. *Journal of Geophysical Research*, **102**, 18267–18288.

Hicks, T., Pine, R., Willis-Richards, J., Xu, S., Jupe, A., and Rodrigues, N. (1996) A thermo-hydro-mechanical numerical model for HDR geothermal reservoir evaluation. *International Journal of Rock Mechanics and Mining Sciences and Geomechanics Abstracts*, **33**(5), 499–511.

Hirschmann, G. (1996) KTB - the structure of a Variscan terrane boundary: seismic investigation - drilling - models. *Tectonophysics*, **264**, 1–4.

Hirschmann, G. (1997) The KTB superdeep borehole: petrography and structure of a 9-km-deep crustal section. *Geologische Rundschau*, **86**, 3–14.

Hoehn, E., Eikenberg, J., Fierz, T., Drost, W., and Reichlmayr, E. (1998) The Grimsel migration experiment: field injection-withdrawal experiments in fractured rock with sorbing tracers. *Journal of Contaminant Hydrology*, **34**(1–2), 85–106.

Holl, H.-G., Moeck, I., and Schandelmeier, H. (2004) Geothermal Well Großschoenebeck 3/90: A Low Enthalpy Reservoir (rotliegend, ne germany). *Proceedings, 66th EAGE Conference F032, Paris, France.*

Holl, H.-G., Moeck, I., and Schandelmeier, H. (2005) Characterisation of the Tectono-sedimentary Evolution of a Geothermal Reservoir - Implications for Exploitation (Southern Permian Basin, Ne Germany). *Proceedings World Geothermal Congress 2005, Antalya, Turkey*, April 2005, 1-5, 24-29.

Hornung, J. and Aigner, T. (1999) Reservoir and aquifer characterization offluvial architectural elements: Stubensandstein, Upper Triassic, Southwest Germany. *Sedimentary Geology*, **129**(3–4), 215–280.

Hornung, J. and Aigner, T. (2002) Reservoir architecture in a terminal alluvialplain: an outcrop analogue study (Upper Triassic, Southern Germany) Part II:Cyclicity, controls and models. *Journal of Petroleum Geology*, **25**(2), 151–178.

Huenges, E., Trautwein, U., Legarth, B., and Zimmermann, G. (2006) Fluid pressure variation in a sedimentary geothermal reservoir in the north German basin: case study gross schoenebeck. *Pure and Applied Geophysics*, **163**, 2141–2152.

Jung, R., Willis-Richards, J., Nicholls, J., Bertozzi, A., and Heinemann, B. (1995) Evaluation of hydraulic test at Soultz European HDR site. *1st Worlds Geothermal Congress*, 18–31 May 1995, Firenze, Italy 2575–2580.

Jung, R. (1991) Hydraulic fracturing and hydraulic testing in the granite section of borehole GPK1, Soulz-sous-Forêts. *Geothermal Science and Technology*, **3**(1–4): 149–198.

Katz, A. and Thompson, A. (1987) Prediction of rock electrical conductivity from mercury injection measurements. *Journal of Geophysical Research*, **92**, 599–607.

Kiryukhin, A., Xu, T., Pruess, K., Apps, J., and Slovtsov, I. (2004) Thermal-hydrodynamic-chemical (THC) modeling based on geothermal field data. *Geothermics*, **33**(3), 349–381.

Klingbeil, R., Kleineidam, S., Asprion, U., Aigner, T., and Teutsch, G. (1999) Relating lithofacies to hydrofacies: outcrop-based hydrogeological characterisation of quaternary gravel deposits. *Sedimentary Geology*, **129**(3–4), 299–310.

Kohl, T. (1992) *Modellsimulation Gekoppelter Vorgänge Beim Wärmeentzug Ausheißem Tiefengestein*. Ph.D. thesis, ETH Zürich.

Kohl, T., Evans, K., Hopkirk, R., Jung, R., and Rybach, L. (1997) Observation and simulation of non-Darcian flow transients in fractured rock. *Water Resources Research*, **33**(3), 407–418.

Kohl, T., Evans, K.F., Hopkirk, R.J., and Rybach, L. (1995) Coupled hydraulic, thermal and mechanical considerations for the simulation of hot dry rock reservoirs. *Geothermics*, **24**(3), 345–359.

Kohl, T. and Megel, T. (2007) Predictive modeling of reservoir response to hydraulic stimulations at the European EGS site Soultz-sous-Forets. *International Journal of Rock Mechanics and Mining Sciences*, **44**(8), 1118–1131.

Kolditz, O. (1995) Modelling of flow and heat transfer in fractured rock: conceptual model of a 3-D deterministic fracture network. *Geothermics*, **24**(3), 451–470.

Kolditz, O. (1997) *Strömung, Stoff- und Wärmetransport im Kluftgestein*, Borntraeger Publisher.

Kolditz, O. (2001) Non-linear flow in fractured rock. *International Journal of Numerical Methods in Fluid and Heat Transport*, **11**(6), 547–576.

Kolditz, O. and Clauser, C. (1998) Numerical simulation of flow and heat transfer in fractured crystalline rocks: application to the hot dry rock site at Rosemanowes (UK). *Geothermics*, **28**(1), 1–26.

Kolditz, O. and Diersch, H.-J. (1993) Quasi-steady-state strategy for numerical simulation of geothermal circulation in hot dry rock fractures. *International Journal of Non-Linear Mechanics*, **28**(4), 467–481.

Kolditz, O., Ratke, R., Diersch, H.J.G., and Zielke, W. (1998) Coupled groundwater flow and transport: 1. verification of variable density flow and transport models. *Advances in Water Resources*, **21**, 27–46.

Kozeny, J. (1927) Über kapillare Leitung des Wassers im Boden. *Akademie der Wissenschaften Wiener*, **136**, 271–306.

Kühn, M. (2004) *Reactive Flow Modeling of Hydrothermal Systems. Lecture Notes in Earth Sciences*, Springer, Berlin/Heidelberg.

Kühn, M., Bartels, J., and Iffland, J. (2002) Predicting reservoir property trends under heat exploitation: interaction between flow, heat transfer, transport, and chemical reactions in a deep aquifer at Stralsund, Germany. *Geothermics*, **31**(6), 725–749.

Kwon, O., Kronenberg, A., Gangi, A., and Johnson, B. (2001) Permeability of wilcoxshale and its effective pressure law. *Journal of Geophysical Research*, **106**, 19–339–19–353.

Legarth, B., Huenges, E., and Zimmermann, G. (2005) Hydraulic fracturing in a sedimentary geothermal reservoir: results and implications. *International Journal of Rock Mechanics and Mining Sciences*, **42**, 1028–1041.

Lehmann, H., Wang, K., and Clauser, C. (1998) Parameter identification and uncertainty analysis for heat transfer at the KTB drill site using a 2-Dinverse method. *Tectonophysics*, **291**(1–4), 179–194.

Lewis, R.W. and Schrefler, B.A. (1998) *The Finite Element Method in the Static and Dynamic Deformation and Consolidation of Porous Media*, 2nd edn, John Wiley & Sons.

Lodemann, M. (1998) On the origin of saline fluids in the KTB (Continental Deep Drilling Project of Germany). *Applied Geochemistry*, **13**(5), 653–671.

Lotz, B. (2004) *Neubewertung des Rezenten wärmestroms im Nordostdeutschenbecken*. FU, Berlin, Dissertation.

Magri, F., Bayer, U., Jahnke, C., Clausnitzer, V., Diersch, H.J., Fuhrman, J., Mö◇ller, P., Pekdeger, A., Tesmer, M., and Voigt, H.J. (2005) Fluid-dynamics driving saline water in the north east German basin. *International Journal of Earth Sciences*, **94**, 1056–1069.

McDermott, C. and Kolditz, O. (2006) Geomechanical model for fracture deformation under hydraulic, mechanical and thermal loads. *Hydrogeology Journal*, **14**(4), 485–498.

McDermott, C.I., Randriamanjatosoa, A.R., Tenzer, H., and Kolditz, O. (2006) Simulation of heat extraction from crystalline rocks: the influence of coupled processes on differential reservoir cooling. *Geothermics*, **35**(3), 321–344.

Megel, T., Kohl, T., and Hopkirk, R. (2006) The potential of the use of dense fluids for initiating hydraulic stimulation. *Geothermics*, **35**(5–6), 589–599.

Morrow, C., Bo-Chong, Z., and Byerlee, J. (1986) Effective pressure law for permeability of westerly granite under cyclic loading. *Journal of Geophysical Research*, **91**, 3870–3876.

Nicol, D. and Robinson, B. (1990) Modelling the heat extraction from the Rosemanowes HDR reservoir. *Geothermics*, **19**(3), 247–257.

Nield, D.A. and Bejan, A. (1999) *Convection in Porous Media. Advances in Water Resources*, Springer, Berlin, Heidelberg, New York, pp. 1–546.

Nilsson, M. (2001) *Thermal Cracking of Young Concrete - Partial Coefficients, Restraint Effects and Influence of Casting Joints.*

Noorishad, J., Tsang, C., and Witherspoon, P. (1984) Coupledthermal-hydraulic-mechanical phenomena in saturated fractured porous rocks: numerical approach. *Journal of Geophysical Research*, **89**, 10365–10373.

O'Sullivan, M.J., Pruess, K., and Lippmann, M.J. (2001) State of the art of geothermal reservoir simulation. *Geothermics*, **30**(4), 395–429.

Pape, H., Clauser, C., and Iffland, J. (1999) Permeability prediction based on fractal pore-space geometry. *Geophysics*, **64**(5), 1447–1460.

Pape, H., Clauser, C., and Iffland, J. (2000) Variation of permeability with porosity in sandstone diagenesis interpreted with a fractal pore space model. *Pure and Applied Geophysics*, **157**, 603–619.

Pebesma, E.J. and Wesseling, C.G. (1998) Gstat: a program for geostatistical modelling, prediction and simulation. *Computers and Geosciences*, **24**(1), 17–31.

Pruess, K. (2008a) On CO2 fluid flow and heat transfer behavior in the subsurface, following leakage from a geologic storage reservoir. *Environmental Geology*, **54**(8), 1677–1686.

Pruess, K. (2008b) On production behavior of enhanced geothermal systems with CO2 as working fluid. *Energy Conversion and Management*, **49**(6), 1446–1454.

Pruess, K. and Spycher, N. (2007) ECO2N – a fluid property module for the TOUGH2 code for studies of CO2 storage in saline aquifers. *Energy Conversion and Managemen*, **48**(6), 1761–1767.

Rautman, C.A. and Treadway, A.H. (1991) Geologic uncertainty in a regulatory environment - an example from the potential Yucca Mountain nuclear waste repository site. *Environmental Geology and Water Sciences*, **18**(3), 171–184.

Rea, J. and Knight, R. (1998) Geostatistical analysis of ground-penetrating radar data: a means of describing spatial variation in the subsurface. *Water Resources Research*, **34**(3), 329–340.

Reinicke, A., Zimmermann, G., Huenges, E., and Burkhardt, H. (2005) Estimation of hydraulic parameters after stimulation experiments in the geothermal reservoir Groß Schönebeck 3/90 (North-German Basin). *International Journal of Rock Mechanics and Mining Sciences*, **42**(7–8), 1082–1087.

Richards, H., Parker, R., Green, A., Jones, R., Nicol, D., Randall, M., Richards, S., Stewart, R., and Willis-Richards, J. (1994) The performance and characteristics of the experimental hot dry rock geothermal reservoir at Rosemanowes, Cornwall (1985–1988). *Geothermics*, **23**(2), 73–109.

Rouhani, S., Srivastava, R., Desbarats, A., Cromer, M., and Johnson, A. (eds) (1996) *Geostatistics for Environmental and Geotechnical Applications*. ASTM.

Rutqvist, J., Barr, D., Birkholzer, J., Chijimatsu, M., Kolditz, O., Liu, Q., Oda, Y., Wang, W., and Zhang, C. (2008) Results from an international simulation study on coupled thermal, hydrological, and mechanical processes near geological nuclear waste repositories. *Journal of Nuclear Technology*, **163**(1), 101–109.

Schindler, M., Baumgartner, J., Fritsch, D., and Graff, J.J. (2008) Von der Reservoirentwicklung zur Stromproduktion im Geothermieprojekt Soultz, *Geothermische Energie*, **59**(2): 5–6.

Schrefler, B.A., Matteazzi, R., Gawin, D., and Wang, X. (2000) Two parallel computing methods for coupled thermo-hydromechanical problems. *Computer-Aided Civil and Infrastructure Engineering*, **15**(3), 176–188.

Shioya, R. and Yagawa, G. (2005) Large-scale parallel finite-element analysis using the internet: a performance study. *International Journal for Numerical Methods in Engineering*, **63**(2), 218–230.

Somerton, W. (1992) *Thermal Properties and Temperature-related Behavior of Rock/fluid Systems*. Elsevier, p. 257.

Span, R. and Wagner, W. (1994) *A New Equation of State for Carbon Dioxide Covering the Fluid Region from Triple-point Temperature to 1100 K at Pressures up to 800 MPa*. Technical Report, Lehrstuhl für Thermodynamik, Ruhr-Universität, Bochum.

Stephansson, O., Hudson, J., and Jing, L. (2004) *Coupled Thermo-Hydro-*

Mechanical-Chemical Processes in Geo-Systems: Fundamentals, Modelling, Experiments, and Applications. Geo-Engineering Book Series, vol. 2, Elsevier.

Stephens, M. (1994) Architectural element analysis within the Kayenta Formation (Lower Jurassic) using ground-probing radar and sedimentological profiling, southwestern Colorado. *Sedimentary Geology,* **90**(3–4), 179–211.

Stober, I. (1986) Strömungsverhalten in Festgesteinsaquiferen mit Hilfe von Pump- und Injektionsversuchen. *Geologisches Jahrbuch,* **C42**.

Su, G., Geller, J., Pruess, K., and Hunt, J. (2001) Solute transport along preferential flow paths in unsaturated fractures. *Water Resources Research,* **37**(10), 2481–2491.

Sultanov, L. (2006) *Application of supercritical CO2 for the extraction of geothermal energy in deep crystalline systems,* MSc Thesis. Technical Report, Center of Applied Geoscience, University of Tuebingen.

Talwani, P., Cobb, J., and Schaeffer, M. (1999) In situ measurements of hydraulic properties of a shear zone in northwestern South Carolina. *Journal of Geophysical Research B: Solid Earth,* **104**(7), 14993–15004.

Tenzer, H., Schanz, U., and Homeier, U. (2000) *HDR Research Programme and Results of Drill Hole Urach 3 to Depth of 4440 m – the Key for Realisation of a HDR Programme in Southern Germany and Northern Switzerland.* Proceedings World Geothermal Congress 2000, pp. 3927–3932.

Tezuka, K. and Watanabe, K. (2000) *Fracture Network Modeling of Hijiori Hot Dryrock Reservoir by Deterministic and Stochastic Crack Network Simulator(d/sc).* Proceedings World Geothermal Congress 2000, pp. 3933–3938.

Topping, B.H.V. and Khan, A.I. (1996) *Parallel Finite Element Computations,* Saxe-Coburg Publications, Edinburgh.

Trautwein, U. and Huenges, E. (2005) Poroelastic behaviour of physical properties in rotliegend sandstones under uniaxial strain. *International Journal of Rock Mechanics and Mining Sciences,* **42**, 924–932.

Tsang, C. (1991) Coupled hydromechanical-thermochemical processes in rock fractures. *Reviews of Geophysics,* **29**(4), 537–551.

Valley, B. and Evans, K.F. (2007) Stress state at Soultz-sous-Forêts to 5 km depth from wellbore failure and hydraulic observations. *Proceedings of the 32nd Workshop on Geothermal Reservoir Engineering, SGP-TR-183,* Stanford University, Stanford, CA, USA, 7 p.

Wagner, W. and Kurse, A. (1998) *Properties of Water and Steam: The Industrial Standard IAPWS-IF97 for Thermodynamic Properties and Supplementary Equations for other Properties,* Springer.

Walsh, R., McDermott, C., and Kolditz, O. (2008) Numerical modeling of stress-permeability coupling in rough fractures. *Journal of Hydrogeology,* **16**(4), 613–627.

Wang, W. and Kolditz, O. (2007) Object-oriented finite element analysis of thermo-hydro-mechanical (THM) problems in porous media. *International Journal for Numerical Methods in Engineering,* **69**(1), 162–201.

Wang, W., Kosakowski, G., and Kolditz, O. (2009) A parallel finite element scheme for thermo-hydro-mechanical (THM) coupled problems in porous media. *Computers and Geosciences,* **35**(8), 1631–1641.

Watanabe, K. and Takahashi, H. (1995) Fractal geometry characterization of geothermal reservoir fracture networks. *Journal of Geophysical Research*, **100**(B1), 521–528.

Watanabe, N., McDermott, C., Wang, W., Taniguchi, T., and Kolditz, O. (2010) *Uncertainty Analysis of Thermo-hydro-mechanical (THM) Coupled Processes in a Hot-dry-rock Geothermal Reservoir*. Computational Mechanics, doi:10.1007/s00466-009-0445-9.

Weidler, R., Gerard, A., Baria, R., Baumgärtner, J., and Jung, R. (2002) Hydraulic and Micro-seismic Results of a Massive Stimulation Test at 5km Depth at the European Hot-Dry-Rock Test Site Soultz, France. *Proceedings, 27th Workshop on Geothermal Reservoir Engineering. Stanford, California*, January 28-30, 2002, pp. 95–100.

Willis-Richards, J. (1995) Assessment of HDR reservoir stimulation and performance using simple stochastic models. *Geothermics*, **24**(3), 385–402.

Willis-Richards, J. and Wallroth, T. (1995) Approaches to the modelling of HDR reservoirs: a review. *Geothermics*, **24**(3), 307–332.

Winter, H., Adelhardt, S., Jerak, A., and Kuchenhoff, H. (2002) Characterization of cataclastic shear-zones of the KTB deep drill hole by regression analysis of drill cuttings data. *Geophysical Journal International*, **150**, 1–9.

Witherspoon, P., Wang, J., Iwai, K., and Gale, J. (1980) Validity of cubic law for fluid flow in a deformable rock fracture. *Water Resources Research*, **16**(6), 1016–1024.

Zheng, C. and Gorelick, S. (2002) *Solute Transport in Flow Fields Influenced by Decimetre-scale Preferential Flow Paths: Implications for Ground Water Remediation*. IAHS AISH Publication 275, pp. 463–470.

Zimmerman, R.W. (1991) *Compressibility of Sandstones (Developments in Petroleum Science)*, vol. 29, Elsevier Publishing Company, p. 173.

Zimmermann, G., Reinicke, A., Brandt, W., Blöcher, G., Milsch, H., Holl, H.-G., Moeck, I., Schulte, T., Saadat, A., and Huenges, E. (2008) Results of Stimulation Treatments at the Geothermal Research Wells in Grossschoenebeck/Germany. *Proceedings Annual Stanford Workshop on Geothermal Reservoir Engineering 2008 Stanford, CA*, 28-30 January 2008.

Zimmermann, G., Reinicke, A., Holl, H.-G., Legarth, B., Saadat, A., and Huenges, E. (2005) Well Test Analysis After Massive Waterfrac Treatments in a Sedimentary Geothermal Reservoir. *Proceedings World Geothermal Congress 2005Antalya, Turkey*, 24-29 April 2005, pp. 1–5.

Zoback, M. and Byerlee, J. (1975) Permeability and effective stress. *American Association of Petroleum Geologists Bulletin*, **59**, 154–158.

Zyvoloski, G., Dash, Z., and Kelkar, S. (1988) FEHM: Finite Element Heat and Mass Transfer Code. Technical Report LA-11224-MS, Los Alamos National Laboratory, Los Alamos.

6
Energetic Use of EGS Reservoirs

Ali Saadat, Stephanie Frick, Stefan Kranz, and Simona Regenspurg

The principle for the energetic use of EGS (enhanced geothermal system) reservoirs is based on the production of a fluid, which carries the geothermal heat, in a production well, the extraction of heat from the fluid on the surface with a heat exchanger, and the reinjection of the cooled fluid into the reservoir in an injection well. The extracted heat can be used for heat, power, or chill provision (Figure 6.1). The different technical aspects and constraints which are related to such plants are the subject of this chapter. The following sections are designed to give a general overview on the energetic utilization options and EGS plant design. The focus will be on typical EGS applications which use formation water as heat carrier, in a temperature range between 100 to about 200 °C.

Other EGS plant concepts which are researched and might be realized at a few sites in the future are not discussed. Such futuristic concepts refer to EGS plants that can assess steam reservoirs and geo-pressured reservoirs or EGS plants using heat carriers other than formation water, such as CO_2.

6.1
Utilization Options

In the following section, the different options for energetic use of EGS reservoirs are outlined with focus on the most important thermodynamic aspects. More details on thermodynamic aspects can be obtained from the literature, for example, Çengel and Boles (2006) and Dinçer and Rosen (2007). The goal of this section is to characterize the different utilization options based on energy efficiency and exergy efficiency considerations.

6.1.1
Energetic Considerations

The form of energy which can be supplied by any heat source mainly depends on the temperature level. For EGS plants, this is the temperature of the reservoir or, more precisely, the temperature of the produced geothermal fluid. Different energy

Geothermal Energy Systems. Edited by Ernst Huenges

ISBN: 978-3-527-40831-3

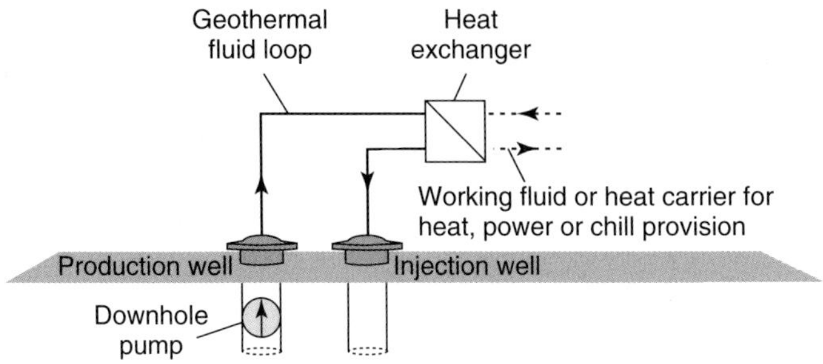

Figure 6.1 Principle for energetic use of EGS reservoirs showing a well doublet, the geothermal fluid loop and a heat exchanger on the surface.

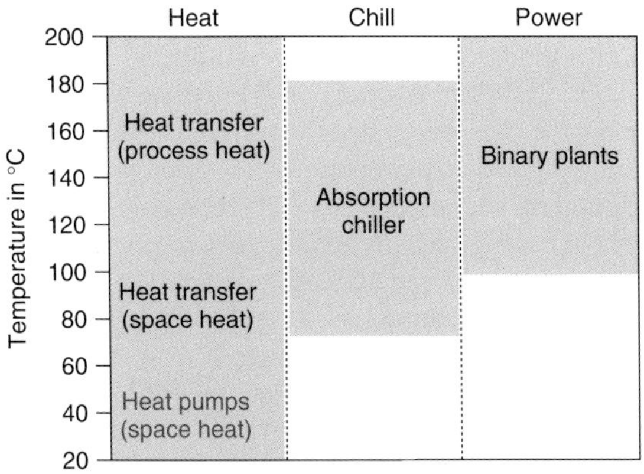

Figure 6.2 Energy provision options for heat source temperature between 20 and 200 °C. (Derived from Lindal, 1973.)

provision options referring to a heat source temperature between 20 and 200 °C are shown in Figure 6.2. From a thermodynamic point of view, the quality of a heat source and therefore the quality of the heat which can be supplied, decreases with decreasing heat source temperature. Since power has a higher quality than heat, it can only be generated (considering ambient conditions) from heat sources with high enough temperatures.

This is a very important aspect when it comes to evaluating the energetic performance of EGS plants, especially when they provide power, because only part of the energy offered by the geothermal fluid can be converted into power (depending on the heat source temperature). An energetic evaluation should therefore differ between the energy and the exergy (the part of the energy which can be converted to power) contained in the geothermal fluid flow. This differentiation is based on the second law of thermodynamics which, in contrast to the first law, does not only consider the conservation of energy in a system but also the

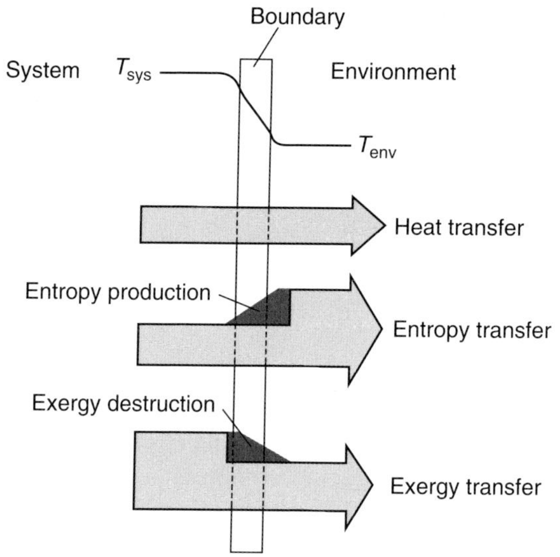

Figure 6.3 Entropy production and exergy destruction for the example of the heat transfer through a wall. (Based on Dinçer and Rosen, 2007.)

irreversibilities of thermodynamic processes due to the production of entropy and the destruction of exergy related therewith. Figure 6.3 shows a simple example of entropy production and exergy destruction for the example of heat transfer.

According to these considerations, different ways to evaluate energetic performance exist. In the following pages, the commonly used approaches are presented and discussed.

The most common way to evaluate the performance of an energy plant or a thermodynamic cycle is measuring the useful energy output relative to the energy input. This ratio is typically referred to as first law efficiency η in case of heat transfer and power plants, and as coefficient of performance (COP) for refrigeration cycles and heat pumps. First law efficiency and COP are based on the first law of thermodynamics and do not account for the quality of the energy.

$$\eta = \text{COP} = \frac{\text{used energy}_{\text{out}}}{\text{energy}_{\text{in}}} \tag{6.1}$$

Comparing two power plants, for example, with different heat source temperatures, the quality of the heat source, however, becomes important information for such a comparison. If the two plants have an identical first law efficiency, this means that the power plant using the heat source with the lower temperature actually uses the available exergy, which is also lower, more efficiently. The first law efficiency, though, is still a very useful measure to compare different energy conversion options or plant concepts for the same heat source or analyze the overall plant performance at a specific site.

In order to account for the energy quality in such an evaluation, the exergy efficiency ε can be used. It typically measures the exergy output $\dot{E}_{out}$ of a plant relative to its exergy input $\dot{E}_{in}$:

$$\varepsilon = \frac{\dot{E}_{out}}{\dot{E}_{in}} \tag{6.2}$$

Another evaluation option in this context is the so-called second law efficiency η_{2nd}. This efficiency is defined as the ratio which compares the first law efficiency of a real plant or cycle η_{real} and a thermodynamically ideal one η_{ideal} operating under the same thermodynamic conditions.

$$\eta_{2nd} = \frac{\eta_{real}}{\eta_{ideal}} \tag{6.3}$$

With reference to the energetic use of an EGS reservoir, resource utilization η_U is another aspect for evaluating plant performance. The utilization can be measured by relating the geothermal heat used by a plant $\dot{Q}_{geo,used}$ to the maximum usable heat available from the geothermal source $\dot{Q}_{geo,\,max}$. The maximum usable geothermal heat can be determined by the ambient temperature as lower temperature limit or a minimum allowable temperature. Comparing two EGS plants, the plant with the higher utilization ratio is the one using the geothermal resource to a larger extent and it reinjects less useful heat into the reservoir:

$$\eta_U = \frac{\dot{Q}_{geo,used}}{\dot{Q}_{geo,\,max}} \tag{6.4}$$

6.1.2 Heat Provision

Depending on their temperature level, EGS reservoirs can be exploited in order to directly supply heat; such as for district heating grids or process heat applications. In this case, the heat of the hot geothermal fluid is transferred to a fluid with lower temperature, such as the return of a district heating system. A typical case of a heat transfer with separately flowing fluids is shown in Figure 6.4.

The heat flow $\dot{Q}$ between hot and cold fluid is derived from the energy balance around the heat exchanger. Since heat losses can be usually neglected for the

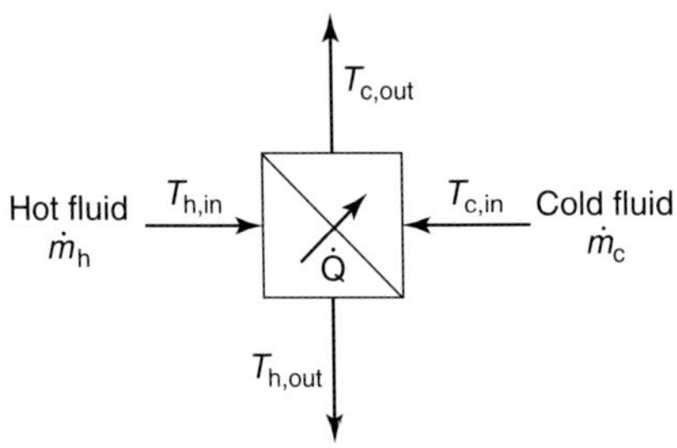

Figure 6.4 Heat transfer between two separated fluids with different temperature.

temperature levels relevant here, and the potential and kinetic energies of the fluids have no influence on the heat transfer, the heat flow $\dot{Q}$ can be calculated either from the heat flow from the hot side $\dot{Q}_h$ or the heat flow to the cold side $\dot{Q}_c$ of the heat exchanger with $\dot{Q} = \dot{Q}_h = \dot{Q}_c$.

In typical EGS applications, the hot geothermal fluid is always in liquid phase. The heat flow from the hot fluid can hence be calculated as follows:

$$\dot{Q}_h = \dot{m}_h \cdot c_{p,h} \cdot (T_{h,in} - T_{h,out}) \tag{6.5}$$

with mass flow rate $\dot{m}_h$, specific heat capacity $c_{p,h}$, and $T_{h,in}, T_{h,out}$ as temperatures of the hot fluid at the inlet of the heat exchanger and outlet, respectively. In special cases, the geothermal fluid might be in the liquid–vapor or the vapor phase. The calculation of the transferred heat is then analogous to the following calculation of the heat flow to the cold fluid.

When calculating the heat flow to the cold fluid, the phase of the cold fluid (i.e., liquid, liquid–vapor, or vapor) and phase changes (i.e., evaporation) have to be considered. Assuming a liquid fluid which is heated, evaporated, and superheated by the geothermal heat, the heat flow $\dot{Q}_c$ is determined as follows:

$$\dot{Q}_c = \dot{m}_c \cdot (\underbrace{c_{p,cl} \cdot (T_{c,in} - T_e)}_{HT_{sensible,l}} + \underbrace{q_v(T_e)}_{HT_{latent}} + \underbrace{c_{p,cv} \cdot (T_e - T_{c,out})}_{HT_{sensible,v}}) \tag{6.6}$$

where mass flow rate of the cold fluid is $\dot{m}_c$, the specific heat capacity of the cold fluid in liquid phase $c_{p,cl}$, its inlet temperature $T_{c,in}$, the evaporation temperature T_e, the evaporation heat $q_v\,(T_e)$, the specific heat capacity of the cold fluid in vapor phase $c_{p,cv}$, and the outlet temperature of the cold fluid $T_{c,out}$. Assigning these different terms to different modes of heat transfer, $HT_{sensible,l}$ is the sensible heat transfer of the liquid phase, HT_{latent} the latent heat transfer of the phase change, and $HT_{sensible,v}$ the sensible heat transfer of the vapor phase.

In case not all modes of heat transfer are relevant for a specific application, the heat flow calculation can be simplified. If the liquid cold fluid is only heated, for example, only the term for the sensible heat transfer of the liquid phase is used with $T_e = T_{c,out}$. If the liquid cold fluid is heated and evaporated to saturated vapor, the sensible heat transfer of the liquid phase and the latent heat transfer with $T_e = T_{c,out}$ are sufficient.

With reference to the energetic evaluation of heat provision, the (first law) efficiency according to Equation (6.1) is determined by

$$\eta = \frac{\dot{Q}_c}{\dot{Q}_h} = 1 \tag{6.7}$$

Since it has been stated above that heat losses are negligible, the efficiency of the heat provision equals one. Therefore, when comparing different EGS heat plants, the efficiency is not useful for a significant energetic evaluation.

In this case, the utilization ratio defined in Equation (6.4) is more applicable. If the maximum usable heat $\dot{Q}_{h_0}$ is determined by the ambient temperature T_0 as lower temperature limit, the utilization ratio η_U can be written according to Equation (6.4).

With reference to the utilization of geothermal heat for heat provision, it has to be remembered that it is not influenced only by the characteristics of the heat resource, but also by the characteristics of the heat use or the heat demand:

$$\eta_U = \frac{\dot{Q}_h}{\dot{Q}_{h0}} = \frac{\dot{m}_h \cdot c_{p,h} \cdot (T_{h,in} - T_{h,out})}{\dot{m}_h \cdot c_{p,h} \cdot (T_{h,in} - T_0)} = \frac{T_{h,in} - T_{h,out}}{T_{h,in} - T_0} \tag{6.8}$$

An EGS plant using a geothermal fluid with a temperature of 150 °C, down to a temperature of 70 °C at an ambient temperature of 20 °C, for example, results in a utilization ratio of 62%. If the heat, in contrast, could be used down to a temperature of 50 °C, the utilization ratio would increase to 77%.

Another possibility for energetic comparison is the calculation of the exergy efficiency according to Equation (6.2). For doing so, the exergy input (exergy content in the hot fluid) and the exergy output (exergy content in the cold fluid) must be first calculated. The exergy content of a fluid mainly depends on the temperature $T_{h,in}$ at which the heat is available and the lower temperature level T_0 down to which the heat can actually be used. Assuming the lower temperature level as ambient temperature and accounting for the finiteness of a geothermal heat source (i.e., the temperature of the geothermal fluid decreases during heat transfer which is accounted for by using the logarithmic average temperature), the exergy content of the hot geothermal fluid is determined as follows:

$$\dot{E}_{in} = \dot{Q}_h \cdot \left(1 - \frac{T_0 \cdot \ln \frac{T_{h,in}}{T_0}}{T_{h,in} - T_0}\right) \tag{6.9}$$

Regarding the exergy output or the exergy content of the cold fluid, phase and phase changes must be considered, analogous to Equation (6.6). For a liquid fluid which has been heated, evaporated, and superheated by the geothermal heat, the exergy output is defined as

$$\dot{E}_{out} = \dot{Q}_c \cdot \left(1 - \frac{T_0 \left(c_{p,cv} \ln \frac{T_{c,out}}{T_e} + \frac{q_V(T_e)}{T_e} + c_{p,cl} \ln \frac{T_e}{T_{c,in}}\right)}{c_{p,cv}(T_{c,out} - T_e) + q_V(T_e) + c_{p,cl}(T_e - T_{c,in})}\right) \tag{6.10}$$

For geothermal fluid temperatures between 100 and 200 °C, heat provision reaches exergy efficiencies up to 80%. This value is significantly higher than for other modes of heat provision such as combustion. This is due to the fact that the primary energy provided from EGS resources is already of relatively low quality so that only a small exergy reduction occurs for direct heat provision.

6.1.3 Chill Provision

Besides the provision of heat for space heating or industrial processes, geothermal energy can also be used for the provision of low temperature heat needed for refrigeration. Vapor compression, adsorption and absorption refrigeration are the most common principles for the provision of chill. Vapor compression refrigeration is the most commonly used principle. In contrast, the heat-driven absorption

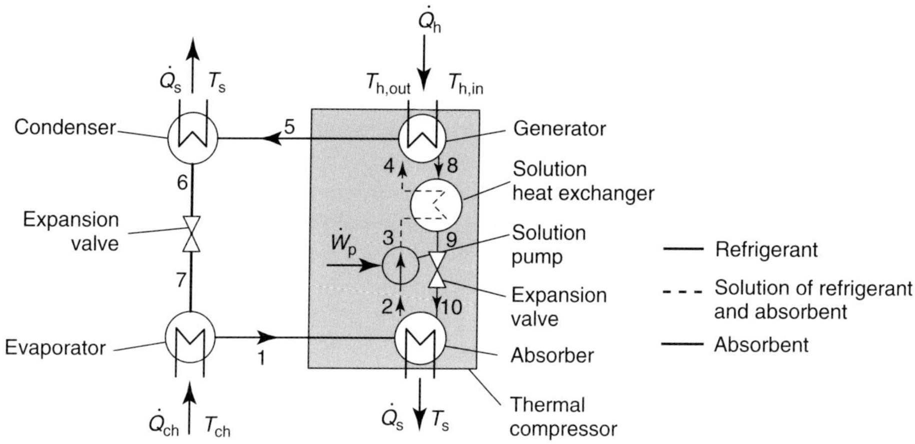

Figure 6.5 Principle of an absorption refrigeration cycle.

refrigeration cycle applied in absorption refrigeration systems (ARSs) is the most applicable for geothermal chill provision. The principle scheme of an ARS is shown in Figure 6.5. A typical ARS uses a mixture consisting of a refrigerant and an absorbent as working fluid and consists of an evaporator, a condenser, a generator, an absorber, a solution heat exchanger, a solution pump, and throttling valves.

Similar to most refrigeration cycles, absorption refrigeration is based on evaporating the refrigerant at low temperature and pressure, compressing the vapor and condensing it at a higher temperature and pressure level. During evaporation, the refrigerant absorbs the heat from the heat source that has to be chilled. During condensation, this heat is transferred to a heat sink at a higher temperature level, such as a water- or air-driven recooling system.

Evaporator, condenser, and throttling valve are similar to the ones used in vapor-compression-refrigeration systems. Due to the use of a mixture instead of a pure fluid as working fluid, ARSs contain a thermal instead of mechanical compressor. Such a thermal compressor is composed of three main elements: absorber, solution pump, and generator (Figure 6.5). In order to obtain a higher efficiency, a solution heat exchanger is used for energy recuperation.

The most widely used fluid combinations are ammonia–water and lithium bromide–water. Ammonia–water is appropriate for cooling and freezing to temperatures below 0 °C. The lithium bromide–water combination is used for air conditioning and chilling applications with temperatures above 4 °C, due to the crystallization of water. Other fluid mixtures that are currently the subject of research include water–lithium-, bromide–lithium-, iodide–lithium- and nitrate–lithium mixtures-, chloride solution-, or ammonia–water–sodium hydroxide.

In the following paragraph, the principle of cycle operation is explained using the example of an ammonia–water ARS (Figure 6.5). The ammonia (refrigerant) vapor exits the evaporator (1) and is absorbed by the water (absorbent) inside the absorber at low pressure (approximately 2.5 bar) and relatively low temperature. The resulting heat of absorption $\dot{Q}_S$ has to be removed from the cycle. The liquid solution (2) contains a relatively high concentration of water. The solution passes the solution pump (2→3) which raises the pressure up to the pressure level of both generator and condenser (approximately 10 bar). Inside the generator (4→5) the ammonia is driven off the water by means of heat input $\dot{Q}_h$, such as from a geothermal fluid. Afterwards the ammonia vapor (5) enters the condenser and the water (8) which contains a low concentration of ammonia, passes the solution heat exchanger (8→9), expands to the low pressure level inside the throttling valve (9→10), and enters the absorber. The solution heat exchanger reduces both, the heat needed in the generator and the one that has to be removed from the cycle. Inside the condenser, the ammonia is cooled until it condenses (5→6). The heat of condensation $\dot{Q}_s$ has to be removed as well. After leaving the condenser (6), the liquid ammonia is expanded by the throttle to the lower pressure level of evaporation (6→7). In the evaporator, the ammonia evaporates completely at a low temperature and pressure level (7→1). The heat needed for evaporation corresponds to the refrigeration capacity of the ARS. After evaporation, the ammonia vapor (1) is redirected to the absorber.

For evaluation of the energetic efficiency of refrigeration systems, the COP according to Equation (6.1) is defined as

$$\mathrm{COP} = \frac{\dot{Q}_{ch}}{\dot{Q}_h + \dot{W}_p} \tag{6.11}$$

where $\dot{Q}_{ch}$ is the heat input to evaporator (chilling capacity), $\dot{Q}_h$ the heat input to the generator, and $\dot{W}_p$ the power needed to drive the solution pump. In comparison to compression vapor cycles, the mechanical work needed to drive the solution pump is very small and thus, negligible. In existing ARSs the work needed for the solution pump amounts to approximately 5–8% in relation to the refrigeration capacity:

$$\mathrm{COP} \approx \frac{\dot{Q}_{ch}}{\dot{Q}_h} \tag{6.12}$$

Since the COP of ASRs depends strongly on the temperature level of input and output thermal energies, it is appropriate to express the COP based on the determining temperatures. According to Ziegler and Alefeld (1987) the COP considering (i) the inlet and outlet temperatures $T_{h,in}$ and $T_{h,out}$ of the hot fluid that passes the generator, (ii) the temperature of the evaporator heat input T_{ch}, and (iii) the temperature of heat output from absorber and condenser T_s results in

$$\mathrm{COP} \approx \frac{T_{ch}}{(T_s - T_{ch})}\left[1 - \frac{T_s}{T_{h,in} - T_{h,out}} ln\left(\frac{T_{h,out}}{T_{h,in}}\right)\right] \tag{6.13}$$

In the case of a sensible heat source such as the geothermal fluid and a sensible chill provision such as the cooling of a liquid fluid flow, the temperatures T_h and T_{ch} represent the corresponding thermodynamic average temperature T_a, considering an incompressible steady-state fluid flow:

$$T_a = \frac{h_{out} - h_{in}}{s_{out} - s_{in}} \tag{6.14}$$

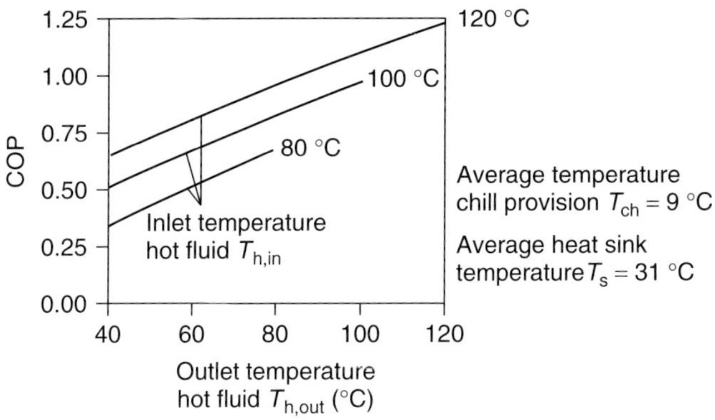

Figure 6.6 Coefficient of performance of absorption chillers according to Equation (6.13) depending on the return temperature and supply temperature of the heat source.

where h_{out} and h_{in} are the specific enthalpy of output and input fluid flow and s_{out}, and s_{in}, the specific entropy.

The relationship between COP and the inlet and outlet temperatures of heat input according to Equation (6.13) is shown in Figure 6.6 for a constant temperature of chill provision and a constant heat sink temperature.

As mentioned above in ARS, the necessary refrigerant compression is obtained by applying a thermal compressor. To drive a vapor compressor, exergy input is needed either from mechanical work and hence electrical power (vapor compression refrigeration systems) or from thermal energy (ARSs). Using thermal energy, the exergy associated with the heat input determines the efficiency of chill production. Since the exergy of the heat input depends on the temperature at which this heat is available, a higher temperature level of the heat input also results in a higher obtainable COP. For that reason, the exergy efficiency for chill provision is also an important benchmark for evaluating the quality. According to Equation (6.2), the exergy efficiency for ARS is defined as

$$\varepsilon_{ARS} = \frac{\dot{E}_{out,ev}}{\dot{E}_{in,gen} + \dot{W}_p} \tag{6.15}$$

where $\dot{E}_{out,ev}$ denotes the exergy output based on the provided chill at the evaporator and $\dot{E}_{in,gen}$ represents the exergy input based on the heat source fluid flow in the generator. Considering a constant temperature of heat input T_h and a constant temperature of chill provision T_{ch}, the exergy efficiency of ARSs can be written as

$$\varepsilon_{ARS} = \frac{|\dot{E}_{q,ch}|}{\dot{E}_{q,h} + \dot{W}_p} \tag{6.16}$$

with

$$\dot{E}_{q,ch} = \dot{Q}_{ch} \cdot \left(1 - \frac{T_s}{T_{ch}}\right) \quad \text{and} \quad \dot{Q}_h \cdot \left(1 - \frac{T_s}{T_h}\right). \tag{6.17}$$

In the case of a sensible heat source, the temperature T_h and T_{ch} must be replaced with the corresponding thermodynamic average temperature T_a (Equation 6.14).

6.1.4
Power Provision

The principle of converting heat into power is based on the heat flow between two heat reservoirs with different temperatures. This heat flow can be used by a working fluid which is performing a thermodynamic cycle. For EGS, closed thermodynamic cycles are the most relevant.

A common method to analyze closed cycles is the Carnot Cycle. This theoretical approach studies an ideal power conversion cycle between an infinite heat source at high temperature and an infinite heat sink at low temperature using a working fluid capable of expansion, such as gas or vapor. This ideal cycle, shown in the temperature–entropy diagram of Figure 6.7a, consists of four changes of thermodynamic state: isentropic pressure increase of the working fluid (1→2), isothermal heat input (2→3), isentropic expansion in an expansion machine (3→4), and isothermal heat removal (4→1). The rectangular area that is marked by the four thermodynamic states represents the work performed by the Carnot Cycle. The efficiency of the Carnot Cycle η_C therefore represents the maximum efficiency which can be achieved in the conversion of heat into power. According to the first law of thermodynamics, it is defined as ratio of work $\dot{W}$ to heat input $\dot{Q}_{in}$ and can be written as

$$\eta_C = \frac{\dot{W}}{\dot{Q}_{in}} = \frac{A_{1234}}{A_{1'234'}} = \frac{(T_h - T_s) \cdot (s_4 - s_1)}{T_h \cdot (s_4 - s_1)} = 1 - \frac{T_s}{T_h} \tag{6.18}$$

with the temperature of the infinite heat source T_h, the temperature of the infinite heat sink T_s, and the entropy difference for the state of change during heat input or heat removal respectively $(s_4 - s_1)$.

An advanced method to analyze closed cycles is the consideration of the finiteness of the heat source (such as geothermal heat sources) which means that the

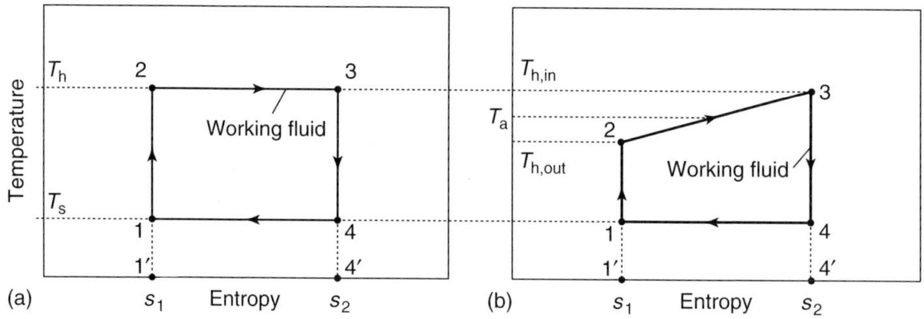

Figure 6.7 Temperature–entropy diagram of (a) the Carnot cycle with infinite heat source and (b) the trilateral cycle with finite heat source.

temperature of the heat source decreases with heat transfer. This ideal cycle is often referred to as the trilateral cycle due to its shape (Figure 6.7b). The efficiency of this cycle can be calculated based on Equation (6.18) using the thermodynamic average temperature of heat input T_a, which is derived from the ratio of enthalpy to entropy difference (Equation 6.14), instead of T_h. Since the average temperature will always be lower than the heat source temperature, the efficiency of the trilateral cycle is lower than the Carnot efficiency for a specific heat source. The influence of a finite heat source on the efficiency thereby increases with higher heat source temperatures (Figure 6.8).

With reference to real thermodynamic cycles, besides the characteristics of the heat source (and heat sink), the characteristics of the working fluid performing the cycle, the working medium, must also be considered. Fluids which can be evaporated are most commonly used. Such thermodynamic cycles are referred to as *Rankine cycles*. Due to the variety of evaporable working fluids, Rankine cycles are versatilely applicable, including for low temperature heat sources. Most other types of thermodynamic cycles used for low temperature heat are modifications of it e.g., Organic Rankine Cycle (ORC) or Kalina. The basic form of the Rankine cycle is shown in Figure 6.9. It consists of the following changes of state:

- pressure increase of the liquid working fluid with a feed pump (1→2);
- heating of the pressurized liquid up to the evaporation temperature (2→3);
- evaporation of the liquid at constant temperature and pressure (3→4);
- expansion of the vapor in an expansion machine (4→5) which produces mechanical work;
- condensing the expanded vapor at constant pressure and temperature (5→1).

For a finite heat source, the evaporation temperature is always lower than that of the heat source. The same considerations can be taken into account for the

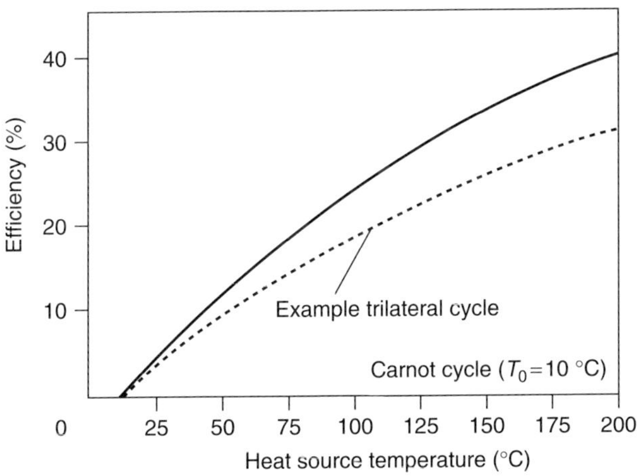

Figure 6.8 Comparison of Carnot efficiency and efficiency of an example trilateral cycle for increasing heat source temperatures at constant heat sink temperature.

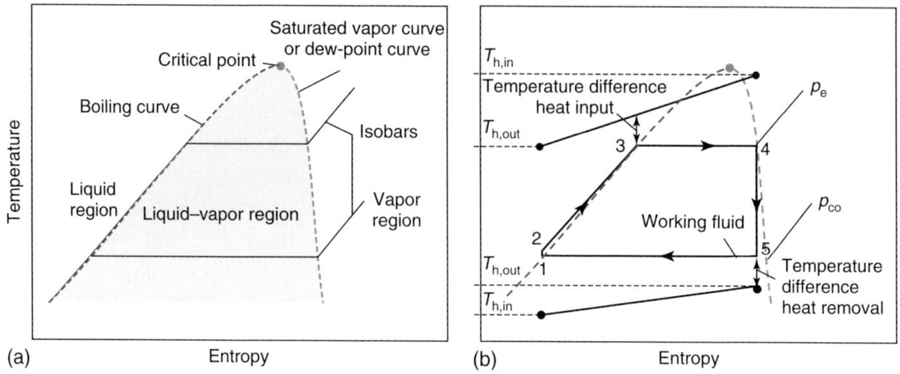

Figure 6.9 Temperature–entropy diagram (*T*–*s* diagram) of an example working fluid showing (a) general notations and (b) a corresponding Rankine cycle with saturated vapor.

condensation temperature in case of a sensible heat sink. The efficiency of the Rankine Cycle is hence always lower than the efficiency of the trilateral cycle using the same heat source. Considering a finite heat transfer area between heat source and working fluid, a minimum temperature difference also called pinch point (Figure 6.9) will remain, resulting in a further decrease of cycle efficiency. Analyzing real cycles, additional irreversibilities occur in technical components, such as friction losses in heat exchangers and in expansion machines. In this context, Figure 6.10 shows the implementation of a Rankine cycle driven by geothermal heat.

The comparison of the energetic performance of these different cycles has so far been based on the first law efficiencies, assuming the same heat source. However, for an energetic evaluation of different cycles using different heat sources, the exergy efficiencies are more significant. The exergy efficiency of a power conversion cycle

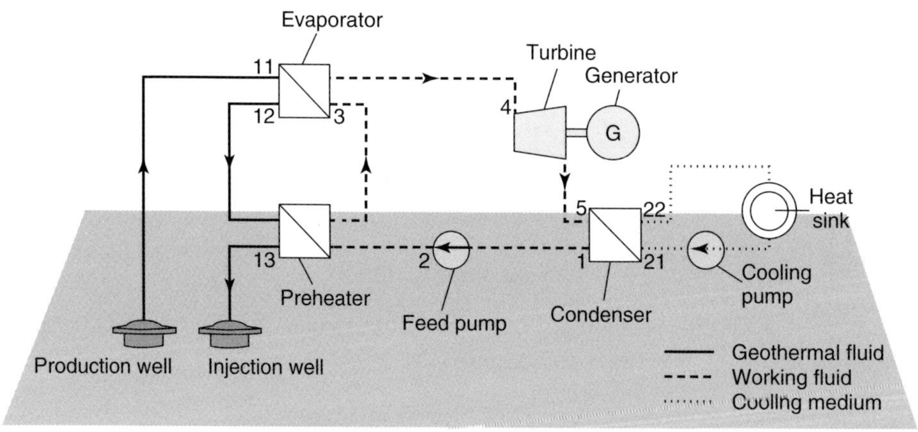

Figure 6.10 Schematic setup of Rankine cycle using geothermal heat as the heat source.

$\varepsilon_{\text{plant}}$ is therefore calculated from the ratio of performed work $\dot{W}_{\text{net}}$ to exergy input $\dot{E}_{\text{in}}$ as shown in Equation (6.19). As addressed above for heat provision, the exergy efficiency of EGS power plants is significantly higher than for other modes of thermal power conversion due to the already relatively low quality of the heat source despite the lower (first law) efficiency:

$$\varepsilon_{\text{plant}} = \frac{\dot{W}_{\text{net}}}{\dot{E}_{\text{in}}} = \frac{\dot{W}_{\text{net}}}{\dot{E}_{11} - \dot{E}_{13}} \tag{6.19}$$

The exergy input for the Rankine cycle shown in Figure 6.10 can be written as the difference between the exergy of the produced and the injected geothermal fluid, $\dot{E}_{11}$ and $\dot{E}_{13}$ respectively. The exergy content at a certain point n of the fluid flow can be derived from the mass flow $\dot{m}_{\text{n}}$ and the specific exergy e_n as

$$\dot{E}_n = \dot{m}_n \cdot e_n \tag{6.20}$$

The specific exergy of a heat flow at a point n can be calculated from the specific enthalpy and entropy at this point h_n and s_n, respectively. Neglecting kinetic and potential energy, the specific exergy can be calculated as follows:

$$e_n = h_n - h_0 - T_0(s_n - s_0) \tag{6.21}$$

with h_0 and s_0 the specific enthalpy and entropy of the fluid at reference conditions (here ambient conditions) and T_0 the reference or ambient temperature.

The exergy analysis can also be used for the thermodynamic evaluation of specific plant components by calculating their exergy efficiency or exergy destruction due to irreversibilities such as shown in Table 6.1. Kanoglu (2002) has shown, for example, that such analyses can be used to detect weak points of a geothermal plant concept.

Exergy analyses are also used in combination with economic considerations for improved project planning. The goal of such exergo-economic analyses is basically to identify the influence of specific plant components on costs and exergy

Table 6.1 Exergy efficiency and exergy destruction for the components shown in Figure 6.10.

Component	Exergy efficiency	Exergy destruction
Preheater	$\varepsilon_{pre} = \frac{\dot{E}_3-\dot{E}_2}{\dot{E}_{12}-\dot{E}_{13}}$	$\dot{E}_{D,pre} = (\dot{E}_{12} - \dot{E}_{13}) - (\dot{E}_3 - \dot{E}_2)$
Evaporator	$\varepsilon_{vap} = \frac{\dot{E}_4-\dot{E}_3}{\dot{E}_{11}-\dot{E}_{12}}$	$\dot{E}_{D,vap} = (\dot{E}_{11} - \dot{E}_{12}) - (\dot{E}_4 - \dot{E}_3)$
Turbine	$\varepsilon_{turb} = \frac{\dot{W}_{turb}}{\dot{E}_4-\dot{E}_5}$	$\dot{E}_{D,turb} = (\dot{E}_4 - \dot{E}_5) - \dot{W}_{turb}$
Feed pump	$\varepsilon_{pump} = \frac{\dot{E}_2-\dot{E}_1}{\dot{W}_{pump}}$	$\dot{E}_{D,pump} = \dot{W}_{pump} - (\dot{E}_2 - \dot{E}_1)$
Condenser	$\varepsilon_{cond} = \frac{\dot{E}_5-\dot{E}_1}{\dot{E}_{22}\ \dot{E}_{21}}$	$\dot{E}_{D,cond} = (\dot{E}_{22} - \dot{E}_{21}) - (\dot{E}_5 - \dot{E}_1)$
Total plant	$\varepsilon_{plant} = \frac{\dot{W}_{net}}{\dot{E}_{11}-\dot{E}_{13}}$	$\dot{E}_{D,plant} = (\dot{E}_{11} - \dot{E}_{13}) - \dot{W}_{net}$ or $\dot{E}_{D,plant} = \dot{E}_{D,pre} + \dot{E}_{D,vap} + \dot{E}_{D,turb} + \dot{E}_{D,pump} + \dot{E}_{D,cond}$

performance. Exergo-economic analysis can also be used to calculate the specific energy costs for heat and power from a combined supply. More details can be obtained; for example, from Bejan, Tastsaronis, and Moran (1996).

6.2 EGS Plant Design

EGS plants can basically be divided into different technical subsystems such as the geothermal fluid loop and an energy provision unit on the surface. The challenge in designing EGS plants is to meet the different design characteristics of these subsystems and manage an efficient and reliable interaction. Hence, boundary conditions and design criteria of an EGS plant and its subsystems vary from site to site.

Based on this variety in plant design, the design optimization of an EGS plant at a specific site requires the comparison of different technical solutions. In this context, simulation tools are an important instrument. So far, modeling of surface installations and reservoir behavior have usually been treated separately, which leads to a strong simplification of the interactions between these systems. Coupled models in contrast, are capable of combining aspects of subsurface and surface modeling and will play an increasingly important role in EGS plant design.

This chapter provides an overview of elementary design aspects of different components of an EGS plant: the geothermal fluid loop, the heat exchanger as an integral part for the utilization of the geothermal energy, and the energy utilization or provision unit on the surface. Regarding the latter, this chapter focuses on direct heat use, power conversion, and a combined energy provision since these options are presently the most relevant. The provision of chill is only briefly addressed, although this utilization option might play a more important role in the future.

6.2.1 Geothermal Fluid Loop

The geothermal fluid loop carries geothermal energy and is therefore, a decisive part of an EGS plant. It needs to integrate different functions such as the production of the geothermal fluid from the reservoir, its transport on the surface, processing, and finally its reinjection into the reservoir.

With reference to the design of the geothermal fluid loop and the components contained, such as production pump, pipelines, heat exchangers, filters, and injection pump, the following aspects need to be considered:

(i) Properties of geothermal fluids differ depending on the site-specific reservoir characteristics and the change of temperature and pressure during energetic utilization. Knowledge of the geochemical composition of the fluid and assessment of its alteration in the geothermal fluid loop or in the reservoir during reinjection must be an integral part of EGS plant design.
(ii) The operational reliability is strongly affected by interaction of the geothermal fluid with the plant materials. In particular, corrosion and scaling are technical

risks which need to be considered while optimizing design and successful operation of the geothermal fluid loop. Adequate measures include selection of durable materials and monitoring of physicochemical conditions of the fluid.

(iii) The production and injection of the geothermal fluid is related to considerable amounts of auxiliary power which depends on the reservoir characteristics and the design of the thermal water loop. To energetically optimize the plant operation, the auxiliary power input must be ameliorated/minimized.

In the following section, these aspects will be discussed in greater detail.

6.2.1.1 Fluid Properties

Geothermal fluid is defined as a heated multiphase substance consisting mainly of gas and liquid, which flows within pores of a geological formation (i.e., bedrock) of a deep geothermal reservoir. As a consequence of temperature and pressure change during uplift and processing, as well as by reinjection of the chilled fluid into the reservoir and by interaction with the surrounding materials (rock, tubing, casing), various chemical and physical processes (e.g., degassing, mineral solution, and precipitation) can be expected to occur within the fluid. These reactions can damage the plant by corrosion and fouling of tubes or clogging of the reservoir pores during reinjection. Thus, for proper design of the geothermal fluid loop, the fluid needs to be characterized in terms of its composition and its ability to alter both, the used materials and the reservoir.

An introduction to fluid types and compositions, as well as the main compounds of the fluids and potential interactions with the plant materials will be given in the following section.

Fluid Composition and Classification The geological source of the fluid bedrock as well as interactions of the fluid with the surrounding rocks during migration determines the chemical composition of geothermal fluids. Roughly, fluids can be divided according to their geological source as sedimentary basin fluids and crystalline rock fluids (Drever, Holland, and Turekian, 2004). Since their original formation, the geothermal fluids have been very mobile and have reacted with the surrounding rocks and minerals. Therefore, they undergo elemental separation and segregation due to microbiological processes, mineral saturation or mineral dissolution, as well as interact with organic matter and minerals. Consequently, with increasing age and depth of the fluid/bedrock, the salinity of a geothermal fluid usually increases. The most important water–rock interactions, which determine the fluid composition and increase the salinity, are the dissolution of halite (NaCl) or mixing of the fluid with halite-derived brines. The latter can be transported over long distances in sedimentary rocks (Kharaka and Hanor, 2004). Even more important is the consumption of water during weathering of minerals such as olivine or feldspar in crystalline rocks; this concentrates the ions in solution (Frape *et al.*, 2004). Further, during fluid migration in the earth's crust, it can mix with fluids of different origin and/or with seawater which would further affect its composition. As an additional group of fluids, one should mention fluids from young/recent volcanic rocks, which can be found, for example, in Iceland (Carvalho,

Fojaz, and Almelda, 2006) or New Zealand (Reyes *et al.*, 2003). They are marked due to their relatively young age by low salinity and due to their location in tectonically active areas by a high geothermal gradient.

Explaining fluid origins and possible evolution paths requires knowledge of characteristic isotopic distributions such as the $^{87}Sr/^{86}Sr$ ratio (McNutt, 2000). More details on isotopic fingerprints can be obtained from subject-specific literature (McNutt, 2000; Kendall and Doctor, 2004). To summarize, the chemical composition since fluid formation has been altered strongly inasmuch the original composition of the fluids does not – unlike shallow groundwater – directly indicate its source, because during geological ages, fluids interact with the surrounding rocks and minerals or mix with different fluids.

Fluids are usually classified according to (i) their major chemical compounds or (ii) according to their salinity:

1) Depending on their dominant ions, fluids are grouped into different types representing the major components of the system. In sedimentary basin fluids, the most frequently occurring compounds are sodium (Na^+), calcium (Ca^{2+}), and chloride (Cl^-) (Figure 6.11a) as well as occasionally, the organic anion acetate (CH_3COO^-). Consequently, the most common geothermal fluids are

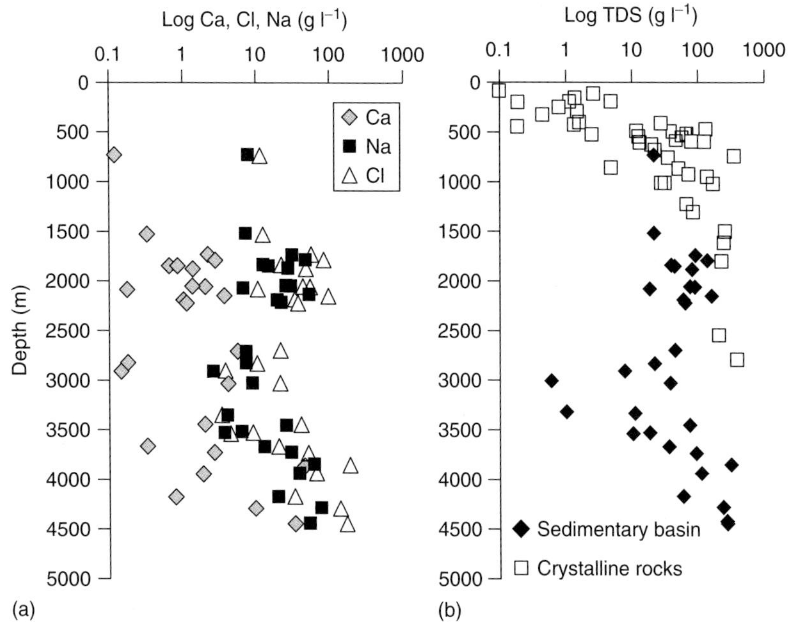

Figure 6.11 (a) Depth distribution of main elements of sedimentary basin fluids plotted versus depth. They represent 78–98 wt.% of total fluid salt content (Kharaka and Hanor, 2004). (b) Depth distribution of dissolved solids (log TDS); data from sedimentary basins (black diamonds; Giese *et al.*, 2002; Kharaka and Hanor, 2004; Wolfgramm *et al.*, 2007 and crystalline rock formations (white squares; Frape *et al.*, 2004.

(cations followed by anions in order of frequency) Ca–Cl, Na–Ca–Cl, Na–Cl, or Na–Cl–CH_3COO^- (Kharaka and Hanor, 2004). In crystalline rocks, besides Na^+, Ca^{2+}, and Cl^-,bicarbonate (HCO_3^-) and sulfate (SO_4^{2-}) might also play a major role (Frape *et al.*, 2004). Depth distribution of main components (log Ca, Na, Cl in g l^{-1}) from sedimentary basin fluids of various origins is shown in Figure 6.11a.

2) Due to the strong variation in salinity from a few milligrams to several hundred grams per liter of total dissolved solid (TDS), fluids are classified as freshwater ($< 1\,g\,l^{-1}$ TDS), brackish water (1–10 g l^{-1}), saline water (10–100 g l^{-1}), and brine ($> 100\,g\,l^{-1}$.) In general, salinity increases with increasing depth. The highest ever measured concentrations in a geothermal fluid are reported from a Ca–Cl brine (TDS $= 643\,g\,l^{-1}$) of the Salina formation of the Michigan Basin (Case, 1945). Most geothermal fluids are brines although less salty waters are less corrosive and thus easier to handle, and would consequently be more suitable for EGS systems. In low enthalpy geothermal systems, lower salinities have only been reported from few, evaporate-free regions. An exception, where hot but less saline fluids are used for geothermal energy production, are fluids from the Malm limestone formation of the South German Molasse Basin (e.g., Unterhaching, 3000–3200 m depth), where carbonate rocks are covered solely by unconsolidated sediments. These Na–Ca–HCO_3–type fluids have a TDS content of only 0.6–1 g l^{-1} (Wolfgramm *et al.*, 2007) and therefore these fluids are particularly suited for geothermal energy use.

Extensive overviews about fluid formation and composition covering a large number of samples from all over the world are given by Hanor (1994) and Kharaka and Hanor (2004) for sedimentary basin fluids, and by Frape *et al.* (2004) for deep fluids from crystalline rocks. Figure 6.11b shows the correlation between depth of the fluid origin and the TDS content in the sedimentary basin as compared to crystalline rock fluid from 76 samples collected from different locations (Giese *et al.*, 2002; Kharaka and Hanor, 2004; Frape *et al.*, 2004; Wolfgramm *et al.*, 2007). It is obvious that less data from deep (>2000 m) crystalline rock fluids (white squares; Figure 6.11b) are available as compared to the sedimentary basin fluids (black diamonds). However, in both fluid types, the TDS content increases considerably with depth. In the graph shown, the mentioned exception of freshwater, Na–Ca–HCO_3 type fluids from Southern Germany stand out due to their exceptionally low TDS content at great depths.

Effect of Fluid Properties on Plant Operation Geothermal fluids represent a mixture of many inorganic ions, dissolved silica, organic materials, and dissolved gases. Some of these compounds are highly dominant in the solution, but of less importance for operating the EGS plant. Others are highly reactive and even trace amounts can have effects on the production process.

Variation of fluid conditions, induced by pressure (p) and temperature (T) change for example, can provoke a change in fluid composition due to degassing or precipitation of minerals because mineral solubility and formation depends strongly on pT conditions. These processes further affect the pH value or induce interaction of the fluid with the plant materials. Thus, knowledge of fluid composition and properties can indicate not only the reservoir type and the origin of the water, but also the potential risk of plant corrosion and scaling (= mineral precipitation).

In the following, an overview of the most relevant properties of a geothermal fluid will be given.

Temperature, pressure, and physicochemical conditions (pH and redox value) The temperature of a fluid from a typical EGS reservoir is around 100–200 °C. During its energetic use on the surface, the temperature is lower than that of the produced solution. Similarly, the pressure decreases from several hundred bars in the reservoir to few bars on the surface where optimally the pressure is kept at around 10 to 15 bar to prevent degassing of CO_2. Both pressure, and in particular, temperature changes strongly affect the solubility of minerals and thus need to be considered in order to assess scaling of minerals and fouling of casing.

At given pT conditions of a solution, the parameters pH value and redox potential determine the dominant species in which a compound (e.g., a metal) occurs in solution. This knowledge is important to estimate if this compound of a certain concentration will remain in solution or precipitate.

The pH value of an aqueous solution defines the transfer of hydrogen ions between chemical species. In geothermal fluids, the pH value usually decreases (i.e., there is an increase of protons in the solution) with increasing chlorinity and depth. It should be noted that most reported pH values were measured at the earth's surface and not at in situ conditions (within the borehole). However, due to changes in pressure and temperature as well as degassing of acid volatiles (mainly CO_2), the pH might increase by 1–2 pH units at the surface (Kharaka, Hull, and Carothers, 1985).

The redox potential of an aqueous solution is the tendency of the solution to either gain or lose electrons when another species is introduced into the system. A solution with a more positive potential than the other species will have a tendency to gain electrons from the new species (i.e., to be reduced by oxidizing the new species) and a solution with a more negative reduction potential will have a tendency to lose electrons to the new species (i.e., to be oxidized by reducing the new species). The transfer of electrons between chemical species determines the reduction potential of an aqueous solution. Geothermal fluids are usually in a slightly reducing state inasmuch some compounds (e.g., Fe) occur predominantly at a more reduced state (as Fe(II)), and others such as sulfur (S), occur mainly in the oxidized form (as S(VI)).

Cations of importance As mentioned above, Na^+ and Ca^{2+} are the highest concentrated cations in geothermal fluids which are mainly enriched due to halite (NaCl) and calcite ($CaCO_3$) dissolution or mineral transformation. In contrast, potassium (K^+) and magnesium (Mg^{2+}) play a less dominant role because they are mainly enriched in the source rock in less soluble compounds. Iron (Fe) and manganese (Mn) are of special importance, because they are ubiquitous elements usually occurring in concentrations of up to several hundred milligrams per liter and form hardly soluble minerals in the presence of oxygen (O_2), carbonates (CO_3^{2-}, HCO_3^-), or sulfide (S^{2-}).

Heavy metals such as lead (Pb), zinc (Zn), copper (Cu), or mercury (Hg) are usually found in very low concentrations in geothermal fluids (< 1 mg l^{-1}), because they usually occur as sulfides which (at anoxic conditions) are hardly soluble minerals. Even the low concentrations found in many geothermal systems can only be explained by their binding to chlorine, bisulfide, or organic matter forming aqueous complexes which would keep the metals soluble despite the presence of sulfide. Similarly, barium (Ba^{2+}) and strontium (Sr^{2+}) are often problematic because in the presence of sulfate, they form hardly soluble barite ($BaSO_4$) or coelestine ($SrSO_4$) minerals.

Few exceptionally rich sources of heavy metals in fluids are known such as red bed fluids (brines) from the central Mississippi Salt Dome Basin (Kharaka *et al.*, 1987) or the European Rotliegend formation (e.g., Gross Schönebeck, Germany, Giese *et al.*, 2002 where Zn and Pb are enriched up to several hundred milligrams per liter at very low S^{2-} concentrations (< 0.02 mg l^{-1}). These highly concentrated heavy metals could represent a problem, because they easily precipitate in the presence of O_2, HCO_3^-, or by changes in redox and pH conditions.

Anions of importance By far, the anion of highest concentration in geothermal fluids is Cl^- which accounts, for example, for an average of 57% TDS (Kharaka and Hanor, 2004) in 27 samples of sedimentary basins, or for 50% TDS as measured in 43 crystalline basin fluids (Frape *et al.*, 2004). This enrichment is a consequence of its frequency in salts and minerals as well as of relatively high solubility of chlorine salts. Since it is known for its corrosive properties (see below), it requires special attention in geothermal plants. Other monovalent anions such as bromide, iodine, or fluoride are important to explain fluid evolution, but due to their lower concentrations, play a minor role in an EGS system.

Sulfur (S) occurs in many redox states in water (−2 to +6), but the common dominating species is the anion sulfate (SO_4^{2-}), whose concentration is usually at most up to 1000 mg l^{-1} and is controlled mainly by anhydrite ($CaSO_4$) solubility. Although SO_4^{2-} occurs at higher concentrations (2700 mg l^{-1}) in seawater, the formation of gypsum ($CaSO_4 \cdot 2H_2O$) and barite ($BaSO_4$) as well as bacterially induced sulfate reduction, lower the SO_4^{2-} content of the fluids. However, the oxidation of sulfides (especially pyrite, FeS_2) or dissolution of sulfate minerals can also increase its concentration. Carbonate and bicarbonate concentration usually occurs below a few hundred milligrams per liter in solutions of salinities > 30 g l^{-1}. This is mainly because the increase of H^+ with increasing salinity shifts the equilibrium toward the carbonic acid ($HCO_3^- + H^+ \rightarrow H_2CO_3$).

Both, sulfate and carbonate represent a problem when the solution becomes oversaturated with respect to sulfates or carbonates and precipitates as minerals such as $CaSO_4$, (gypsum), $BaSO_4$ (barite), $FeCO_3$ (siderite), $MnCO_3$ (rhodochrosite), or $CaCO_3$ (calcite or aragonite).

Organic compounds The majority of organic acid anions are generated by thermal alteration of kerogen (i.e., chemically transformed residues of natural organic substances such as plant residues) in the source rock. Most organic anions are mono-and di-carboxylic acids such as acetate, propionate, butyrate, malonate, and oxalate (Kharaka, Hull, and Carothers, 1985; Kharaka, Lundegard, and Giordano, 2000) as well as phenols (Lundegard and Kharaka, 1994). Concentrations range between few and several thousand milligrams per liter. They play an important role for many chemical reactions inasmuch they can act as proton donor for pH dependent reactions (especially for microbiologically induced reactions), as pH-, Eh-buffering agents, and as complexing agents for many (heavy) metals, thus keeping the complexed metal in solution.

Measured data usually represent minimum values since many organic compounds are degraded during processing by bacteria and by thermal decarboxylation, resulting in an increase of dissolved CO_2.

Dissolved silica The concentration of dissolved SiO_2 is mainly controlled by the solubility of quartz and usually ranges between 10–200 mg l^{-1} in formation waters. However due to slow precipitation kinetics of quartz, many fluids are not in equilibrium with quartz and are highly oversaturated (Land and Macpherson, 1992). Quartz equilibrium is usually achieved when temperatures exceed 70–120 °C (Bjorlykke *et al.*, 1995; Bazin, Brosse, and Sommer, 1997). Silica precipitation is especially undesirable in plants, because it can hardly be removed and only the addition of HF, which is highly risky to handle, can dissolve these precipitates.

Dissolved gases Fluids contain various amounts of dissolved gases (0.05–1 l gas per liter fluid) which consist mainly of a mixture of nitrogen (N_2) and carbon dioxide (CO_2) as well as occasionally, hydrogen sulfide (H_2S), methane (CH_4), and traces of helium (He). Of special interest for plant design, is the content of noncondensable gases (NCGs) such as CO_2 and N_2, which are not easily condensed by fluid cooling and thus may not be simply injected back into the reservoir. Moreover, NCGs at a given field also vary over time; this can cause problems with the fluid production equipment. Depending on local flow conditions, the NCG will then accumulate in the installations of the geothermal fluid loop (such as heat exchangers and injection well). To avoid scaling and corrosion induced by reaction of gas with plant material, a removal system (vacuum equipment) is necessary. Of special concern is the gas hydrogen sulfide (H_2S), because sulfide can react with metals such as Pb, Cu, or Fe to form various sulfide minerals which will clog the pipes, whereas the free protons (H^+) induce their corrosion. Carbon dioxide and methane are greenhouse gases and hence, measures have to be taken to avoid their escape into the atmosphere.

The amount and composition of gases can vary dramatically from reservoir to reservoir and from well to well within the same reservoir. Consequently, it is very difficult to establish accurate design bases for the gas fraction in geothermal plants.

6.2.1.2 Operational Reliability Aspects

The reliability of the geothermal fluid loop is based on the choice of suitable materials and component layouts according to the characteristics of the geothermal fluid and local fluid properties which are expected during operation. Improper design can lead to deterioration of EGS plant performance or even plant failure if components of the geothermal fluid loop have to be replaced.

For the operation of a geothermal plant, corrosion of plant materials and clogging of the pipes (scaling) are the two risks which strongly affect plant reliability. They will be discussed in the following sections. In Table 6.2 an overview on the most significant/damaging compounds of a geothermal fluid and their special reactivity in an EGS system is given.

The reliability of the fluid production technology is another important aspect in this context and will be addressed in Section 6.2.1.3.

Table 6.2 Relevant fluid properties and parameters responsible for scaling and corrosion.

Relevant parameter	Scaling	Corrosion	Process
O_2	+ +	+ +	Oxidation of plant material (steel/iron) (corrosion) and of dissolved compounds in solution (scaling of oxides)
H^+ (pH value)	+	++	Proton promoted material dissolution (corrosion); or shift of pH in the Pourbaix diagram which moves the compound saturation thus controlling the scaling (most minerals precipitate at higher pH value)
Redox change	+	+	Shift in the Pourbaix diagram; see pH
T-decrease	++	–	Decrease of mineral solubility at decreasing temperature and pressure, resulting in oversaturation of many compounds
p-decrease	+	–	
Fe, Mn	++	+	Facilitated precipitation as oxide/hydroxide, carbonate, sulfide (scaling)[a]
Pb, Zn, Cu	+	+	Precipitation as sulfide or elemental metal (scaling)[a]
Sulfide	+	+	Scaling with metals[a]
Sulfate	+		Scaling with alkaline earths (Ca, Mg, Ba, Sr); especially $BaSO_4$ is hardly soluble[a]
Cl^- concentration	–	+	Corrosion (damage of the passivation film of materials)
Dissolved silica	+	+	Scaling as (amorphous) silica; hardly soluble[a]
Dissolved gases: degassing	–	+	Alteration of the pH (see pH effect above)
Dissolved gases: H_2S	+	++	Interaction of plant materials with protons (corrosion) and of fluid compounds and plant materials with sulfide (scaling)

[a] Corrosion at the scaling–pipe interface.
"+" Signifies a slight effect, "++" a strong effect, and "–" no direct, or very little effect.

Corrosion Corrosion is the destruction of a material due to chemical reactions with its surrounding environment (gas or liquid), for example, by reaction with water and oxygen. Corrosion occurs under a variety of conditions which involve two basic mechanisms: electrochemical corrosion and (hot gas) oxidation (Cornell and Schwertmann, 1996). In the presence of oxygen in air, thin films of oxide form on the surface of steel or iron; these increase in thickness with increasing temperature due to thermodynamically favored conditions toward iron oxides (e.g., $2\ Fe^{2+} + 1\frac{1}{2}\ O_2 \rightarrow Fe_2O_3$).

Electrochemical corrosion describes the oxidation/dissolution of iron in an aqueous solution or at air humidity >60% (Cornell and Schwertmann, 1996): $Fe \rightarrow Fe^{2+} + 2\,e^-$. This anodic half reaction requires an additional cathodic reaction such as the reduction of oxygen ($H^+ + \frac{1}{4}\ O_2 + e^- \rightarrow \frac{1}{2}\ H_2O$) or hydrogen ($H^+ + e^- \rightarrow \frac{1}{2}\ H_2$).

Consequently, several properties/compounds of the geothermal fluids show significant corrosive effects.

- **pH value (H^+ concentration):** As mentioned above, at low pH values (high H^+ concentration) iron dissolution can be expected. A decrease of the fluid pH value can be induced, for example, by CO_2 degassing or by precipitation (scaling) of certain minerals, which results in a local change of conditions (especially beneath the scaling).
- **Oxygen (O_2):** As long as geothermal fluids are not in contact with the atmosphere or mix with O_2 containing surface water or shallow ground water, O_2 gas dissolved in water is not expected to occur in geothermal fluids. However, if O_2 enters the fluid, it will quickly react to oxidize the reduced species of the fluid such as Fe(II) or Mn(II) to precipitate as Fe(III) or Mn(IV) oxide/hydroxide. Similarly, the tubing material (Fe- or Mn-containing steel) could be oxidized by O_2, resulting in material corrosion. A good indicator to guarantee an O_2-free environment is the presence of hydrogen sulfide whereupon even traces are sufficient because they cannot both – O_2 and S^{2-} – be present in the solution. However, after transport of the geothermal fluid to the surface, prevention of O_2 contamination is extremely difficult, especially if pumps other than down-well submersible pumps are used to move the geothermal fluid. Even though the fluid system may be maintained at overpressure, air infiltration at the pump seals is likely, particularly in the light of the level of maintenance likely in many installations.
- **Redox potential (Eh value):** The redox potential is the tendency of the solution to either gain or lose electrons. In a geothermal fluid, redox reactions might occur during pumping, degassing or scale formation. Together with the pH, the redox value determines strongly the speciation of compounds in a solution. The Pourbaix diagram (pH–Eh predominance area diagram) in Figure 6.12 gives an example on the effect of the redox/and pH value on iron speciation and precipitation, respectively. It displays a simple aqueous system containing only Fe, O, and H: While at high pH and high positive Eh values, the (solid) Fe_2O_3 species would form predominantly, whereas reducing conditions would favor iron to occur as dissolved Fe^{2+}. Elemental (native) Fe would not form in an aqueous solution, because its stability line is below the one of water (Figure 6.12).

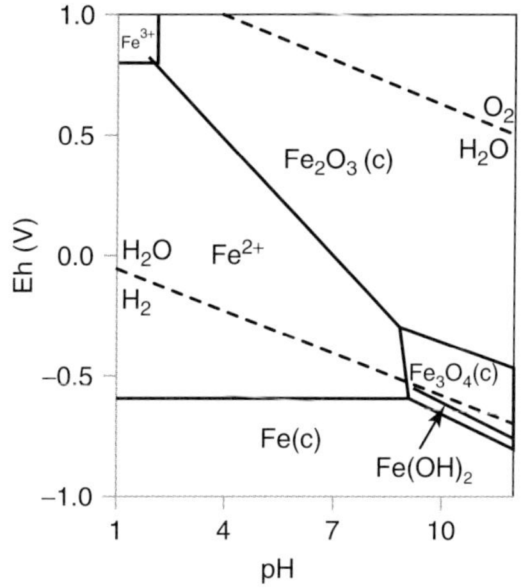

Figure 6.12 Predominance area Eh-pH diagram (Pourbaix) for the Fe(II)/Fe(III)- redox couple as calculated with HYDRA/MEDUSA software (Puigdomenech, 2004) of a 0.01 mM Fe(II) solution, at ionic strength of 1 M and at atmospheric conditions (25 °C). Dashed lines indicate stability area of water and solid lines of respective Fe species; (c) indicates a solid species.

- **Highly concentrated salt (chloride):** Very high concentrations of anions such as chloride can interfere with a given alloy's ability to re-form a passivation film. These thin and hard films occurring on steel and alloy surfaces usually consist of metal oxide or nitride and will, due to their low reactivity, inhibit corrosion. However, local fluctuations induced by high chloride concentrations will prevent the film formation inasmuch the oxide film will be degraded at a few critical points which can then be greatly amplified and can cause corrosion pits of several types, depending upon the conditions.
- **Sulfide:** Susceptible alloys, especially steels, react with hydrogen sulfide, forming metal sulfides and elementary hydrogen. The hydrogen diffuses into the metal matrix and continues damaging the material (refer to effects of pH value). This form of corrosion is usually described as sulfide stress cracking (SSC), or sulfide stress corrosion and can be enhanced in the presence of sulfate-reducing bacteria which catalyze sulfide production. High content of nickel in the steels greatly improves their resistance to SSC. This type of corrosion is worst at temperatures around 80 °C. (Chen *et al.*, 2005).

While some efforts to reduce corrosion merely redirect the damage into less visible, less predictable forms, controlled corrosion treatments such as passivation and chromate conversion will increase a material's corrosion resistance.

Forms of corrosion Corrosion can be concentrated locally to form a pit or crack, or it can extend across a wide area to produce general deterioration. This uniform corrosion, which involves a uniform thinning of the metal, is the most common form but can be controlled relatively easily with appropriate engineering design. In contrast, nonuniform corrosion such as crevice attack is more difficult to control because it has a number of causes. It occurs due to discontinuities of the materials localized, for example, at holes, joints, and bends in pipes. If two different metals are joined together, bimetallic corrosion occurs (galvanic corrosion). Pitting corrosion describes the local corrosion which occurs when the protective coating of a metal is defective. Intergranular corrosion characterizes the attack along grain boundaries in the iron causing whole grains to fall out. Finally, the effect of mechanical stress often combined with local electrochemical corrosion can cause rapid cracking in equipment such as pipes and reactors.

Due to the heterogeneous and often aggressive composition of the geothermal fluids, corrosion is likely to be expected in the tubing of the geothermal loop system. Thus, depending on the fluid chemistry, the design of the geothermal fluid loop requires careful engineering.

Since corrosion depends strongly on the kind of material used, separate corrosion tests of potential materials should be performed. Moreover, *in situ* monitoring of corrosion rates as well as of the mentioned corrosion parameters should be performed in the geothermal fluid loop to assess corrosion or the potential risk of corrosion.

As products of corrosion, the formation of all major iron hydroxides can be expected. Depending on corrosion type and conditions, certain compounds are formed which could point to the source of their formation. Most frequently occurring minerals are magnetite (Fe_3O_4), lepidocrocite (FeOOH), and green rust ($Fe(OH)_2$) (Cornell and Schwertmann, 1996).

Corrosion prevention To minimize corrosion, either the cathodic species (O_2, H^+) need to be removed from the solution or (potentially rusting) materials such as iron or steel should be protected by providing a barrier against the reacting species in the fluid. In most cases, this implies the formation of a uniform, nonporous, adherent, protective film. The formation of these films can be achieved either by alloying the iron or by addition of inhibitors into the aqueous phase. Anodic inhibitors would deactivate the anodic site of the metal and bring the potential into the passive region of the Pourbaix diagram (Figure 6.12). Usually, nitrite or chromate, two strongly oxidizing agents are used as inhibitors which would rapidly form a protective film of Fe(III)oxide in case of nitrite, or a mixed $Cr_2O_3-Fe_2O_3$ on the surface of the iron or steel in case of chromate. Cathodic inhibitors such as phosphoric acid would form a protective layer of mixed Fe(II)/Fe(III) phosphate.

Similarly, anticorrosive paints containing chemically active pigments can form protective coatings at the paint–metal interface which consist, for example, of iron oxides or iron phosphates.

Best corrosion prevention, of course, is achieved by the choice of appropriate materials to be used for the tubes of an EGS plant in a certain environment; this depends mainly on fluid properties. In general, alloying the iron with nickel or chromium increases the corrosion resistance, because a higher electromotive force (emf) would be required for corrosion to take place. Corrosion resistant alloy (CRA) development is a broad field in material sciences and cannot be discussed here in detail. However, the latest alloy developments for materials in geothermal systems support the use of ruthenium enhanced titanium alloys such as Ti–6%Al–4%V–Ru, which exhibit long-term corrosion resistance to sulfide- and chloride-induced stress corrosion cracking in anoxic, aggressive brines over a wide pH range and at high temperatures (up to 210 and 330 °C, respectively; Schutz and Watkins, 1998; Schutz, Porter, and Horrigan, 2000).

Scaling

Scale formation Scaling characterizes the formation of a solid within a solution as the product of a redox reaction or as consequence of oversaturation of the solution with a certain salt. The concentration of salts to be saturated depends on the chemical compounds as well as on the pH value, flow dynamics, temperature, and pressure of the solution. In natural systems, the kinetics of solid precipitation is often slow and thus even when a solution is oversaturated, precipitation cannot be observed. However, the presence of solid surfaces or microorganisms can strongly catalyze precipitation reactions.

Within the tubing of the geothermal fluid loop, scaling effects can be crucial due to solid precipitation directly on surfaces such as a heat transfer surface or a pipe wall, and thus cause clogging of the pipes. Another effect is the formation of small (colloidal) particles which remain suspended in the fluid. These can then be transported over large distances in the well, as well as in the fluid loop, and accumulate somewhere else in the system thereby causing clogging or interaction with the tubing material (corrosion).

The nature of the scaling/precipitate depends mainly on the composition of the fluid. In carbonated solutions the formation of carbonates, especially $CaCO_3$ (calcite, aragonite) and $FeCO_3$ (siderite), represents a major problem; whereas in sulfate-rich waters, gypsum ($CaSO_4$) and barite ($BaSO_4$) form the most common precipitates. In high silica waters, the formation of (amorphous) silica scales represents a major problem due to the low solubility of these materials, which means that removal of silica scales will be an extremely elaborate process.

Similarly the presence of sulfide in solution, which reacts with all kind of heavy metals often used as tubing material, will lead to the formation of hardly soluble precipitates such as galena (PbS), sphalerite (ZnS), or various copper and iron sulfides. These scalings are frequently observed directly on the production pump, because in this area, the redox conditions vary locally and strongly. Consequently metals precipitate as sulfides or even in their elemental (native) form (especially Cu or Pb; Figure 6.13).

Another factor inducing massive scaling is the introduction of O_2 into the system; this has to be avoided by all means. Especially in the presence of Fe(II) ions, the O_2 is at once consumed to produce Fe(III) oxides which can precipitate over a wide range of pH.

The changes of fluid temperature and pressure shift the chemical equilibrium and thus potentially induce scaling. Moreover CO_2 gas, which is often dissolved in the geothermal liquid, can be released due to changes in pT conditions or when flashing a liquid to produce steam in separators. A CO_2 release a shifts the pH to

Figure 6.13 Lead scales near the pump in the Groß Schönebeck well (Germany).

higher values resulting in the potential precipitation of certain minerals, especially hydroxides.

As mentioned above, H_2S (gas) is of special concern in the plants due to the reaction of sulfide (S^{2-}) with metals such as lead (Pb), nickel (Ni), copper (Cu), or iron (Fe) to form various sulfide minerals which precipitate in the tubes of the geothermal fluid loop. Simultaneously, the precipitation of the sulfides increases the H^+ concentration in the system which in turn increases material corrosion. The amount of hydrogen sulfide in an EGS plant will vary considerably depending on the characteristics of the reservoir(s), the design of the plant, and the local regulatory requirements.

Scaling prevention and removal Scaling can be dealt with in a variety of ways. To avoid scale formation, a developer can reduce the heat captured from the geothermal fluid (thereby reducing plant efficiency), because a less drastic heat reduction results in less mineral precipitation since the solubility of a mineral increases with increasing temperature. However, for this measure, caution is essential if the fluids are rich in calcium and carbonate, because calcite precipitation is increased at decreasing temperature. Other options are the addition of scaling inhibitors to the fluid (usually polymers which prevent the formation of sulfide minerals). This is often applied for oilfield wells or the acidification of the fluid to increase mineral solubility. Further, biocides or disinfectants can be added to the fluids to prevent microbiologically catalyzed mineral precipitation. Depending on the kind of scaling and on the location where scaling occurs, several mechanical or chemical methods can be applied for its removal. Mechanical cleaning involves the application of either mud and water or the use of a "Rotating Control Head Preventer," as it was successfully applied to remove calcite scaling from a Turkish geothermal well (Yenice and Dünya, 2007). Other mechanical methods as applied in well water regeneration that might also work for geothermal wells, are the removal of scaling by brushing, pumping out, blasting (with dynamite) or exposure to ultrasound (Houben and Treskatis, 2003). In particular, carbonates and low crystalline iron oxides can be dissolved by weak acids. The dissolution of iron and manganese precipitates can be achieved by adding complexing agents such as organic ligands which bind with the metals to form aqueous complexes. Certain precipitates such as amorphous SiO_2 are extremely poorly soluble and removal from the pipes of an EGS system would probably be possible only by shutdown of the plant and cleaning by reaction with hydrofluoric acid (very aggressive).

More complex methods of scale control have been improved in recent years, with technologies such as the Crystallizer–Reactor–Clarifier (which turns the superheated fluid into steam while removing solids from it) or pH modifications which are now successfully used at geothermal facilities.

Monitoring some relevant chemical parameters of the fluid and calculating the mineral precipitation with adequate chemical simulation codes at given pT conditions can help to estimate the potential risk of mineral precipitation. Attempts to predict mineral scale formation in geothermal systems by modeling have been made for example, by Garcia *et al.* (2005) and Garcia *et al* (2006) who used the UNIQUAC model to predict sulfate and carbonate scaling or by Duan *et al.* (1996) to predict scaling with GEOFLUIDS and TEQUIL. However, these codes are usually strongly simplified due to the high complexity of these systems (usually extremely high ionic strength, very high pT conditions, and involving several phases) and consequently lack adequate equilibrium constants. For simple, one-dimensional transport modeling of chemical equilibrium in multiphase geothermal systems, the code PhreeqC can be used with the llnl database (Parkurst and Apello, 1999). The code TOUGHREACT (Xu and Pruess, 1998, 2001) also enables modeling of reactive chemical transport in porous media such as the interaction of geothermal fluids with primary minerals by considering fluid flow, mass transport, and chemical reactions in systems with large temperature variations as shown by Kiryukhin *et al.* (2004).

6.2.1.3 **Fluid Production Technology**

Under typical geological conditions, the pressure in EGS reservoirs will not overcome the hydrostatic pressure of the water head which builds up in the well. In this case, fluid production technology to lift the geothermal fluid from the reservoir is needed. In designing EGS plants, the reliability of the fluid production technology and the auxiliary power to run the equipment are a crucial aspect. The most important aspects regarding the design of downhole pumps are briefly addressed below. More information on downhole pumps are given, for example, in Economides and Ungemach (1987), Culver and Rafferty (1998), Legarth (2003) and Aksoy (2007). Flashman and Lekic (1999) give a brief overview of other technical fluid production options.

In EGS plants, the use of downhole pumps installed below the fluid level in the production well are most common. Downhole pumps are distinguished depending on their mode of drive in line shaft pumps or electrical submersible pumps. Line shaft pumps are driven by a shaft and an electrical motor located aboveground. This limits their application to vertical wells and installation depths up to 600 m (Aksoy, 2007). In contrast, electrical submersible pumps are driven by an electrical motor located in the production well and have therefore, a wider range of application (Figure 6.14).

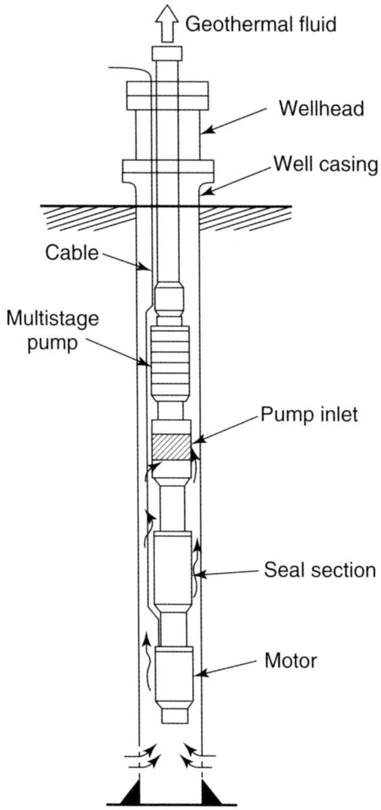

Figure 6.14 Scheme of an electrical submersible pump.

The physical life of geothermal downhole pumps is typically limited to several years. Depending on the site and the operating conditions in the production well, the physical life can however be affected. It has to be considered that the replacement of a downhole pump can take up significant amount of time. One main aspect of design and operation of downhole pumps must therefore be to maximize the physical life of the pumps under the given circumstances. As for the design of other components in the geothermal fluid loop, aspects of corrosion, scaling, and thermal expansion must be considered for design and operation of downhole pumps. A further limiting condition for the physical life of such pumps is the high temperature of the geothermal fluid. This is a problem especially for electrical submersible pumps. Since the cooling of the motor must be realized by the streaming hot geothermal fluid, overheating of the motor is an issue. Another aspect, which can limit the physical life of downhole and especially electrical submersible pumps, is the aggressiveness of the produced fluid which can be a problem, for example, for seals. Unintentional degassing of the geothermal fluid in the pump due to potential local pressure drops results in material stresses and may lead to material failure.

Other production technologies do exist but presently have less relevance for EGS. Gas lift based on inert gases, for example, is not considered a solution for continuous fluid production due to high costs. The turbopump technology, which consists of a turbine-driven submersible pump supplied by high-pressured water from the surface, has only been installed at one site so far due to its low efficiency (Boissavy and Dubief, 1995).

Depending on the reservoir pressure, the production well is filled with geothermal fluid up to a certain level which is called the *static fluid level* and is usually measured in terms of distance from the surface. If geothermal fluid is produced from the reservoir, the fluid level in the production well lowers. The fluid level during production mainly depends on the produced flow rate and the reservoir characteristics and is called *dynamic fluid level*. Compared to the static fluid level, the dynamic fluid level is lower and decreases with increasing flow rate. With reference to the surface, the dynamic fluid level h_{DFL} can be calculated as follows:

$$h_{\text{DFL}} = h_{\text{SFL}} + \frac{\dot{V}_{\text{geo}}}{\text{PI} \cdot \rho_{\text{geo}} \cdot g} \tag{6.22}$$

with the static fluid level in the production well h_{SFL}, the volume flow of the geothermal fluid $\dot{V}_{\text{geo}}$, the productivity index of the reservoir PI, the density of the geothermal fluid ρ_{geo}, and the gravity constant g.

Accordingly, the installation depth of the downhole pump must be adapted to the design flow rate and the fluid level drawdown related therewith. Furthermore, it has to be considered that downhole pumps need a minimum intake pressure (approximately 10–20 bar); so, the pump must be installed at a sufficient depth with reference to the lowest dynamic fluid level which occurs during operation in order to ensure this intake pressure with the overlying water head (Figure 6.15). With state-of-the-art electric submersible pumps, installation depths of up to 3000 m can be realized (Legarth, 2003).

The effort P_{prod} to produce the geothermal fluid from the reservoir is derived from the flow rate, the pressure increase applied by the downhole pump Δp_{prod}, and the efficiency of the pump η_{p} (Equation 6.23). Downhole pumps typically have an efficiency of approximately 50–75% depending on the temperature conditions and the contents of dissolved gases in the produced fluid. The pressure increase of the downhole pump must be regulated according to the pressure to overcome the difference in height between the dynamic fluid level in the production well and the surface, and the necessary wellhead pressure p_{wh}. The latter is set in order to avoid degassing and to ensure the needed injection wellhead pressure (in cases where there is no injection pump). The hydraulic conditions in the production well and the pipes of the geothermal fluid loop and therefore pressure losses due to friction Δp_{loss} must also be considered.

$$P_{\text{prod}} = \dot{V}_{\text{geo}} \cdot \Delta p_{\text{prod}} \cdot \frac{1}{\eta_{\text{p}}} = \dot{V}_{\text{geo}}(-\rho_{\text{geo}} \cdot g \cdot h_{\text{DFL}} + \Delta p_{\text{loss}} + p_{\text{wh}}) \cdot \frac{1}{\eta_{\text{p}}} \tag{6.23}$$

While designing the fluid production, not only the correlations shown in Equation (6.23), but also technical restrictions such as a maximum installation depth must

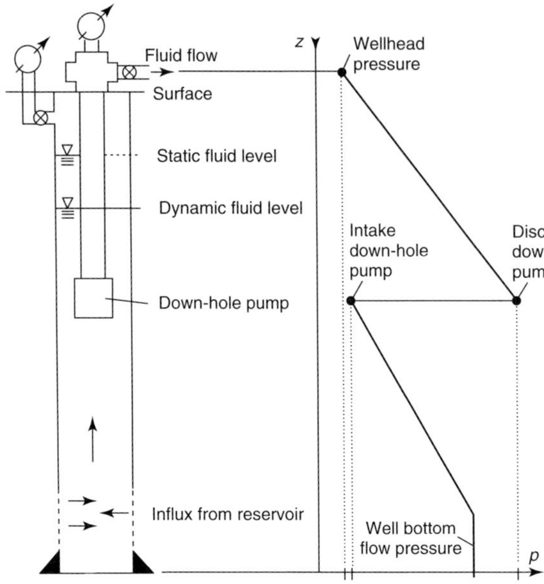

Figure 6.15 Scheme of a production well showing down-hole pump, static fluid level and dynamic fluid level (left) and the pressure curve during fluid production (right).

be considered. According to the present state-of-the-art downhole pumps, the maximum producible fluid flow is approximately 600 $m^3\ h^{-1}$ and the maximum motor capacity is about 1200 kW (Legarth, 2003).

From an energetic point of view, another very important aspect in the design of fluid production is the increase of the production effort with increasing flow rate. The pumping of a higher flow rate results in a larger drawdown of the fluid level and a larger difference in height for which the downhole pump has to compensate. The production effort therefore shows a quadratic dependence on the flow rate (Figure 6.16). The influence of an increasing flow rate on the fluid production effort is thereby larger for lower reservoir productivities.

At EGS sites, where the pressure at the bottom of the injection well resulting from the water head in the injection well and the wellhead pressure is not sufficient to overcome the reservoir pressure, an injection pump must be installed on the surface before the injection well. This is the case when the injectivity of a site is insufficient. The effort for the fluid injection is then calculated from geothermal flow rate, the efficiency of the injection pump, and the necessary pressure increase in order to realize a sufficient pressure at the bottom of the injection well.

6.2.2
Heat Exchanger

Heat exchangers used in EGS plants are essential components for producing power as well as providing heat and/or chill. The general purpose of heat exchangers

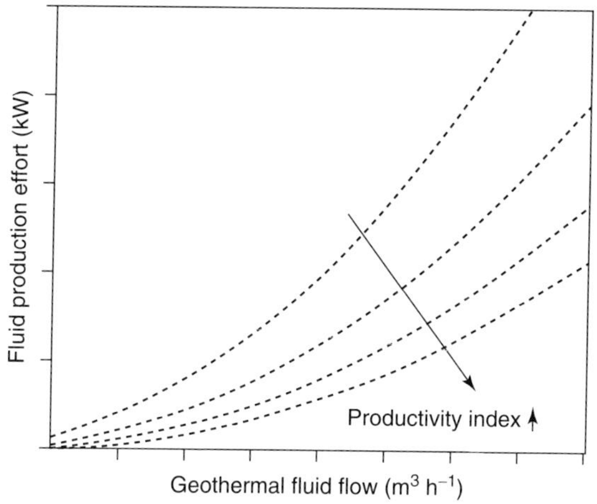

Figure 6.16 Qualitative development of fluid production effort as a function of geothermal fluid flow for different reservoir productivities.

is to transfer heat from one fluid to another, in the most reliable way for any given application. In geothermal applications, various types of heat exchangers are needed. For direct use they basically serve as liquid/liquid heat exchangers, for example, transferring the heat from the geothermal fluid to a district heating system. In binary power plants, they are used as preheaters, evaporators, superheaters, condensers, and recuperators and the fluids which are present may be liquids or mixtures of gases and liquids or even mixtures of liquids and solids. Because of the different boundary conditions in geothermal applications, no off-the-shelf heat exchangers will meet all requirements and almost every heat exchanger has to be designed separately.

6.2.2.1 Heat Exchanger Analysis – General Considerations

Analysis and design of heat exchangers are quite comprehensive and only a general discussion can be given here. Further information is given, for example, in Incropera and Dewith (2007), Bejan and Kraus (2003), Smith (2006), and Cengel (2007).

When considering a heat exchanger with two fluid streams, the rate of heat transferred from one fluid to the other is expressed as follows for each stream:

$$\text{Cold fluid: } \dot{Q}_c = \dot{m}_c(h_{c,out} - h_{c,in})$$
$$\text{Hot fluid: } \dot{Q}_h = \dot{m}_h(h_{h,in} - h_{h,out})$$

with $\dot{m}_c$, $\dot{m}_h$ indicating mass flow rates, h_c, h_h the specific enthalpy, and T_c, T_h the temperatures of the hot and cold streams, respectively.

If no heat losses occur, the energy balance of the heat exchanger can be written as

$$\dot{Q} = \dot{m}_c(h_{c,out} - h_{c,in}) = \dot{m}_h(h_{h,in} - h_{h,out}) \quad (6.24)$$

Considering a heat exchanger with two fluid streams and no phase change during heat transfer (i.e., evaporation or condensation), the rate of heat transferred from the hot to the cold fluid is expressed according to Equations (6.5) and (6.6) as follows:

$$\dot{Q} = \dot{m}_c \cdot c_{p,c} \cdot (T_{c,out} - T_{c,in}) = \dot{m}_h \cdot c_{p,h} \cdot (T_{h,in} - T_{h,out}) \tag{6.25}$$

This equation does not include any indication of the heat exchanges size needed to transfer the prescribed rate of heat which is an important design aspect.

The heat exchanger size can be derived from the following equation:

$$\dot{Q} = U \cdot A \cdot \Delta T_m \tag{6.26}$$

where U indicates the overall heat transfer coefficient, A is the surface area, and ΔT_m denotes the driving or mean temperature difference. Assuming a counter or parallel flow heat exchanger, the driving temperature difference is represented by the log mean temperature difference LMTD as

$$\Delta T_m = \text{LMTD} = \frac{\Delta T_1 - \Delta T_2}{\ln\left(\frac{\Delta T_1}{\Delta T_2}\right)}. \tag{6.27}$$

The temperature difference ΔT_1 and ΔT_2 refer to the temperature difference at the cold end and the hot end of a specific heat exchanger. The temperature profiles for different heat exchanger modes are shown in Figure 6.17.

For multipass or cross-flow heat exchangers, as in most applications, the driving temperature difference results from the LMTD complemented by a correction factor F. The correction factor takes into consideration the real flow arrangement and can be determined according to Bejan and Kraus (2003) and VDI-Gesellschaft Verfahrenstechnik und Chemieingenieurwesen (GVC) (2006), for example

$$\Delta T_m = F \cdot \text{LMTD} \quad 0 \leq F \leq 1. \tag{6.28}$$

The LMTD method is applicable when all temperatures at the inlet and outlet are known or can be determined from the energy balance. If the overall heat transfer coefficient and the mass flow rates are known, the required surface area can be calculated.

In case the outlet temperatures of a heat exchanger are not known and must be calculated from the mass flow rates, inlet temperatures, and heat transfer rate for a specific heat exchanger design, the LMTD method is less useful since recursive

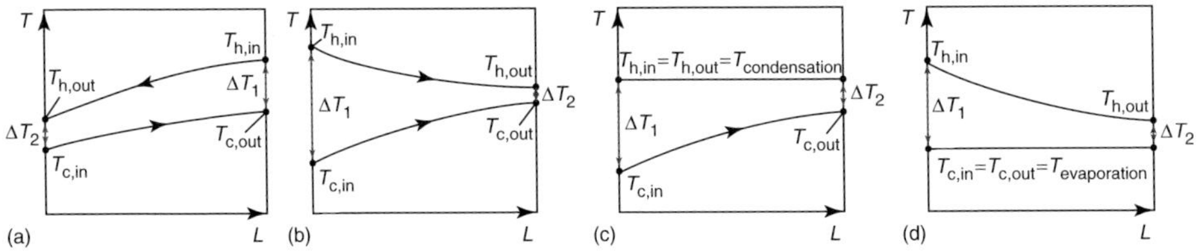

Figure 6.17 Temperature profiles along the length L of heat exchangers with (a) counterflow, (b) parallel flow, (c) condensation, and (d) evaporation.

estimation methods have to be applied. Here, other methods are recommended such as the effectiveness-NTU (number of heat transfer units) method (Incropera and Dewith, 2007; Bejan and Kraus, 2003).

The calculation of the overall heat transfer coefficient depends on the fluid flow conditions, shape of heat transfer area (pipe, wall), thermophysical properties of fluids and materials, and the heat transfer regime such as boiling, condensation, or heat transfer without phase change. In general, the following equation can be applied considering a plane wall as heat transfer area:

$$U = \left(\frac{1}{h_1} + \frac{\delta}{\lambda} + \frac{1}{h_2}\right)^{-1} \tag{6.29}$$

where h_1 and h_2 are the heat transfer coefficients between medium and heat transfer surface at each medium side measured in Watts per square meter per Kelvin, δ is the thickness of the plane wall, and λ is the thermal conductivity of the wall material given in Watts per meter per Kelvin. For other heat transfer areas such as pipes, the equation has to be adapted. In order to estimate the heat transfer coefficients for a specific application, several approaches can be found in, for example, Bejan and Kraus (2003) and VDI-Gesellschaft Verfahrenstechnik und Chemieingenieurwesen (GVC) (2006).

6.2.2.2 **Selection of Heat Exchangers**

The selection and the design of heat exchangers depend on several factors, described in the following paragraph.

The *heat transfer rate* of a heat exchanger, which is the heat transferred per unit time, is based on the preliminary planning of the geothermal plant. With reference to the binary power plant, the necessary heat transfer rate of a specific heat exchanger is defined by the binary cycle design calculation. When focusing on the heat exchanger design, the heat transfer rate is determined by the overall heat transfer coefficient and the driving temperature difference between the fluid streams. Both factors determine the required heat transfer surface area.

The *materials* used in the construction of a heat exchanger for geothermal applications can be an important issue. In particular, the chemical properties of the geothermal fluid as well as the fluid on the secondary side of the heat exchanger, such as the working fluid in a binary cycle, require corrosion-resistant materials such as stainless steel and titanium, or coatings.

Due to the differing requirements of heat transfer applications, many *types of heat exchangers* are available and therefore, a wide range of design approaches have been developed.

A general distinction is made between the flow arrangements. In *parallel flow*, both fluids enter the heat exchanger at the same end and flow in the same direction (Figure 6.17a). In *counterflow*, both fluids enter the heat exchanger at opposite ends and flow in opposite directions (Figure 6.17b). The third flow arrangement is called *cross-flow*. Here the fluids move in a perpendicular direction through the heat exchanger. Because of the different flow arrangements, different temperature profiles occur within the heat exchanger. The flow arrangements mentioned rarely

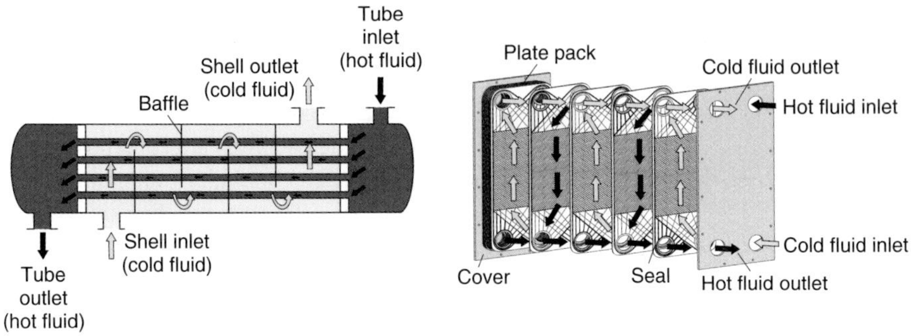

Figure 6.18 Simple illustration of (a) a shell-and-tube heat exchanger and (b) a plate heat exchanger.

appear in their pure form in real applications. In most cases, a combination of these arrangements occurs.

With reference to construction-related classifications, shell-and-tube heat exchangers and plate heat exchangers such as those shown in Figure 6.18 are commonly used in geothermal applications. Plate heat exchangers are characterized by a high surface-to-volume ratio which leads to a compact construction form. Once installed, they can easily be expanded when more heat capacity or heat exchange surface is needed. Plate heat exchangers are commonly used as preheaters (Culver and Rafferty, 1998). Their use as evaporators or condensers is also possible as is the case for chilling technologies. However, the design of plate heat exchangers is based on more detailed calculations as compared to the design of shell-and-tube heat exchangers (Zhua and Zhangb, 2004). In comparison, shell-and-tube heat exchangers have a larger surface-to-volume ratio and larger realizable temperature gradients. Furthermore, fouling can be a larger issue than for plate type heat exchangers (Chandrasekharam and Bundschuh, 2002). The design of such heat exchangers is based on a wide range of theoretical and experimental experience and is therefore very reliable (Costa Andre and Queiroz Eduardo, 2008). Shell-and-tube heat exchangers can be used for almost all geothermal applications. However, they are predominantly applied as evaporators and condensers.

Heat exchangers are designed for special applications and a specific *operational mode* depending on inlet and outlet temperatures and mass flow rates. When operating a geothermal plant, particularly a binary power plant, different operational modes occur due to changing parameters and boundary conditions. For example, the ambient temperature alternates seasonally and can affect the condensing temperature in the binary cycle. When considering a combined heat and power provision, the temperatures and the heat capacity of the district heating system will affect the operating conditions. For that reason, the heat exchanger design has to be focused on full load as well as part load conditions. The part load behavior becomes a crucial factor when operation off the design point is predominant.

Heat exchanger design must consider *fouling*, that is, precipitation of minerals such as silica and deposits which accumulate on the heat transfer surface. In geothermal plants, this is a very important issue due to the changes of chemical equilibrium of geothermal fluids which result from the transport and cooling of the fluid, and the pressure and temperature changes related therewith (Section 6.2.1.2). The layer of deposits represents an additional heat transfer resistance between the hot and the cold fluid. This effect can be considered as a fouling factor which has to be incorporated in the calculation of the overall heat transfer coefficient. Different fouling mechanisms have been identified: precipitation fouling, corrosion fouling, chemical reaction fouling, biological fouling, particulate fouling, and freezing fouling (Bejan and Kraus, 2003). Other than the latter, all fouling mechanisms are relevant in geothermal applications. Precipitation and corrosion fouling are the most important mechanisms. The determination of the fouling factor is somewhat indefinite and is based on experiences with different geothermal fluids. In addition, the definition of the fouling factor suggests steady-state conditions, whereas fouling occurs in a time-dependent manner.

6.2.2.3 Specific Issues Related to Geothermal Energy

Designing heat exchangers for geothermal applications requires accurate knowledge of *thermophysical fluid properties* such as specific heat capacity, viscosity, and density. With reference to geothermal fluids, the manifold chemistry may hinder the calculation of properties with the required accuracy. Methods for calculating density, heat conductivity, and viscosity depending on temperature, pressure, and salinity are given for example, in McDermott *et al.* (2006). Methods for calculating the specific heat capacity are given for example, in Thomsen (2005) and Millero, F. (2009). In case methods suitable for calculating these properties are not available, the data need to be measured.

An additional issue is the need for low *temperature differences* between the geothermal fluid and the fluid on the secondary side, particularly the working fluid in binary cycles. The small temperature difference is important for decreasing the exergy destruction during heat transfer. It must be considered however, that low temperature differences lead to an increasing heat transfer surface.

Focusing on evaporation, the boiling heat transfer coefficient h_b depends, apart from thermophysical fluid properties, on the temperature difference. Here the difference between the heat transfer surface temperature and the saturation temperature of the fluid is important. This temperature difference is called the *excess temperature* ΔT. The general dependency of the boiling heat flux q on the excess temperature is shown in Figure 6.19. Because of the necessarily low temperature differences in geothermal binary cycles the excess temperature while evaporation is low as well. This leads to a combination of nucleate and natural convection boiling regimes (Figure 6.19) with low heat transfer coefficients. The nucleate boiling heat transfer coefficient depends on the temperature difference as well (Cengel, 2007) and decreases with decreasing temperature differences. As a result, the necessary heat transfer area increases with decreasing temperature differences in an order

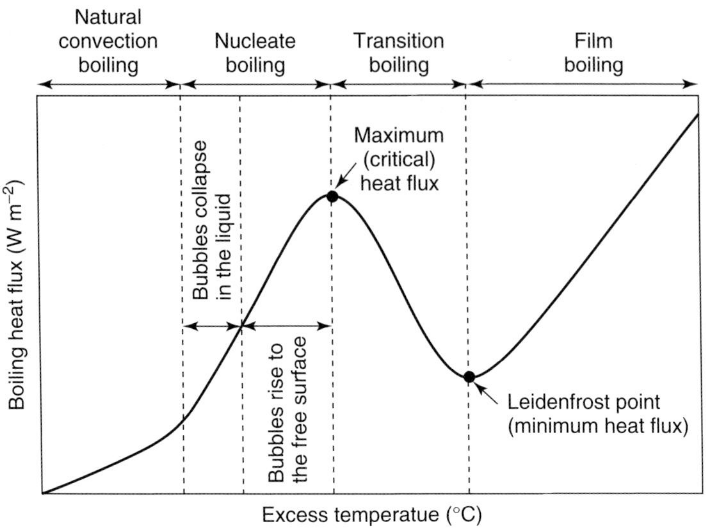

Figure 6.19 Principle boiling curve showing the different boiling regimes for pool boiling.

higher than 1 (equation 6.30).

$$A = f(\Delta T^m) \text{ with } m > 1 \tag{6.30}$$

In contrast, the necessary heat transfer area without phase change is linearly dependent on the temperature difference.

In order to reduce the additional heat transfer area caused by low temperature differences, the boiling can be enhanced by means of specially formed microsurfaces on the outside or inside of tubes (Bejan and Kraus, 2003).

6.2.3 Direct Heat Use

The direct use of the heat from an EGS reservoir is usually realized by transferring the geothermal heat to a water-operated heating network or a heat carrier used in process heat applications. In contrast to electrical power grids, which are widely developed networks in many countries due to the good physical transport properties of electricity, heating networks refer to local supply structures. Typical district heating networks are fed by a small number of heating stations, which are preferably located close to the heat customers in order to avoid large energy losses during transportation. The design of EGS plants for direct heat use therefore does in most cases, not only depend on the site-specific reservoir conditions but also on the structure of the heat customers within a heating network. The characteristics of heat consumption and their influence on EGS plant design are outlined in the following section.

A schematic setup of an EGS heat plant feeding a district heating system is shown in Figure 6.20. At some sites, heat plants must also include a backup system (for example, in case the EGS plant is the only heat provider of the district heating

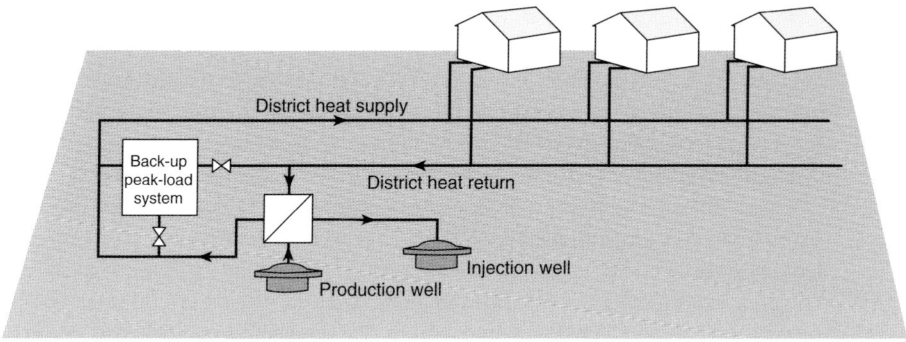

Figure 6.20 Scheme of a well doublet with backup/peak-load system feeding a district heating grid.

system) or a peak-load system (in case the EGS heat plant is designed to cover the base-load demand). The heat demand of a heating network is defined by its supply and return temperature and the demanded heat capacity, which is the sum of the simultaneous demands of the connected heat customers. The temperature level of a single heat customer depends on the use of the heat, such as for space heating, warm water production, or also industrial use such as that given in Lindal (1973).

Existing geothermal heat plants typically feed district heating networks which are mainly used for space heating. Since space heating represents large part of the worldwide heat demand, the supply of district heat will also be typical for EGS heat plants. The supply of district heat is characterized by supply temperatures between 50–90 °C and return temperatures between 30–70 °C. In contrast to an annually constant heat demand for many industrial processes, the demand for space heating is characterized by a variable heat demand during the year. Figure 6.21 shows the annual variations of the space heat demand according to a varying ambient temperature.

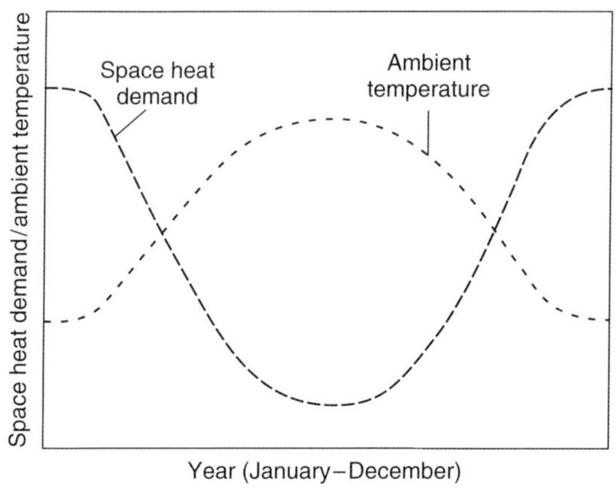

Figure 6.21 Annual space heat demand depending on the ambient temperature.

With reference to such heat demand characteristics, it must be considered that a geothermal heat source is not optimally utilized because with a variable heat demand, it can only partly be used over a year. If the return temperature of a district heating system is above the temperature at which the geothermal heat can be used, the heat source is also not utilized to its full extent. Based on these considerations, different design aspects of EGS plants are discussed in the following section.

With a given heat demand, EGS plant design including borehole and reservoir stimulation concept, dimensioning of the heat exchanger and other surface installations, aims to fully cover the demand by geothermal energy. In case the given heat demand varies during the year, the energetic potential of the EGS reservoir is only used to a small part. As shown in Figure 6.22a, the geothermal full load hours of such a plant are low. Another possibility is to design the geothermally driven part of the heat plant for a part of the heat supply which can, for example, be the base-load heat demand (Figure 6.22b). For the hours of the year where this base-load demand is exceeded, a peak-load system such as another renewable heat technology gadget or a conventional boiler is installed. This way, the geothermal full-load hours can be increased and the use of the EGS resource can be improved significantly.

Another approach to EGS plant design is to adapt the heat demand side to the continuous characteristics of the EGS resource. In the summer months, where the demand for space heat is lower, the geothermal heat can, for example, be used for the supply of chill. This can be done directly by supplying chill (which requires a cooling grid) or connecting consumers who use the heat for decentralized chill provision. Presently, a more common way is to use the geothermal heat available off the heating period, to produce power (Figure 6.23).

Besides the "design" of the heat consumer structure in terms of a more continuous heat demand throughout the year, the cooling of the geothermal fluid is also an important aspect. By cascading heat consumers with different temperature levels, the utilization of the geothermal resource can be optimized.

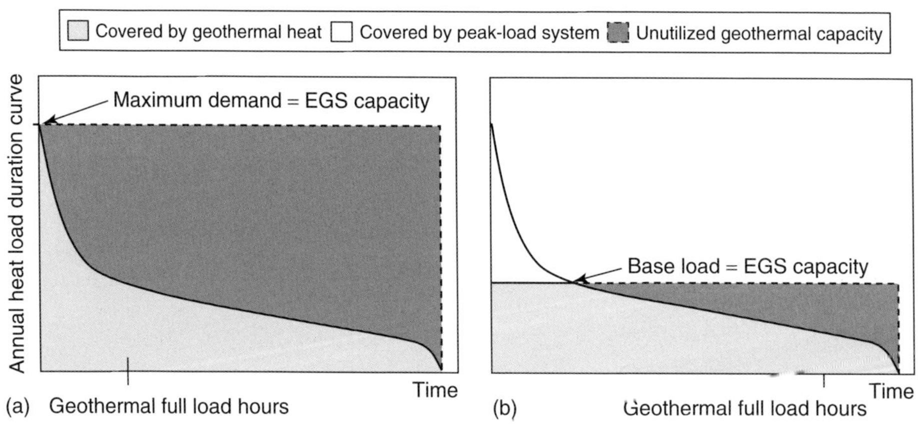

Figure 6.22 Annual heat load duration curves and corresponding EGS plant design for heat plants with (a) low and (b) high geothermal full load hours.

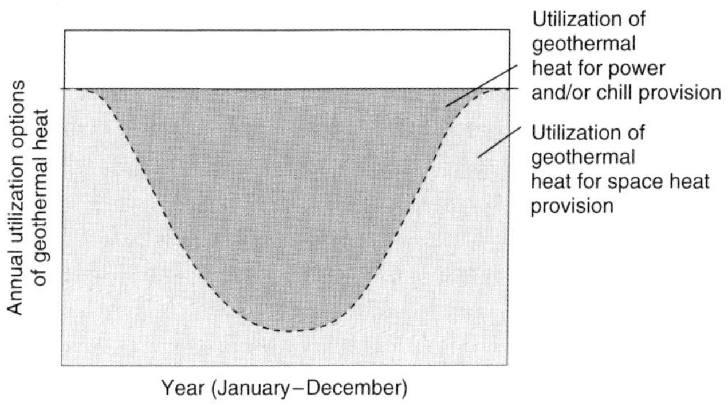

Figure 6.23 Seasonal utilization options for continuous geothermal heat use.

Design aspects which are related to a combined supply of different energy demands will be addressed in Section 6.2.5.

6.2.4 Binary Power Conversion

For the conversion of heat from typical EGS reservoirs (as defined at the beginning of this chapter) into power, the heat is transferred to another working fluid contained in a so-called binary power unit. As shown in the schematic setup of a geothermally driven binary cycle in Figure 6.10, p. 312. This working fluid is preheated and evaporated. The generated vapor is expanded and the released enthalpy is used by an expansion machine. In most applications, turbines are used as expansion machines due to their wide application range and widespread availability. For small power ratings, screw or piston expanders are also available. The expansion machine is connected to a generator in order to convert the mechanical work into electricity. The exhaust vapor from the turbine is condensed. The cooled working fluid is pressurized in the feed pump and recycled to the preheater. The power which is produced in a binary conversion cycle depends on the heat source and, on the binary side, on component-specific parameters (such as efficiency of the turbine, generator, and feed pump; temperature differences in heat exchangers), and binary cycle design.

Existing binary conversion cycles generally use different working fluids and cycle layouts. With reference to geothermal resources, the Organic Rankine Cycle (ORC), which uses an organic working fluid, is the most common type. At the moment more than 150 binary units with an average capacity between 1 and 3 MW are installed world wide (DiPippo, 2008). Another frequently-discussed but so far rarely installed type is the Kalina cycle. One Kalina plant has been installed in Húsavik, Iceland (Maack and Valdimarsson, 2002). A second plant is operated in Unterhaching, Germany (Knapek and Kittl, 2007). The Kalina cycle is characterized by an ammonia–water mixture and an absorption and desorption process within the cycle. However, Rankine Cycles with inorganic working fluids such as pure ammonia or other mixtures are also possible. A detailed overview of existing geothermal binary power plants is given in DiPippo (2008).

Geothermally driven binary cycles are designed according to the temperature level and the characteristics which are encountered at a specific site. Site-specific cycle design has to aim at a high utilization of the geothermal heat with an efficient and reliable power conversion technology.

In the following, the most important design approaches regarding general cycle design and working fluid, their influence on utilization ratio and conversion efficiency, and some other aspects will be addressed. Binary power units using heat sources with temperatures up to approximately 200 °C are discussed in more detail, for example, in Köhler (2005), Saleh *et al.* (2007), and DiPippo (2008). Attention is also paid to the design of the recooling of a binary cycle because power conversion from EGS has to deal with large amounts of waste heat and engineering of such cooling systems must consider site-specific preconditions.

6.2.4.1 General Cycle Design

The decisive aspect for power generation using expansion machines, is the conversion of working fluid enthalpy into mechanical (and simultaneously into electrical) energy. The electrical power obtained from a binary cycle P_{el} can therefore be calculated as follows:

$$P_{el} = \dot{m}_{WF} \cdot (h_{in} - h_{out}) \cdot \eta_T \cdot \eta_G \tag{6.31}$$

from the specific enthalpy difference over the expansion machine (h_{in} - h_{out}), the working fluid mass flow rate $\dot{m}_{WF}$, and the mechanical turbine and generator efficiency η_T and η_G.

The usable enthalpy difference in the expansion machine is determined by the design of the conversion cycle. In conventional thermal conversion power plants, the energy input can be controlled and therefore, can be designed to obtain a defined electrical power output. Geothermal binary cycles, in contrast, must be designed according to the given characteristics of an existing heat source. The design parameters of a basic binary conversion cycle will be addressed in the following section. Some design aspects of other cycle setups are also briefly outlined.

Basic Design Parameters A basic binary conversion cycle with saturated vapor is shown, for example, in the temperature–entropy diagram of Figure 6.9b, p. 312. In this case, the working fluid is preheated and evaporated. The heat input ends at the right side of the dew-point curve where the working fluid becomes saturated vapor. The design parameters to define such a cycle are the working fluid, evaporation temperature, and condensation temperature. With these parameters, the mass flow rate of the working fluid can be derived from the energy balance around the heat input to the conversion cycle.

In order to realize a maximized absolute enthalpy difference and to optimize the electric power obtained from a geothermally driven binary cycle, the design parameters must be chosen in such a way that a high utilization of the geothermal heat and also a high efficiency of its conversion is achieved. The conversion efficiency of a binary cycle depends on the thermodynamic average temperatures of heat input and removed heat, the temperature differences of the heat exchangers, the working fluid, and component efficiencies. Based on these considerations, the electrical power obtained from a geothermally driven binary cycle can be written as

$$P_{el} = \dot{m}_h \cdot c_{p,h} \cdot (T_h - T_{h,0}) \cdot \eta_U \cdot \eta_{bin} \tag{6.32}$$

with the hot fluid representing the geothermal fluid, the utilization ratio η_U calculated according to Equation (6.4), and η_{bin} representing the conversion efficiency of the binary cycle.

The choice of the working fluid enables the adaptation of the binary cycle to the heat source since different shapes of the dew-point curve and different evaporation characteristics can be realized with different media and mixtures. The choice of a suitable working fluid is a crucial step in designing binary plants since more than just thermodynamic properties need to be considered. Section 6.2.4.2 will therefore deal in detail with different design aspects regarding working fluids.

The choice of the evaporation temperature is decisive for the average temperature of heat input. The cycle efficiency will therefore increase with higher evaporation temperatures, but the evaporation temperature is limited by the temperature and finiteness of the geothermal source and the quality of the heat exchanger (realizable pinch point). The evaporation temperature also influences the cooling of the geothermal fluid. The utilization of the geothermal fluid thereby typically decreases with higher evaporation temperatures as shown in Figure 6.24a. For the maximum power output, the optimum evaporation temperature has to be found (Figure 6.24b). The development of cycle efficiency and utilization ratio against the evaporation temperature thereby depends on the characteristics of the working fluid.

The choice of the condensation temperature is decisive for the average temperature of heat removal because a lower condensation temperature leads to higher cycle efficiencies. Since the utilization ratio also increases with decreasing condensation temperatures, the maximum power output is reached at the minimum

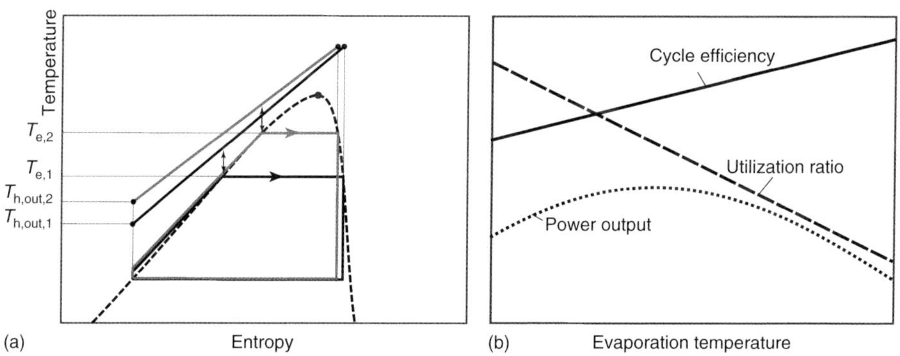

Figure 6.24 *T–s* diagram showing (a) the correlation of evaporation temperature and geothermal outlet temperature and (b) development of power output as a function of the evaporation temperature.

condensation temperature. The minimum condensation temperature depends on the heat sink quality at a specific site, the chosen working fluid, and the realizable temperature difference between cooling medium and working fluid. Even though the produced power is increased with lower condensation temperatures, it must be considered that the demand for recooling is increased. Depending on the cooling system, this can result in a significantly higher auxiliary power demand. Since this is a very important aspect of the provision of net power, the different modes of cooling will be discussed in greater detail in Section 6.2.4.3.

Cycle Setups Apart from the general design parameters discussed above, the performance of a binary cycle can also be influenced with different cycle setups. This section will give an overview of the most common and important layouts.

Binary cycle with superheating In cycles with superheating, the heat input to the working fluid goes beyond the right side of the dew-point curve. The superheated working fluid then has a temperature higher than the evaporation temperature, and a higher average temperature of heat input than a saturated vapor cycle. Depending on the working fluid and the shape of the dew-point curve, superheating of the working fluid increases the cycle efficiency as is known from the design of conventional water vapor cycles. However, the dew-point curve can also be shaped in such a way that the increase of the average temperature of the heat input is compensated for by a concomitant increase in the average temperature of the heat removed. This can be seen in Figure 6.25, which shows the T–s diagram for a wet working fluid (a) and a dry working fluid (b).

Even though superheating of wet working fluids results in higher conversion efficiencies, it has to be remembered that the utilization of the geothermal fluid decreases as discussed above, according to the choice of evaporation temperature. From an engineering point of view, superheating might however be necessary in case the saturated vapor cycle leads to excessive vapor wetness in the turbine. Usually a maximum vapor wetness of 85% is allowed in order to avoid damaging the turbine.

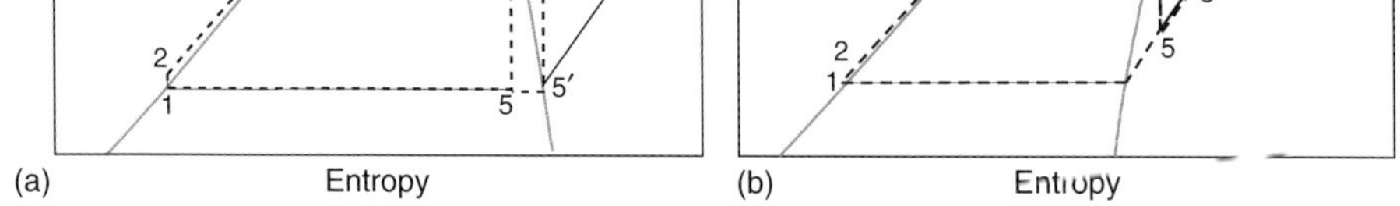

Figure 6.25 T–s diagrams of binary cycles (a) with superheating for a wet working fluid and (b) a dry working fluid.

Supercritical cycles If the evaporation temperature lies above the dew-point curve and the critical temperature of a working fluid, binary cycles are termed to be *supercritical* cycles. As discussed for a basic binary cycle, such an increase in evaporation temperature leads not only to an increase of conversion efficiency, but also to a lowering of the utilization ratio. The advantage of a supercritical cycle, however, is the almost triangular shape of this process and therefore, the adaptation to the cooling-off curve of the geothermal fluid (Figure 6.26). Depending on the working fluid, the increase of cycle efficiency can exceed the decrease of the utilization ratio so that the power output can be improved with such cycle setups. Since the corresponding upper process pressures of supercritical binary cycles are typically low, their design and operation is less problematic from an engineering point of view than in the case of supercritical water vapor cycles. However, supercritical cycles are, so far, only executed at the laboratory scale.

Internal heat recuperation If the exhaust vapor at the outlet of the turbine is superheated, such as is shown in Figure 6.27, the vapor must first be de-superheated

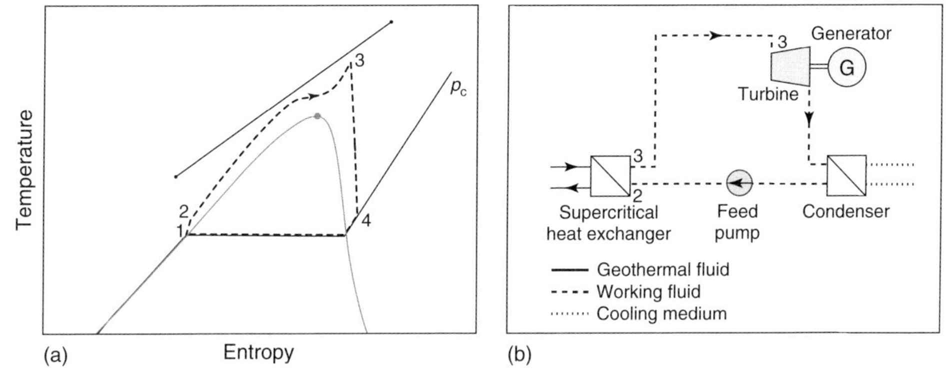

Figure 6.26 (a) T–s diagram and (b) schematic setup of a supercritical binary cycle.

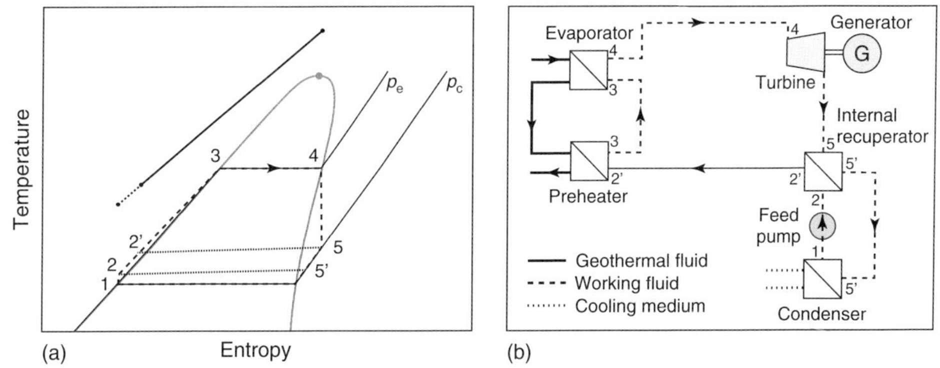

Figure 6.27 (a) T–s diagram and (b) schematic setup of a binary cycle with internal heat recuperation.

before it can be condensed, that is, heat removal should occur until the dew-point curve is reached. Part of this heat from de-superheating (Figure 6.27a; 5→5′) can be used internally in the binary cycle for the preheating of the working fluid (2→2′). In case of wet working fluids, superheated exhaust vapor occurs only when the working fluid is superheated during the heat input. Dry working fluids, in contrast, offer this opportunity of internal heat recuperation also in a saturated vapor cycle. In the example in Figure 6.27, it can be seen that the internal use of the exhaust vapor substitutes a part of the geothermal heat which is used for preheating the working fluid. This means that no additional power is produced. However, internal heat recuperation can, for example, be advantageous if the cooling of the geothermal fluid is limited due to geochemical restrictions or due to further use of the geothermal heat such as for the additional supply of district heat. Another aspect is that with internal heat recuperation, the demand for recooling can be decreased.

Multistage cycles Another possible approach to a triangular cycle is to design a binary cycle as a multistage cycle in which the evaporation is split into several processes with different temperature levels. Two possibilities are to evaporate one working fluid at different evaporation pressures or to use different working fluids in serial cycles. The most common multistage cycles are dual-pressure cycles (DiPippo, 2008; Drescher, 2008). In such cycles, the working fluid is first preheated to the evaporation temperature of the low pressure stage (Figure 6.28a). One part of the working fluid mass flow is evaporated and drives the turbine. The other part of the working fluid mass flow is put to a higher pressure level and evaporated at this pressure. Dual-pressure cycles can be designed with a dual-admission turbine, or, due to the usually small size of binary cycle turbines and practical considerations, two separate turbines are foreseen (Figure 6.28b).

Kalina cycle The Kalina cycle is a binary cycle in which a water-ammonia mixture is used as a zeotropic working fluid (Section 6.2.4.2). Besides the use of a zeotropic

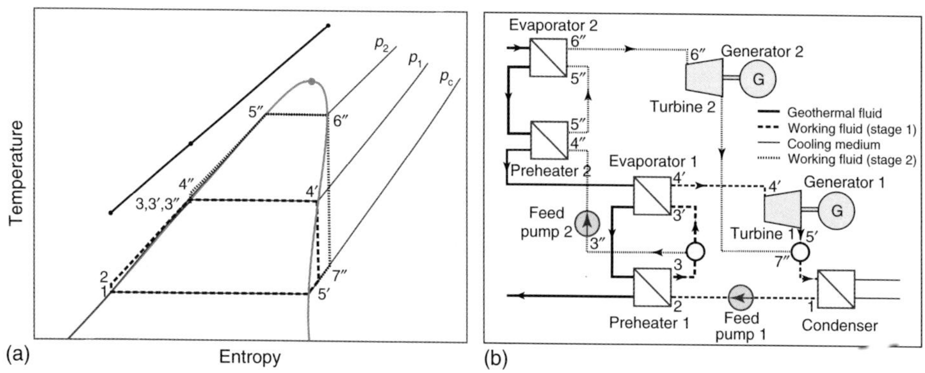

Figure 6.28 (a) T–s diagram and (b) schematic setup of a multistage binary cycle with different evaporation pressures and separate turbines.

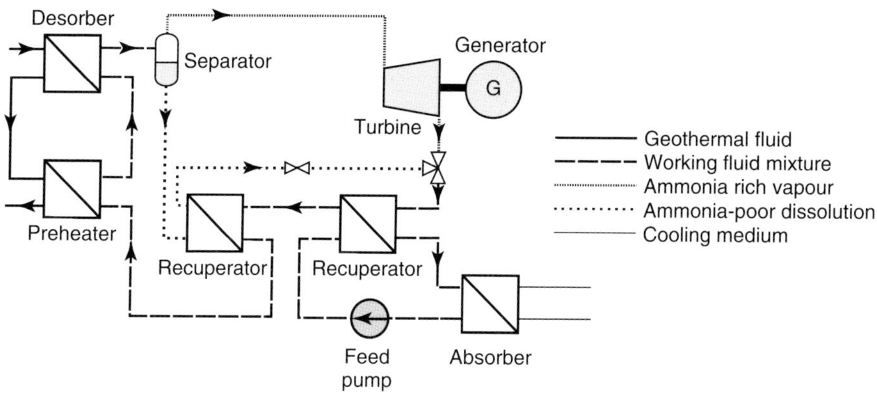

Figure 6.29 Schematic setup of a Kalina cycle according to the KCS34. (After Leibowitz and Mlcak, 1999.)

mixture, the setup of a Kalina cycle is characterized by the use of a desorber and absorber which replace the evaporator and the condenser, respectively. Since the Kalina cycle is the most commonly discussed and realized binary cycle setup using mixed working fluids, it will be briefly described here. For the application of the Kalina cycle different plant setups have been developed. Representative for the use of heat sources between 100 and 200 °C is the cycle shown in Figure 6.29 (Leibowitz and Mlcak, 1999). In the desorber, the heated working fluid is separated in ammonia-rich vapor and ammonia-poor dissolution, which increases the usable enthalpy difference in the turbine compared to the use of a vapor-mix in the turbine. The ammonia-poor fluid is used to preheat the working fluid mixture in an internal heat exchanger. The cooled ammonia-poor fluid and the ammonia-rich exhaust vapor are mixed and the released absorption heat is used for internal preheating of the working fluid mixture. The full absorption is realized in the absorber, where the heat is transferred to a cooling medium.

6.2.4.2 **Working Fluid**

The working fluid should basically ensure a high conversion efficiency and allow a high utilization of the available geothermal heat. The selection of the working fluid for binary cycles is crucial for plant efficiency since it mainly depends on the thermodynamic characteristics. Also important are aspects of material compatibility between working fluid and plant components and therefore, plant reliability. Furthermore, economic aspects such as availability, and environmental considerations such as potential health risk and unintentional emission of greenhouse gases at the plant play a role. The accuracy of thermophysical properties is essential for cycle and component design. Possible working fluids are related to alkane hydrocarbons, alcohols, and heterocyclic compounds. In addition, mixtures of pure fluids may act as working fluids, particularly mixtures that exhibit a nonisothermal evaporation. Consequently, it is possible to approach the temperature profile of the geothermal fluid with a suitable working fluid regarding

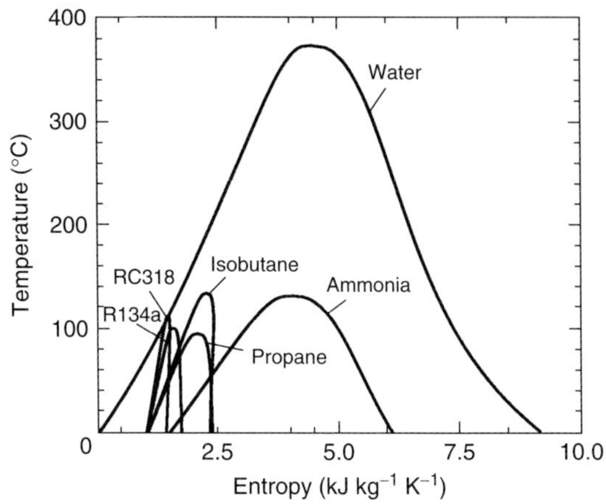

Figure 6.30 Temperature–entropy diagram showing the saturation vapor curves of different working fluids.

the heat transfer. These mixtures are called *zeotrope mixtures*. Ammonia–water or propane–butane mixtures are zeotrope fluids.

Thermodynamic Aspects In general, working fluids suitable for binary cycles offer a lower boiling point than water as basically used in fossil power plants. The lower boiling point is needed to generate vapor from heat sources with temperatures below 200 °C.

A general characterization of different working fluids can be expressed according to the saturation vapor curve in the temperature–entropy diagram (Figure 6.30). A "wet" fluid shows a negative slope of the saturation vapor curve. When expanding saturated vapor in an expansion machine, the state obtained is always located in the two-phase region. Water and ammonia, for example, are typical "wet" fluids. A "dry" fluid, such as isobutene, shows a positive slope of the saturated vapor curve. After expanding saturated vapor from a dry fluid, its state is always located in the superheated region. If the saturated vapor curve does not show any slope, the fluid is called isentropic.

In selecting a suitable working fluid, the temperature of the heat source is decisive. A rule of thumb is given in Köhler (2005) and Invernizzi and Bombarda (1997) where the critical temperature of the fluid should be close to, or slightly lower than the temperature of the geothermal fluid. The ratio of the liquid working fluid's specific heat capacity to the enthalpy of vaporization should be as large as possible. Therefore, the temperature profile of the working fluid during heat absorption comes closer to the triangle shape so that exergy destruction can be reduced. Furthermore, it is recommended that a working fluid is chosen with an almost isentropic saturated vapor curve in order to avoid wet turbine exhaust vapor or to limit the exhaust vapor wetness which can cause damage in the turbine. When using a "dry" working fluid, the specific heat capacity of the superheated vapor should be low to limit the heat removal during de-superheating.

A wide range of possible working fluids are investigated in Saleh *et al.* (2007) with reference to the achievable cycle efficiency and considering a minimum and maximum cycle temperature of 100°C and 300 °C, respectively. Saleh *et al.* thereby consider subcritical as well as supercritical cycle designs.

Mixtures as Working Fluid Mixtures as working fluids can lead to a better match of the temperature profile between the geothermal fluid and the working fluid within the evaporator and therefore, to a higher cycle efficiency. In general, mixtures can be distinguished according to their vapor–liquid equilibrium behavior. When evaporating a zeotropic mixture its temperature increases during isobaric evaporation for any proportion of the constituents (nonisothermal evaporation). During evaporation, the composition of the vapor and the liquid phase changes continuously. The temperature slide allows for better temperature matching in the evaporator (Figure 6.31).

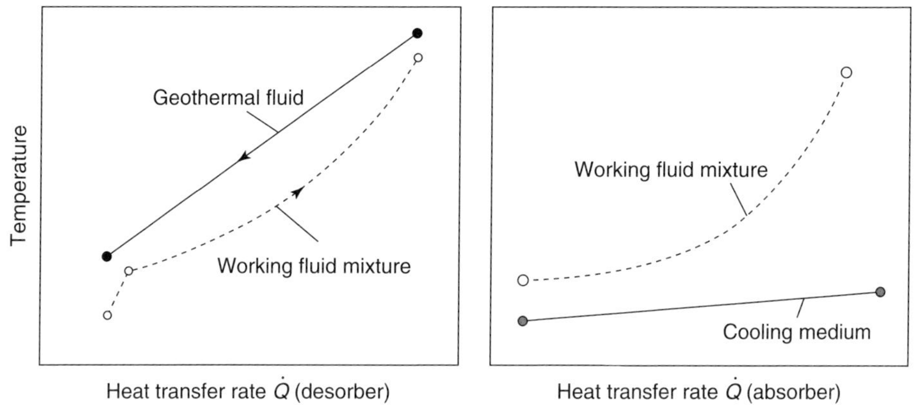

Figure 6.31 T–Q diagram showing the temperature profiles in different heat exchangers.

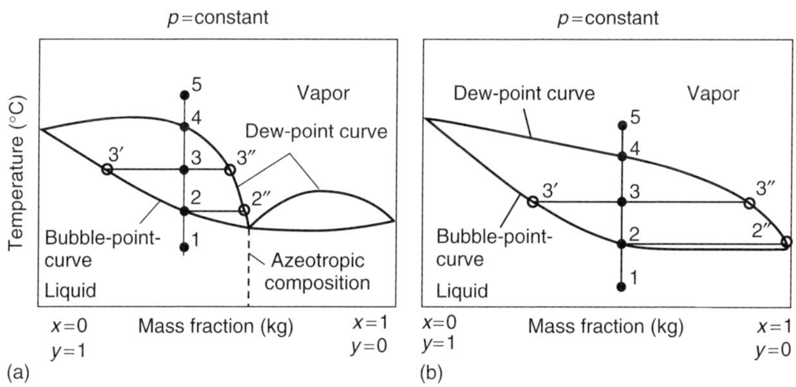

Figure 6.32 Temperature–composition diagram of (a) an azeotropic mixture and (b) a zeotropic mixture.

For azeotropic mixtures, in contrast, a composition exists, at which the temperature remains constant during isobaric evaporation. This composition is called the *azeotrope* and it shifts with changing pressure (Figure 6.32a).

The zeotropic behavior of both kinds of mixtures can be illustrated with the temperature–x,y fraction diagram in Figure 6.32b showing the composition of a mixture during its isobaric evaporation. Starting at point 1, the liquid mixture is subcooled. With heat input, the temperature rises and the mixture reaches point 2 at the bubble-point curve. Here, the mixture starts to evaporate. At point 3 the mixture is in the liquid–vapor phase. The saturated vapor (point 3″) has a higher mole fraction of the more volatile component x than the basic mixture. On the other hand, the saturated liquid (point 3′) has a higher mole fraction of the less volatile component y in relation to the basic mixture. With additional heat input, the less volatile constituent concentrates in the vapor phase until reaching the dew-point curve at point 4. At the dew-point curve, the vapor reaches the basic composition and with further heating, the vapor is superheated (point 5).

Zeotropic mixtures are of interest for an application as working fluids. The most common mixture is ammonia–water used in the Kalina Cycle (Section 6.2.4.1). Other mixtures are also possible. In Gawlik and Hassani (1997), for example, isobutane and propane-based mixtures were investigated with a focus on achieving levelized electricity cost reduction when applied in ORCs.

Other Aspects In selecting working fluids, additional important aspects have to be taken into account. The working fluid should be thermally stable in the long-term within the range of operational temperatures. With reference to the components of a binary cycle, the fluid needs to be noncorrosive and compatible with the materials used, for example, sealing materials and other possible substances such as lubricating oil.

With regard to safety, the fluid should be nonflammable or at least of low flammability and nonexplosive. Furthermore, the fluid should be nontoxic and should have a high threshold limit value (TLV) that allows humans to be close to the plant for a long time without adverse health effects. From an environmental point of view, the selection of fluids is governed by environmental legislation in many countries. In any case, the fluid should have no ozone depletion potential and the greenhouse potential should be low.

Apart from thermodynamic, health-related and environmental aspects, the market availability, and the costs of the working fluid need to be considered.

Working Fluid Effect on Component Design The working fluid affects the thermodynamic cycle performance as well as the design and performance of each component. The following paragraphs provide a few indications of the influence that the working fluid has on several selected components.

Considering the fact that every binary cycle is specially designed to utilize the geothermal heat as efficiently as possible, almost no standard turbine will meet these site-specific requirements. That means that every turbine has to be specially designed or at least modified. The design boundary conditions for the turbine are given by the basic cycle design. It defines the working fluid, the mass or volume flow rate, the change of enthalpy inside the turbine and therefore, the pressure drop.

The most common turbine type in low temperature binary cycles is the axial flow turbine. However, advantages of the use of radial flow turbine in low temperature conversion cycles are discussed in Marcuccilli and Zouaghi (2007). Besides the mechanical, geometric, and manufacturing constraints, the volume flow rate at the inlet and outlet determines the turbine size, for example, the rotor and blade dimensions. The change of enthalpy determines the vapor velocity and therefore the blade speed and its rotation speed. The design of blades and nozzles is based on fluid flow calculations. In particular, the sonic velocity of the working fluid affects its geometry. Apart from design conditions, the viscosity and density contribute to the losses which occur in the turbine. More information on turbine design is presented for example, in Balje (1981) and Invernizzi, Iora, and Silva (2007).

Heat exchanger design is affected in such a way that the heat transfer–governing properties of the working fluid have an essential impact on the required heat transfer surface. Depending on the boiling regimes, these properties are the heat conductivity, surface tension, and viscosity. Most of the woring fluids suitable for binary cycles show considerably less satisfactory heat transfer behavior than water. Therefore, the heat exchanger has to be designed with care and the more accurate the thermophysical data are, the more reliable the design is. With regard to mixtures and their evaporation, the temperature can be of convex or concave shape in the temperature–heat diagram. As a consequence, the LMTD method is not applicable and a discrete calculation is needed (Incropera and DeWitt, 2007; Bejan and Kraus, 2003).

Provision of Thermophysical Data for Working Fluids The availability of sufficiently accurate data on thermophysical properties is crucial in designing binary cycles and their components. In general, thermophysical properties can be classified into equilibrium or thermodynamic properties and nonequilibrium or transport properties. A large variety of methods were developed in recent decades to estimate or determine the thermophysical properties of fluids in the liquid–vapor region. The required fluid properties for power plant design are shown in Table 6.3.

Table 6.3 Thermophysical properties required for binary power plant design.

Equilibrium thermodynamic properties	Transport properties
T – temperature	η – dynamic viscosity
p – pressure	ν – kinematic viscosity
h – specific enthalpy	λ – thermal conductivity
s – specific entropy	–
ρ – density	–
c_p – isobaric specific heat	–
c_v – isochoric specific heat	–
w – speed of sound	–

A suitable estimation method, for engineering purposes, should be as accurate as required and convenient to use in the single and two-phase regions. Commonly, thermodynamic property tables and diagrams or different empirical equations that are obtained from measurements are used. These equations are called equations of state (EOS) and a wide range of EOS were developed and published. Most of the suitable property estimation methods use equations based on incomplete theory with empirical correlations of parameters that are not considered by the theory. Because of the numerous suitable working fluids and the different requirements depending on the boundary conditions of the geothermal site, no general estimation method can be given here. However, some useful references suitable for engineering purposes are given in the following paragraph.

REFPROP-database (REFerence fluid PROPerties) is developed by the National Institute of Standards and Technology (US). It provides tables and plots of the thermodynamic and transport properties of industrially important fluids and their mixtures, with an emphasis on refrigerants and hydrocarbons (National Institute of Standards and Technology, 2002). In addition, special emphasis is given to properties of natural gas systems. REFPROP implements three models for calculating the thermodynamic properties of pure fluids: the equations of state explicitly as Helmholtz energy, the modified Benedict–Webb–Rubin equation of state, and an extended corresponding states (ECS) model. The calculation of mixtures is also possible. These calculations employ a model that applies mixing rules to the Helmholtz energy of the mixture components. Viscosity and thermal conductivity are modeled with either fluid-specific correlations, an ECS method, or in some cases, the friction theory method.

FluidCal has been developed by the chair of thermodynamics at Ruhr University, Bochum (Germany), for calculating thermodynamic properties from equations of state (fundamental equations in the form of the Helmholtz energy) for research and technical applications (Ruhr-Universitat Bochum, 2009). FluidCal includes equations of state that were established at the chair and accurate equations of other authors. Hence, the program enables the user to calculate more than 20 different thermodynamic properties of more than 60 substances. For a large number of substances, the most common transport properties can also be calculated. As parameters, every possible combination of the properties such as temperature T, pressure p, density ρ, specific enthalpy h, and specific entropy s can be chosen.

Furthermore in Sengers *et al.*, (2000), Poling *et al.*, (2004), a comprehensive database of methods for calculating thermodynamic and transport properties is given. These books provide the theoretical basis and the practical use of each type of equation. Their strengths, weaknesses, and limits are discussed.

6.2.4.3 **Recooling Systems**

In order to "close" a conversion cycle and enable the working fluid to reabsorb the geothermal heat after its use in the turbine, the exhaust vapor at the outlet of the turbine must be condensed. If the exhaust vapor is superheated (i.e., state of the exhaust vapor is to the right of the dew-point curve), it also must be de-superheated before its condensation. For removing the waste heat in the condenser from the

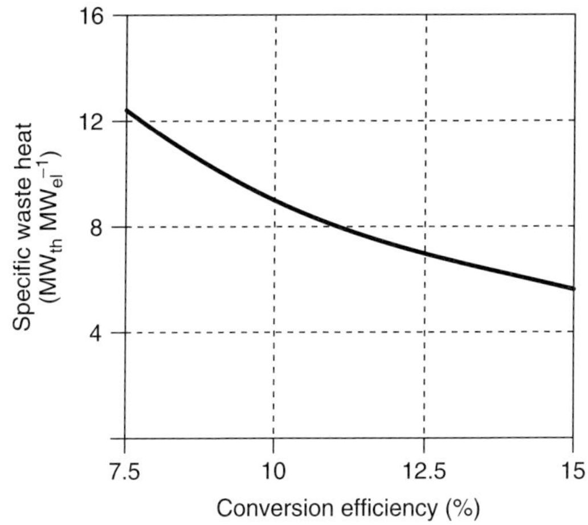

Figure 6.33 Specific waste heat to be removed from a conversion cycle as a function of conversion efficiency.

working fluid, a suitable heat sink is necessary. In power plant engineering, surface water, water from groundwater wells or ambient air is used as the heat sink. In conventional power plant engineering, usually water cooling is preferred to air cooling due to the possibility of realizing lower condensation temperatures and therefore, a larger enthalpy difference in the turbine.

At many EGS sites however, the precondition of sufficient supply of cooling water can, if at all, only be met with additional technical and energetic effort. The design of the recooling system in EGS power plants will therefore oftentimes be a compromise between technical and energetic aspects. In addition, environmental aspects need to be considered depending on the used recooling systems. Another difference from conventional plant engineering is the relatively larger amount of waste heat due to the low conversion efficiencies (Figure 6.33). This results in a larger influence of recooling on the net power output.

In the following, different recooling systems which are relevant for use in EGS plants and their most important design aspects will be outlined. The information is summarized from different sources, for example, Klenke (1970), Recknagel and Schramek (1995), and Integrated Pollution Prevention and Control (IPPC) (2001). As basic systems, once-through cooling, cooling with wet-cooling towers and air cooling (also called dry cooling) are distinguished (Figure 6.34). Hybrid cooling towers, which combine wet and dry cooling, will also be addressed briefly.

Once-through Cooling Systems Recooling systems with once-through cooling discharge the heat from the condenser to a water body. The condenser is thereby typically fed by surface water such as from rivers, lakes, or the sea. For smaller cooling capacities, groundwater is also used. The used and heated water is typically discharged back into the water source.

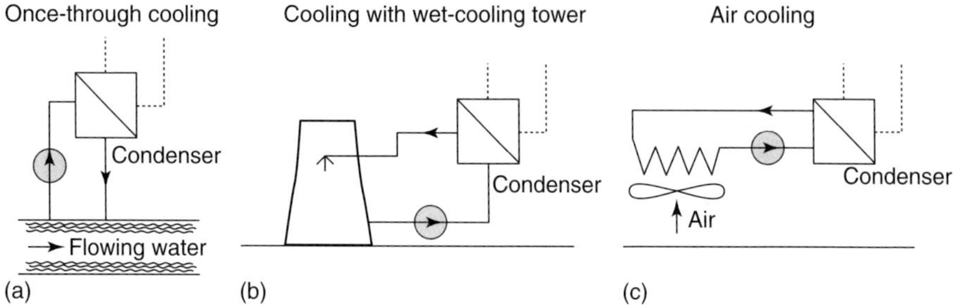

Figure 6.34 Simplified setup of different recooling systems relevant for EGS plants.

Once-through cooling systems can realize low condensation temperatures, depending on the cooling water mass flow, the inlet temperature of the cooling water, and possible limitation of the outlet temperature (such as avoiding environmentally adverse heat pollution of surface and groundwater). They cover a broad range of cooling capacities from less than 10 kW_{th} up to more than 1000 MW_{th} (Integrated Pollution Prevention and Control (IPPC), 2001).

Once-through cooling systems can only be installed at sites where sufficient water is available without negative impacts on the environment. Subject to the realizable temperature difference between cooling water inlet and outlet temperature, the specific cooling water flow ranges between $16 - 48\ kg\,s^{-1}\ MW_{th}^{-1}$ for a cooling range (temperature difference between cooling water inlet and outlet) of 15–5 K respectively.

When designing once-through cooling systems at a specific site, impacts such as heat emissions when discharging the used cooling water, possible disturbance of fish stocks or lowering of the groundwater level can play a role and must be considered properly. Since once-through cooling is a widely applied technology that has been used for a long time, such environmental aspects are known and usually regulated by local authorities. Regarding the reliability of the cooling system, the quality of the cooling water and its possible effects on materials (such as corrosion) and components (such as fouling in heat exchangers) must be considered. If additives are used to avoid fluid–material interactions, the emissions of these materials into the surface water source or the ground water must be handled with care.

Concerning the auxiliary power demand, pumps for water intake and circulation need to be considered. Depending on the cooling water flow rate, the pumping pressure, and the efficiency of the pump, the power consumption related to the removed waste heat typically ranges between $5 - 25\ kW_{el}\ MW_{th}^{-1}$ (Integrated Pollution Prevention and Control (IPPC), 2001).

Wet-cooling Towers Recooling systems with wet-cooling towers transfer the waste heat from the conversion process to the air by evaporation and convection. The condenser is thereby fed with water coming from the wet cooling tower. The used and heated cooling water is recirculated. In the following text, open and closed type wet-cooling towers with mechanical draught are distinguished.

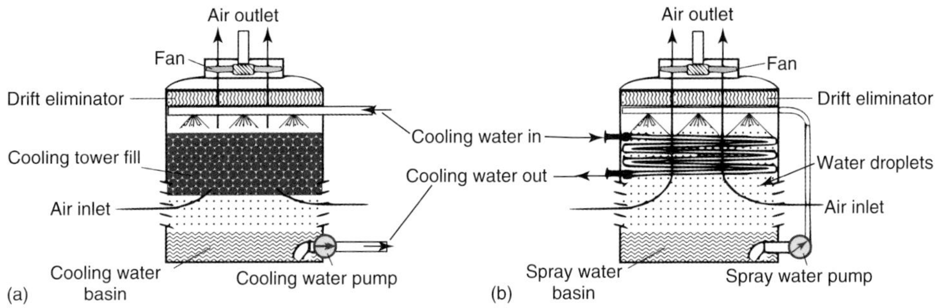

Figure 6.35 Schematic setup of (a) open and (b) closed induced draught wet-cooling tower.

In an open wet-cooling tower (Figure 6.35a), the heated cooling water is distributed in fine droplets (e.g., sprayed) on the cooling tower fill and is cooled by the up-streaming passing air flow, which is generated with fans. Cooling towers with forced draught have fans at the base and induced draught at the top of the tower. The cooling tower fill serves to enlarge the contact area between water and air. Since the air flow through the cooling tower is not saturated with water vapor, a part of the cooling water is evaporated which leads to cooling of the remaining water due to the removed evaporation heat (latent heat transfer). Another small part of the heat in the cooling water is transferred to the air stream by convection (sensible heat transfer).

A closed wet-cooling tower (Figure 6.35b), in contrast, carries the cooling water in a coil and the coil is sprayed with water. The cooling water transfers the heat to the spray water which is cooled due to evaporation and convection by the up-streaming air flow. Besides carrying the cooling water in the coil, it is also possible to circulate the working fluid in the coil and remove the waste heat from the conversion process directly inside the cooling tower.

The minimum theoretical cooling water temperature which can be realized in wet-cooling towers is the wet-bulb temperature. The wet-bulb temperature is achieved (in an adiabatic system) at the contact surface between an unsaturated air flow and an infinite water surface area in which case, water will be evaporated until the air flow is saturated with water vapor (which correlates to a relative air humidity of 100%). From the Mollier diagram for humid air at an ambient pressure of 1 bar (Figure 6.36), it can be seen that the wet-bulb temperature is always below the air temperature and that depending on both, the air temperature and relative humidity, water is evaporated until the point of saturation.

Since the contact area in a wet-cooling tower is finite, the achievable cooling water temperature is above the wet-bulb temperature. This temperature difference is usually called approach. In open wet-cooling towers, the minimum approach is approximately 3 to 5 K, whereas in closed cooling towers it is higher due to the additional heat transfer from the cooling water to the spray water.

The installed cooling capacity of wet-cooling towers can vary from less than 100 kW_{th} to more than 1000 MW_{th} (Integrated Pollution Prevention and Control

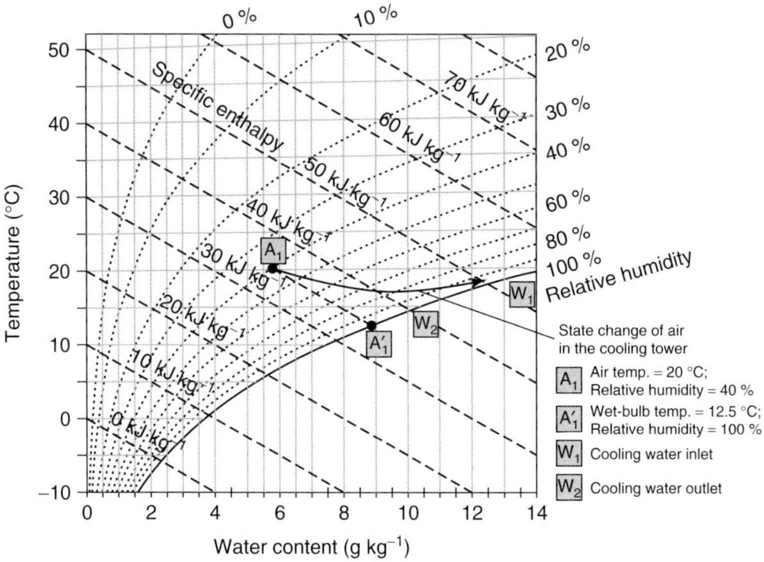

Figure 6.36 Mollier diagram for humid air at 1 bar showing the state change of air and cooling water in an example wet-cooling tower.

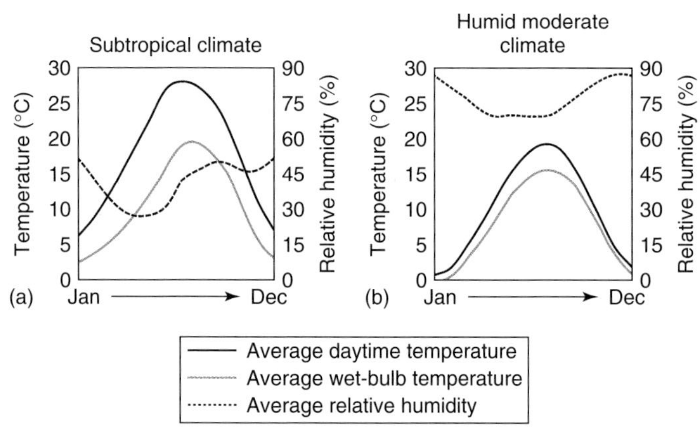

Figure 6.37 Annual climate data for the example of subtropical and humid moderate climate.

(IPPC), 2001). However, regarding their operation, it must be considered that the cooling performance can considerably differ from the installed capacity throughout the year due to the alternating ambient conditions at a plant site. Figure 6.37 shows the annual variations of air temperature, relative humidity, and corresponding wet-bulb temperature for two different climate types.

Even though the cooling or spray water used in a wet-cooling tower is recirculated, cooling water losses occur. Cooling tower design needs to consider these losses by providing sufficient make-up water. Cooling water losses are related to losses due to evaporation and blowdown of water from the cooling water basin in order to maintain the required water quality and limit suspended solids and impurities. In open wet-cooling towers, a make-up water amount of about 1–5% (Integrated Pollution Prevention and Control (IPPC), 2001) of cooling or spray water mass flow or between $0.3 - 1\ \text{kg}\,\text{s}^{-1}\ \text{MW}_{\text{th}}^{-1}$ must be available at the plant site. The necessary amount of make-up water for closed wet-cooling towers is lower.

Regarding the design of a wet-cooling tower for a specific site, different aspects must be considered. One main aspect is the adaptation to the climatic conditions and their annual alternation. Fans, for example, can vary in diameter, blade size, and operational mode (one- or multispeed operation). The cooling system can be operated with one wet-cooling tower or a modular cooling system with several towers used in parallel. If ambient temperatures below 0 °C occur, the freezing of the water in the cooling tower during standstill must be avoided, for example, with basin heating. In order to avoid freezing of the cooling tower during operation, cooling water temperatures need to be above 1 °C.

Another main aspect is the design of the water treatment because a certain water quality is needed to avoid fouling on the coil of a closed wet-cooling tower and in particular, the fill of an open wet-cooling tower. In some cases, closed systems might be advantageous due to the smaller amount of water which needs to be treated.

Depending on the environment of the plant site (such as sites near residential areas), other aspects which can play a role in designing wet-cooling towers are noise emissions and plume formation.

The auxiliary power demand of wet-cooling towers results from the operation of the cooling and/or spray water pump and the fans. The power consumption of the pumps is in the same range as for the cooling water pumps in once-through cooling systems and varies between $5 - 20\ \text{kW}_{\text{el}}\ \text{MW}_{\text{th}}^{-1}$ (Integrated Pollution Prevention and Control (IPPC), 2001). The power consumption of the fans basically depends on the ventilated air flow and lies approximately between 5–10 $\text{kW}_{\text{el}}\ \text{MW}_{\text{th}}^{-1}$ (Integrated Pollution Prevention and Control (IPPC), 2001). Since the necessary air flow must be increased when realizing lower cooling water temperatures, the power demand for the fans increases. From an energetic point of view, this has to be considered for the design of an EGS plant. Due to the alternating ambient conditions, the auxiliary power demand (especially of the fans) can significantly vary at a site throughout the year.

Air Cooling In air cooling or dry cooling systems, a cooling medium (water or refrigeration substances) or the working fluid of the binary cycle is circulated through coils and tubes which are cooled by the passing air, typically in cross-flow. For recooling applications below 200 MW_{th}, the air flow is generated by fans. The direct cooling of the working fluid can also be realized in an air-cooled condenser (ACC; Figure 6.38a). For an indirect cooling of the working fluid, the condenser is supplied by a cooling medium which is circulated through air-cooled coils and tubes in a liquid cooler (Figure 6.38b).

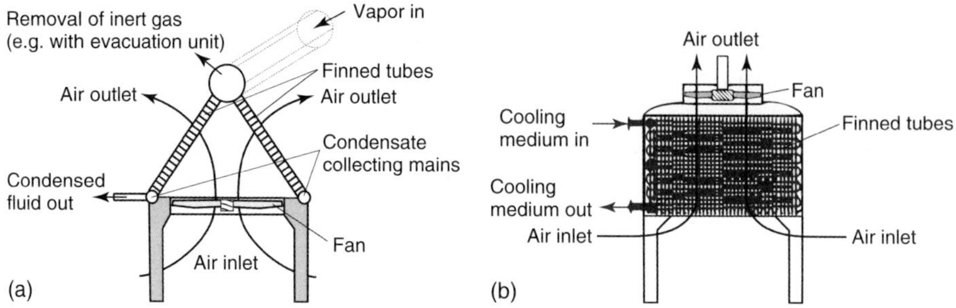

Figure 6.38 Schematic setup of (a) air-cooled condenser (ACC) with forced draught and (b) air-cooled liquid cooler with induced draught.

Since the only heat transport phenomena in dry cooling is convection, the minimum condensation temperature (direct cooling) or the minimum cooling medium temperature (indirect cooling) which can be achieved theoretically is the temperature of the ambient air. In practice however, the heat exchange area is limited and the heat transfer resistance must be taken into account. Due to the low heat capacity of air and the low heat transfer coefficients for air cooling, a larger heat exchange area is required than with water cooling for the same cooling capacity. In order to enlarge the surface of air-cooled heat exchangers, fins are often placed on the coils and tubes. With these measures, a temperature difference of 10–20 K between passing air flow and cooled medium can be achieved. The cooling capacity of ACCs and liquid coolers has a range of up to more than 500 MW_{th}.

As for wet-cooling towers, adaptation of the dry cooling system to variable ambient conditions is mainly achieved with adequate fan design (such as size and mode of operation) and typically, a modular setup of parallel air coolers. For the operation of dry cooling systems, cleaning intervals of the heat exchanger surface must be foreseen in order to avoid larger accumulations of airborne debris which impair the heat transfer.

Regarding site-specific preconditions, dry cooling is advantageous at sites with insufficient cooling water availability or where plume formation might be a problem. However, dry cooling is related to considerable noise emissions during operation and large space requirements (due to the necessarily large heat exchange area); these need to be considered when designing the implementation of the cooling system at the plant site.

Depending on the ambient conditions at a site and the cooling temperature, the auxiliary power consumption of the fans can range from 10–50 $kW_{el}\ MW_{th}^{-1}$ (Integrated Pollution Prevention and Control (IPPC), 2001). The influence of the cooling temperature on the auxiliary power demand is thereby larger than for wet-cooling towers for which, the necessary increase of the air flow rate in order to realize lower cooling temperatures has already been addressed. As mentioned above, this is a very important aspect of the overall design of an EGS plant.

Hybrid Cooling Towers Hybrid cooling towers have been developed to combine the advantages of wet-cooling towers and dry coolers. Hybrid cooling towers are therefore characterized by lower plume formation (plume-free or plume-reduced types) and lower make-up water consumption as compared to wet-cooling towers and a better cooling performance than dry coolers. The idea behind hybrid cooling is the flexibility to vary the proportions of evaporative and dry cooling, according to particular operating conditions. Hybrid cooling towers can be advantageous at sites where cooling water availability and/or plume formation is only a seasonal problem so that for these periods the cooling tower is operated in dry or hybrid mode. However, this flexibility in operation is followed by a larger technical effort regarding construction, operational control, and cooling water treatment.

An open hybrid cooling tower can be operated in wet and in hybrid mode depending on the ambient conditions. Wet operation is the same as in an open wet-cooling tower. During hybrid operation, a part of the cooling medium passes through the dry section of the cooling tower where heat is removed by an induced air flow. The cooling medium is then conducted through the wet section. In the upper part of the open hybrid cooling tower, the heated air from the dry section is mixed with the vapor from the wet section. This leads to a decrease in the relative humidity of the discharged air and a reduced plume formation.

The closed hybrid cooling tower with wetted cooling elements (such as finned coils, tube bank, or plated bank) can be operated in dry mode, similar to an air cooler. For wet operation, its cooling elements can be wetted with a water film which runs down. The cooling water that runs off the cooling elements is collected in a basin at the bottom of the tower and is then recirculated.

6.2.5 Combined Energy Provision

Besides the provision of heat, chill, and power from separate plants, EGS reservoirs also offer the possibility of a combined energy supply. This is possible with cogeneration which is known from conventional power plant design. Using EGS reservoirs, a combined energy supply in more or less independent plant parts which are connected in parallel or serial is often applicable. EGS plants can also combine cogeneration, serial, and parallel connection in one plant concept so that a variety of plant setups are possible depending on the demand characteristics at a specific site. The main design aspects which have to be considered for supplying heat and/or power and/or chill with these different concepts are outlined in the following section. An overview of existing geothermal plants providing power and heat is given by Lund and Chiasson (2007).

6.2.5.1 Cogeneration

Cogeneration or combined heat and power supply (CHP) is known from conventional power plant engineering. In this setup the waste heat from the conversion process, which otherwise would be discharged to the environment, is used for heat supply (Figure 6.39). Cogeneration is therefore characterized

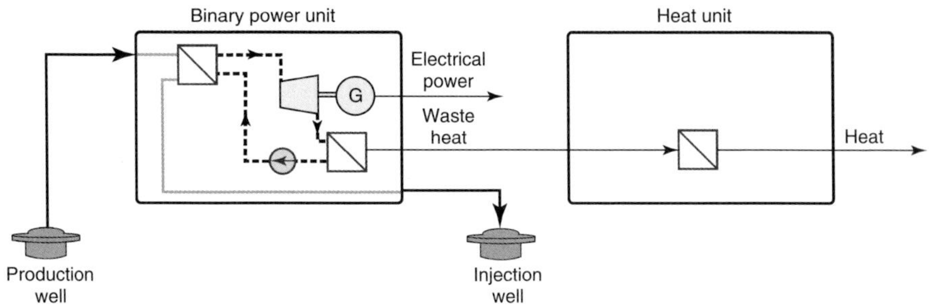

Figure 6.39 Schematic setup of an EGS plant with conventional cogeneration of power and heat.

by a simultaneous supply of power and heat from the same unit and leads to a significantly higher utilization of the heat source driving the conversion process.

The heat contained in the exhaust vapor at the outlet of the turbine of conventional power plants can, for example, be used to supply district or process heat with supply temperatures above 100 °C or to convert this heat and provide chill. In EGS power plants, in contrast, the temperature of the exhaust vapor is normally lower than in conventional power plants. The temperature correlates with the geothermal fluid temperatures. For geothermal fluid temperatures of up to 170 °C, the exhaust vapor temperature typically lies below 60 °C; for higher fluid temperatures of up to 200 °C, an exhaust vapor temperature of about 80 °C can be expected (Köhler, 2005). It is generally possible to raise the exhaust vapor temperature though this measure has limited influence and leads to a considerable decrease in efficiency of the conversion process.

Cogeneration in EGS plants is therefore dependent on a suitable heat consumer structure characterized by low supply temperatures. With reference to the relatively large waste heat amounts of EGS power plants, cogeneration is however, also interesting because the technical use of the waste heat can reduce the demand for recooling.

6.2.5.2 **Serial Connection**

Geothermal fluids offer a large temperature difference, which can be used for energy supply. In many cases, the fluid's single use for power, heat, or chill provision does not utilize this potential. When providing power from geothermal heat for example, the geothermal fluid at the outlet of the binary unit still carries heat at a usable temperature level, which is higher than the temperature of the exhaust vapor.

By connecting different energy-providing units in series, the utilization of the geothermal fluid is possible at different temperature levels for different modes of energy provision (Figure 6.40). The main design criterion which has to be considered for such a connection is that the temperature of the geothermal fluid at the inlet of each downstream unit has to be sufficiently high. When designing a serial connection, this precondition can be achieved by adapting the design of

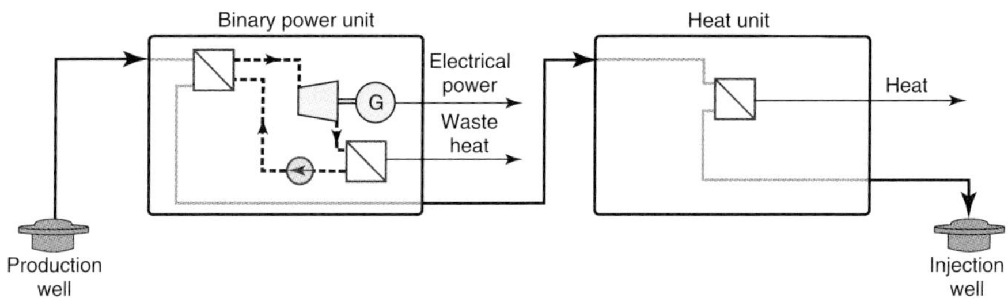

Figure 6.40 Schematic setup of an EGS plant with power and heat supply in serial connection.

the downstream unit to a particular inlet temperature. If the downstream unit is dependent on a certain inlet temperature, the outlet temperature of the upstream unit must be adapted accordingly. At some sites, all units might be optimized for their use in serial connection, for example in the case of a binary power unit which is connected serially with a chill provision unit. In contrast to cogeneration, which refers to power and heat or chill provision at the same time, serial connections can also be designed for a varying energy provision mix, for example, due to an annually varying heat demand.

A typical geothermal serial connection is represented by a unit providing power and a downstream unit providing heat. If a binary unit optimized for power output has a geothermal outlet temperature of 70 °C, the heat unit can, for example, feed a low temperature heating grid with a supply temperature of 60 °C and a return temperature of 40 °C. If the heat unit has to provide 70 °C instead, the power conversion cycle needs to be adapted in order to cool the geothermal fluid to only about 75–80 °C, which leads to a lower power output. Assuming a typical heat demand which varies throughout the year, this reduction in the power output is relevant only for times of heat demand.

Based on these considerations, a serial connection is especially advantageous when the inlet and outlet temperatures of the connected units closely match the optimum temperatures of their single application.

6.2.5.3 **Parallel Connection**

In a parallel connection of different energy supply units, the geothermal mass flow is split between these units so that the inlet temperatures of all units are the same (Figure 6.41). The main design criterion is the split-up of the geothermal mass flow. Each energy supply unit is optimized for the geothermal fluid temperature and a certain mass flow. In comparison to units connected in series, units connected in parallel can be designed more independently. Parallel connections can also be designed easily for a varying energy provision mix. In case of a variable mass flow, the part load behavior of a unit needs to be considered.

Parallel connections are advantageous when the necessary inlet temperatures of different energy supply units are at the same temperature level, especially when these temperatures are close to the geothermal fluid temperature. An EGS plant

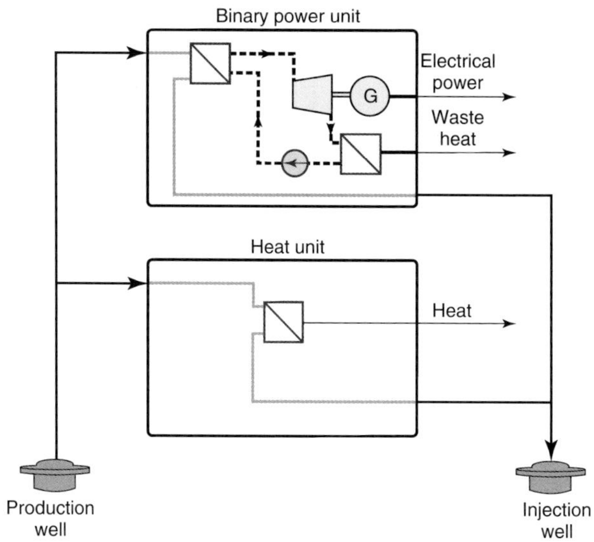

Figure 6.41 Schematic setup of an EGS plant with power and heat supply in parallel connection.

with parallel supply of power and heat, for example, uses all of the geothermal mass flow for heat provision in times of large heat demand. In times of lower heat demand (but still with a supply temperature close to the geothermal fluid temperature), only a part of the geothermal mass flow is used for heat supply so that the remaining mass flow can be used in a binary power unit.

6.3 Case Studies

By means of case studies, two examples will be given for holistic geothermal plant design. The case studies analyze the provision of power, and the provision of power and heat from an EGS plant. For each case study, a specific objective is defined. According to this objective, a practical design approach following the ideas of holistic plant design will be discussed. The goal of this section is to show the relevance of an integrated design for EGS which combines different engineering disciplines.

The results of the case study on power provision indicate that EGS plants are typically characterized by a considerable auxiliary power demand. This is especially due to the energetic effort required for the production of the geothermal fluid and the recooling of the binary unit. Furthermore, it will be shown that a holistic design approach can lead to a considerable improvement of the net power output based on existing technologies. From the results of the case study on power and heat provision, it can be seen that the applicability of a specific mode of connection depends on the amount and the temperature of the required heat.

Since the reservoir and site preconditions, and operational characteristics vary from site to site, holistic power plant design is a site-specific approach. At some sites nontechnical aspects, such as noise emission thresholds or land use, can also be issues which are integrated into holistic approaches. It is important to note that the success of this approach is based on interdisciplinary collaboration.

6.3.1 Power Provision

6.3.1.1 Objective

The contribution of geothermal binary power plants to the energy system is based on their provision of net power. The objective in the following case study is therefore to maximize the net power output of an EGS binary power plant at a specific site. Two different approaches are compared. Based on the implementation of existing and reliable technology for both approaches, one approach aims at the maximization of the installed capacity or gross power and the other one directly maximizes the net power by means of a holistic plant design where the different characteristics of the auxiliary power demand are considered.

The reference reservoir of the following case study is assessed with a doublet. On the surface, the produced geothermal fluid is used in a binary unit. The conversion unit is based on a Rankine Cycle with a pure working fluid and is cooled by an induced-draught cooling tower. The reservoir and the ambient conditions are defined as follows:

Reservoir temperature	150 °C
Reservoir depth	4000 m
Specific heat capacity geothermal fluid	3.5 kJ kg^{-1} K^{-1}
Geothermal fluid density	1.147 kg m^{-3}
Productivity index	30 m^3 (h MPa)$^{-1}$
Injectivity index	30 m^3 (h MPa)$^{-1}$
Pore pressure gradient	10.7 bar/100 m
Average ambient temperature	15 °C
Average relative air humidity	0.75

6.3.1.2 Design Approach

The parameters which are varied in order to optimize the system are the working fluid, evaporation temperature (corresponding to a certain reinjection temperature of the geothermal fluid), condensation temperature, and geothermal fluid flow rate.

General considerations in every plant design also refer to component-specific aspects such as size, efficiencies of turbines and pumps, temperature differences in heat exchangers, or cooling tower characteristics. Further design aspects of the

EGS plant refer to the allowable operating conditions in the geothermal fluid loop, such as the intake pressure of the downhole pump or the wellhead pressure in the production well. For the comparison of the different design approaches these design considerations will not be addressed in detail. The following parameters are therefore assumed for both design approaches:

Geothermal fluid loop	
Well head pressure production well	10 bar
Efficiency downhole pump	0.75
Intake pressure downhole pump	10 bar
Relative roughness riser tube	0.003
Relative roughness casing	0.008
Diameter riser tube	5 in.
Diameter casing	8.5 in.
Binary conversion cycle	
Heat exchanger temperature differences	5 K
Pressure loss per heat exchanger	0.1 bar
Efficiency feed pump	0.8
Isentropic efficiency turbine	0.75
Mechanical efficiency turbine	0.95
Mechanical efficiency generator	0.95
Wet-cooling tower	
Approach to wet-bulb temperature	3 K
Cooling range	6 K
Cooling tower constant	0.8
Specific heat capacity water	$4.2\,kJ\,kg^{-1}\,K^{-1}$
Installation height cooling tower fill	1.5 m
Water-sided pressure losses	1 bar
Pressures increase fans	0.002 bar
Cooling pump efficiency	0.8
Cooling fan efficiency	0.8

6.3.1.3 Gross Power versus Net Power Maximization

The results of the case study are shown in Table 6.4. It can be seen that under the conditions assumed for the example site, the net power varies between 1.13–1.27 MW and that fluid production and recooling have a significant auxiliary power demand which ranges between 39–44% and 10–16%, respectively. Comparing the different design approaches, the maximization of the net power results in a 12% higher net power output as compared to the plant based on maximizing the gross power. The installed capacity or gross power for the plant with optimized net power, in contrast, is 16% lower. With net power maximization, the ratio of net power output to gross power is therefore increased by 33%. These improvements are based on the differently chosen design parameters which result in a decrease of the auxiliary power demand for recooling and geothermal fluid production. From

Table 6.4 Results of the case study.

	Gross power maximization	Net power maximization
Working fluid	Isobutane	Isobutane
Reinjection temperature (°C)	65	65
Evaporation temperature (°C)	100	100.5
Condensation temperature (°C)	25.7	28
Geothermal fluid flow rate (kg s^{-1})	100	88
Gross power (MW)	3.5	3.0
Auxiliary power binary cycle (MW)	0.28	0.24
Auxiliary power cooling (MW)	0.57	0.29
Auxiliary power geothermal fluid loop (MW)	1.50	1.27
Net power (MW)	1.13	1.27

the following table it can be seen that all design criteria, except for the working fluid, depend on the design objective.

The difference of the optimum evaporation temperatures between gross and net power maximization is small. Regarding the optimum condensation temperature, in contrast, the design approach has a significant influence. The working fluid in the binary cycle of the plant designed for maximum gross power is condensed at the minimum condensation temperature depending on the ambient conditions. This results in a specific auxiliary power demand for recooling of 19 kW_{el} $MW_{th}{}^{-1}$. When maximizing the net power in contrast, a higher condensation temperature is chosen so that the specific recooling effort is reduced to 11 kW_{el} $MW_{th}{}^{-1}$ which results in a 50% decrease of the auxiliary power demand for recooling.

A significant influence of the design approaches can also be seen for the optimum geothermal fluid flow rate. Maximizing the power plant for gross power, the maximum flow rate with respect to technical restrictions (e.g., installation depth, volume flow, pump capacity) is produced from the reservoir. For the assumed reservoir conditions, the maximum flow rate is limited to 100 kg s^{-1}. Applying a holistic design approach, the optimum flow rate is 88 kg s^{-1}. Figure 6.42 shows that the optimum flow rate for a specific reservoir depends on the gross power output and the auxiliary power demand. Under more favorable ambient conditions, for example, more gross power is produced and less auxiliary power consumed for recooling so that the optimum flow rate is at a higher value. Producing a net power optimized flow rate, the auxiliary power consumption of the downhole pump is reduced by 24%.

In contrast, the reduction of the auxiliary power consumption of the feed pump in the binary cycle is only 12%. This improvement is based on a lower pressure increase in the feed pump due to the higher condensing pressure and the lower working fluid flow rate.

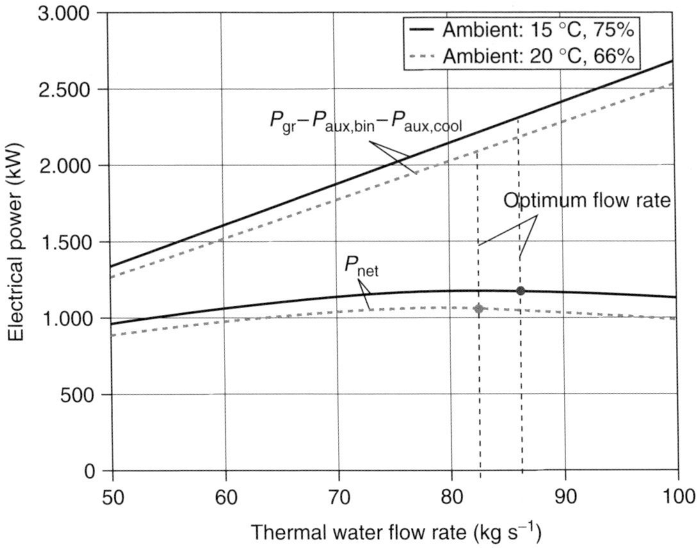

Figure 6.42 Influence of ambient conditions on optimum geothermal fluid flow rate and maximized net power output.

6.3.2 Power and Heat Provision

6.3.2.1 Objective

Since EGS reservoirs can be used for the supply of power and heat, an integrated or holistic approach can also refer to the design of the energy supply concept. The design of the provision of power and heat from one EGS plant offers many degrees of freedom. Based on the case study for power provision, the objective in the following is the maximization of the net power output at the same site at which a defined heat demand has to be met. The goal is to find the appropriate connection for the power and heat provision to a low temperature grid, and for the power and heat provision to a medium temperature grid. The heat demand is characterized as follows:

Characteristics of the low temperature (LT)-heating grid	
Supply temperature (°C)	70
Return temperature (°C)	50
Thermal capacity (MW)	5–9
Characteristics of the medium temperature (MT)-heating grid	
Supply temperature (°C)	90
Return temperature (°C)	70
Thermal capacity (MW)	5–9

6.3.2.2 Design Approach

The focus of the following design approach is to analyze different plant connections for power and heat supply that allows the supply of the demanded heat and realize the maximum possible net power output. Due to the temperature of the reservoir, the waste heat from the power conversion process is insufficient to provide the demanded temperature level for the heat supply. Therefore, the analyzed connections refer to the use of the thermal energy in the geothermal fluid. In the following, a parallel connection which splits the geothermal flow rate between power generation and heat provision, and a serial connection which uses the geothermal heat first for the power conversion and then for the heat provision, are compared.

The site-specific conditions and general parameters are taken from the previous case study. In this case study, a geothermal flow rate of $88\,kg\,s^{-1}$ is produced. For the supply of district heat, a temperature difference of 5 K between geothermal fluid and energy carrier in the heating grid was assumed. The parameters on which the following discussion will focus on is the connection of power and heat supply, referring to the conduction and cooling of the geothermal fluid. For each mode of connection, the working fluid, evaporation temperature, and degree of internal heat recuperation have been optimized.

6.3.2.3 Serial versus Parallel Connection

The results of this case study are shown in Figure 6.43. It can be seen that questions regarding the optimum mode of connection cannot be answered in a

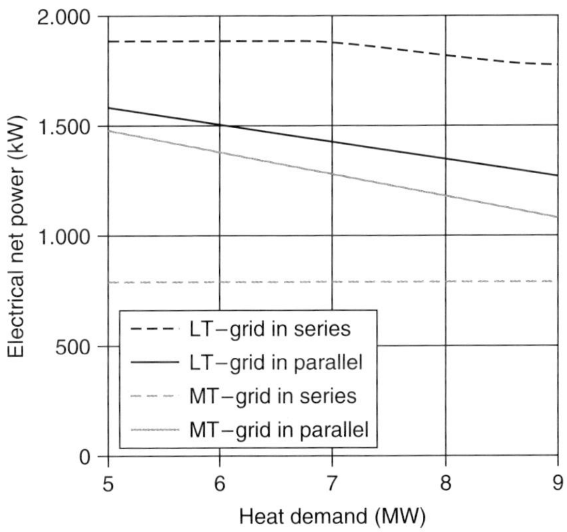

Figure 6.43 Net power for serial and parallel connection depending on heat demand and supply and return temperatures of the low temperature(LT)-heating grid and the medium temperature(MT)-heating grid.

general manner. Under the conditions assumed in this case study, the design of the connection depends on the temperature level of the heat demand.

In case of the supply of low-temperature heat, the serial connection results in a higher net power output. This is due to the fact that at this temperature level of heat demand, decreasing the temperature difference available for the power conversion results in lower power production losses than splitting the geothermal fluid flow rate as in the case of a parallel connection. Up to a heat demand of 7 MW, the serial connection results in a net power of 1.8 MW. In this case, a geothermal fluid temperature at the outlet of the binary unit 5 K higher than the supply temperature of the heating grid is sufficient to meet the heat demand. For a heat demand above 7 MW, this is no longer the case because the geothermal flow rate is too low to supply the demanded heat. For a higher heat demand, the geothermal fluid temperature after the binary unit therefore needs to be raised so that the net power production decreases due to the smaller temperature difference which can then be used for the power conversion. In a parallel connection, the net power production always decreases linearly with increasing heat demand due to the mass flow which is needed for the heat supply.

In contrast, for the supply of medium-temperature heat, the parallel connection is more appropriate. In this case, splitting the geothermal fluid flow rate has less impact on the net power production than decreasing the temperature difference available for the power conversion in case of a serial connection.

References

Abel, H.P. (2005) Argentina country update. Proceedings World Geothe Rmal Congress 2005, April 24–29, 2005, Antalya.

Aksoy, N. (2007) Optimization of downhole pump setting depths in liquid-dominated geothermal systems: a case study on the Balcova-Narlidere field, Turkey. *Geothermics*, **36**, 436–458.

Balje, O.E. (1981) *Turbomachines: A Guide to Design, Selection, and Theory*. John Wiley & Sons, Inc., New York.

Bazin, B., Brosse, E., and Sommer, F. (1997) Chemistry of oilfield brines in relation to digenesis of reservoirs: 1. Use of mineral stability fields to reconstruct *in situ* water composition. Example of Mahakam basin. *Marine and Petroleum Geology*, **14**, 481–495.

Bejan, A. and Kraus, A.D. (2003) *Heat Transfer Handbook*, John Wiley & Sons, Inc.

Bejan, A., Tastsaronis, G., and Moran, M. (1996) *Thermal Design and Optimization*, John Wiley & Sons, Inc., New York.

Bjorlykke, K., Aagaard, P., Egeberg, P.K., and Simmons, S.P. (1995) Geochemical constraints from formation water analyses from the North Sea and the Gulf Coast basins on quartz, feldspar, and illite precipitation in reservoir rocks. *Geological Society Special Publication*, **86**, 33–50.

Boissavy, M.C. and Dubief, R. (1995) New technologies in the geothermal well. Proceedings World Geothermal Congress, 1995.

Carvalho, M.R., Fojaz, V.H., and Almelda, C. (2006) Chemical composition of deep hydrothermal fluids in the Ribeira Grande geothermal field (Sao Miguel, Azores). *Journal of Volcanology and Geothermal Research*, **156** (1-2), 116–134.

Case, L.C. (1945) Exceptional silurian brine near Bay City, Michigan. *American Association of Petroleum Geologists Bulletin*, **29**, 567–570.

Cengel, Y.A. (2007) *Introduction to Thermodynamics and Heat Transfer*, McGraw-Hill Higher Education.

Çengel, Y. and Boles, M. (2006) *Thermodynamics: An Engineering Approach*, 5th edn, McGraw-Hill Higher Education, 988 p.

Chandrasekharam, D. and Bundschuh, J. (2002) *Geothermal Energy Resources for Developing Countries*, A A Balkema.

Chen, Y.Y., Liou, Y.M., and Shih, H.C. (2005) in Stress corrosion cracking of type 321 stainless steels in stimulated petro-chemical process environments containing hydrogen sulphide and chloride, *Materials Science and Engineering*, **A407**, 114–126.

Cornell, R.M. and U. Schwertmann (1996) in *The Iron Oxides* (eds R.M. Cornell and U. Schwertmann), VCH, New York.

Costa Andre', L.H. and Queiroz Eduardo, M. (2008) Design optimization of shell-and-tube heat exchangers. *Applied Thermal Engineering*, **28**, 1798–1805.

Culver, G. and Rafferty, D.K. (1998) Well pumps, in *Geothermal Direct-Use, Engineering and Design Guidebook*, 3rd edn (eds J.W. Lund, J.P. Lienau, and B.C. Lunis), Oregon Institute of Technology, Geo-Heat Center, Idaho, pp. 211–239.

Dinçer, I. and Rosen, M. (2007) *Exergy: Energy, Environment and Sustainable Development*, Elsevier, 454 p.

DiPippo, R. (2008) *Geothermal Power Plants, Principles, Applications, Case Studies and Environmental Impact*, 2nd edn, Elsevier, Ltd, Oxford, p. 493.

Drescher, U. (2008) *Optimierungspotenzial des Organic Rankine Cycle für Biomassebefeuerte und Geothermische Wärmequellen, Thermodynamik: Energie – Umwelt – Technik*, Bd. 14 (in German), D: Logos Verlag, Berlin.

Drever, J.I., Holland, H.D., and Turekian, K.K. (2004) *Treatise on Geochemistry*, Surface and Groundwater, Weathering, and Soils, Vol. 5, Elsevier Science Technology Book, 626 p.

Duan, Z., Moller, N., DeRocher, T., and Weare, J.H. (1996) Prediction of boiling, scaling and formation conditions in geothermal reservoirs using computer programs TEQUIL and GEOFLUIDS. *Geothermics*, **25** (6), 663–678.

Economides, M. and Ungemach, P. (1987) *Applied Geothermics*, John Wiley & Sons, Inc.

Flashman, R. and Lekic, H.O. (1999) Artificial lift for high-volume production. *Oilfield Review*, Spring **11** (1), Schlumberger, 49–63.

Frape, S.K., Blyth, A., Blomkvist, R., McNutt, R.H., and Gascoyne, M. (2004) Deep fluids in the continents: II. Crystalline rocks, in *Treatise on Geochemistry*, Surface and Groundwater, Weathering, and Soils, Vol. **5**, Elsevier Science Technology Book, (eds J.I. Drever, H.D. Holland, and K.K. Turekian), pp. 541–580.

Garcia, A.V., Thomson, K., Erling, H., and Stenby, E.H. (2005) Prediction of mineral scale formation in geothermal and oilfield operations using the extended UNIQUAC model Part I. Sulfate scaling minerals. *Geothermics*, **34**, 61–97.

Garcia, A.V., Thomsen, K., and Stenby, E.H. (2006) Prediction of mineral scale formation in geothermal and oilfield operations using the Extended UNIQUAC model. Part II. Carbonate-scaling minerals. *Geothermics*, **35**, 239–284.

Gawlik, K. and Hassani, V. (1997) Advanced binary cycles: optimum working fluids. Proceedings of the 32nd Intersociety of Energy Conversion Engineering Conference (IECEC), Vol. 3, pp. 1609–1814.

Giese, L.B., Seibt, A., Wiersberg, T., and Pekdeger, A. (2002) Geochemistry of the Formation Fluid. GFZ Scientific Technical Report No. STR02/14, Deutsches GeoforSchungszentrum, 145–169.

Hanor, J.S. (1994) Origin of saline fluids in sedimentary basins, in *Geofluids: Origin, Migration and evolution of Fluids in Sedimentary Basins* (ed. J. Parnell), Geological Society Special Publication No 78, Geological Society, pp. 151–174.

Houben, G. and Treskatis, C. (2003) *Regenerierung und Sanierung von Brunnen*, Oldenburg Industrieverlag, München, 280 p.

Incropera, F.P., and DeWitt, D.P. (2007) *Fundamentals of Heat and Mass Transfer*, 6th edn, Vol. **25**, John Wiley & Sons, Inc., p. 997.

Integrated Pollution Prevention and Control (IPPC) (2001) Reference Document on the Application of Best Available Techniques in Industrial Cooling Systems, European Commission.

Invernizzi, C. and Bombarda, P. (1997) Thermodynamic performance of selected HCFs for geothermal applications. *Energy*, **22**, 887–895.

Invernizzi, C., Iora, P., and Silva, P. (2007) Bottoming micro-Rankine cycles for micro-gas turbines. *Applied Thermal Engineering*, **27**, 100–110.

Kanoglu, M. (2002) Exergy analysis of a dual-level binary geothermal power plant. *Geothermics*, **31**, 709–724.

Kendall, H. and Doctor, D.H. (2004) Stable isotope applications in hydrologic studies, in *Treatise on Geochemistry*, Surface and Groundwater, Weathering, and Soils, Vol. **5**, Elsevier Science Technology Book, (eds J.I. Drever, H.D. Holland, and K.K. Turekian), pp. 499–540.

Kharaka, Y.K. and Hanor, J.S. (2004) Deep fluids in the continents: I. Sedimentary basins, in *Treatise on Geochemistry*, Surface and Groundwater, Weathering, and Soils, Vol. **5**, Elsevier Science Technology Book, (eds J.I. Drever, H.D. Holland, and K.K. Turekian), pp. 499–540.

Kharaka, Y.K., Hull, R.W., and Carothers, W.W. (1985) Water rock interactions in sedimentary basins, in *Relationship of Organic Matter and Mineral Digenesis: Society of Economic Patrolytes: American Journal of Science*, Vol. **288**, Society of Economic Paleontologists, Tulsa, Okla, (eds D.L. Gautiers, Y.K. Kharaka, and R.C. Surdam), pp. 19–98.

Kharaka, Y.K., Lundegard, P.D., and Giordano, T.H. (2000) Distribution and origin of organic ligands in subsurface waters from sedimentary basins. *Reviews in Economic Geology*, **9**, 119–132.

Kharaka, Y.K., Maest, A.S., Carothers, W.W., Law, L.M., Lamothe, P.J., and Fries, T.L. (1987) Geochemistry of metal-rich brines from central Mississippi Slat Dome Basin, USA. *Applied Geochemistry*, **2**, 543–561.

Kiryukhin, A., Xu, T., Pruess, K., Apps, J., and Slovtsov, I. (2004) Thermal–hydrodynamic– chemical (THC) modeling based on geothermal field data. *Geothermics*, **33**, 349–381.

Klenke, W. (1970) Zur einheitlichen Beurteilung und Berechnung von Gegenstrom- und kühltürmen, Kältetechnik-Klimatisierung 22. Jahrgang, Heft 10/1970, pp. 322–330.

Knapek, E. and Kittl, G. (2007) Unterhaching power plant and overall system. Proceedings European Geothermal Congress 2007, 30 May-1 June 2007, Unterhaching, pp. 1–6.

Köhler, S. (2005) Geothermisch angetriebene dampfkraftprozesse – analyse und vergleich binärer kraftwerke. Dissertation, Technische Universität Berlin, Berlin, 184 p.

Land, L.S. and MacPherson, G.L. (1992) Geothermometry from brine analysis – lessons from the Gulf Coast, USA. *Applied Geochemistry*, **7** (4), 333–340.

Legarth, B. (2003) Erschließung sedimentärer speichergesteine für eine geothermische Stromerzeugung. Dissertation, Technische Universität Berlin, Berlin.

Leibowitz, H.M. and Mlcak, H.A. (1999) Design of a 2 MW Kalina cycle binary module for installation in Husavik, Iceland. Proceedings Geothermal Resources Council 1999 Annual Meeting, October 17–20, 1999, Reno, pp. 75–80.

Lindal, B. (1973) Industrial and other applications of geothermal energy, in *Geothermal Energy: Review of Research and Development*, LC No. 72-97138, UNESCO, Paris, pp. 135–148.

Lund, J.W. and Chiasson, A. (2007) Examples of combined heat and power plants using geothermal energy. Proceedings European Geothermal Congress 2007, 30 May-1, June 2007, Unterhaching.

Lundegard, P.D. and Kharaka, Y.K. (1994) Distribution of organic acid in subsurface waters, in *Organic Acids in Geological Processes* (eds E.D. Pittman and M.D. Lewan), Springer, Berlin, pp. 40–69.

Maack, R. and Valdimarsson, P. (2002) Operating Experience with Kalina Power Plants, VDI- Berichte 1703 Geothermische Stromerzeugung, Potsdam 17. und 18. October 2002.

Marcuccilli, F. and Zouaghi, S. (2007) Radial inflow turbines for Kalina and organic Rankine cycles. Proceedings European Geothermal Congress 2007, 30 May-1, June 2007, Unterhaching.

McDermott, C.I., Randriamanjatosoa, A.R., Tenzer, H., and Kolditz, O. (2006) Simulation of heat extraction from crystalline rocks: the influence of coupled processes on differential reservoir cooling. *Geothermics*, **35** (3), 321–344.

McNutt, R.H. (2000) Strontium isotopes, in *Environmental Tracers in subsurface Hydrology* (eds P. Cook and A.L. Herczeg), Kluwer Academic, Boston, pp. 234–260.

Millero, F. (2009) Thermodynamic and kinetic properties of natural brines. *Aquatic Geochemistry*, **15**, 7–41.

National Institute of Standards and Technology. (2002) NIST Reference Fluid Thermodynamic and Transport Properties Database (REFPROP): Version 8.0 *http://www.nist.gov/srd/nist23.htm*.

Parkurst, D.L. and Apello, C.A.J. (1999) User's Guide to Phreeqc (Version 2) – A Computer Program for Speciation, Batch-Reaction, One-Dimensional Transport, and Inverse Geochemical Calculations. Water-Resources Investigations Report No. 99-4259.

Poling, B.E., Prausnitz, J.M., and O'Connell, J.P. (2000) *The Properties of Gases and Liquids*, 5th edn, McGraw-Hill Professional.

Puigdomenech, I. (2004) HYDRA and MEDUSA Chemical Equilibrium Software. Software and Documentation, *http://web.telia.com/~u15651596/*.

Recknagel, H., Sprenger, E. and Schramek E.R (1995) *Taschenbuch für Heizung und Klimatechnik*, Oldenburg Verlag, 1011–1556 and 1663–1796.

Reyes, A.G., Trompetter, W.J., Britten, K., and Searle, J. (2003) Mineral deposits in the Rotokawa geothermal pipelines, New Zealand. *Journal of Volcanology and Geothermal Research*, **119** (1-4), 215–239.

Ruhr -Universität Bochum. (2009) FluidCol: Software for the Calculation of Thermodynamic Properties for a great number of Substances, *http://www.ruhr-uni-bochum.de /thermo/Software /Seiten/Fluidcal-eng.htm*.

Saleh, B., Koglbauer, G., Wendland, M., and Fischer, J. (2007) Working fluids for low-temperature organic Rankine cycles. *Energy*, **32**, 1210–1221.

Schutz, R.W., Porter, R.L., and Horrigan, J.M. (2000) Qualifications of Ti-6%Al-4%V-Ru alloy production tubulars for aggressive fluoride-containing Mobile Bay well service. *Corrosion*, **56** (11), 1170–1179.

Schutz, R.W. and Watkins, H.B. (1998) Recent developments in titanium alloy application in the energy industry. Conference on Thermomechanical Processing and Metallurgy of Titanium Alloys, JUL 07–11, 1997 Univ Wollongong, Wollongong, Australia in *Materials Science and Engineering A-Structural Materials Properties, Microstructure and Processing*, **243** (1-2), 305–315.

Sengers, J.V., Kayses, R.F., Peters, C.J., and White, H.J. (2000) *Equations of State for Fluids and Fluid Mixtures*, 1st edn, Elsevier Science, pp. 896.

Smith, E.S. (2006) *Advances in Thermal Design of Heat Exchangers: A Numerical Approach: Direct-sizing, Step-wise rating, and Transients*, John Wiley & Sons, Inc.

Thomsen, K. (2005) Modeling electrolyte solutions with the extended universal quasichemical (UNIQUAC) model. *Pure and Applied Chemistry*, **77** (3), 531–542.

Thorolfsson, G. (2005) Maintenance history of a geothermal plant: Svartsengi Iceland. Proceedings World Geothermal Congress 2005, April 24–29, 2005, Antalya.

VDI-Gesellschaft Verfahrenstechnik und Chemieingenieurwesen (GVC) (eds) (2006) *VDI – Wärmeatlas*, 10th edn, VDI-Buch, Reihe, 1500 p.

Wolfgramm, M., Bartels, J., Hoffmann, F., Kittl, G., Lenz, G., Seibt, P., Schulz, R., Thomas, R., and Unger, H.J. (2007) Unterhaching geothermal well doublet: structural and hydrodynamic reservoir characteristic; Bavaria (Germany). Proceedings European Geothermal Congress 2007; 30 May-1, June 2007, Unterhaching, pp. 1–6.

Xu, T. and Pruess, K. (1998) Coupled Modeling of Non-isothermal Multiphase Flow, Solute Transport and Reactive Chemistry in Porous and Fractured Media: 1. Model Development and Validation. Report No. LBNL-42050, Lawrence Berkeley National Laboratory, Berkeley.

Xu, T. and Pruess, K. (2001) Modeling multiphase non-isothermal fluid flow and reactive geochemical transport in variability saturated fractured rocks: methodology. *American Journal of Science*, **301**, 16–33.

Yenice, H. and Dünya, H. (2007) Mechanical cleaning of calcite scaling with rotating control head preventer in a geothermal well 2007. *Energy Exploration and Exploitation*, **25** (6), 451–465.

Ziegler, F. and Alefeld, G. (1987) Coefficient of performance of multistage absorption cycles. *International Journal of Refrigeration*, **10**, 285–296.

Zhua, J. and Zhangb, W. (2004) Optimization design of plate heat exchangers (PHE) for geothermal district heating systems. *Geothermics*, **33**, 337–347.

7
Economic Performance and Environmental Assessment

Stephanie Frick, Jan Diederik Van Wees, Martin Kaltschmitt, and Gerd Schröder

7.1
Introduction

Besides the technical feasibility of the energy provision from enhanced geothermal system (EGS), its deployment depends on the establishment on the energy market. This means that the characteristics of national and international energy markets need to be understood to improve EGS's classification therein.

Energy markets are not driven only by demand and supply. The provision of energy addresses also social and political issues for the present and future generations. On one side, today's living standard is directly linked to the use of energy. And to maintain and improve this living standard energy must be securely available and affordable. On the other side, today's energy provision depletes finite fossil resources and emits substances, which are damaging the local environment as well as the global climate.

Sustainable energy markets, avoiding unwanted effects resulting from the provision and the use of energy, therefore need to manage the three dimensions: security of supply, economic affordability, and environmental compatibility (Figure 7.1) and the conflicts resulting thereof. Since different players with varying priorities are involved in the energy markets, the governments have to set an energy-political framework in order to ensure the development of energy markets, which can meet the mentioned sustainability criteria.

Against this background, a goal defined in many energy-political frameworks is to increase the share of renewable energies contributing to the mitigation of greenhouse gas emissions and the reduction in the consumption of finite energy resources within the energy system. But the costs for energy provision from renewable energy sources are not yet fully competitive to the prices in the market (even though these prices do not reflect the full lifecycle costs related to the use of fossil energy resources). Furthermore, the integration of renewable energies into the existing structures of the energy system is not always easy to realize. Therefore, promotion measures have been implemented in many countries over the last

Geothermal Energy Systems. Edited by Ernst Huenges

ISBN: 978-3-527-40831-3

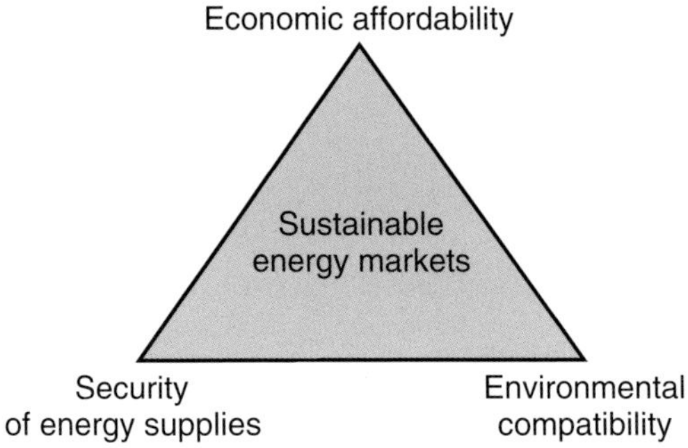

Figure 7.1 Strategic triangle of sustainable energy markets (based on Erdmann and Zweifel, 2008).

decade. Typical instruments are feed-in tariffs, quota models, or fiscal instruments (Bürger *et al.*, 2008; Fouquet and Johansson, 2008).

In general, economic affordability can be measured by the costs for a product in relation to the amount that the purchaser is able or willing to pay. For the provision of heat and/or electricity based on EGS, this means that the average energy provision cost, typically referred to as *levelized cost of energy* (LCOE), needs to be competitive to the price level in the energy markets. From this viewpoint, EGS plants are not yet competitive in most cases. However, EGS is still at the beginning of the market introduction and thus at the beginning of the learning curve. This means that with increasing deployment and accumulating experience, possibly stimulated by supporting measures, a reduction of the LCOE such as experienced for other energy technologies can be expected (IEA/OECD, 2000; Barreto, 2001). The potential development of EGS including research and referring to installed capacity and LCOE has been discussed, for example, by Ledru (2008) who summarized the results and experiences from the EU project "Enhanced Geothermal Innovative Network for Europe (ENGINE)" (Figure 7.2). It is hence important to develop the market in order to accumulate experience and to reduce LCOE.

Regarding environmental aspects, all options for the provision of energy are to a certain extent connected with an impact on the environment. An energy provision option with no impact on the environment does not exist. Against this background, all impacts of EGS projects on the local and global environment need to be analyzed precisely and must be assessed in relation to the locally given circumstances.

In this context, the chapter first outlines economic aspects of EGS projects. Secondly, potential environmental impacts are addressed. By means of representative case studies, the most important influencing factors will be identified and discussed. On the basis of the results, a better understanding of the economic and environmental characteristics to be considered in the planning and realization of EGS projects can be achieved.

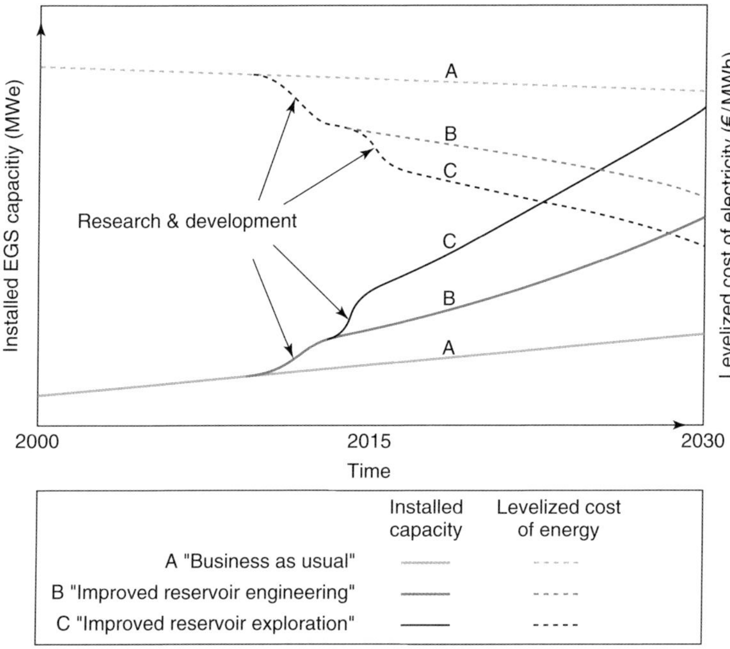

Figure 7.2 Potential development of the installed capacity and levelized cost of energy for EGS plants over time for different scenarios indicating the influence of research and development (derived from Ledru, 2008).

7.2 Economic Aspects for Implementing EGS Projects

The planning of EGS projects and especially the decision to realize a project is based on the estimation of the costs and revenues, which are related to a project. Since EGS projects are characterized by a long planning period, large initial investments, and a long technical lifetime, estimating prospective costs and revenues involves uncertainties and risks. This is true because no reliable statements on market development, detailed geologic site conditions, or technological problems can be made at the beginning of a project. In order to minimize existing risks, cost influences must be known and risks must be analyzed.

Analyzing the LCOE of an EGS plant, the general cost structure, the influencing factors, and the cost reduction potentials will be outlined. The section on decision and risk analysis then discusses how the planning of EGS projects can deal with the existing risks.

7.2.1 Levelized Cost of Energy (LCOE)

The LCOE is calculated based on the total costs throughout the overall economic lifetime of a plant related to the provided energy. The calculation of the LCOE is therefore an important and commonly applied approach in economic evaluation because it allows the comparison of different energy provision technologies or options with each other and with the prices that are paid on the energy markets, (e.g., IEA/OECD 2005; Tester *et al.*, 2006 and Ahmed 1994). In order to analyze economic aspects of EGS plants, the calculation of the LCOE is based on the estimation of the prospective costs (and revenues from possible by-products) since no commercial plants exist so far. The prospective costs thereby depend on the site- and project-specific conditions. Accordingly, LCOE, which are representative for all EGS sites cannot be declared. However, the general cost structure and influencing factors should be similar for different projects.

Within the following explanations, the methodological approach to calculate the LCOE and the basic investments for EGS plants for the provision of power and/or heat are discussed. The indicated costs thereby represent only orientation values, which are derived from different publications and own experiences, and can deviate for each site. By means of case studies, the LCOE are calculated for representative EGS power and heat plants. On the basis of these studies, the determining factors and the potentials to reduce the LCOE are addressed.

7.2.1.1 Methodological Approach

The methodology to calculate the LCOE is a commonly applied approach that is described in detail, for example, in Ahmed (1994) and IEA/OECD (2005). It accounts all payments of an EGS project in the monetary value of the reference year, which include the following:

- costs of capital related to the investments;
- operation costs such as for service and personnel;
- cost for consumables such as for supplies and auxiliary power;
- other costs such as for insurance and taxes; and
- revenues for by-products such as heat in case of power and heat supply.

Using the annuity method, these costs are converted in a series of constant annual payments. Equation 7.1 distinguishes between annual costs for consumables, operation and other expenses, annual revenues, and annualized capital user costs. The latter refers to constant annual payments which are the repayments and interest payments for the used capital. Equation 7.2 derives the annualized capital user costs from the investments for installing and maintaining the plant, to which an annuity factor is applied.

$$\mathrm{LCOE} = \frac{A_{\mathrm{total}}}{E_{\mathrm{a}}} = \frac{O_{\mathrm{a}} + I_{\mathrm{a}} - R_{\mathrm{a}}}{E_{\mathrm{a}}} \tag{7.1}$$

where,

LCOE = levelized cost of energy

A_{total} = annualized overall payments

E_{a} = annually provided energy

O_{a} = annual costs for consumables, operation and miscellaneous

I_{a} = annualized capital user costs

R_a = annual revenues for by-products

$$I_a = a \cdot I_{total} = \frac{i(1+i)^L}{(1+i)^L - 1} I_{total} \tag{7.2}$$

where,

a = annuity factor
I_{total} = total capital investments
i = imputed interest rate
L = number of annual periods within economic lifetime

7.2.1.2 **Cost Analysis**

The total costs of an EGS project are dominated by the investments at the beginning of the project. These investments mainly consist of costs for

- reservoir exploration;
- well drilling and completion;
- reservoir engineering measures;
- installation of the geothermal fluid loop; and
- construction of the plant on the surface for power and/or heat provision.

Further investments can include exploration measures, project planning, risk insurances, or replacement purchases during operation.

The variable costs during the operating phase are mainly caused by the salaries for personnel, and supplies to run and maintain the subsurface and surface installations. Additionally, payments for the consumption of auxiliary power need to be considered in case the required power to produce the geothermal fluid from the reservoir is not provided by the plant itself (e.g., in case of EGS heat plants).

Quantifying the investment costs of an EGS project is crucial since the costs depend on different conditions, which cannot be generalized for all EGS sites and must be estimated with large uncertainties. Additionally, the costs for accessing and developing the reservoir, which dominate the overall investments, are strongly related to the site-specific geologic conditions and the actually consumed time for drilling, well completion, and reservoir engineering. The investments for installing the surface part, including the geothermal fluid cycle and the plant unit on the surface, depend on technological specifications. The estimation of the investments in an EGS project must therefore be based on site-specific key geological and technical parameters.

In the following, the influence of site-specific conditions and technological specifications on the costs of an EGS project are analyzed in detail. The costs, which are presented in this section refer, unless not indicated, to the monetary value of 2008 and must be seen as indicative of the order of magnitude. It must be considered that the costs of specific EGS plants can differ from the indicated values due to locally given circumstances.

Reservoir Access and Development The costs for accessing and developing the reservoir represent the largest part of the investments in EGS projects. A general assessment of these costs is difficult due to varying geological conditions influencing the drilling, completion, and reservoir stimulation. When estimating the cost previous to the operational activities, the geological conditions are only roughly predictable, especially when a site is located in an unexploited geological area. In case of the reservoir engineering measures, the limited experience available so far is determining the uncertainties in cost estimation.

Well costs Cost estimations at the beginning of a project are usually based on existing cost data for already drilled and completed wells. However, getting access to such data is not always possible because the data are often confidential and/or not reported in a comparable way of documentation.

Detailed cost calculations use geographical and geological site information. Such calculations usually consider also a supplemental charge for unforeseen troubles, such as stuck pipes or hole-stability problems. This charge typically lies between 10 and 20%, but can also be higher in unknown geologic areas.

The drilling and completion costs can be split into the following cost items:

- **Rig rent:** The rig rent is usually paid on an hourly or daily rate for the time the drilling rig is used. The rate for a particular drilling rig thereby depends on its specifications such as hook load and depth capacity. A drilling rig with larger hook load results in a higher rate but, on the other side, can realize a faster drilling progress and a decrease of the term of lease. From an economic viewpoint, the choice of the drill rig is therefore a compromise between rig capacity and drilling progress.
- **Material costs:** The material costs basically include the expenditures for casings, drilling mud, and drilling bits. These costs depend on the borehole design, such as diameter, depth, and well course, as well as on the site-specific stratigraphy, which for example, determines the casing material and its insulation thickness.
- **Energy costs:** The energy costs refer to the power to drive the drilling rig and the drilling mud pumps and depend on the means of energy provision (e.g., energy provision from electricity grid or diesel electric rig drive).
- **Service costs:** The service costs include quantity-dependent service costs, which are borehole-related services (such as installation of the casings, cementation, and logging) and drilling site–related activities (such as installation and dismantling of the drilling rig and drilling site preparation). Time-dependent service costs contain costs for core barrels, jars, stabilizers, surveys, and drilling mud treatment.

Depending on the site and well design, the composition and the total amount of the borehole costs can significantly vary such as shown in Figure 7.3. The borehole costs thereby increase overproportionally with the depth. This is mainly related to the decreasing drilling progress with larger depth. This can be seen from Figure 7.4, which shows the development of the well costs and depth versus time for a specific site. The costs increase nearly linearly with increasing time, whereas

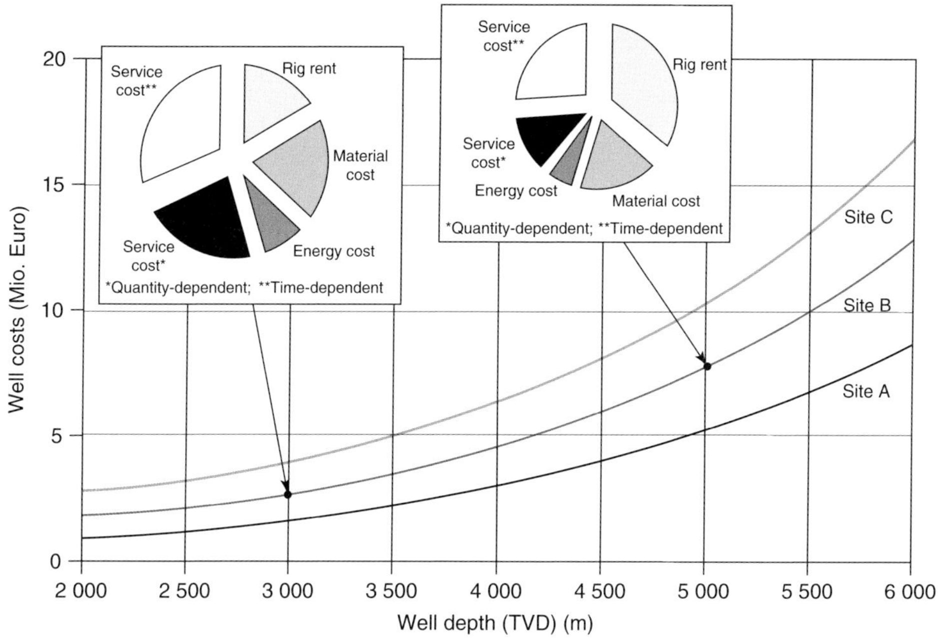

Figure 7.3 Development of well costs versus well depth for the example of different sites (= different geological conditions and borehole design) and typical compositions of well costs at a depth of 3000 and 5000 m (based on Legarth, 2003).

the drilling progress decreases with larger depths because, particularly, the round trip times increase.

Besides geological and technical influences, the situation on the drilling rig and the commodity markets has a decisive impact on the well costs. The price development in the steel markets, for example, and the increased demand on drilling rigs from the oil and gas industry (which usually can offer the drilling rig operator a better capacity utilization) have resulted in significantly increasing well costs. Figure 7.5 shows the results of a study done by Tester *et al.* (2006), which represent the development of the borehole costs compared to the year 1977 based on a drilling cost index. Thus, the drilling costs are very much influenced by the energy prices and basically by the prices of crude oil.

Reservoir engineering costs The cost for reservoir engineering measures cannot be based on existing cost data. Such data are not or only scarcely existent. Existing data is furthermore valid only for particular sites with comparable characteristics referring to petrophysical and rock mechanical properties of the reservoir, and therefore a comparable technical effort for the stimulation measures. Further factors that influence the technical effort are skin damage caused by drilling and the targeted increase of transmissibility.

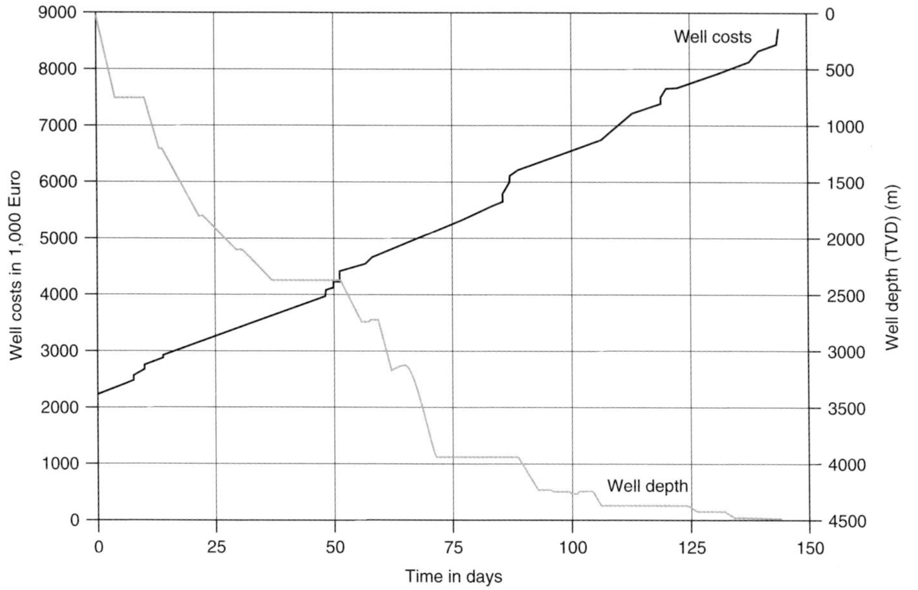

Figure 7.4 Development of well costs and well depth versus time for the example of the well GrSk 4/05 drilled in 2005 at the Groß Schönebeck site.

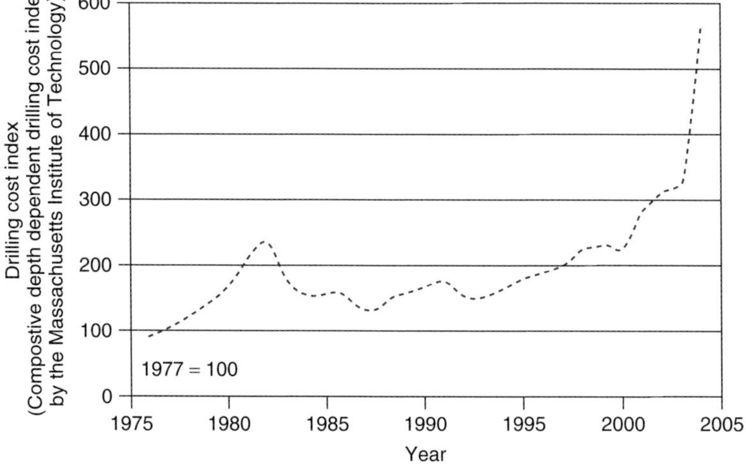

Figure 7.5 MIT depth dependent drilling cost index using average cost per well from Joint Association Survey on Drilling Costs for well depths between 400 and 6000 m (Tester *et al.*, 2006).

On the basis of existing experiences and publications it can however, be derived that the stimulation costs can also be split into equipment rent, material cost, energy cost, and service cost. The equipment rent includes the cost for fluid pumps, blenders for mixing the frac-fluid, and miscellaneous peripheral equipment. The material costs depend on the used fluid and additives and the amounts, which are injected into the reservoir. The energy cost refers to the power that is consumed by the injection pumps, for example. The required power is determined by the injection pressure, the flow rate, and the duration of the stimulation measures.

There is no common approach to estimate the stimulation costs. In Tester and Herzog (1990) the estimation of the cost for basement rock stimulation is based on a defined reservoir area, which needs to be stimulated to obtain 1 kW electric power output. The declared stimulation cost thereby varies between €360 and 1200 $kWel^{-1}$, depending on the size of the reservoir. Legarth (2003), in contrast, estimates the stimulation costs per well at a specific site based on the frac-fluid volume and amount of proppants that are used for a specific reservoir, and comes to total costs between €0.32 and 0.64 million (excluding permanent or temporary installations such as injection tubes and packer). More recent publications such as Heidinger, Dornstädter, and Fabritius (2006) and Sanyal *et al.* (2007) assume fixed costs of €0.36–0.71 million per well or €0.58 million per reservoir, respectively.

Surface Installations The investment costs for the surface part of an EGS plant include the cost of the geothermal fluid cycle and cost of the plant unit. The estimation of these costs can be based on cost estimations for the single components and peripheral equipment and for their installation at the plant site.

Geothermal fluid loop The investments for the geothermal fluid cycle contain the cost for the equipment such as pumps, pipes, filter, and slop systems to produce and circulate the geothermal fluid. At some sites, injection equipment is also used.

In most EGS projects, the production pump will represent the main cost factor because of the technical requirements, which this component has to meet. The pump must be capable of supplying the necessary pressure increase to produce the geothermal fluid from the reservoir and to handle temperature, chemical composition, and gas content for the proposed technical lifetime.

Furthermore, the production at many EGS sites is based on the use of downhole pumps. Since these pumps are installed in the production well, design constraints such as installation depth and diameter, mechanical drive, and energy provision are also further technical restrictions. Depending on the site- and plant-specific conditions, the costs for downhole pumps range between €2000 (e.g., Legarth, 2003; Rogge, 2004) and 7500 $(m^3/h)^{-1}$ according to experiences from the Groß Schönebeck site and Heidinger, Dornstädter, and Fabritius (2006). Regarding the cost composition, such as shown in Figure 7.6, this range mainly results from varying costs for the frequency converter and the motor, which both depend on the pump capacity.

The costs for other components such as pipes, filters, and slop systems mainly depend on the flow rate of the geothermal fluid and the necessary overpressure

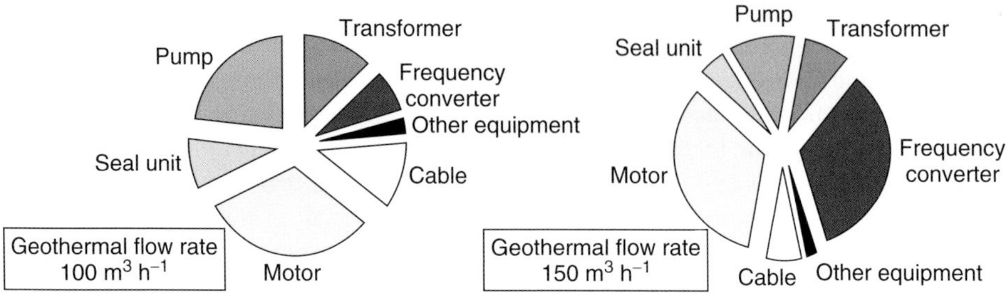

Figure 7.6 Typical compositions of downhole pump costs for different flow rates (based on Legarth, 2003).

in the fluid cycle. Further influencing parameters are chemical composition, gas content, and temperature of the geothermal fluid, which determine the material choice. The costs for the pipes also depend on the length and, in case of long-distance pipelines, on the laying (i.e., surface or subsurface). The cost for the geothermal fluid cycle, excluding the production pump, can vary between €75 and 600 m^{-1} (Kaltschmitt *et al.*, 2007 Rogge, 2004).

Power plant unit The costs for a binary power unit are generally related to the installed capacity, whereby the specific investments typically decrease with larger capacity due to economy of scale. The main cost factors of a binary plant are the turbine and generator unit, the heat exchangers, and the cooling unit. With regard to EGS projects, the influence of the geothermal fluid temperature and site-specific conditions, which determine the mode of the installed cooling system (water or air cooling), are additional important factors. For installing the same capacity at a site with a low geothermal fluid temperature, for example, will be more expensive due to the larger heat exchange area which is required compared to a site with higher geothermal fluid temperature. Referring to the installation of the cooling system at a specific site, the realization of air cooling is in many cases more expensive than using wet-cooling towers. Furthermore, the complexity of the power conversion cycle (e.g., basic Rankine cycle, Rankine cycle with two pressure levels or Rankine cycle with working fluid mixture) influences is the cost. The characteristics of the geothermal fluid need to be considered for determining the material and layout of the respective heat exchangers. According to Köhler (2005), the specific binary plant investment is approximately between €1400 and 2300 kW^{-1} for an installed capacity in the range from 500 to 2000 kW (Figure 7.7).

Heat plant unit When referring to a geothermally based heat supply without additional heat generators as backup or peak-load units, the investments for the heat plant unit mainly include the cost for the heat exchanging equipment. Depending on the layout and design of the heat exchangers, the specific cost can range between €10 and 100 kW^{-1} thermal capacity. Investments for the construction of a heating grid are not considered in this section.

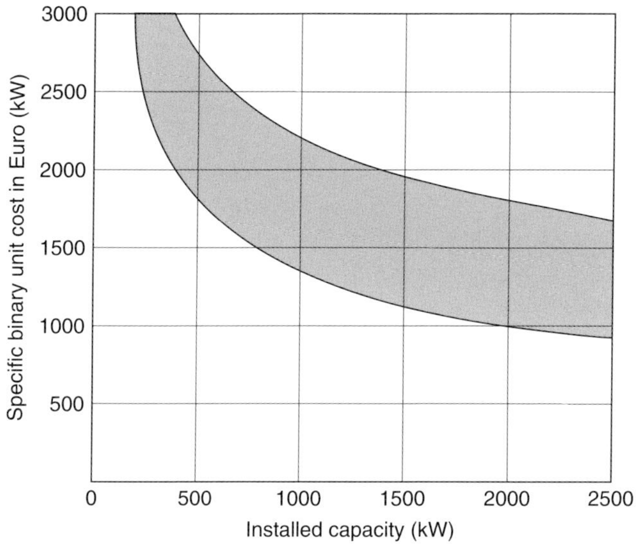

Figure 7.7 Range of specific binary plant cost (Köhler, 2005).

Operational and Other Costs The annual operating cost of EGS plants mainly include the costs for personnel, consumption material (e.g., lubrication oil), overhaul, and maintenance. In many cases, EGS plants can operate without continuous supervision so that the annual personnel cost is low. The costs for consumption material, overhaul, and maintenance are usually estimated with a small percentage of the investments for the subsurface and the surface part. If the auxiliary power demand cannot or is not provided by the EGS plant itself, then the cost for auxiliary power consumption also needs to be considered.

Depending on the project, potentially additional investments need to be considered. Project planning, for example, can be a complex and time-intensive part of realizing successful EGS projects including among others feasibility studies, siting, permitting, and coordinating the engineering design of the wells and the surface installations. Project planning can take up to 10% of the overall investments. Depending on the available geological information for a site, further information on the subsurface is needed. Subject to the planned program, exploration can take up to €1 million (Heidinger, Dornstädter, and Fabritius, 2006). Further investment costs can also arise if a project is carried out close to housing areas, where noise protection, such as the erection of sound insulating walls, is necessary. Insurances, such as for covering the geological risk, are further cost factors which might need to be considered when estimating the costs for EGS projects.

7.2.1.3 **Case Studies**

In the following, the LCOE for different representative EGS plants from which power and/or heat is provided, are analyzed. By means of case studies general

Table 7.1 Example EGS plants for power production.

	Power plant 1	Power plant 2
Reservoir[a]		
Depth (km)	4.0	4.0
Temperature (°C)	125	165
Physical life (a)	20	20
Geothermal fluid cycle		
Geothermal flow rate ($m^3\,h^{-1}$)	250	120
Distance between wells (m)	500	500
Auxiliary power need relating to flow rate kW ($m^3\,h^{-1})^{-1}$	1.3	1.3
Binary plant unit		
Installed capacity (MW)	1.6	1.6
Conversion efficiency (%)	10[b]	12[b]
Full load hours ($h\,a^{-1}$)	7 500	7 500
Auxiliary power need relating to installed capacity (%)	10	10

[a]Geothermal fluid circulation with downhole pump at a pressure increase of 35 bar and a downhole pump efficiency of 75% based on Legarth (2003).
[b]Assumed conversion efficiency depending on geothermal fluid temperature, injection temperature 70 °C, geothermal fluid heat capacity 4.2 kJ $(kg\,K)^{-1}$ based on Köhler (2005).

correlations between the plant data and the LCOE, referring to electricity or heat, are derived. Based on these correlations, the potentials which exist in each EGS project to reduce the LCOE will be discussed.

The case studies are not designed to replace cost estimations for the realization of EGS projects. Site- and country-specific conditions such as taxes, subsidies, and special depreciation possibilities are not considered.

LCOE Electricity Provision Two representative EGS sites for power production are investigated. Both reservoirs are assessed with two deep wells of a depth of 4000 m. Due to different temperature gradients, the reservoirs differ in temperature. The reservoirs are engineered with the same technical effort but reach a different transmissibility because of differences in the natural reservoir conditions. The geothermal fluids are produced from the reservoirs using submersible pumps in the production wells. On the surface, the geothermal energy is converted to power within binary power plant units. The necessary data to define both plants are summarized in Table 7.1.

Power plant 1 produces the geothermal fluid with a temperature of 125 °C and a flow rate of 250 $m^3\,h^{-1}$. For power plant 2, the temperature is 165 °C and the flow rate 120 $m^3\,h^{-1}$. On the surface, the fluid is on each site transported in a closed fluid cycle. The geothermal heat is transferred to the binary plants which contain a low-boiling working medium and the turbine-generator unit. After the heat transfer, the cooled geothermal fluid is transferred to the respective injection well, which is located 500 m away from the production well and returned into the reservoir with an assumed temperature of 70 °C.

The installed capacity of both power plants is 1.6 MW. The auxiliary power needed for circulating the geothermal fluid is assumed with 1.3 kW $(m^3\,h^{-1})^{-1}$ for both sites. The auxiliary power need in the binary cycle (e.g., for the feed-pump and cooling) is defined with 10% of the installed capacity. Both power plants operate 7500 full load hours per year which corresponds to an availability of 86%. The total technical operating time is 20 years.

Table 7.2 shows the economic data for the defined EGS power plants. For both plants, the investment costs sum up to €23 million or a specific investment of about €15 000 kW^{-1}. The well costs represent with 69% the largest cost factor of the overall investments. The next largest investment, with a share of 12%, is the binary plant unit. Within the other costs, additional costs such as for exploration activities or project planning and management are considered with 10% of the total investments. The costs for the reservoir engineering measures are assumed with €0.75 million per well (i.e., a share of 6% in the investments). The costs for the geothermal fluid loop are about 3%.

Based on an assumed interest rate of 6% and an economic lifetime equal to the technical life time, the overall investments result in an annuity of about €2 million a^{-1}, which represents 80% of the annual payments. The remaining 20% are the annual operating costs, which include the cost for personnel, consumption material, overhaul, and maintenance and also include the replacement costs for components with a technical lifetime below the lifetime of the EGS plant. Overhaul and maintenance costs of the subsurface and surface installations are estimated with 1.5% of the corresponding investments and 6% respectively.

With these data, LCOE of €0.32 $(kW\,h)^{-1}$ for power plant 1 and €0.26 $(kW\,h)^{-1}$ for power plant 2 are calculated. The difference of 19% in the LCOE indicates that they do not only depend on the investments for a specific installed capacity since these costs only differ by 2% comparing power plants 1 and 2. Therefore, the plant-specific net-power output is the crucial factor, because power plant 2 has a lower auxiliary power consumption due to the lower flow rate in the geothermal fluid loop.

LCOE Heat Provision In the following, the LCOE for two EGS heat plants at different representative sites are compared. Analog to the analysis of the power provision, at both sites well doublets tap the reservoirs which are located in the same depth but differ in regard to temperature and natural conditions. The necessary data to define the heat plants are summarized in Table 7.3.

Heat plant 1 is characterized by a geothermal fluid temperature of 100 °C and a flow rate of 210 $m^3\,h^{-1}$. In heat plant 2, a geothermal fluid with 130 °C and a flow rate of 120 $m^3\,h^{-1}$ is produced. On the surface, the geothermal heat is transferred to a low temperature district heating grid with a supply temperature of 70 °C and

Table 7.2 Cost and economic data of the defined EGS power plants.

Economic data	Power plant 1	Power plant 2	Comment
Investment costs in €1000			
Well doublet	16 000	16 000	€8 million per well
Reservoir engineering	1 500	1 500	€0.75 million per well
Downhole pump	500	240	€2 000 $(m^3\,h^{-1})^{-1}$
Geothermal fluid loop	250	250	€500 m^{-1}
Binary plant unit	2 800	2 870	€1 800 kW^{-1}
Other costs[a]	2 339	2 318	10% of total investments
Operating costs in €1000 a^{-1}			
Personnel and consumables	50	50	One-man operation
O&M[b] subsurface	263	263	1.5% of subsurface investments
O&M[b] surface	213	202	6% of surface investments
Annuity calculation			
Imputed interest rate	6%	6%	–
Economic lifetime	20[a]	20[a]	–
Annuity investments in €1000 a^{-1}	2 039	2 021	–
Annual operating costs in €1000 a^{-1}	526	514	–
LCOE (€ (kW h)$^{-1}$)	***0.32***	***0.26***	–

[a]Other costs such as for project planning, exploration, and so on.
[b]Overhaul and maintenance.

a return temperature of 50 °C. The cooled geothermal fluid is reinjected with a temperature of 60 °C considering a counterflow heat transfer with a temperature difference of 10 K. It is assumed that the heat plants are only driven by geothermal energy. Therefore, no additional equipment, such as backup or peak-load systems, needs to be considered in this case study.

Based on these data, a thermal capacity of 9.8 MW is installed at both sites. The auxiliary power, which is needed for circulating the geothermal fluids, is assumed with 1.3 kW $(m^3\,h^{-1})^{-1}$. Based on a typical seasonal demand for district heat, the heat plants operate 3100 full load hours per year. The total technical operating time is 20 years.

Table 7.4 shows the economic data of these EGS heat plants. The total investments are about €16 million or €1600 kW^{-1} thermal capacity. The well costs are responsible for 74% of the overall investments and are the predominant cost factor. Reservoir engineering is declared with €0.5 million per well (i.e., 6% of the investments). The influence of the heat unit on the surface on the total capital investment is in the same order of magnitude. The investments for the geothermal fluid cycle sum up to 4% of the total budget. Other costs including exploration

Table 7.3 Example EGS plants for heat provision.

	Heat plant 1	Heat plant 2
Reservoir[a]		
Depth (km)	3.0	3.0
Temperature (°C)	100	130
Physical life (a)	20	20
Geothermal fluid loop		
Distance between wells (m)	500	500
Geothermal flow rate ($m^3\,h^{-1}$)	210	120
Auxiliary power need ($kW\,m^{-3}\,h^{-1}$ flow rate)	1.3	1.3
Heat plant unit		
Installed capacity (MW)	9.8	9.8
Full load hours ($h\,a^{-1}$)	3.100	3.100

[a]Geothermal fluid circulation with downhole pump at a pressure increase of 35 bar and a downhole pump efficiency of 75%.
[b]Feed temperature 70 °C and return temperature 50 °C, brine cooling temperature 60 °C, brine heat capacity 4.2 kJ (kg K^{-1}).

activities, project planning and management are assumed with 10% of the total investments.

With an assumed interest rate of 6%, and an economic lifetime of 20 a, the annuity of the capital investments sums up to €1.4 million a^{-1}. The overall annual payments therefore consist to approximately 78% of the investment annuity and to 22% of the operating costs. The annual operating costs contain the cost for personnel, consumables, overhaul, maintenance, and the annual consumption of auxiliary power.

According to these assumptions, the LCOEs are calculated with €0.061 $(kW\,h)^{-1}$ for heat plant 1 and €0.059 $(kW\,h)^{-1}$ for heat plant 2, which result in a difference of 4% between the two plants. Since the provided amount of district heat is the same for both plants and overall investments only differ by 2%, this difference mainly reflects the different operating costs, which are larger for heat plant 1, due to the larger geothermal fluid flow.

Cost Reduction Potentials Based on the defined EGS plants, the potentials to reduce the LCOE are discussed in the following section. These potentials refer to technical improvements which can be expected with the further development of EGS (i.e., learning curve effects), and also to project design and planning approaches which allow more cost-effective EGS projects. The most effective way to reduce the LCOE is achievable if all reduction potentials can be addressed in an EGS project.

Table 7.4 Cost and economic data of the defined EGS heating plants.

Economic data	Heat plant 1	Heat plant 2	Comment
Investment costs in €1 000			
Well doublet	12 000	12 000	€6 million per well
Reservoir engineering	1 000	1 000	€0.5 million per well
Downhole pump	420	240	€2 000 $(m^3\,h^{-1})^{-1}$
Geothermal fluid loop	630	360	€400 m^{-1}
Heating plant unit	980	980	€100 kW^{-1}
Other costs[a]	1 622	1 602	10% of total investments
Operating costs in €1 000 a^{-1}			
Personnel and consumables	50	50	One-man operation
Auxiliary power consumption	84	48	€0.10 $(kW\,h)^{-1}$
O&M[b] subsurface	195	195	1.5% of subsurface investments
O&M[b] surface	96	85	6% of surface investments
Annuity calculation			
Imputed interest rate (%)	6	6	–
Economic lifetime (a)	20	20	–
Annuity investments in €1 000 a^{-1}	1 414	1 397	–
Annual operating costs in €1 000 a^{-1}	341	330	–
LCOE(€(kW h)$^{-1}$)	0.061	0.059	–

[a]Costs for miscellaneous such as project planning, exploration, and so on.
[b]Overhaul and maintenance.

Technical improvements Due to learning curve effects, further technical developments and the deployment of EGS will lead to a reduction of the investment costs. This will also result in a decrease of the LCOE, since EGS plants are characterized by large investments. Also, cost improvements regarding overhaul and maintenance (e.g., such as cost reductions for the exchange of the downhole pump), can be expected and will have a positive effect. Figure 7.8 shows, for the example of power plant 1, how a reduction of the main cost factors influences the LCOE.

According to Entingh and McVeigh (2003), a cost reduction of 50% is considered for each reservoir stimulation measures, binary units, and overhaul and maintenance. In case of power plant 1, such improvements lead each to a reduction of the LCOE of 3–9%. Regarding the well costs, in contrast, the same cost reduction could lead to a decrease of the LCOE by more than 30%. However, the existing cost-saving potential for well drilling and completion is limited. Even if technical improvements or a better knowledge of specific geologic areas result in a more efficient drilling and completion process, the price increases of the drill rig and within the steel market can compensate these improvements (cf. Figure 7.5). Considering

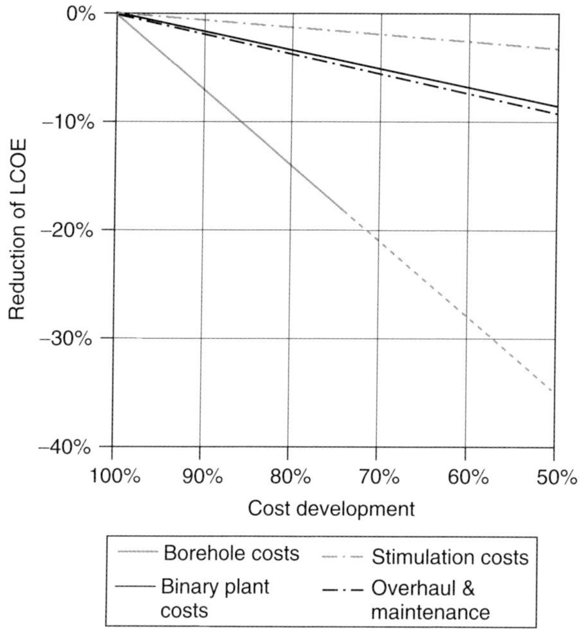

Figure 7.8 Influence of lower investment costs on the LCOE using the example of power plant 1.

a 25% decrease of the well costs as realistic value, the LCOE of power plant 1 are reduced by 17%.

Increased net-power output by improved project design Besides the reduction of investment and overhaul and maintenance costs, the LCOE from EGS plants can also be improved by increasing the net-power output at a specific site. In this context, Figure 7.9 shows the influence of an increased flow rate on the LCOE based on an enhancement of the reservoir productivity for power plant 1. With an improvement of the existing reservoir engineering measures, the same technical effort (and therefore the same costs for stimulation) will result in a higher reservoir productivity or transmissibility. Therefore, significantly higher flow rates can be produced with the same specific effort for pumping. An increase of the flow rate by 50% reduces the LCOE by 27%. If the flow rate can be doubled, a cost improvement of 39% is achieved.

Besides possible improvements in reservoir engineering, the enhancement of the reservoir productivity with a larger technical and therefore monetary effort can also reduce the electricity production costs since higher cost of stimulation has a comparatively small impact (cf. Figure 7.8). Even if the increase of the cost for reservoir engineering is five times higher than the realized increase of the flow rate, the LCOE is still reduced. For power plant 1, a doubling of the flow rate will in this case result in 25% lower production costs.

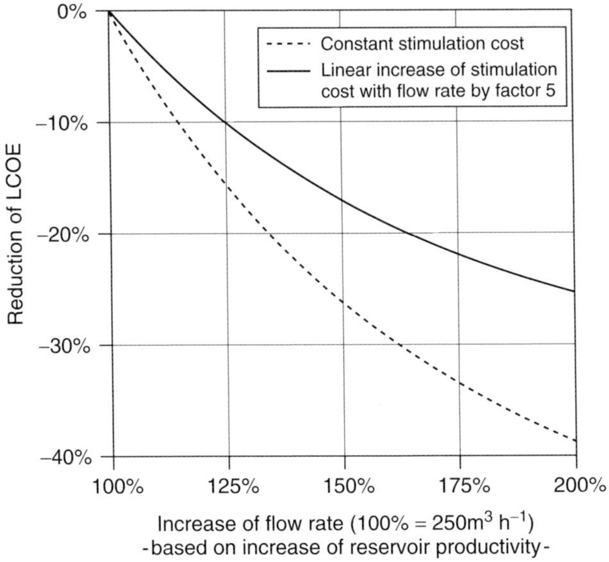

Figure 7.9 Influence of flow rate increase with additional or more efficient reservoir engineering on the LCOE using the example of power plant 1.

Another possibility is to enhance the net-power output by modifying the binary power conversion unit in order to realize higher conversion efficiencies. According to Figure 7.10, increasing the conversion efficiency of power plant 1 by 1%-point or 10% results in 10% lower LCOE – if this improvement is achieved by improved technology which can be installed for the same specific investments as presented in Table 7.2. However, even if higher conversion efficiencies are realized with more complex technology resulting in a doubling of the specific investments for the binary unit, the LCOE are decreased by 6%.

Other opportunities to increase the net-power production are technological modifications to decrease the auxiliary power demand by choosing more efficient cooling systems or installing more efficient fluid production technology in the geothermal fluid loop. In the case of EGS heat plants, a more efficient production of the geothermal fluid (due to a more productive reservoir or downhole pumps with improved efficiency) will result in the reduction of annual costs for auxiliary power consumption.

Increased or additional heat supply by improved project design A further approach to reduce the LCOE is the increase of the utilization ratio of the heat contained in the geothermal fluid.

In EGS power plants, the utilization ratio can be increased, for example, by using the heat in the geothermal fluid, apart from the heat transfer to the binary power unit, also for the supply of district heat. In Figure 7.11a the influence of

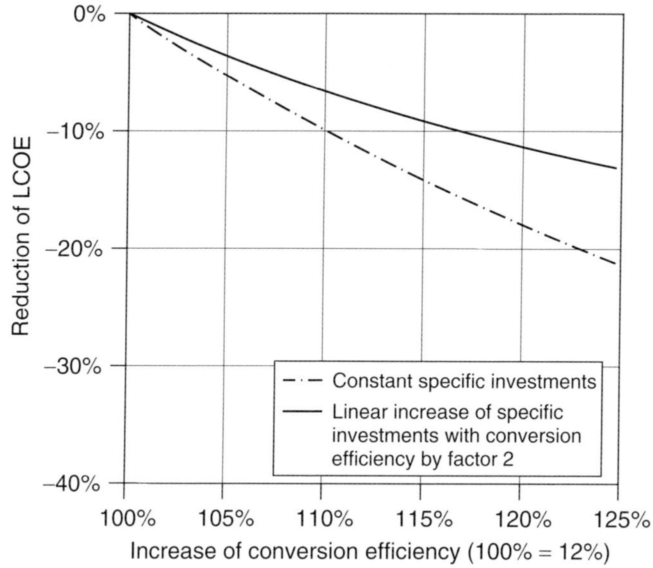

Figure 7.10 Influence of conversion efficiency increase with improved conversion technology on the LCOE using the example of power plant 1.

such a combined power and heat supply is shown for power plant 1. With the heat supply to a low temperature district heating system, a reduction of the LCOE of up to 55% can be realized depending on the supply and return temperature in the heating grid and the reimbursement rate of the heat.

In EGS heat plants, the utilization ratio can be improved with the development of an appropriate heat consumer structure so that a continuous heat demand with high thermal full load hours and a large temperature difference in the heating grid are met. Using the example of heat plant 1, Figure 7.11b shows the influence of increasing thermal full load hours and a larger temperature difference on the LCOE. These measures offer for heat plant 1 a cost reduction potential of up to 60%.

Improved project planning Since EGS plants are characterized by large capital investments and a long operation period, the economic lifetime and the interest rate have a significant influence on the LCOE. As shown in Figure 7.12 using the example of power plant 1, a reduction of the interest rate to 3% would decrease the production cost by 18%. If the economic and also technical lifetime is increased from 20 to 30 years, a cost saving potential of 13% can be seen for power plant 1. Cost-effective EGS projects therefore, need to be based on appropriate financing models, which adequately consider the high capital investments and the long operating period.

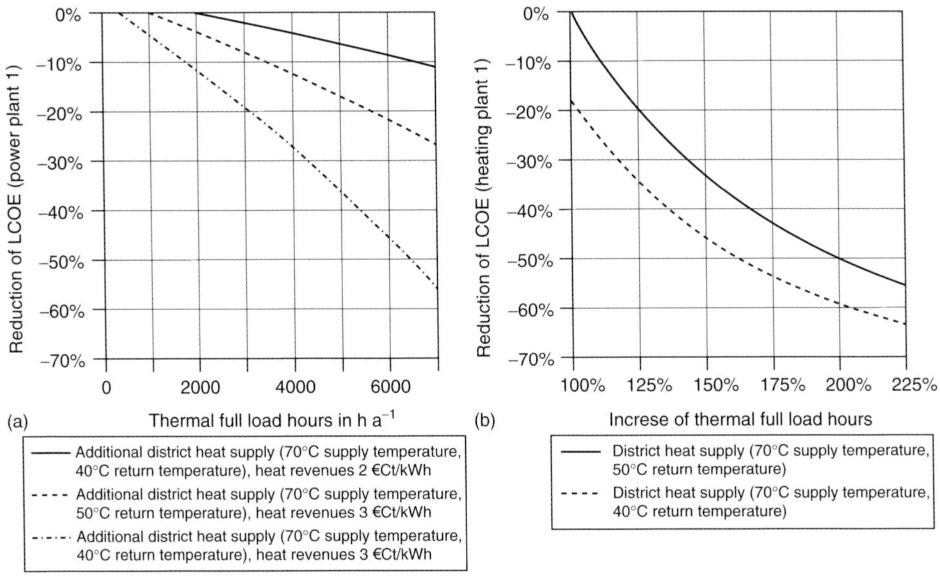

Figure 7.11 Influence of an increase of the geothermal utilization with larger district heat supply on the LCOE using the example of power plant 1 (a) and heat plant 1 (b).

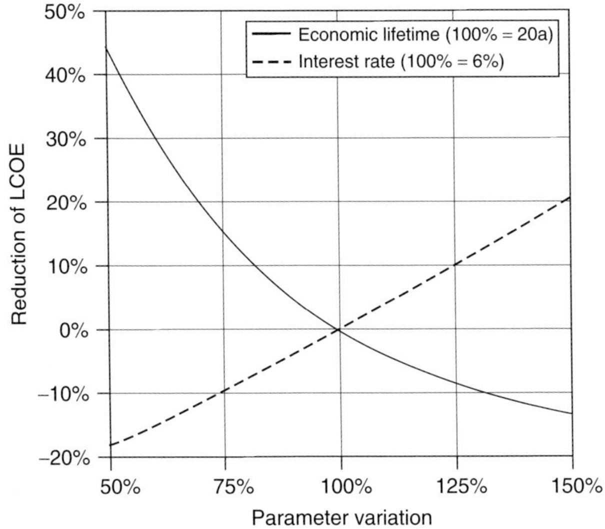

Figure 7.12 Influence of economic lifetime and interest rate on LCOE using the example of power plant 1.

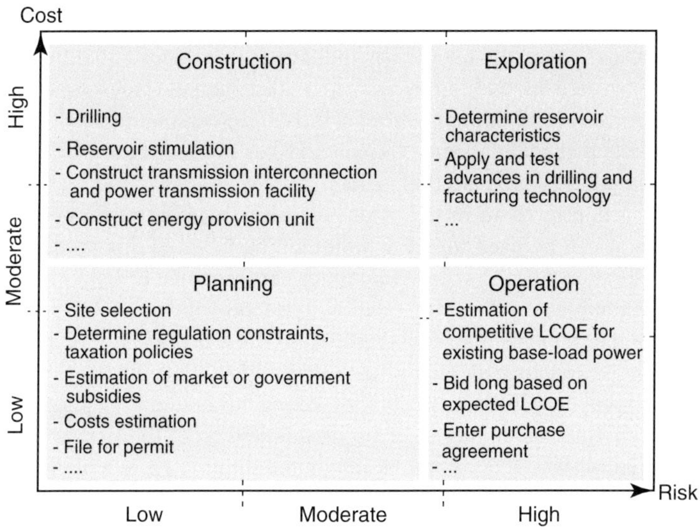

Figure 7.13 Classification of costs and risks in EGS projects. (Based on Tester *et al.*, 2006.)

7.2.2
Decision and Risk Analysis

In the previous section, performance calculations for LCOE have been discussed for various types of geothermal projects. In addition, the sensitivity of LCOE for possible future improvements, regarding technical, and engineering options have been addressed. Such improvements deal with factors which are under human control and enhance economic performance.

In addition to factors under human control, there are many parameters which are largely uncertain and uncontrollable. Such parameters can have a large influence on the economic performance. Such factors, among others, deal with subsurface uncertainty such as reservoir temperature, and productivity, unforeseen delays, as well as legal or technical hurdles, and economical uncertainties especially in terms of drilling costs and unforeseen changes in the energy prices. Tester *et al.* (2006) have presented an overview of the different stages of EGS projects and the corresponding costs and risks. Their classification is shown in Figure 7.13.

In the capital-intensive oil and gas industry, it has been long recognized since decades that many exploration and production (E&P) projects failed to deliver the performance they promised as a result of the underestimation of risks (Bratvold and Begg, 2008). Studies by Merrow (2003) showed that one in eight E&P projects with capital expenditure ranging from \$1 million to \$3 billion US dollars were disasters. Such a disaster is defined in terms of the project failing on two out of the following three metrics:

- not more than 40% cost growth,
- not more than 40% time slippage, and

- first year operability at least 50% of plan.

Based on these experiences, the oil and gas industry has developed quantitative frameworks for better decision making regarding the allocation of resources, when faced with uncertainty beyond human control (Floris and Peersmann, 2002; Bos, 2005). Evidence exists that adopting such frameworks has improved the economic performance of companies (Jonkman *et al.*, 2000).

Therefore, this section gives an introduction to such a quantitative framework and describes how it can be used for EGS projects. The focus of this section is thereby on the exploration stage at which stage much can be gained by risk analysis and improved decision making under uncertainty. The tools which have been used in this section are also published as deliverable from the EU project "Enhanced Geothermal Innovative Network for Europe (ENGINE)" (ENGINE, 2008).

By using a case study it will be shown that by taking advantage of quantitative risk and decision analysis techniques, a better understanding of the risks will be achieved and risk mitigation actions can be integrated into the project planning. In this context, the potential positive influence of staged exploration strategies, in particular, is discussed. Optimized and novel exploration technologies can considerably help in minimizing risk and increasing economic success. Another important aspect is sharing the costs of risk mitigation among a group of projects with comparable geologic preconditions. Dependencies may not only be related to critical characteristics shared between nearby project sites, but can also be related to conceptual breakthroughs in pilot projects.

Apart from improved exploration techniques, technological breakthroughs for production technology, energy conversion, and improved project planning are also likely to increase the expected return in the future as it has been shown in Section 7.2.1.

7.2.2.1 **Methodology**

Uncertainties in a project's technical and cost specification also result in uncertainties when estimating LCOE at the beginning of a project. If, as a result of technical uncertainty (e.g., a lower than expected flow rate), LCOE is higher than expected, it is to be considered a risk. However, positive consequences of uncertainty are also observed through a lower LCOE, for example, when the temperature and flow rate is higher than expected. For economic performance assessment of projects prior to the exploration activity and decision making, the LCOE is usually not convenient. A project may be aborted at an early stage at certain costs without having delivered any power or heat. In order to take this effect into account the *net present value* (*NPV*) is generally preferred to be used over LCOE in decision making.

NPV is defined as the total present value of a time series of outgoing and incoming cash flows related to a project. It is a standard method for using the time value of money to appraise long-term projects. Used for capital budgeting, and widely throughout economics, it measures the excess or shortfall of cash flows, in present value terms. To calculate NPV, all cash flows are discounted back to its present value and are summed into the cumulative discounted cash flow (CDF). Assuming that all capital investments are converted in a series of constant annual payments and the monetary value of a project after its economic lifetime is zero, the CDF can be calculated according to Equation (7.3):

$$\text{CDF} = \sum_{t=1}^{T} \frac{R_t}{(1+i)^t} \tag{7.3}$$

where

t = the time (year) of the cash flow relative to the start of the project
T = the period under consideration (economic lifetime) in years
R_t = the net cash flow (the amount of cash according to inflow minus outflow) at time t
i = the imputed interest rate (Section 7.2.1.1).

For a realized project, the NPV equals the CDF at the end of the project for $t = T$. Figure 7.14 gives an example of the calculation of cash flow and NPV for an EGS project during its lifetime.

From Figure 7.14 it can also be seen that the NPV of a project which is aborted before it reaches the operating stage turns out to be negative. The financial risk is therefore defined by the so-called downside, which equals the average returns below the target of NPV ≥ 0 (cf. Markowitz, 1952; Sharpe, 1964).

For financial risk analysis and decision making, different instruments, which are based on the calculation of NPV of a project, are used with the general aim to lower the financial risk, and at the same time, increase the expected return. Integrating capable planning instruments (such as probabilistic models, sensitivity analysis, decision trees, and portfolio analysis) in the project development, risk mitigating actions, and their staging in the project execution can be optimized and associated decision tollgates and project milestones can be implemented effectively. These tools

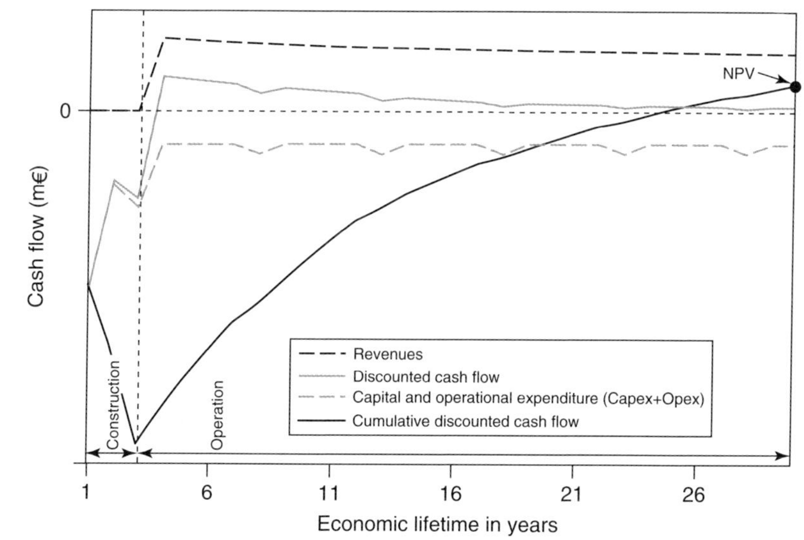

Figure 7.14 Cash flow calculation and net present value of an EGS project.

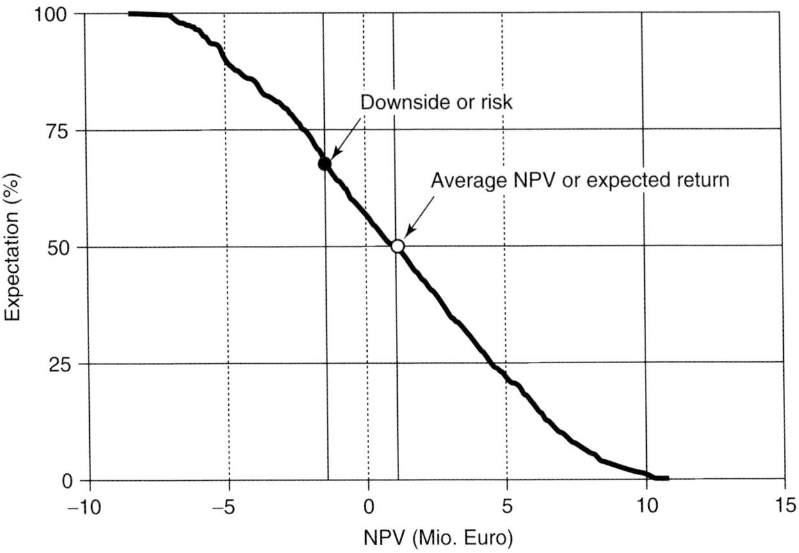

Figure 7.15 Expectation curve of a NPV distribution and expected NPV considering subsurface uncertainties generated through Monte Carlo sampling.

are briefly addressed in the following paragraphs. Their application as well as their integrated use will be described in the following section by means of a case study.

Probabilistic Models The probability of a negative project outcome can be related to the probability density function of forecasted NPV prior to a project-phase execution. Integrated value-chain probabilistic models allow one to assess the impacts of uncertainties in technical and economic parameters on the expected distribution of NPV, with multiple runs and using Monte Carlo sampling. Figure 7.15 gives an example of an NPV distribution adopting various subsurface uncertainties. The average performance, which is also considered the expected NPV, is positive, but due to the uncertainties there is a probability of more than 40% that the forecasted NPV is less than zero.

Sensitivity Analysis In contrast to probabilistic models, sensitivity analyses study the impact of single or various selected uncertainties on the NPV. In doing so, not only the overall uncertainty or risk of a project but also the weight of different parameters is analyzed qualitatively or quantitatively. Sensitivity analyses can be based on the variation of single parameters or Monte Carlo sampling of a set of selected parameters. By being aware of sensitivities, one is able to identify the key parameters and appropriate actions to reduce the uncertainties and financial risks.

Decision Trees Knowing the key parameters to reduce financial risk, which have been derived from sensitivity analyses, for example, the decision framework of a project can be adapted (Floris and Peersmann, 2002). A main aspect in realizing projects with large capital investments at the beginning and high financial risk is to establish decision making, in regard to project continuation, modification, or abortion, depending on milestones. Decision tree presentations (cf. Koller, 2000; Bratvold and Begg, 2008) are in this context, used to represent and calculate quantitatively the effects of alternative decisions and adjusted workflows.

In order to support decision makers, results need to be delivered quickly. This cannot be accomplished effectively by using classical, comprehensive, and rigorous models of all the components that contribute to an investment decision evaluation. These models tend to be too comprehensive and the detail, which they can reproduce, is usually not necessary. Further, they lack integration of decision logic and dependencies over their boundaries. It has been argued by many experts that for optimized decision making, a holistic techno-economic integration which is capable of analyzing all decision options to the fullest is preferred over model precision (Bos, 2005).

Portfolio Analysis The Portfolio theory (Markowitz, 1952; Sharpe, 1964) allows one to rationally benchmark prospect performance under uncertainty. In this approach, projects are plotted in a portfolio plot with risk on the horizontal axis and the average NPV as expected return on the vertical axis (Figure 7.16). Economically promising projects are thereby marked by a large expected return at low risk. The economically best projects are located on an imaginary line which is called the efficient frontier. When multiple projects are plotted, these can be ranked relative to each other and to the efficient frontier. When different alternative workflow options of one project, which can be derived from decision trees, for example, are plotted, portfolio plots also allow to quantitatively benchmark the robustness of the project development strategy.

7.2.2.2 Case Study

The following case study deals with an example for analyzing economic risks of a planned EGS project and shows how to mitigate or minimize these risks by means of an effective decision framework. In the example, focus is on dealing with various subsurface uncertainties.

The techno-economic value-chain model used in this case study has been derived from Heidinger, Dornstädter, and Fabritius (2006) and builds on experiences from the EGS project in Soultz-sous-Forêt. A detailed model description can be obtained from ENGINE (2008). Here, only the most important aspects for the risk analysis are outlined. In the model by Heidinger, Dornstädter, and Fabritius (2006), preexisting natural fractures zones appear to play a key role in fluid flow transport within the reservoir and in economic performance measured by NPV. Predicting the formation of natural fracture zones is, however, related with different uncertainties. The natural fracture zones have a large areal extent in the order of 3 km^2 surface area and are capable of connecting injector and producer wells. These fractures are relatively thin, in the order of few centimeters, resulting in progressive cooling of the produced water. However, as shown by tracer tests by

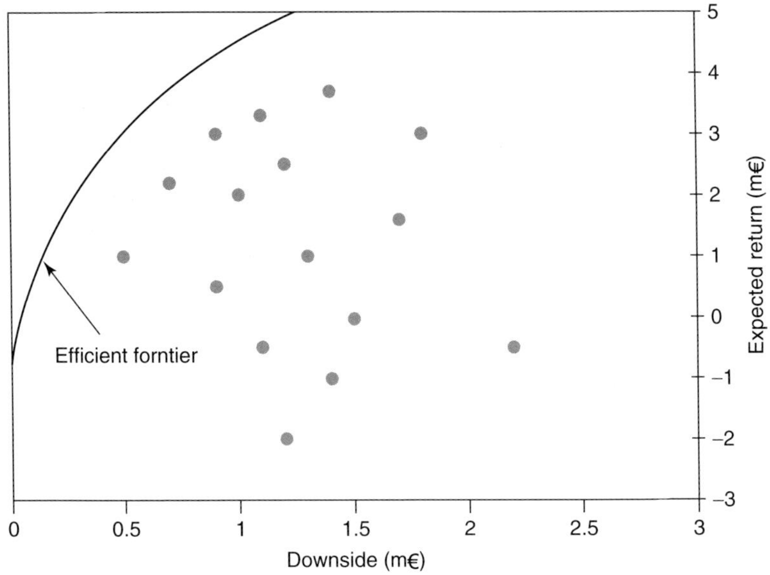

Figure 7.16 Example of a portfolio plot.

Sanjuan *et al.* (2006), only a limited portion (approximately 30%) of the produced water originates from fluid flow from the injector well. Consequently, the cooling effect on the production water is limited to this fraction.

Predicting the formation of the natural fracture zones, it is assumed that significant uncertainty exists regarding the sustainable flow rate (80–120 l s^{-1}), the number of fracture zones (1, 2, or 3), the percentage of inflow from the fractures which will originate from the injection well (30–50%), and the initial reservoir temperature (205–210 °C) (Table 7.5). Based on these uncertainties, the installed electrical capacity of the power plant unit on the surface will approximately lie between 5 and 8 MW_{el}. The cost and economic data which are needed to forecast the economic performance of the EGS plant are summarized in Table 7.6.

Analyzing Economic Risk of the Default Project The work flow assumed for the example project is to decide on the development of a well doublet based on the given uncertainties listed as subsurface uncertainties in Table 7.5. These uncertainties therefore, represent the information available at the project beginning, prior to drilling activities at a planned site. The listed subsurface uncertainties are not solely dependent on the effect of continuous distributions since the number of fractures connecting the injection and production well is discrete.

Such discrete uncertainties are represented in a decision tree with different branches, which means that in case of the decision for the development of the doublet, a reservoir with different numbers of natural fracture zones can be

Table 7.5 Key subsurface parameters with uncertainty and their assumed distribution for the Monte Carlo sampling.

Parameter	Uncertainty	Distribution
Sustainable flow rate	80–120 l s^{-1}	Uniform
Number of connecting fractures	1, 2, or 3	Discrete with predicted probability of 0.7, 0.15, and 0.15, respectively
Inflow from connected fractures	30–50%	Uniform
Initial reservoir temperature	205–210 °C	Uniform

Table 7.6 Cost and economic data of the EGS example case (for further details see ENGINE, 2008).

Cost and economic data	Value	Comment
Investment costs in € 1 000		
Well doublet	17 000	€8.5 million per well
Stimulation	2 500	–
Power plant unit	7 500–12 000[a]	€1 500 kW_{el}^{-1}
Geothermal fluid loop	2 200	–
Operating costs in € 1 000 a^{-1}		
Fixed O&M	1 022–1 180	3.5% of investments
Variable O&M	500–800	€100 kW_{el}^{-1}
NPV calculation		
Imputed interest rate	5%	–
Electricity price	€15 Cts $(kW\,h)^{-1}$	–
Economic lifetime	30 a	–

[a] According to the uncertainties in Table 7.5, the installed electrical capacity of the EGS plant varies between 5 and 8 MW.

tackled leading to different expected NPV (Figure 7.17). The predicted probability of the number of fractures is calculated from a merger of samples from the various branches underlying the event node, in which the selected number of branch-specific samples is weighted for the probability of the particular branch. Varying the sustainable flow rate, the percentage of inflow from the connected fractures and the reservoir temperature, the NPV distribution, and expected NPV can be calculated for each branch as shown in Figure 7.18. The merged NPV distribution is derived from the weighted sum (e.g., considering the different branch probabilities) of the NPV distribution in the branches. The merged distribution is marked by an expected NPV of –€1.07 million and a downside of approximately €3 million. Consequently, the decision should be not to develop the project.

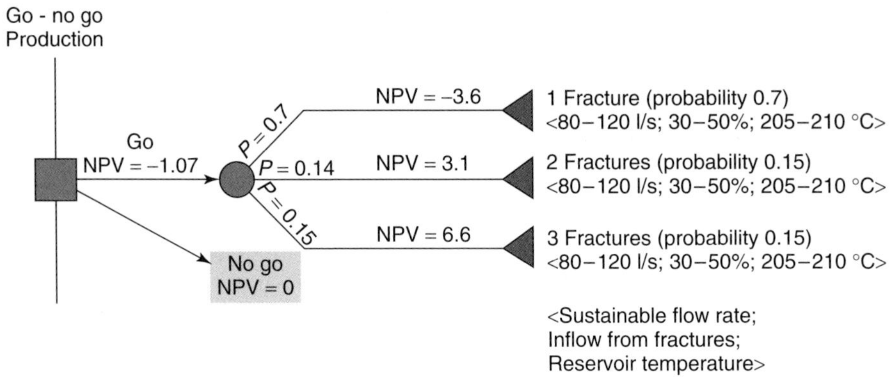

Figure 7.17 Default decision tree of the example case showing not to execute doublet development and adjacent production under the given subsurface uncertainties due to negative expected NPV.

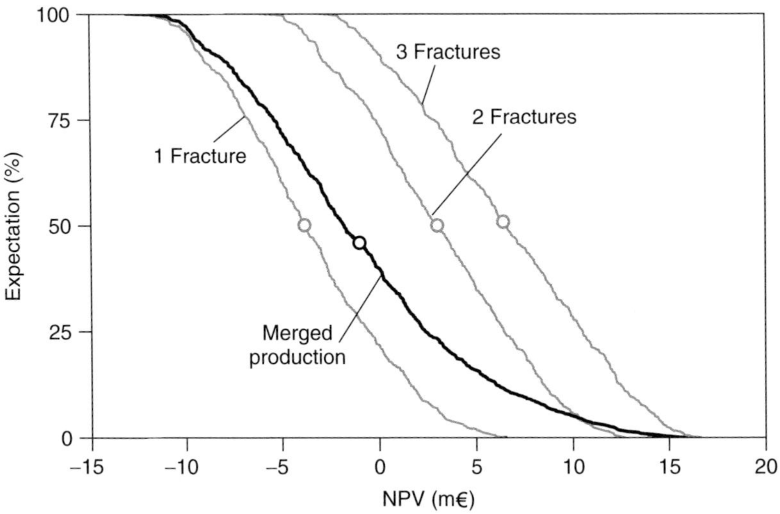

Figure 7.18 Expectation curve of the NPV for different numbers of connected fractures (end nodes Figure 7.17) and merged distribution at the event node.

Even if the NPV distribution of the default project is marked by a negative NPV and expected return, a considerable part of the NPV distribution is positive (Figure 7.18). If the downside could be mitigated, the project's performance would be improved. This could, for example, be accomplished through exploration activities which reduce the uncertainties of the subsurface parameters.

A sensitivity analysis allows targeting exploration activities to obtain more information on the parameters which have the largest impact on NPV. From the

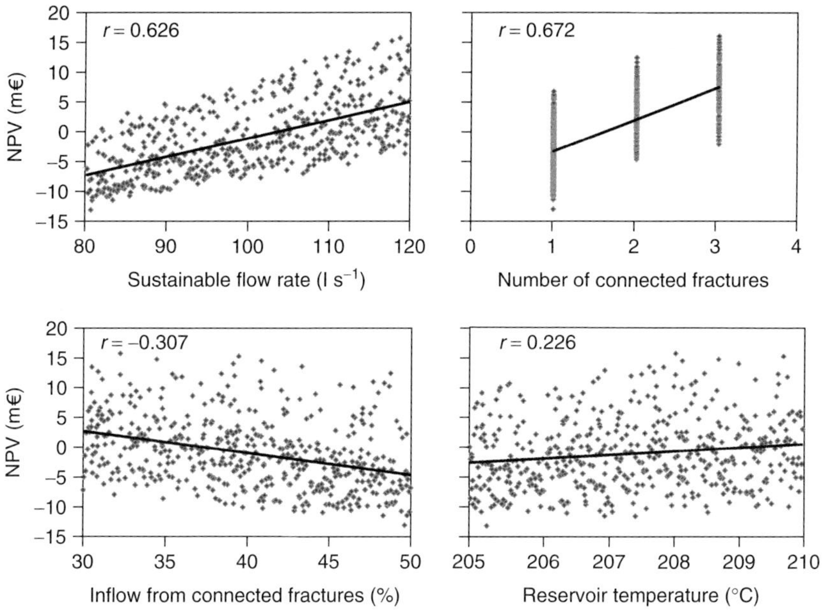

Figure 7.19 Correlation diagrams showing the correlation between subsurface uncertainty parameters and NPV represented by the correlation factor *r*.

Monte Carlo samples correlation factors are produced of the input parameters and NPV, which are shown in Figure 7.19. The stronger the correlation, which is represented by the absolute value of the correlation factor r, the more significant is the impact on the project's NPV. Consequently, these parameters are to be further explored to reduce uncertainty. In this case study, exploration measures focusing on the detection of natural fracture zones would result in a significant uncertainty reduction.

Exploration Approach to Minimize Downside and Maximize Expected Return As it has been shown by the sensitivity analysis, the "one fracture" scenario is marked by strongly negative NPV ruining the economic success of the example project in case the decision framework foresees doublet development as the first event. However, the decision framework can be modified to a staged decision based on additional exploration activities. Here, more information (data and/or analysis) is gathered to reduce uncertainty. This is the case for drilling an exploration well first, which can detect the "one fracture" scenario at an earlier project stage so that the project can be aborted at lower costs since no doublet has been drilled. The additional costs of an exploration well, to the investor, are assumed to be €2 million.

The decision tree accounting for making the additional go-no go decision on the project is displayed in Figure 7.20. The first decision to be made is to drill or not to drill an exploration well. The exploration has a 70% chance for a negative

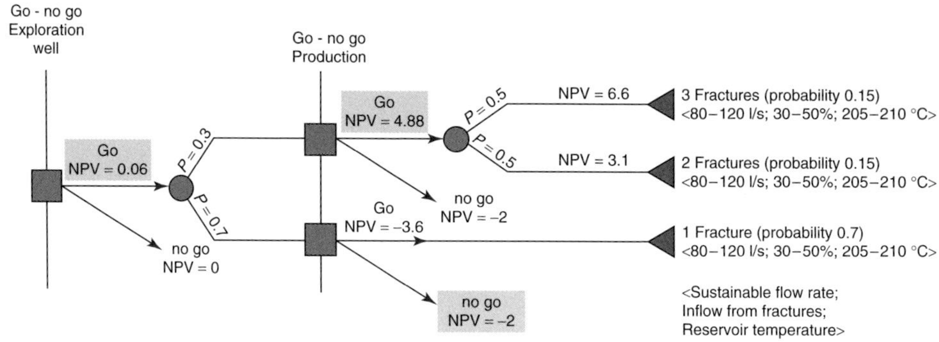

Figure 7.20 Modified decision tree with two go–no go decisions: the drilling of an exploration well at the start of the project and the decision after exploration to develop the doublet for production.

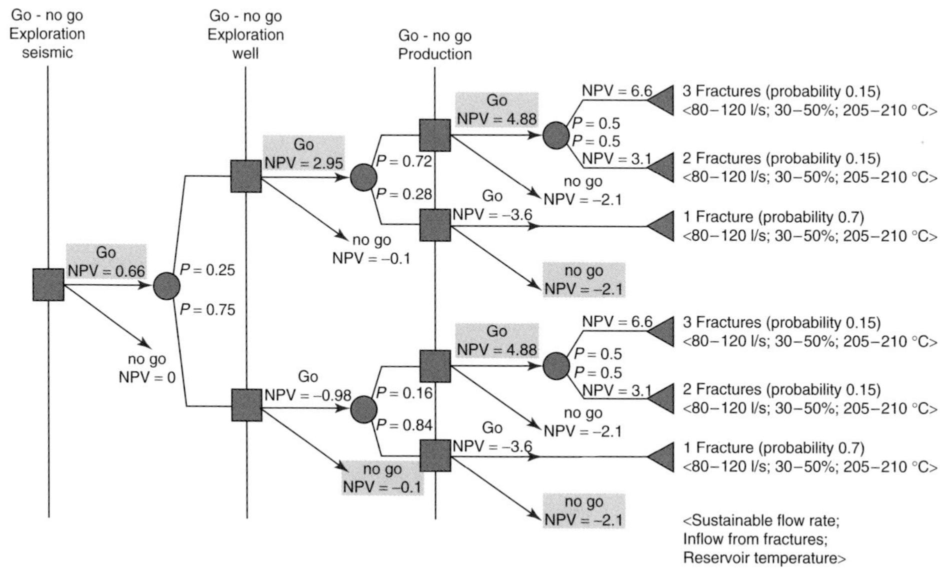

Figure 7.21 Modified decision tree with three go–no go decisions: seismic exploration activity at the start of the project, the second decision to drill an exploration well, and the third to develop the doublet for production.

"one fracture" scenario and a 30% chance for a positive result proving a two or three fractures scenario. Since a significant portion of the downside of the default project has been mitigated by avoiding the high costs of doublet development (Figure 7.22), consequently the expected return of the project is increased to €0.06 million, however, with a still significant downside of €1.5 million.

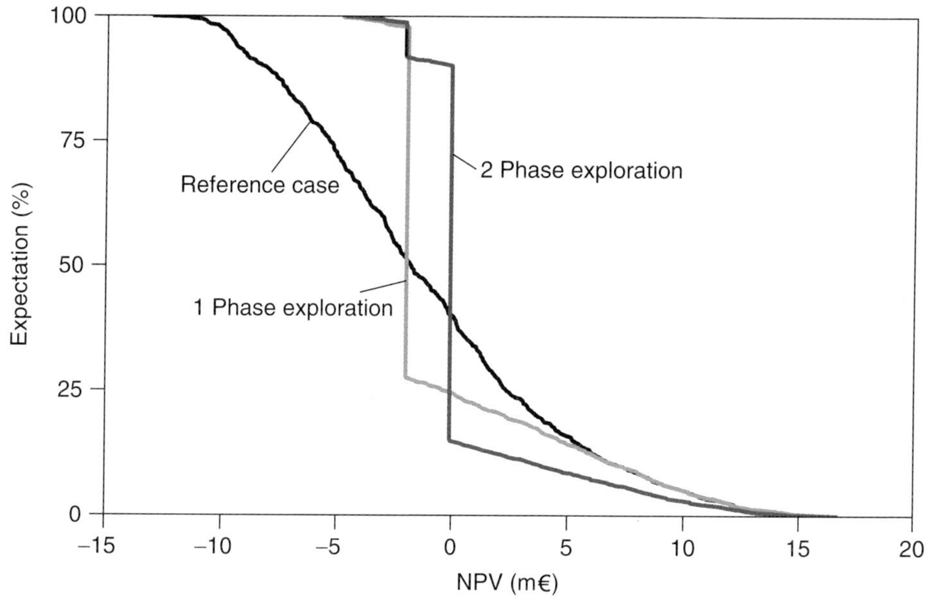

Figure 7.22 Expectation curve of the NPV of executing the project at the first go–no go decision in the default tree (Figure 7.17), the exploration well tree (Figure 7.20), and the exploration seismic tree (Figure 7.21).

The downside can be reduced by further staged decisions and associated exploration phases. This is demonstrated through extending the exploration phase with exploration techniques from the surface. Here, it is assumed that, for example, well-seismic in a nearby existing oil and gas well can be applied which can give additional information prior to drilling the geothermal exploration well (Figure 7.21). The seismic can give an indication for the number of natural fracture zones in the rock (cf. Cuenot *et al.*, 2007 – Soultz-sous-forêts). The costs of these seismic exploration phases is assumed to be €0.1 million. The costs for project abortion after seismic exploration and exploration well drilling are then €2.1 million. The probabilities of the subsequent exploration drilling phase to detect or exclude a particular fracture scenario are adapted accordingly. This imperfect information is generally modeled with conditional probabilities based on experience with similar exploration cases (Koller, 2000; Van Wees *et al.*, 2008). In this case we assume that in 90% of the cases in which the "one fracture" scenario is indicated by seismic exploration, the fracture is really found. For the "two fracture" and "three fracture" scenarios, in contrast, it is less perfect, capable of proving both in 40% of the cases. From the conditional probabilities, the probabilities for the outcomes of the seismic exploration phase and the exploration well are calculated as well as the probabilities for the underlying one, two, and three fractures scenario outcomes. The sum of the one, two, and three fracture scenario probabilities over the various branches in the tree equate back to the original probabilities of 0.7, 0.15, and 0.15, respectively.

Evaluation of this modified decision tree demonstrates the capability to increase the probability of a positive outcome (e.g., assessing a two or three fracture scenario) at the stage of drilling the exploration well up to 72% at relative little costs (€0.1 million). This has a strongly positive effect on the NPV expectation curve for the project, which is marked by a less pronounced downside (Figure 7.22). The expected return of the project is now increased to €0.66 million and the downside has decreased to €0.31 million. This increase of NPV relative to the exploration with exploration well is called *value* of *information* (VOI) which is €0.6 million for including the seismic exploration.

This decision tree approach moves the prospect toward the efficient frontier (Figure 7.23) due to an inexpensive exploration study prior to doublet drilling. The NPV of the prospect can possibly be further enhanced by additional decision tollgates and exploration phases.

If critical factors for risk of different individual prospects are shared (e.g., regional factors such as flow rates for natural fault zones, temperature gradient) it is also possible to mitigate the economic risk for a group of prospects at the costs of a single exploration well.

The positive effect of this prospect dependency can be demonstrated simply by considering the development of a group of two prospects in which the natural fractures are shared. The evaluation of such a decision tree is displayed in Figure 7.24. The expected return of the project has increased to €1.59 million,

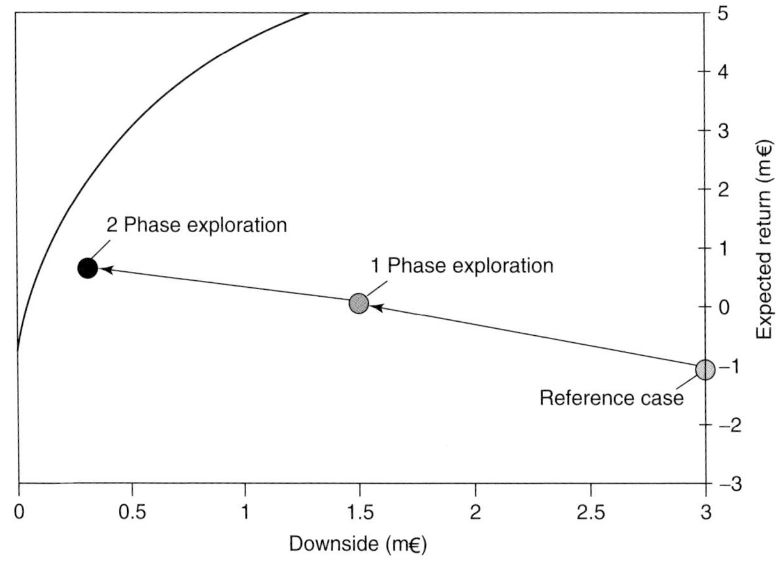

Figure 7.23 Portfolio plot showing progressive improvement of staged exploration approach.

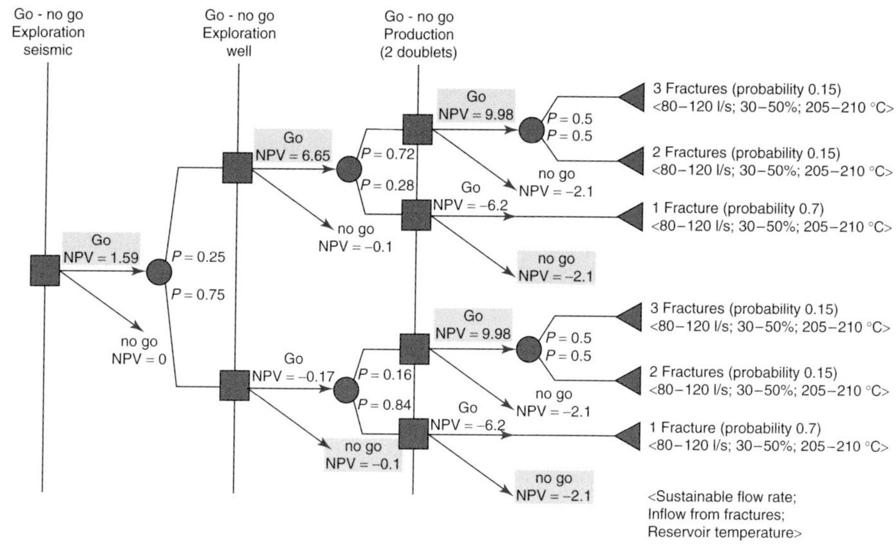

Figure 7.24 Modified decision tree with two go–no go decisions in which exploration results are used for the development for two production doublets in a comparable geologic area.

which is much higher than the cumulative value of two independent prospects adding up to €1.32 million (cf. Figure 7.21).

7.3 Impacts on the Environment

Like all other options of energy provision, EGS has impacts on the global and local environment. Therefore, it is important to identify and evaluate any impact which results from the implementation of an EGS plant at the beginning of a project. The goal must be to avoid or minimize negative impacts on the environment during all stages of an EGS project (e.g., construction, operation, and deconstruction) and to meet the objectives and requirements of climate and environment protection, nature conservation, and preservation of finite resources.

In the following, the impacts on the environment related to EGS plants are discussed. Firstly, the methodology of life cycle assessment (LCA) is introduced and applied for representative EGS plants. Based on this approach, the emission of pollutants, which add to the anthropogenic greenhouse effect and the acidification and eutrophication of natural eco systems, as well as the consumption of finite energy resources are quantified. Secondly, impacts on the local environment, by which an EGS plant is surrounded, are analyzed qualitatively.

The goal of this chapter is to give an overview on potential impacts, which need to be considered in planning environmentally sound EGS plants. The relevance and extent of the addressed impacts can vary from site to site.

7.3.1
Life Cycle Assessment

Even if EGS plants are not related to (continuous) gaseous emissions during operation (due to the carrying of the geothermal fluid in a closed pipeline system on the surface), environmental impacts such as airborne emissions or the consumption of finite energy resources (such as steel used for well completion or fuel for drilling rig operation) occur during other life cycle stages. Therefore, all life cycle stages need to be considered in order to analyze the environmental performance of an EGS plant.

In this context, LCA is a widely applied approach to evaluate and compare specific environmental impacts of different products or technologies. The idea is to carry out a detailed analysis of the life cycle of a product or a duty emerged in response to increased environmental awareness of the public, the industry, and governments. With the LCA, an instrument has been developed to help, for example, manufacturers to analyze and improve their products.

An LCA involves two main stages: the collection of data, related to the product or duty and relevant for the environment, and the interpretation of the collected information. For transparency and traceability of LCA results, standards, such as ISO 14040, ISO 14044[1)] have been developed.

In the following, the methodological approach of an LCA is presented and applied for representative EGS plants. Based on this approach, aspects, which influence the environmental impact during the life cycle, and parameters, which need to be considered in the planning of environmentally sound EGS plants, can be identified.

7.3.1.1 Methodological Approach

The LCA methodology is based on the fact that the environmental impacts of a product (such as the power generation from geothermal energy) are not limited to the production process itself (i.e., the power conversion process). Substantial environmental impacts may also occur within the prechains such as the production and transportation of material needed for the production of the analyzed product (i.e., diesel fuel for running the drilling rig, steel for the completion of the wells). Therefore, within an LCA the overall life cycle of a product is investigated from "cradle to grave." For EGS plants, this is true for all environmental impacts directly and indirectly related to the construction, operation, and deconstruction of the plant.

According to given standards, the LCA is carried out in four steps:

1) **Goal and scope definition:** The goal of LCA is to assess selected environmental effects in the different life cycle stages as well as throughout the whole life cycle

1) Standard of the International Organization for Standardization: ISO 14040:2006 and ISO 14044:2006. Environmental management – Life cycle assessment – Principles and framework and Requirements and guidelines.

of EGS plants. The environmental effects are related to 1 kW h of net-electricity output (functional unit for EGS power plants) or thermal energy output (functional unit for EGS heat plants). According to Chapter 6, *net-electricity* is thereby defined as the difference of produced electricity (e.g., gross electricity) and the electricity needed to produce the geothermal fluid from the reservoir (e.g., pumps for pumping the geothermal fluid) and to run the power plant (e.g., cooling devices). The time reference is the year 2006. The environmental effects which will be analyzed in LCA are the cumulated demand of finite energy resources, the contribution to the anthropogenic greenhouse effect, as well as the acidification and the eutrophication effects on natural eco systems.

2) **Inventory analysis:** In this step, the material and energy flows for all products and processes required to provide 1 kW h net-electricity from an EGS reservoir are quantified and assigned to different process chains. Energy flows include, for example, the electricity needed to operate the drilling rig which is provided by diesel generators. The diesel fuel is produced from different types of crude oil. Its use at the drill site as well as the provision of the fuel to the site results in airborne emissions. The schematic setup of an inventory analysis can be seen in Figure 7.25.
3) **Impact analysis:** In order to quantify the environmental effects, all inventoried material and energy flows are transformed to different impact indicators based on the conversion factors shown in Table 7.7.
4) **Interpretation:** The interpretation of the results from the impact analysis is realized qualitatively by separately discussing the single impact categories.

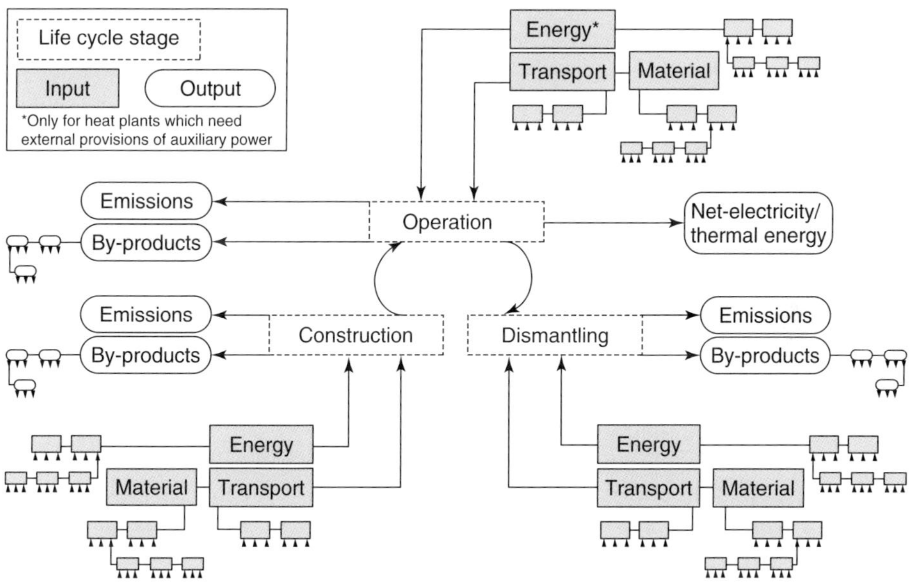

Figure 7.25 Typical scheme of an inventory analysis within an LCA.

Table 7.7 Analyzed environmental effects, impact indicators and characterization factors (Ecoinvent, 2006).

Environmental effect	Impact indicator	Inventoried inputs/outputs	Characterization factors
Cumulated demand of finite energy resources	Finite energy resources	Crude oil	1 MJ/MJ
		Hard coal	1 MJ/MJ
		Lignite	1 MJ/MJ
		Natural gas	1 MJ/MJ
		Nuclear power[a]	10 908 MJ/kWh
Global warming potential	CO_2-equivalent (time horizon 100 years)	CO_2	1 $kg_{equiv}/kg_{pollutant}$
		CH_4	23 $kg_{equiv}/kg_{pollutant}$
		N_2O	296 $kg_{equiv}/kg_{pollutant}$
		SF_6	22 200 $kg_{equiv}/kg_{pollutant}$
		CF_4	5 700 $kg_{equiv}/kg_{pollutant}$
		C_2F_6	11 900 $kg_{equiv}/kg_{pollutant}$
Acidification potential	SO_2-equivalent	SO_x as SO_2	1 $kg_{equiv}/kg_{pollutant}$
		NO_x as NO_2	0.7 $kg_{equiv}/kg_{pollutant}$
		NH_3	1.88 $kg_{equiv.}/kg_{pollutant}$
		HCl	0.88 $kg_{equiv}/kg_{pollutant}$
		HF	1.6 $kg_{equiv}/kg_{pollutant}$
		H_2S	1.88 $kg_{equiv}/kg_{pollutant}$
Eutrophication potential	PO_4-equivalent	NO_x as NO_2	0.13 $kg_{equiv}/kg_{pollutant}$
		NH_3	0.35 $kg_{equiv}/kg_{pollutant}$

[a] Net-electricity from nuclear power plants.

7.3.1.2 **Case Studies**

In the following case study the LCA method is applied to the power plants defined in Table 7.1 (Section 7.2.1.3). The LCA results are discussed in general and also in relation to a reference electricity mix. The main goal of this section is to show the general correlations which need to be considered to design environmentally sound EGS plants, so that these considerations are also applicable to EGS heat plants. A more comprehensive study on LCA of geothermal binary plants providing power and heat is presented in Frick *et al.*, (2010).

In order to identify the data which need to be factored in for the inventory analysis of an EGS plant, the general life cycle stages are described briefly.

Before using EGS reservoirs for energy provision, deep wells have to be drilled. For the material- and energy inputs for drilling and completing deep wells, value ranges can be indicated. The used amounts of material and energy strongly depend on the depth and diameter of the wells. Additionally, geologic conditions have an influence since they determine the composition of the cementation and the drilling mud or the necessary thickness of the casing wall (and therefore, the amount of steel). For example, the amount of diesel to drive the drilling rig and the mud pumps can roughly vary between 6 and 8 GJ per drilled meter. The amount of drilling mud, mainly consisting of water, ranges under normal geologic conditions between 700 and 1000 kg m^{-1}. For the completion of 1 m open hole, approximately 80–120 kg of steel for the casing and 45–65 kg cement to seal the casing can be estimated.

For the reservoir stimulation, it is not possible yet to derive representative bandwidths for the necessary material- and energy inputs. For the two representative plants the needed energy is estimated with 3000 GJ per well and the material is assumed by 260 000 tons of water per well.

On the surface, the geothermal fluid cycle and the binary power unit are installed. The basic parts of the geothermal fluid cycles are the submersible pump, the pipeline to connect the production and injection well, and the heat exchanger which transfers the geothermal heat to the binary unit. The data for such components can be derived from existing technical literature (e.g., Heck, 2004) as well as from the industry. Material and energy inputs for these parts are mainly determined by the flow rate of the geothermal fluid and with regard to the pipeline, additionally, by its length and the mode of construction. The amount of material needed for the heat exchanger depends on its thermal capacity.

Regarding the binary power unit, the working fluid and the system components, such as the heat exchangers, the turbine, the generator, and other peripheral equipment have to be considered in the inventory analysis. The main influence for these components is the installed electrical capacity. In case of using the heat of the geothermal fluid after the binary plant unit for the supply of district heat, an additional heat exchanger needs to be taken into account.

During the operation, the geothermal fluid cycle is operated as a closed loop so that no direct gaseous pollutants are emitted and only the exchange of the submersible pump needs to be included in the inventory analysis.

After the use of the plants, the wells are filled with gravel and cement. The surface installations are disposed or recycled.

Based on these assumptions and the data presented in Frick *et al.* (2010), the life cycle calculations for power plants 1 and 2 are carried out using the Ecoinvent Database which provides the more general data for the inventory analysis. The resulting impact indicators of the case studies and a reference electricity mix are shown in Table 7.8.

Under the circumstances outlined above power plants 1 and 2 show comparable environmental impact indicators. Power plant 2 has however, slightly lower impacts and so the larger energy- and material input for developing the deeper reservoir is compensated by a higher net-electricity output.

To identify more determining factors, which are responsible for the differences in the LCA results of the analyzed EGS plants, Figure 7.26 relates the impact indicators to the different life cycle stages of the power provision. It can be seen that the construction of the EGS plants is responsible for most of the environmental

Table 7.8 Environmental impact indicators for power plant 1, power plant 2, and a reference electricity mix.

Impact indicator	Power plant 1	Power plant 2	Reference electricity mix[a]
Finite energy resources (kJ (kW $h_{el})^{-1}$)	872	831	8 910
CO_2 equivalent in (g (kW $h_{el})^{-1}$)	57.5	54.9	566
SO_2 equivalent (mg (kW $h_{el})^{-1}$)	471	453	1083
PO_4 equivalent (mg (kW $h_{el})^{-1}$)	64.1	62.0	60

[a]breakdown of provided net electricity according to Frick *et al.* (2010): 26% lignite coal, 26% nuclear power, 24% hard coal, 12% natural gas, 4% hydropower, 4% wind power, 1%, crude oil, 3% other fuels

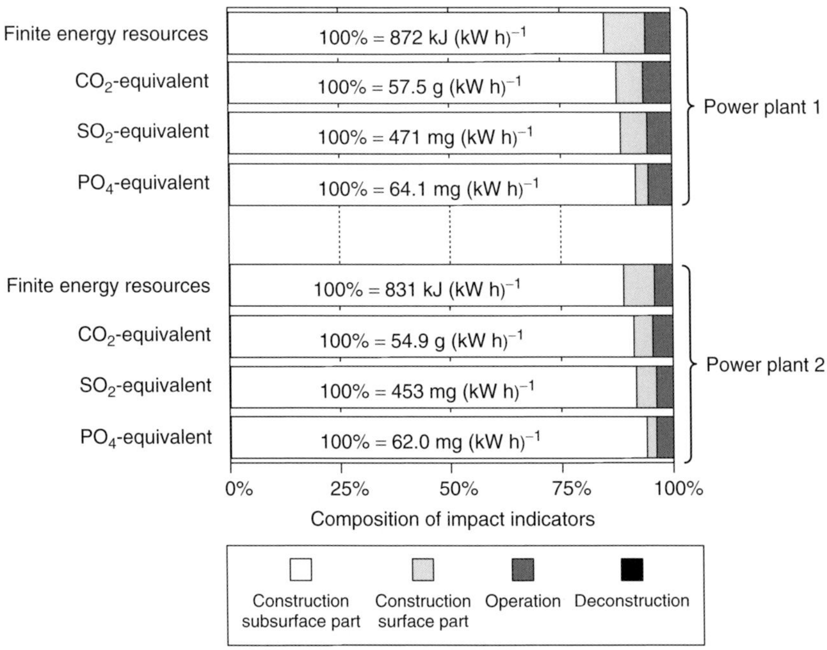

Figure 7.26 Composition of impact indicators for power plant 1 and power plant 2.

effects analyzed here. The construction of the part, which is located underground, causes 85–94% of the analyzed environmental impacts. It has therefore, the largest influence of all life cycle stages. For power plant 2, this influcnce is slightly larger compared to plant 1 because of the larger effort in locking up the deeper reservoir. In comparison to the part of the power plant located underground, the contribution of the part located above ground is with 2–9% of the analyzed environmental impact indicators comparatively low. The operation phase of power

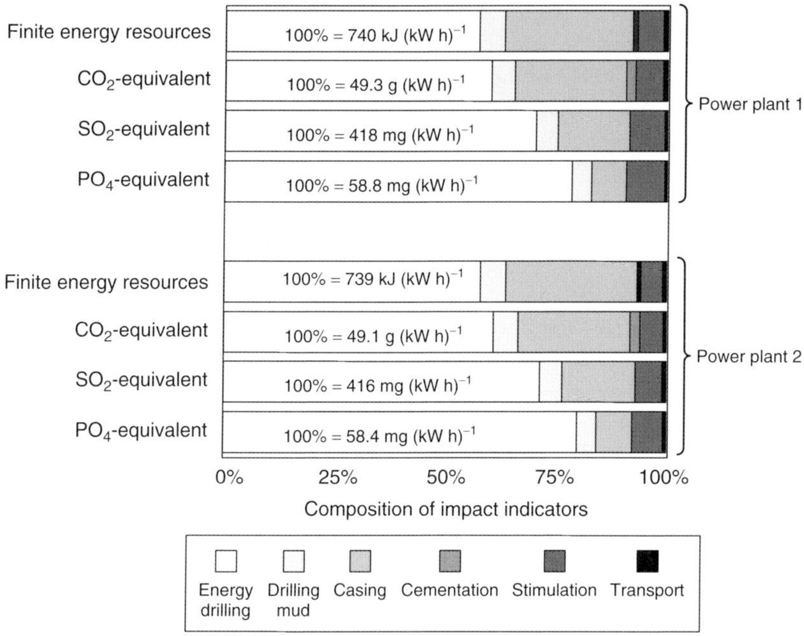

Figure 7.27 Composition of impact indicators for the construction of the subsurface part for power plant 1 and power plant 2.

plants 1 and 2 influences the environmental impact indicators with 4–6% only. This small influence is mainly caused by the replacement of the downhole pump. The influence of the deconstruction on the environmental impacts investigated here is negligible.

According to Figure 7.27, the environmental impact resulting from the subsurface part is dominated by the energy needed to drill the deep wells and the material (i.e., steel) for the casing. By contrast, the influence of the material- and energy input for reservoir stimulation is significantly lower.

Compared to the reference electricity mix, the analyzed EGS plants are characterized by a significantly lower consumption of finite energy resources and emissions of CO_2-equivalent pollutants. But also considerably less SO_2^- equivalent emissions are observed. The PO_4^- equivalents, in contrast, are in the same order of magnitude. Based on this comparison, power plants 1 and 2 can be rated as environmentally sound.

Considering other EGS sites, the geological conditions can significantly vary. Based on the presented results it can, however, be derived that sites which need larger amounts of material and energy for the construction of the subsurface part and thereof especially drilling and completing the deep wells, can also be environmentally promising; the precondition is that the provided energy compensates the larger mass- and energy flows. This can be realized basically by the same measures as addressed for the LCOE reduction potential in Section 7.2.1.3. These measures are an increased net-electricity provision due to enhancement of the reservoir productivity, improvement of the conversion efficiency in the binary unit, and/or reduction of the auxiliary power demand of the overall system. Another possibility is the supply of additional district heat.

7.3.2
Impacts on the Local Environment

Regarding the development of conventional geothermal resources, several local environmental changes have been reported in the last 40 years. Some of them have been significant and even severe. In part of Wairakei field (New Zealand), for example, the withdrawal of fluid from the reservoir led to a subsidence of about 15 m. The degradation of geothermal reservoirs was observed at several sites like Larderello (Italy), The Geysers (USA), and Wairakei (New Zealand). Today, it is known that these problems mainly occurred due to much larger production than injection rates and so adapted reservoir management systems and strategies were developed in order to avoid further impacts (Hunt, 2001).

Regarding construction, operation, and deconstruction of EGS plants, such negative experiences have to be avoided. Therefore, the impacts on the local environment of an EGS plant have to be analyzed precisely. This has to be done by preliminary studies and also on-site monitoring. These aspects must be considered in every EGS project, and also in regulatory frame works, since the protection of the environment is a defined political goal in most countries throughout the world.

In the following sections, an overview on potential environmental impacts is presented. Later, the general course of such an environmental impact assessment (EIA) including legal aspects is outlined.

7.3.2.1 Local Impacts

EGS plants are related to different impacts on diverse parts of the local environment and different characteristics regarding their duration (temporary or continuous), their reversibility, and their degree of probability. The potential impacts on the local environment are briefly outlined below, while also addressing issues which were discussed already in more detail in other chapters. Since the relevance and impact of the issues addressed below varies from site to site, this section can only give a rough overview on aspects which need to be considered from an environmental viewpoint.

Many of the impacts on the local environment are related to assessing the reservoir. Based on the experiences from the oil and gas exploration, most of these impacts are known and technically controllable. Negative impacts due to reservoir exploitation can be avoided with proper reservoir monitoring and management. Environmental impacts from constructing and operating the surface facilities are comparatively low. Some environmental impacts have to be considered however, in connection with the power production from EGS due to the relatively large amounts of waste heat.

Drilling Operations Drilling operations have a large impact on the environment and are related to different risks. Since EGS well drilling uses mainly the same processes as for gas and oil exploration, the different environmental impacts and risks are known and technical measures and safety precautions do exist in order to avoid or minimize them. The following environmental aspects and risks need to be considered in connection with drilling operations; all of them are limited to the period of drilling operations and therefore, last only a couple of weeks or months:

- **Drillings site preparation:** The drilling site needs to be prepared in such a way that drilling operations can be carried out safely, both for persons working at the drilling site and the environment. An important aspect is the covering of the soil in order to prevent its pollution with material which is mounted and handled at the site (such as drilling mud or fuel). In order to do this, the preparation of the drilling site usually includes destruction of local vegetation. The use of land is however, locally restricted to an area of about 2000 m^2 and can be largely recultivated after drilling operations are finished (Kaltschmitt *et al.*, 1999).
- **Noise emissions:** Drilling operations are related to the emission of noise. Operating the drilling rig, handling of casings, or simply communication of the drilling crew causes a noise level between 70 and 125 dB(A) (Kristmannsdóttir and Armannsson, 2003). Depending on the proximity to housing areas and animal habitat and also topographical and meteorological site conditions, this can be an environmental impairment. By using mufflers or implementing proper noise insulation measures (such as sound insulating walls), these emissions can generally be reduced to a tolerable minimum. In many countries, allowed noise thresholds for specific areas are defined and need to be considered in order to get a drilling permission.
- **Subsurface emissions:** The normal drilling process involves local emissions of drilling mud in the subsurface surrounding of a drilled well. Intersecting geologic anomalies such as fault zones also more extensive emissions are possible. For environmental but also safety reasons (e.g., open hole stabilization), the drilling mud composition is adapted to the geological conditions. The emission of drilling mud into already drilled areas or causing a hydraulic short circuit between different aquifers (which can both lead to the pollution of groundwater bearing formations) is a known environmental risk. It is prevented by immediate casing (and cementing) after finishing a drill section. In many countries, the composition of the drilling mud and the closing-off of different drill sections is a matter of legal regulations.
- **Airborne emissions:** Besides the possible creation of fumes and dust during drilling operations, gaseous emission can occur when gas components (such as CH_4, CO_2, H_2S) dissolved in the drilling mud are released at the surface. This can happen when formations with gas-bearing fluids are intersected. Hazardous concentrations for the environment, however, are normally not expected and with the use of gas separators, such emissions can be avoided. Furthermore, small amounts of gaseous emissions related to well testing can also occur.
- **Water usage:** For the supply of drilling mud water is needed, which is usually taken from nearby surface or groundwater bodies. The needed amount differs

from site to site and mainly depends on the geological conditions and the requirements for the drilling mud. Even if this water usage is temporarily terminated, a sufficient availability of water must be checked also from an environmental viewpoint. Otherwise, water must be delivered.

- **Waste disposal:** Related to drilling, large amounts of waste such as disused drilling mud and cuttings must be handled and disposed properly. Whereas the disposal of disused drilling mud (and cuttings) on the basis of freshwater is not very critical, drilling mud on the basis of saltwater or oil has to be handled with care. With proper drilling operation management, the disposable amounts can be reduced, for example, by reusing drilling mud for other wells or using cuttings as filling material for civil engineering.
- **Visual impact:** The installed drilling rig and nocturnal lighting can be an inconvenience for the surrounding environment. However, this visual impact is temporary.

Other incidents which can possibly lead to adverse impacts on the local environment are severe geomechanical changes triggered by the drilling process and uncontrolled well blow outs. However, such geomechanical changes are only possible in tectonically active regions. Regarding well blow outs, the use of blow-out-preventers at the wellhead is state-of-the-art.

Reservoir Stimulation Hydraulic stimulation measures are based on inducing geomechanical alternations in the subsurface in order to enhance the natural transmissibility of a formation. With regard to geomechanical changes, two phenomena have to be differed. On one hand, geomechanical changes can directly result from the injection of stimulation fluids under high pressure (which exceeds the minimum principle stress). On the other hand, they can occur due to sudden reduction of naturally built up stress (which can lead to much larger geomechanical changes). Depending on the surrounding rock formations and their magnitude, geomechanical changes can be followed by build-up and dispersion of microseismic activities up to the surface. The stimulation of sediments, graben and karst structures so far did not cause any microseismic events of significant magnitude. Even massive stimulation measures in the Genesys project in Hannover (Germany) only led to events in the magnitude of 0 (scale of Richter). During research activities in Groß Schönebeck (Germany), no microseismicity could be measured at the surface so far. In contrast, the stimulation of basement rocks has caused several microseismic events with a magnitude around 3 (in the Richter scale) such as the one in Basel (Switzerland) or in Soultz-sous-Forêts (France).

Even though geomechanical changes can lead to damage of buildings and even be hazardous for human beings and animals, the earth tremors caused by EGS reported so far can be categorized as sensible but not as adverse impacts on the environment. For the occurrence of damaging events, large enough prestressed fractions have to exist, which allow much larger deformation energy (between 2 magnitudes, the deformation energy increases by factor 30). However, based on the present state of knowledge, larger impacts cannot be totally excluded since the knowledge about the stress situation and the development of larger microseismic events in the subsurface is still insufficient (Majer *et al.*, 2007).

Reservoir Exploitation The exploitation or operation of an EGS reservoir can lead to different alternations in the reservoir and the surrounding subsurface. Impacts on the local subsurface environment such as hydraulic and thermal alternations as well as circulation losses need to be considered for sustainable reservoir management, but are not considered as adverse environmental impacts.

Large geomechanical changes in the reservoir due to reservoir exploitation, in contrast, can potentially lead to environmental impacts on the surface. Soil subsidence, for example, can theoretically occur due to reservoir cooling and significant pressure drops, also depending on the overlaying formations. Magnitudes of several meters such as those reported for the exploitation of conventional geothermal systems (Hunt, 2001) are not to be expected (Tester *et al.*, 2008) since the largest subsidence from conventional geothermal systems was related to an uneven mass balance due to missing reinjection. Geomechanical changes can potentially also result in microseismic events due to pressure changes in the reservoir while reinjecting the geothermal fluid. Adverse impacts on the environment are however, not to be expected.

When operating artificial fracture systems in the subsurface, circulation losses can occur and these then must be replaced by water from overlaying aquifers or surface water bodies. The extent of such water losses must be assessed site specifically after testing the reservoir. In Soulz-sous-Forêts, for example, tests of the local reservoir could be carried out without circulation losses.

The exploitation of an EGS reservoir is typically related to circulating the geothermal fluid. This means that typically no environmental impacts are related to improper disposal of the geothermal fluid since it is reinjected into the reservoir. During normal operation, the geothermal fluid is carried in a closed loop on the surface, so that no gaseous emissions, such as from dissolved gases in the circulated fluid, occur. Regarding the circulation of the fluid, however, the proper handling of filter residues needs to be considered from an environmental viewpoint. Depending on the geology, also naturally occurring radioactive material (NORM) can occur such as those known from oil and gas exploitation. NORM can lead to toxicological risks during revision operations; radiation protection however, is not relevant.

The circulation of the geothermal fluid can lead to a temperature increase of the soil in the vicinity of the boreholes or pipelines. The thermal influencing is locally limited and occurs normally without environmentally adverse temperature gradients. However, the activity of soil biota can be influenced by temperature changes.

The environmental risk from material emissions due to broken casings and cementation in the deep wells or pipes in the geothermal fluid cycle is little and can be minimized by proper design.

Installation and Operation of Surface Facilities The installation of the surface part of an EGS plant such as the geothermal fluid cycle, heat plant unit, or binary power unit is related to general environmental impacts which come with construction work such as noise emissions and dust creation.

Regarding the operation phase of an EGS plant, some issues relevant for the surface installations have already been addressed in correlation with reservoir exploitation.

Land usage of an operating EGS plant is caused by the geothermal fluid cycle (and therefore mainly depends on the distance between production and injection well) and the need of space for the heat and/or power provision unit. Land usage can be minimized, for example, by using directional drilling from one site or the construction of pipelines along already existing infrastructure. In a qualitative comparison to other energy technologies, EGS is characterized by a comparatively small land usage. On the one hand, EGS plants are located directly at the energy source and do not need additional land for fuel provision (such as transport infrastructure, mining, or biomass cultivation). On the other hand, EGS can continuously provide energy so that the provided energy per square meter is much larger compared to wind power or photovoltaic.

Further environmental impacts are mainly restricted to EGS plants with power production. One aspect during normal plant operation is the emission of noise resulting from generators, transformers, and especially components in the cooling system such as fans and wet-cooling towers. Such emissions can be minimized by proper plant design. Another, very important aspect from an environmental viewpoint is the handling of the relatively large waste heat amounts. Depending on the installed mode of cooling, different environmental impacts need to be considered. A very effective way to avoid or minimize impacts related to the cooling system of an EGS power plant is to find and integrate the technical use of the waste heat from the binary cycle in the plant concept.

Using dry cooling, typically, noise emission and the demand for space needs to be taken into account. Wet-cooling towers use considerable amounts of cooling water in order to make up for evaporation and elutriation losses. The dissipation of the waste heat to the environment locally increases the temperature and humidity of the ambient air and can influence the microclimatic conditions. Depending on the site-specific meteorological conditions (such as ambient temperature, humidity, and inversion formation), wet-cooling towers can lead to vapor emissions (which can also be a visual impact), locally increased rainfall and dewfall, and a decrease of the annual solar radiation. This can be avoided with the (partial) implementation of dry cooling. In case of wet-cooling towers, furthermore, the disposal of elutriation products and also noise emissions have to be considered. Once-through cooling systems need a nearby streaming water body. The heated cooling water is then returned to the stream. Depending on the mass and temperature ratio of cooling water and streaming water, thermal pollution can occur in the stream. A temperature increase of stream water can directly affect the habitat of fishes. This must be avoided by defining and following a maximum temperature or temperature increase. In case additives are used with once-through cooling systems to manipulate the cooling water quality, the additives are also emitted into the streams.

With regard to the binary cycle, gaseous emissions can be caused by the used working fluid due to unavoidable leakages or malfunction. Negative environmental impacts (in this case mainly referring to working personnel) can be avoided with the use of innoxious fluids and by implementing appropriate constructive and safety-related components (such as extraction systems).

Dismantling Besides general environmental aspects which have to be considered with the dismantling of infrastructure, in EGS plants the decommissioning of the deep wells is special because abandoned wells are a potential source for material emissions to the subsurface. With proper sealing and filling of the wells, this can be avoided.

7.3.2.2 Environmental Impact Assessment

Today, most countries include environmental issues in their legislation, also based on the Rio Declaration on Environment and Development which was recorded in the 1992 United Nations "Conference on Environment and Development" (UNCED). Typical Regulations or Acts involve the preservation of the natural environment including water bodies, vegetation and animals, preservation of cultural heritage, pollution control of air and water bodies, noise control, natural resource use, conservation and recovery, and human safety. These issues can be dealt with separately in single Acts or are combined in superordinate Directives. In some countries, also geothermal-specific Acts do exist, such as in United States, Philippines, New Zealand, Italy, or Iceland (Hunt, 2001; Hongying, 2000; Kristmannsdóttir and Armannsson, 2003).

For the development and realization of EGS projects therefore, different environmental regulations, standards, and permission procedures are binding depending on the country. This can also include the collaboration with different legal authorities of a country, such as mining authority (since *geothermal heat* is in many countries defined as a natural resource such as gas, oil, and coal), environmental authority, or water authority. The main features of getting permission for an EGS plant from an environmental viewpoint are however, similar.

A widely used instrument in this context is the EIA. It has first been established in the United States in the 1970s and is used today in many countries, which have set up their own procedures (Hongying, 2000). An EIA is carried out to ensure that all possible impacts on the environment of a planned project are identified and assessed before a decision is made on whether the proposed project should be allowed to proceed. This means that the most environmentally favorable option, or at least an environmentally acceptable option, can be identified at an early stage. With this tool, projects designed to avoid or minimize impacts are realizable.

In order to point out elementary steps, which are also needed in EGS projects to properly deal with impacts on the environment, the most important steps of a typical EIA process are briefly outlined below based on Figure 7.28.

The first step of an EIA process is based on the project description. The project description should concisely describe the project's ecological, geographic, social, and temporal context. This step is focused on general considerations, which can be

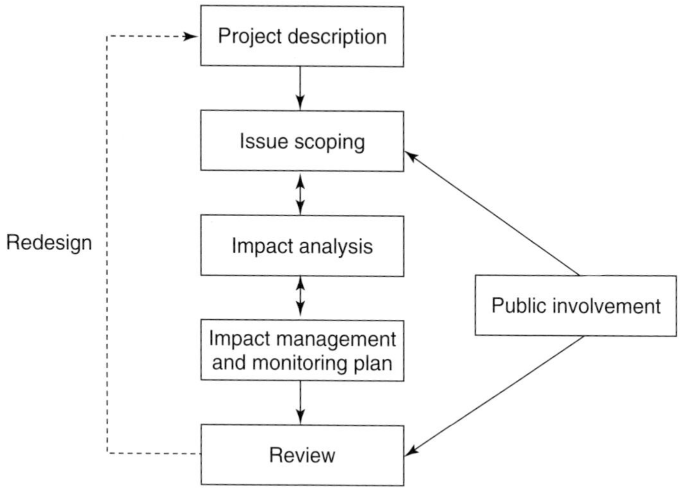

Figure 7.28 Important steps in an environmental impact assessment process.

derived from already available information sources and local experience, rather than detailed analysis. Standard mitigation of known environmental impacts according to best practice (such as best practice for drilling) must be an integral part of EGS projects right from the beginning.

In order to identify the key issues of environmental concern for the described project, issue scoping has to clarify the applicable legislation and regulation. In this context, it should be considered that, in an EGS project, project development might have to deal with different authorities, many of them probably inexperienced in dealing with EGS technology. Therefore, issue scoping of EGS projects should involve professional judgment, such as from environmental agencies, and the collaboration with responsible authorities. Experience has shown that in particular, legislation and regulations with a large range of interpretation of geothermal technology and EGS can be a hindrance for project realization (Kontoleontos, 2007). On the level of project development, this can be avoided by establishing an open dialogue with the authorities. On the level of lobbying, improvement and adaptation of existing legislation in some countries to the most important geothermal particularities should be worked on. Another important aspect of issue scoping is the involvement of the public. Many EGS projects are carried out very close to the public or are at least of large public interest. The lack of information can cause distrust and public fears. Ignoring such facts can cause major problems up to complete failure of a project. Therefore, an open-minded communication and information strategy based on long-term considerations is imperative to initiate public understanding.

Based on the key issues of concern, the impact analysis has to identify and predict the environmental effects (including the already foreseen standard mitigation) of the project and evaluate their likelihood and significance. The impact analysis comprises a more detailed description of the project and its positive and negative influences on the environment within the defined assessment boundaries (such as spatial, temporal, environmental, and administrative scope). The significance of the environmental impacts varies depending on their magnitude, geographical extent, duration, frequency, irreversibility, secondary effects, and cumulative impacts. The likelihood of an environmental impact depends on the probability of its occurrence and the scientific knowledge or uncertainty about this effect.

For environmental impacts which have been identified to be significant, even with the already planned mitigation measures (further) impact or mitigation management measures must be planned to be implemented. It must be considered however, that such measures are also related to additional cost so that mitigation might not only change a proposed project from a technical viewpoint. For some environmental impacts, a monitoring plan must also be established. The purpose of monitoring is to observe the compliance of allowable or defined thresholds and to be able to notify adverse changes in advance and react to them. Monitoring should be carried out during the construction and operation phases of a project.

With the review of the elaborated impact management and monitoring plan and the analyzed impacts of the (redesigned) project, it is possible to make a decision whether or not a project can be realized or whether it must be redesigned from an environmental viewpoint. The public should be involved in the review process.

References

Ahmed, K. (1994) Renewable Energy Technologies – A Review of the Status and Costs of Selected Technologies, Technical Report No. 240, World Bank Technical Paper, Washington.

Barreto, L. (2001) Technological learning in energy optimisation models and deployment of emerging technologies, PhD thesis, Swiss Federal Institute of Technology Zurich.

Bos, C.F.M. (2005) A framework for uncertainty quantification and technical-to-business integration for improved investment decision making: SPE Paper 94109, Society of Petroleum Engineers Europec/European Association of Geoscientists and Engineers Annual Conference, Madrid, Spain.

Bratvold, R.A. and Begg, S.H. (2008) I would rather be vaguely right than precisely wrong: A new approach to decison making in the petroleum exploration and production industry. *AAPG Bulletin*, **92**,(10), 1373–1392.

Bürger, V., Klinski, S., Lehr, U., Leprich, U., Nast, M., and Ragwitz, M. (2008) Policies to support renewable energies in the heat market. *Energy Policy*, **36** (8), 3150–3159.

Cuenot *et al.* (2007) How to optimize drilling strategies and reservoir management: lessons learned from the Soultz EGS project. From Simmelink E., Van Wees J.D. & Bonté D.(eds.) 2007, in Actes/Proceedings of the Engine Workshop 7 "Risk analysis for the development of geothermal energy" 7-9 November 2007, Leiden, The Netherlands. ISBN: 978-2-7159-2992-0. Orleans BRGM Edition. Collection Acts/Proceedings. ISSN: 1773-6161

Ecoinvent (2006) Swiss Centre for Life Cycle Inventories. Ecoinvent Data v2.0.

ENGINE (2008) Performance Assessment Tool, ENGINEPA.XLS, ENGINE DSS, Deliverable No~51 of the ENGINE Coordination Action (ENhanced Geothermal Innovative Network for Europe), September 2008, Utrecht. Article presenting the risk evaluation for the development of geothermal energy.

Entingh, D.J. and McVeigh, J.F. (2003) Historical improvements in geothermal power system costs. *Geothermal Resources Council Transactions*, **27**, 533–537.

Erdmann, G. and Zweifel, P. (2008) *Energieökonomik – Theorie und Anwendung*, Springer.

Floris, J.T. and Peersmann, R.H.E. (2002). Integrated scenario and probabilistic analysis for asset decision support: *Petroleum Geoscience*, **8**, 1–6.

Fouquet, D. and Johansson, T.B. (2008) European renewable energy policy at crossroads–focus on electricity support mechanisms. *Energy Policy*, **36** (11), 4079–4092.

Frick, S., Kaltschmitt, M., and Schröder, G. (2010) Life cycle assessment of geothermal binary power plants. *Energy*, submitted in March 2010.

Heck, T. (2004) Wärme-Kraft-Kopplung, in *Sachbilanzen von Energiesystemen: Grundlagen für den ökologischen Vergleich von Energiesystemen und den Einbezug von Energiesystemen in Ökobilanzen für die Schweiz* (ed. R. Dones *et al.*), Paul Scherrer Institut Villigen, Swiss Centre for Lice Cycle Inventories, Dübendorf, Final Report, Econinvent 2000 No. 6-V (in German).

Heidinger, P., Dornstädter, J., and Fabritius, A. (2006) HDR economic modelling: HDRec software. *Geothermics*, **35** (5-6), 683–710.

Hongying, L. (2000) Geothermal environmental impact assessment studies in Hebei province, China, in *Reports of the United Nations University Geothermal Training Programme* (ed. L.S.Georgsson), United Nations University, pp. 231–266.

Hunt, T. (2001) Five Lectures on Environmental Effects of Geothermal Utilization, Lectures on environmental studies given in September 2000, United Nations University, Geothermal Training Programme, Reykjavík (Island).

IEA/OECD (2000) Experience curves for energy technology policy.Technical report, International Energy Agency/Organisation for Economic Co-operation and Development, Paris.

IEA/OECD (2005) Projected costs of generating electricity – 2005 update. Technical report, International Energy Agency/Organisation for Economic Co-operation and Development/Nuclear Energy Agency, Paris.

Jonkman, *et al.* (2000) *Best practices and methods in hydrocarbon resource estimation, production and emissions forecasting, uncertainly evaluation and decision making:* SPE Paper 65144, Society of Petroleum Engineers European Petroleum Conference, Paris, France, October 24-25.

Kaltschmitt, M., Huenges, E., Wolff, H., Baumgärtner, J., Hoth, P., Jung, R., Kayser, M., Lux, R., Sanner, B., Schallenberg, K., Scheytt, T., Kaltschmitt, M., Huenges, E., and Wolff, H. (eds) (1999) *Energie aus Erdwärme*, DVG Deutscher Verlag für Grundstoffindustrie.

Kaltschmitt, M., Streicher, W., Frick, S., Huenges, E., Jorde, K., Jung, R., Kabus, F., Kehl, K., Laing, D., Lewandowski, I., Ortmanns, W., Rau, U., Sanner, B., Sauer, U., Schneider, S., Schröder, G., Seibt, P., Skiba, M., Weinrebe, G., and Wiese, A. (2007) *Renewable Energy – Technology, Economics and Environment*, Springer.

Köhler, S. (2005) Geothermisch angetriebene Dampfkraftprozesse–Analyse und Vergleich binärer Kraftwerke, PhD thesis, Technische Universität Berlin.

Koller G.R. (2000) *Risk modeling for determining value and decision making:* London, New York, Washington, Boca Raton. Chapman & Hall/CRC p.321.

Kontoleontos, E. (ed.) (2007), Increasing policy makers'awareness and the public acceptance. Workshop Abstracts of the Engine Workshop 6 13–14 September 2007, Athens, Greece.

Kristmannsdóttir, H. and Ármannsson, H. (2003) Environmental aspects of geothermal energy utilization. *Geothermics*, **32** (4-6), 451–461.

Ledru, P. (2008) The final conference of the ENGINE coordination action: a milestone towards EGS demonstration projects. From Calcagno, P. and Sliaupa, S. (eds), Actes/Proceedings of the ENGINE Final Conference "Enhanced Geothermal Innovative Network for Europe", February 12–15, 2008, Vilnius, pp. 11–12.

Legarth, B. (2003) Erschließung sedimentärer Speichergesteine für eine geothermische Stromerzeugung, PhD thesis, Technische Universität Berlin.

Majer, E.L., Baria, R., Stark, M., Oates, S., Bommer, J., Smith, B., and Asanuma, H. (2007) Induced seismicity associated with enhanced geothermal systems. *Geothermics*, **36** (3), 185–222.

Markowitz, H.M. (1952) Portfolio Selection, *Journal of Finance*, **7** (1), 77–91.

Merrow, E. (2003) Taking on the cult of mediocrity. Upstream, May 23, 2003.

Rogge, S. (2004) Geothermische Stromerzeugung in Deutschland –Ökonomie, Ökologie und Potenziale. PhD thesis at the Technische Universität Berlin, Berlin (in German).

Sanjuan, B. *et al.* (2006) Tracer testing of the geothermal heat exchanger at Soultz-sous-forêts (France) between 2000 and 2005. *Geothermics*, **35**, 622–653.

Sanyal, S.K., Morrow, J.W., Butler, S.J., and Robertson-Tait, A. (2007) Cost of electricity from enhanced geothermal systems.Proceedings of the Thirty-Second Workshop on Geothermal Reservoir Engineering, Standford University.

Sharpe, W.F. (1964) Capital asset prices: A theory of market equilibrium under conditions of risk, *Journal of Finance*, **19** (3), 425–442.

Tester, J., Anderson, B., Batchelor, A., Blackwell, D., DiPippo, R., Drake, E., Garnish, J., Livesay, B., Moore, M., Nichols, K., Petty, S. ,Toksöz, M. Jr., and Veatch, R. (2006) The future of geothermal energy – impact of enhanced geothermal systems (EGS) on the United States in the 21st century. Technical report, Massachusetts Institute of Technology. pp.358.

Tester, J. and Herzog, H. (1990) Economic Predictions for Heat Mining: A Review and Analysis of Hot Dry Rock (HDR) Geothermal Technology, Final Report for the U.S. Department of Energy, Massachusetts Institute of Technology.

Van Wees, J.D., *et al.* (2008) A new method for assessing the impact of exploration prospect inter-dependency: an application to predicted gas discoveries in the Netherlands. *AAPG Bulletin.*, **92** (10), 1315–1336.

8
Deployment of Enhanced Geothermal System Plant and CO_2 Mitigation

Ernst Huenges

8.1
Introduction

The provision of energy addresses societal and political issues for the present and future generations. A goal, which is defined in many energy-political frameworks, is to increase the share of renewable energies in the provision of energy, in order to mitigate green house gas emissions and reduce the consumption of finite energy resources. One future option, which has to be considered in this context, is the use of enhanced geothermal systems (EGS). They are supposed to make a large contribution to a sustainable energy mix, in the future. This means that, besides improving technical aspects, EGS must be realizable with a climate friendly life cycle and competitive energy production costs. Regarding the sustainability, competitiveness does not refer to break even with the energy costs from conventional energy carriers, because the prerequisite of the environmental compatibility is not included in this comparison. In integrating the environmental aspect in such a comparison, competitiveness must rather be referred to CO_2-mitigation costs.

For a worldwide deployment of EGS substituting mainly coal-fired power plants, the CO_2-mitigation costs will be derived from estimating the specific energy production costs and the probable CO_2 emissions of EGS plants using the results from life cycle assessment calculations.

8.2
CO_2 Emission by Electricity Generation from Different Energy Sources

EGS plants usually require deeper boreholes than conventional geothermal plants and significantly more effort to engineer the reservoir. However, EGS plants are operated in closed systems at the surface. The thermal water is reinjected after utilizing its heat at the heat exchanger. There exists no gas release during the operation. The comparison of the results of emission studies on conventional systems in the United States (see Bloomfield, Moore, and Neilson, 2003), and life cycle analyses of EGS as given in Figure 8.1 show that still lower CO_2 emissions are referred to EGS.

Geothermal Energy Systems. Edited by Ernst Huenges

ISBN: 978-3-527-40831-3

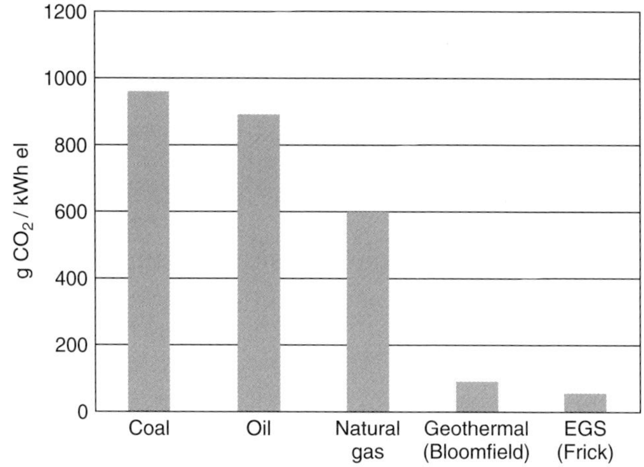

Figure 8.1 Comparison of CO_2 emission from electricity generation from different energy sources. Data from Bloomfield, Moore, and Neilson (2003) in the USA and from Frick *et al.* (2010).

Coal-fired power plants are widely spread all over the world and responsible for several gigatonnes of CO_2 emissions every year. In addition, there are huge programs all over the world now to build up new coal-fired power plants every year, with the result of further impact on the climate change due to their operation. An emission of 960 g $(kW\,h)^{-1}$ for operation of coal-fired power plants (as given in Figure 8.1) was reported by Bloomfield; other studies give numbers somewhat above 1000. Thus, substituting coal-fired power plants by EGS plants to fulfill a given task of energy supply lead to a CO_2 mitigation at least of about 800 g $(kW\,h)^{-1}$.

During the discussion, providing a report on renewable energy in the framework of the IPCC (Intergovernmental Panel on Climate Change) process, scenarios were made for a reasonable development of the capacity of geothermal plants worldwide. Figure 8.2 is based on such a scenario from a 2010 capacity of 10 GW to an extension of the capacity of 140 GW by 2050, with a yearly contribution of about 1000 TW h. Half of the future capacity is expected to be contributed by EGS plants. The substitution of coal-fired power plants by extended geothermal energy provision, which can be reached in the year 2050, mitigates every year more than 1 Gt CO_2 emissions worldwide.

8.3 Costs of Mitigation of CO_2 Emissions

Society is highly engaged at the moment to develop strategies for mitigation of CO_2 emissions. A pronounced role plays, nowadays, the development of technologies of capture CO_2 emissions mainly from coal-fired power plants, transport them, and store the CO_2 in the underground (i.e., CCS (carbon, capture and storage)). The IPCC report from 2005 (IPCC 2005) presented costs between €25 and 55 per

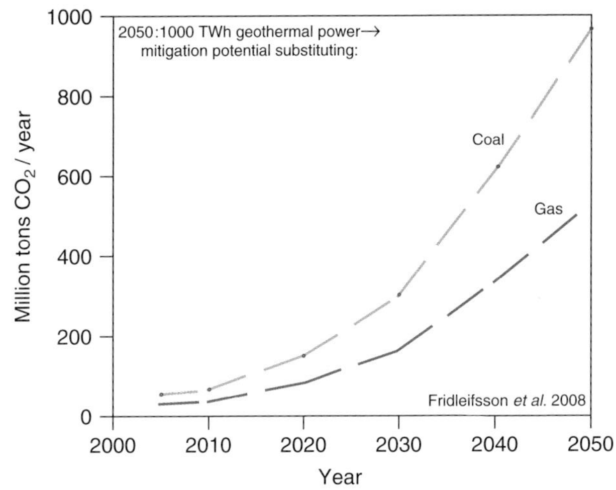

Figure 8.2 CO_2-mitigation calculation based on a forecast of the development of installed capacity of geothermal electricity up to about 140 GW providing about 1000 TWh from about 10 GW providing about 70 TWh in the year 2010.

tonne CO_2 for CCS including the sequestration in deep reservoirs. It is important to note that other authors assume higher costs, because costs for storage and monitoring are poorly presented in the report due to the lack of experiences from given field experiments. Given these cost estimates, which would increase costs for the generation of electricity from coal-fired power plants, the question is, if the EGS concept is cost-competitive under these circumstances.

Numbers from studies at representative sites given by Frick *et al.* (2010) lead to the following conclusion, which is represented here for clarification with rounded figures. The costs for power supply from EGS today allow provision of 200 kW h for the €50 that are estimated for the mitigation of 1 t CO_2 by CCS. Taking into account the estimated effect of geothermal technology development, even about 600 kW h could be delivered in the future with the same efforts. As mentioned above, substituting coal-fired power plants by EGS plants would lead to a CO_2 mitigation of about 800 g $(kW\,h)^{-1}$. Therefore, for €50, about 600 kW h electrical power can be supplied and a mitigation of 600 times 800 g CO_2 that results in about 500 kg CO_2 by substituting coal-fired power plants is possible. In summary, double of the currently assumed costs for CCS, which may have also some cost reduction following a learning curve, deliver the same CO_2 mitigation and additional power supply from EGS. The interaction of CO_2 sequestration with other utilization of underground reservoirs and the probable prevention of these is not yet transferred to costs. Fischedick and Esken (2007) calculated that the costs of low carbon emission coal-fired power plants (i.e., conventional power plant + efficiency loss due to carbon capture + transport and storage of CO_2) are comparable to the costs of renewables.

All these approaches are related to sites with normal geothermal gradients and can become much more favorable for EGS in preferred geothermal environments. Therefore, it can be concluded that the deployment of geothermal energy is the way for provision of base load energy as part of future renewable energy supply and part of the CO_2-mitigation strategy by substituting fossil fuels. Furthermore, the costs of an extended installation of EGS power plants is highly competitive with other CO_2-mitigation strategies.

8.4 Potential Deployment

Overall, the geothermal–electricity market appears to be accelerating, as indicated by the trends in both the number of new countries developing geothermal energy and the total of new megawatts of power capacity under development. It is, however, difficult to predict future rates of deployment, because of the numerous variables involved. With the present engineering solutions, an increase from the current value of 10 GWe installed capacity, up to about 50–70 GWe, should be easily achievable by 2050 (Fridleifsson, 2008). The gradual introduction of new technology improvements is expected to boost the growth rate, with exponential increments after 10–20 years, thus reaching an expected global target of ~140 GWe by 2050. Some of the new technologies (for example, binary plant) have already been proven and are now rapidly deploying, whereas others are entering the field demonstration phases to prove commercial viability (EGS), or early investigation stages to test practicality (supercritical temperature and offshore resources).

Low temperature power generation with binary plants has opened up the possibilities of producing electricity in countries that do not have high temperature resources. EGS technologies (deep drilling, stimulation, and pumping) are being developed to access resources in this setting. Supercritical and offshore resources are also under investigation. If these technologies can be proven economical at commercial scales, the geothermal market potential could be limited only by the size of the grid or load in many countries of the world. It is anticipated that, by 2050, approximately half of the deployed capacity could come from these new technologies.

Direct use of geothermal energy for heating is currently commercially competitive, using accessible hydrothermal resources. A moderate increase is expected in the future development of such hydrothermal resources for direct use, mainly because of dependence on resource proximity, and therefore on local economic factors.

8.5 Controlling Factors of Geothermal Deployment

8.5.1 Technological Factors

Direct heating technologies, district heating, and EGS methods are available. These have different degrees of maturity. The direct use of thermal fluids from deep aquifers, and heat extraction using EGS, have costs and risks, which can be reduced by further technical advances associated with accessing and engineering fractures in the geothermal reservoirs. The latter requires a better knowledge and measurement of the subsurface stress fields. For EGS, further remaining challenges are drilling, well completion, brine management, mitigation of induced seismicity, reliability of system components, and mitigation for corrosion and scaling. Knowledge acquired while developing geothermal reservoirs will lead to better practices and standards and increased deployment confidence.

Geothermal power generation technologies also have different degrees of maturity. Reducing subsurface exploration risks will contribute to more efficient and sustainable development . The drilling of high temperature reservoirs requires advance technologies to prevent reservoir damage by drilling mud; an example is the use of balanced drilling procedures. Improved utilization efficiency requires better auxiliary energy use and improved performance of surface installations. Better reservoir management, with improved simulation models, will optimize reinjection strategy, avoid excessive depletion, and plan future make-up well requirements, to achieve sustainable production.

The quality of the heat extracted, and its potential diversity of use, increases with heat source temperature. Improvements in energy utilization efficiency from cascaded use of geothermal heat are an important deployment strategy. Evaluating the performance of geothermal plants, including heat and power EGS installations, will consider heat quality of the fluid by differentiating between the energy and the exergy content (that part of the energy that can be converted to power).

8.5.2 Economic and Political Factors

Distributions of potential geothermal resources vary from being nearly site-independent (e.g., EGS) to site-specific (for hydrothermal sources). The distance between electricity markets or centers of heat demand and geothermal resources is a factor in the economics of power generation and direct use. When making development choices, there is sometimes a trade-off between the quality of hydrothermal resources and their remoteness from secure grid connections or demand centers. The renewable, reliable, and cost-competitive nature of geothermal energy has, in the past, attracted some energy-intensive industries (e.g., aluminum smelting, pulp and paper, timber drying) to colocate with geothermal resources to attain a comparative commercial advantage. In the context of mandates for increased use of renewable energy and for reductions in GHG (green house gas) emissions, this colocation trend is expected to increase. At present, growth in direct-use demand is dominated by single building applications for shallow ground-source heat pump systems, where the cost of energy distribution is not an issue. The direct use of heat from hydrothermal systems and EGS projects can satisfy the demand of district heating systems and industrial heating more effectively, but only where the politics, economics, and infrastructure of heat distribution are favorable.

The deployment of all technologies relies on the availability of skilled installation and service companies. For deep geothermal drilling and reservoir management, such services tend to be concentrated in a few countries only. For district heating, there is also a correlation between local availability and awareness of service companies, and technology uptake. For enhanced global deployment, such services would be better distributed worldwide. Larger deployment is generally facilitated by establishing insurances to cover drilling, development, and production risks. Therefore, project risk management is another requirement for financing, installing, and operating large geothermal installations. Prior knowledge and expertise within the local banking and insurance industries generally assist in accelerating local deployment rates.

Geothermal deployment will be supported, politically, by a CO_2-mitigation strategy, through establishing incentives for market penetration of geothermal energy supply technologies. These incentives can include, for example, subsidies, guarantees, and tax write-offs to cover the risks of initial deep drilling. Policies to attract energy-intensive industries (e.g., aluminum smelting) to known geothermal resource areas can also be useful. Feed-in tariffs with confirmed geothermal prices have been very successful in attracting commercial investment in some countries (e.g., Germany). However, feed-in tariffs for direct heating are difficult to arrange. Therefore, direct subsidies for building heating and for district heating systems may be more successful. Subsidy support for refurbishment of existing buildings with geothermal energy will open a much greater market for future deployment.

Policy support for research and development is required for all geothermal technologies, but especially for EGS. Public investment in geothermal research drilling programs should lead to a significant acceleration of EGS deployment. Support is also needed for programs to educate and enhance public acceptance of geothermal energy use, and to conduct research toward the avoidance or mitigation of potential induced hazards and adverse effects.

References

Bloomfield, K.K., Moore, J.N., and Neilson, R.N. (2003) Geothermal energy reduces greenhouse gases. *Geothermal Resources Council Bulletin*, **32**, 77–79.

Fischedick, M. and Esken, A. (2007) *RECCS: Strukturell-ökomisch-ökologischer Vergleich Regenerativer Energietechnologien (RE) mit Carbon Capture and Storage (CCS)*, Bundesministerium für Umwelt, Naturschutz und Reaktorsicherheit, Berlin.

Frick, S, van Wees, J.D., Kaltschmitt, M., and Schröder, G. (2010) Economic performance and environmental assessment, in *Enhanced Geothermal Energy Systems, Geothermal Technology for Resource Assessment, Exploration, Field Development, and Utilization* (ed. E. Huenges), John Wiley & Sons, Ltd, in press.

Fridleifsson, I.B., Bertani, R., Huenges, E., Lund, J. W., Ragnarsson, A., and Rybach, L. (2008) The possible role and contribution of geothermal energy to the mitigation of climate change, in *IPCC Scoping Meeting on Renewable Energy Sources, Proceedings* (eds O. Hohmeyer and T. Trittin), Lübeck, January 20–25, 2008, pp. 59–80.

IPCC (2005) Carbon Dioxide Capture and Storage. A Special Report of Working Group III of the Intergovernmental Panel on Climate Change, pp. 1–443.

Color Plates

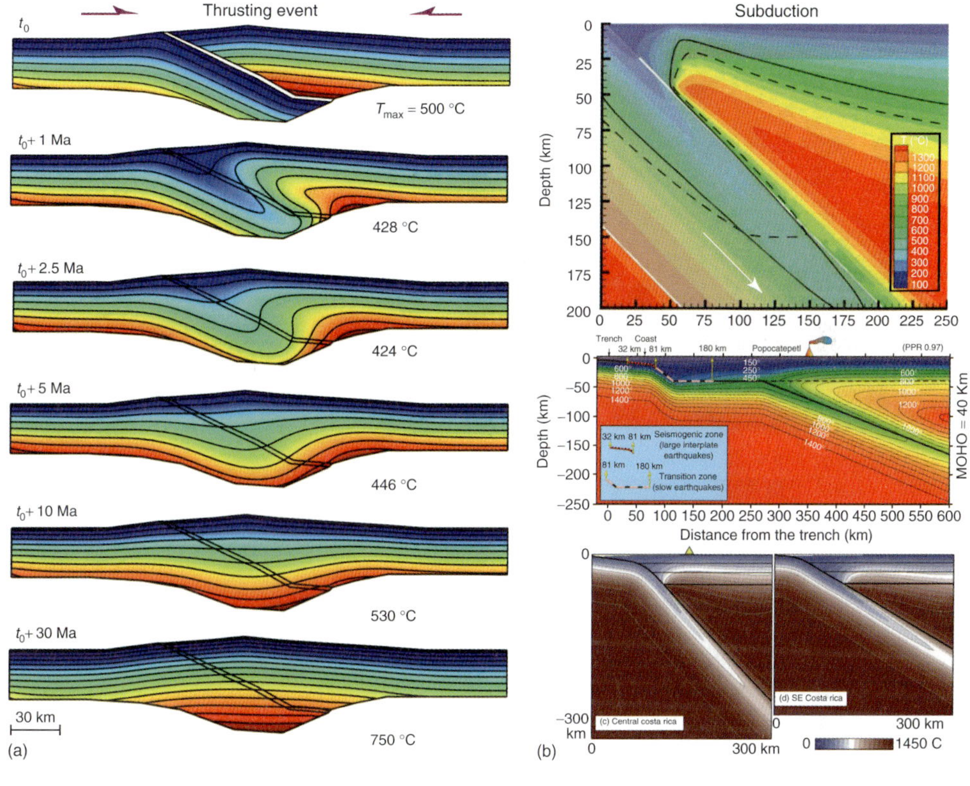

Figure 1.6 Examples of large-scale transient and steady-state thermal perturbations: (a) thrusting event resulting in a thickened and more radiogenic crust and (b) distinct models of thermal fields around subduction zones, where slab dip angle and plate velocities differ from one case to the other. (After, from top to bottom, Cagnioncle, Parmentier, and Elkins-Tanton, 2007; Manea *et al.*, 2004; Peacock *et al.*, 2005.) (This figure also appears on page 11.)

Geothermal Energy Systems. Edited by Ernst Huenges

ISBN: 978-3-527-40831-3

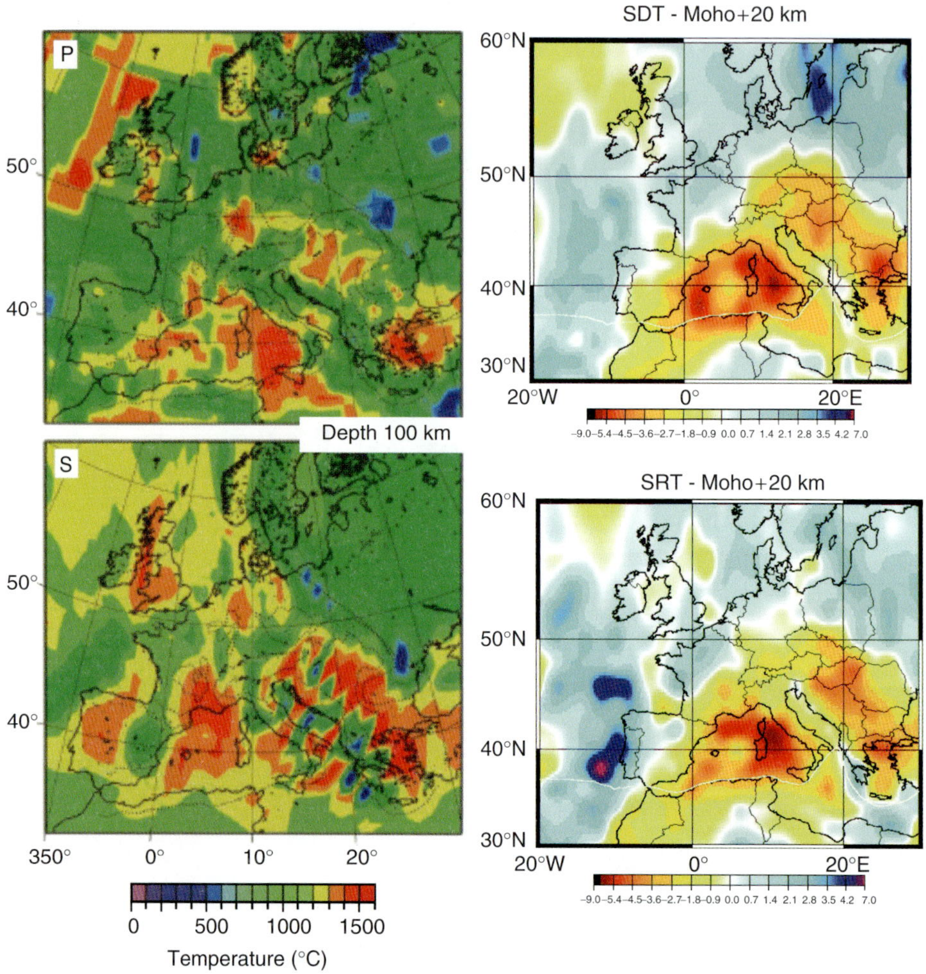

Figure 1.9 Left: Temperatures at 100 km depth estimated from the P and S velocity anomalies. (After Goes *et al.*, 2000.) Right: Tomographic models extracted from an upper mantle shear velocity model (Shapiro and Ritzwoller, 2002); top: diffraction tomography, bottom: ray tomography. (This figure also appears on page 16.)

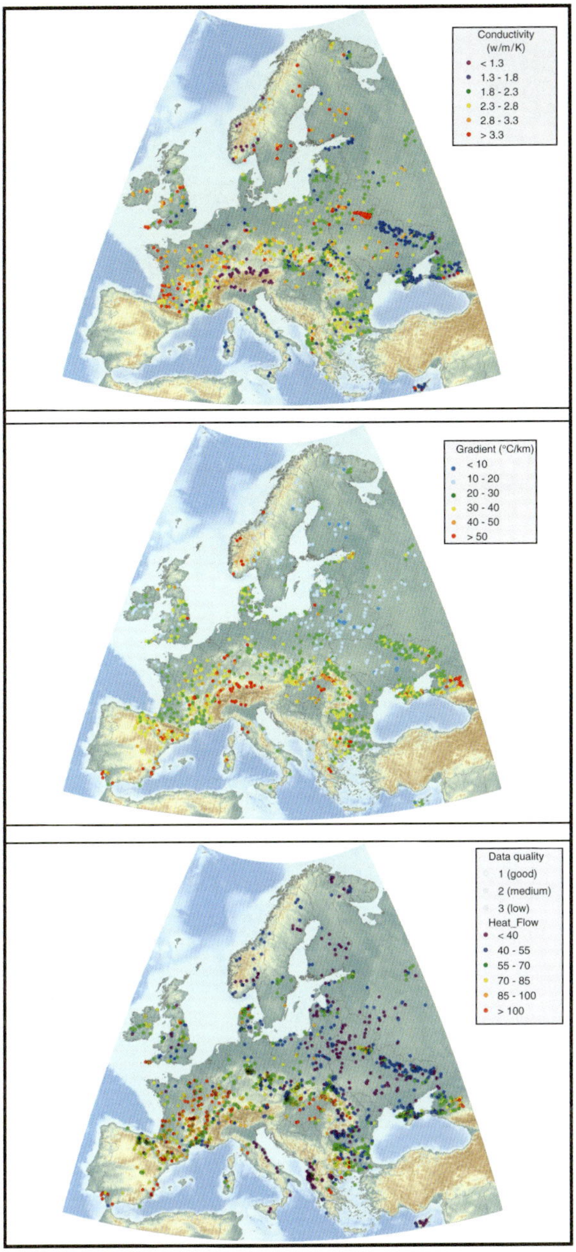

(a) (b) (c)

Figure 1.10 Thermal conductivity (a), temperature gradient (b), and heat flow data (c) as compiled from this study. Each color is assigned a range of values, and for heat flow data, a quality criterion is added (see text). (This figure also appears on page 19.)

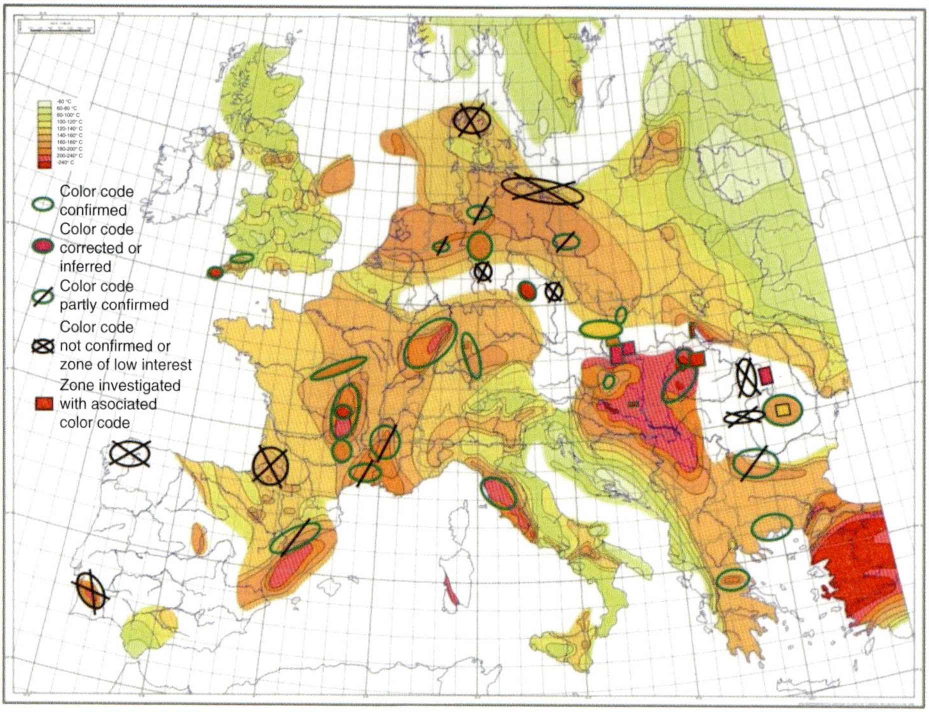

Figure 1.11 Map of temperature at 5 km depth, as inferred from unavailable (confidential) BHT measurements (Hurtig *et al.*, 1992; EIEG, 2000) and critical analysis by Genter *et al.* (2003) from published thermal data (see text). (This figure also appears on page 20.)

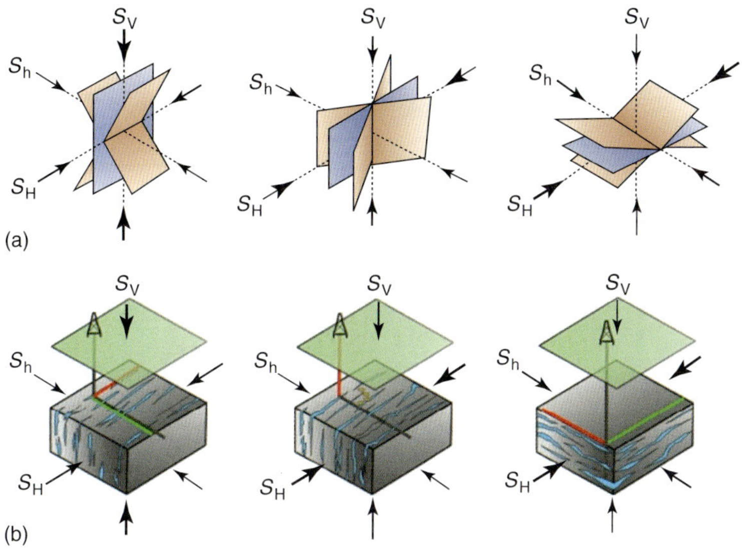

Figure 2.2 (a) Geometrical relation between stress axes, stress regimes, and fracture planes. Brown: shear fractures; blue: tensile fractures. Stress regimes from left to right: normal faulting, strike-slip faulting, and reverse faulting. (b) From left to right, orientation of tensile fractures in normal faulting, strike-slip faulting, and reverse faulting regime. Red drill path is least stable; green drill path is most stable. In strike-slip regimes, the most stable drill path depends on the stress ratios of S_V and S_H. (This figure also appears on page 46.)

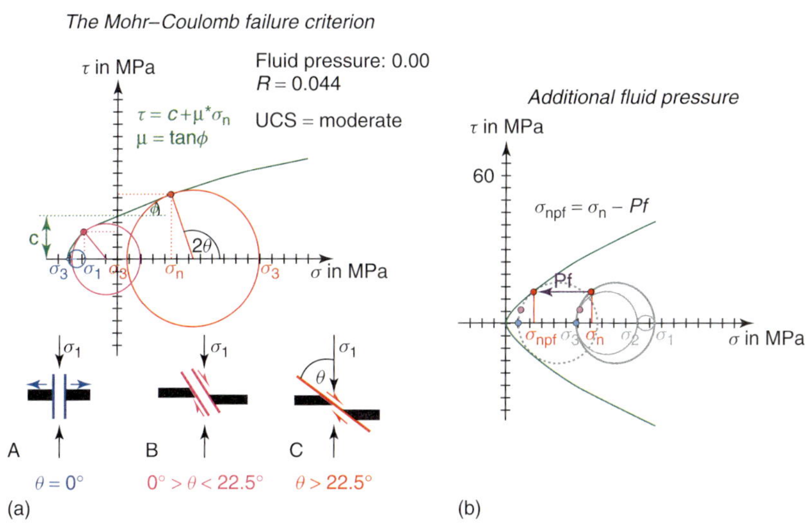

Figure 2.3 The Mohr–Coulomb failure criterion (see text). (This figure also appears on page 47.)

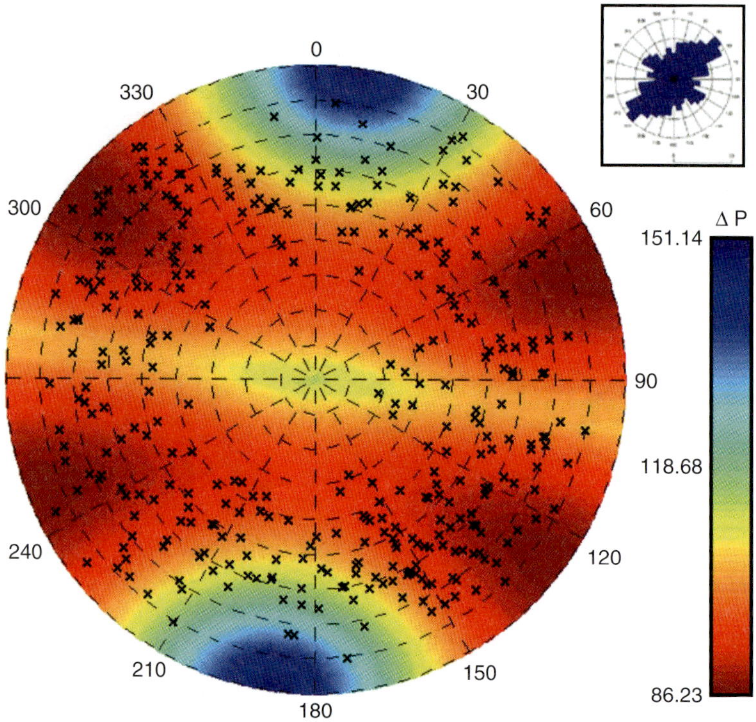

Figure 2.4 Fracture susceptibility diagram. The amount of pore pressure increase, ΔP_p needed to cause failure of a fracture with a given orientation is indicated by the color scale shown at the right edge of the image. Fracture orientations observed from image logs or oriented cores can be plotted as planes to poles. If they lie in the red areas of the diagram, ΔP_p is relatively low and fractures are more likely to fail and to be conductive than in the blue areas (after Mildren, Hillis, and Kaldi, 2002). In the example shown here, steeply dipping fractures striking NW–SE or NE–SW are much more likely to be conductive than a steeply dipping fracture striking E–W. (This figure also appears on page 49.)

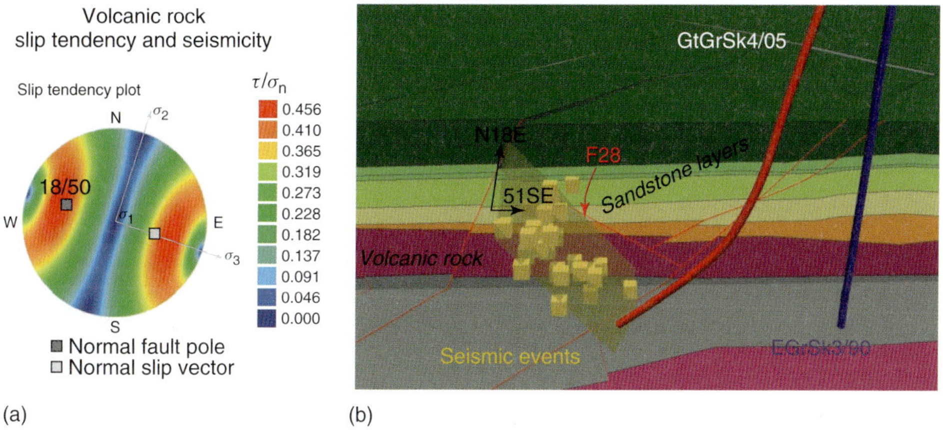

Figure 2.5 (a) Slip tendency plot of the lower Permian volcanic rocks in the Groß Schönebeck field. The pole of plane represents the mean plane as derived from microseismicity. (b) Mean plane of recorded seismic events together with a spatial distribution of recorded seismicity (yellow boxes) together with least-square fitted plane (transparent yellow). The distribution of seismicity fits the orientation of the F28 fault plane within the reservoir (Moeck, Kwiatek, and Zimmermann, in press). (This figure also appears on page 52.)

Figure 3.26 "Bit sample" of casing material. (This figure also appears on page 161.)

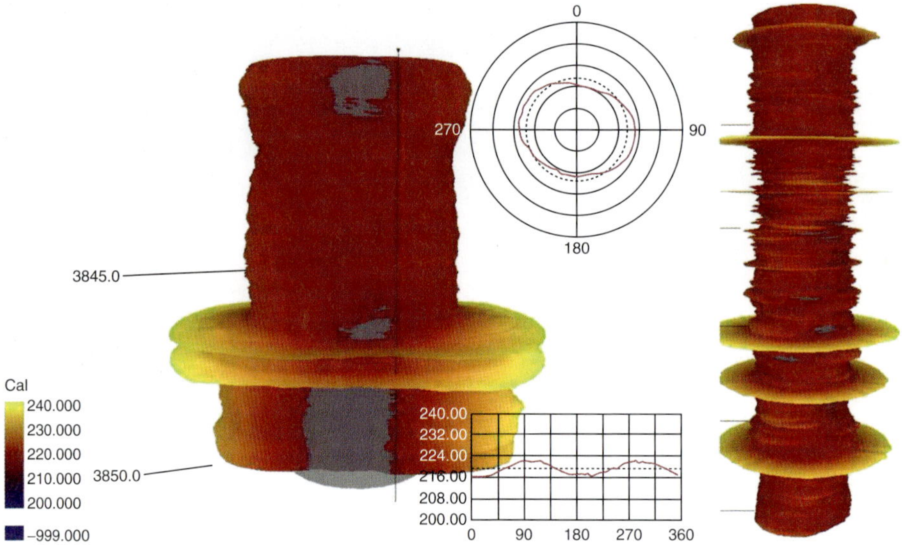

Figure 3.27 Example of multifinger caliper results in the damaged casing section. (This figure also appears on page 162.)

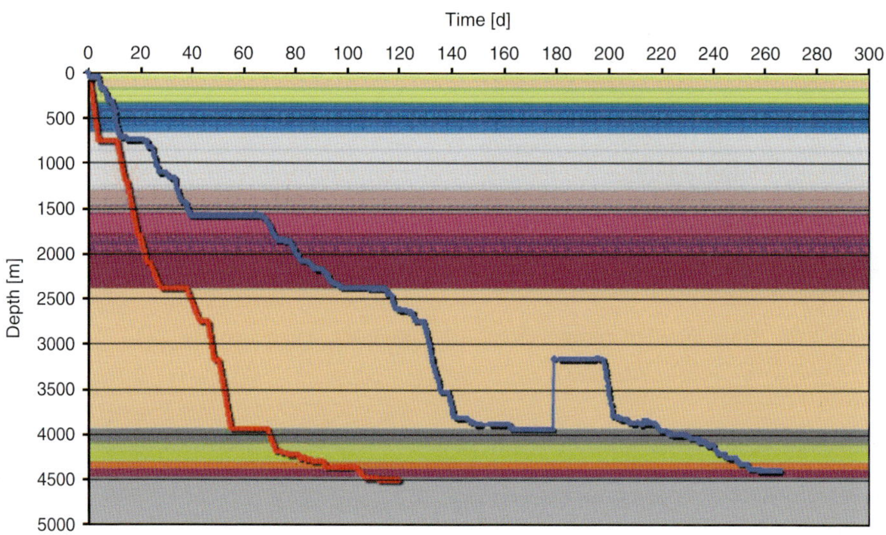

Figure 3.29 Depth versus time (red – planned, blue – real, colors indicate lithostratigrafic members). (This figure also appears on page 164.)

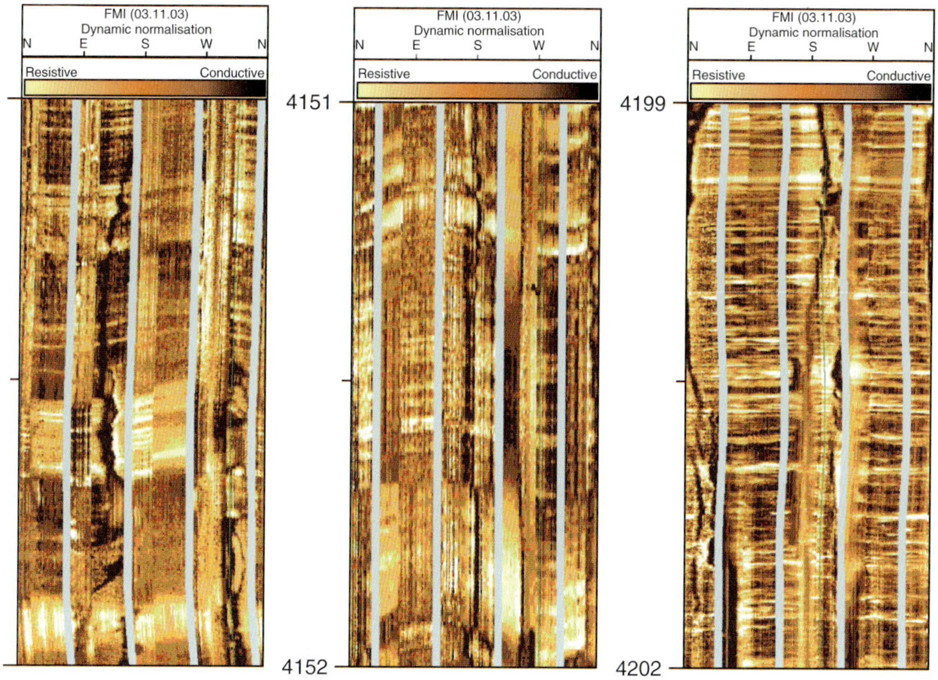

Figure 4.3 Formation MicroImaging measurement in the borehole Groß Schönebeck 3/90 exhibiting a "roll out" wall of the borehole. The dark color shows a vertical fissure in the rough direction south-north which was opened by a massive waterfrac treatment. The color scale runs from high electrical resistance (light) to low resistance (dark). (This figure also appears on page 180.)

Figure 4.6 Main equipment necessary to perform a hydraulic stimulation treatment (left pump aggregats, middle tanks, left container for proppants) (from treatments in Groß Schönebeck 2007). (This figure also appears on page 187.)

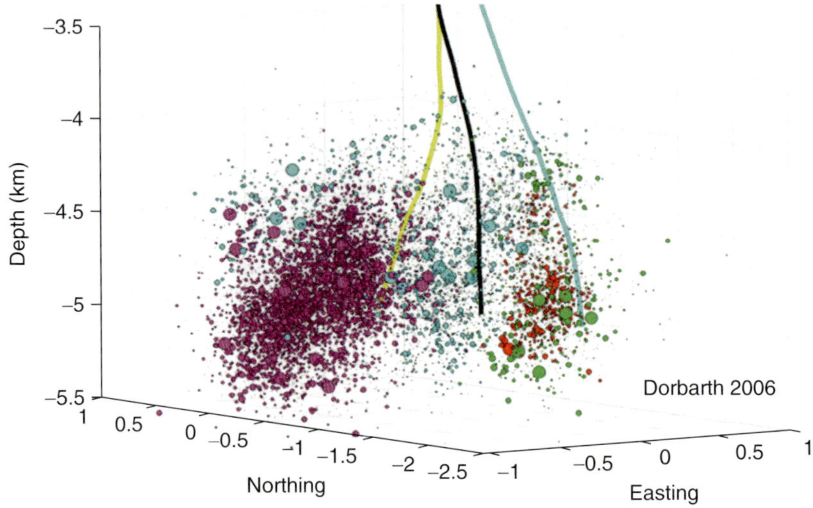

Figure 4.8 Relocation of induced seismicity event related to a stimulation treatment in Soultz sous Forets (Dorbath, 2006). (This figure also appears on page 190.)

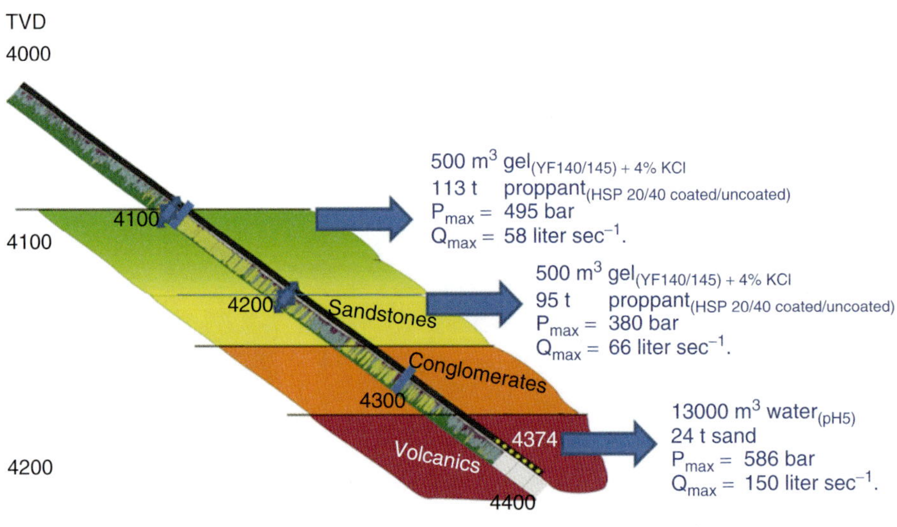

Figure 4.12 Summary of three stimulation treatments in the deviated well GtGrSk4/05 in Groß Schönebeck (see text). (This figure also appears on page 214.)

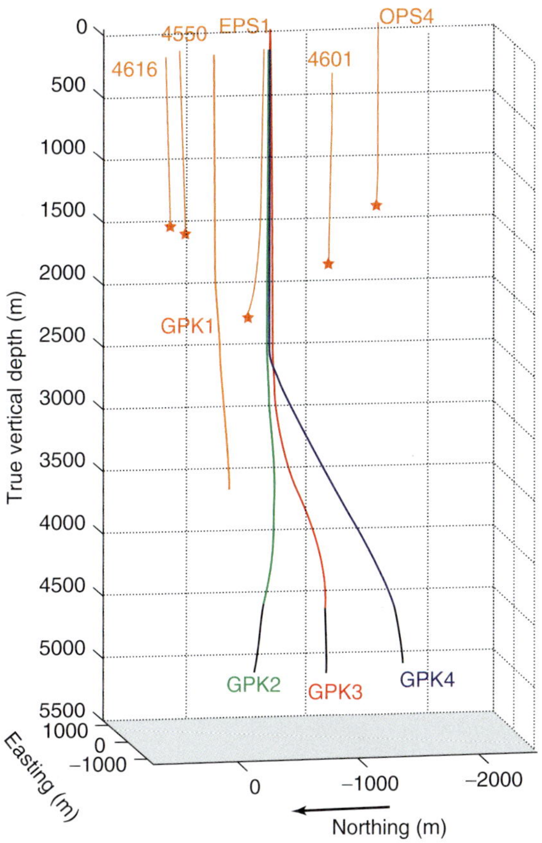

Figure 4.19 Sketch of the three deep wells and the surrounding seismic observation wells. The observation wells are orange, the stars indicate the depth level of the installed downhole seismic sensors. (This figure also appears on page 222.)

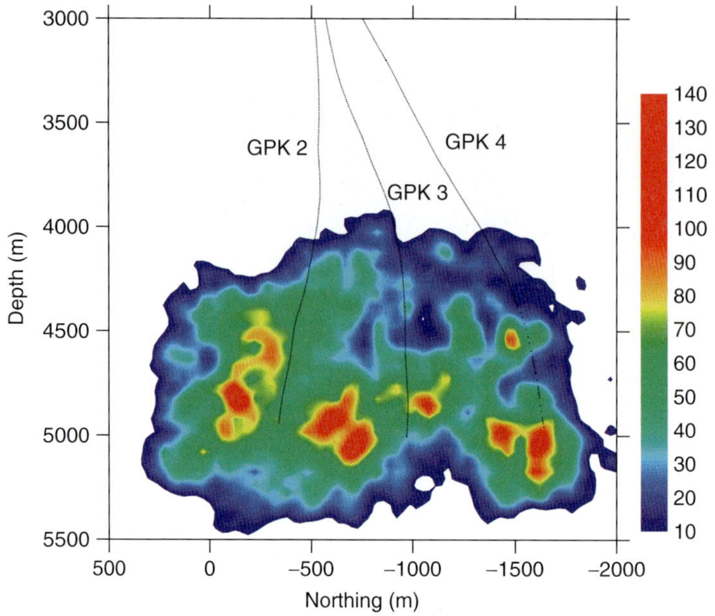

Figure 4.20 Color contour plot of the event density in 50 × 50 m cells in the plane of the graph. Perpendicular to this plane the cells are unlimited. All events localized during the stimulations 2000, 2003, 2004, and 2005 were included. (This figure also appears on page 223.)

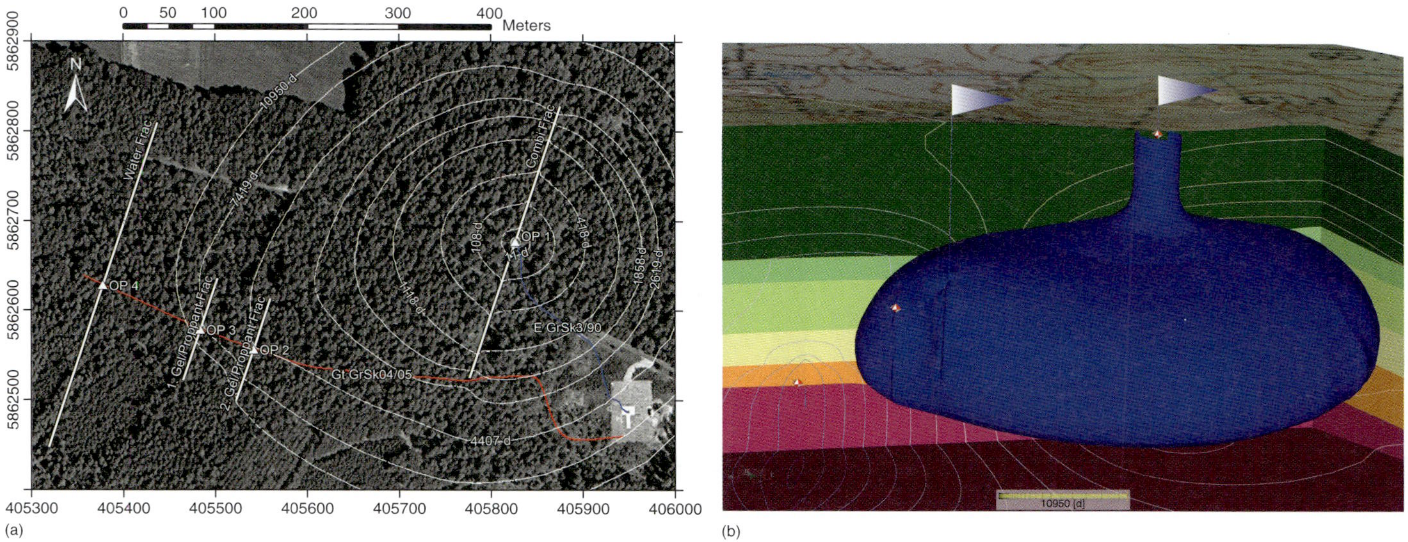

Figure 5.12 Propagation (a) and final stage (b) of simulated 130 °C isosurface around the injection well during 30 years of production. The propagation is illustrated by a horizontal cut through the Elbe base sandstone II unit (b). (This figure also appears on page 266.)

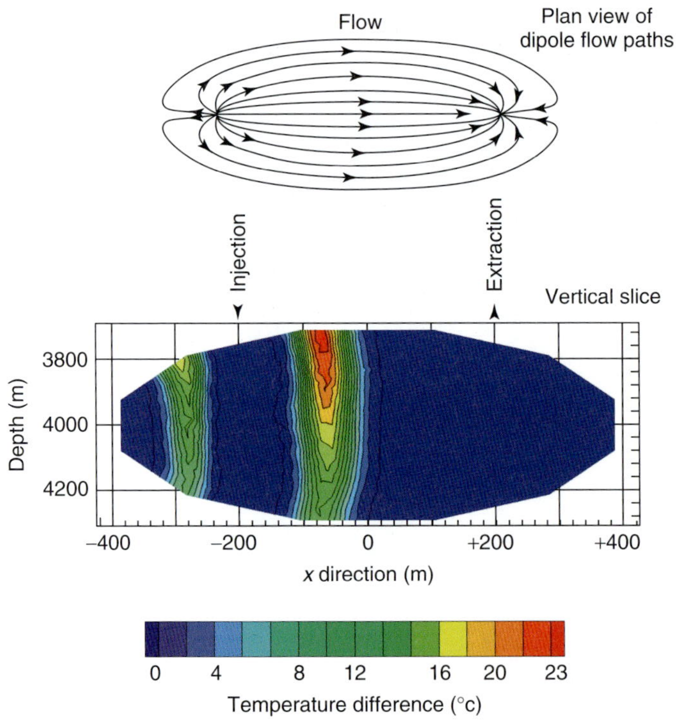

Figure 5.21 Hydromechanical coupling, the reservoir is more permeable under the lower stress conditions at the top of the reservoir than the higher stress conditions at the base of the reservoir. (This figure also appears on page 276.)

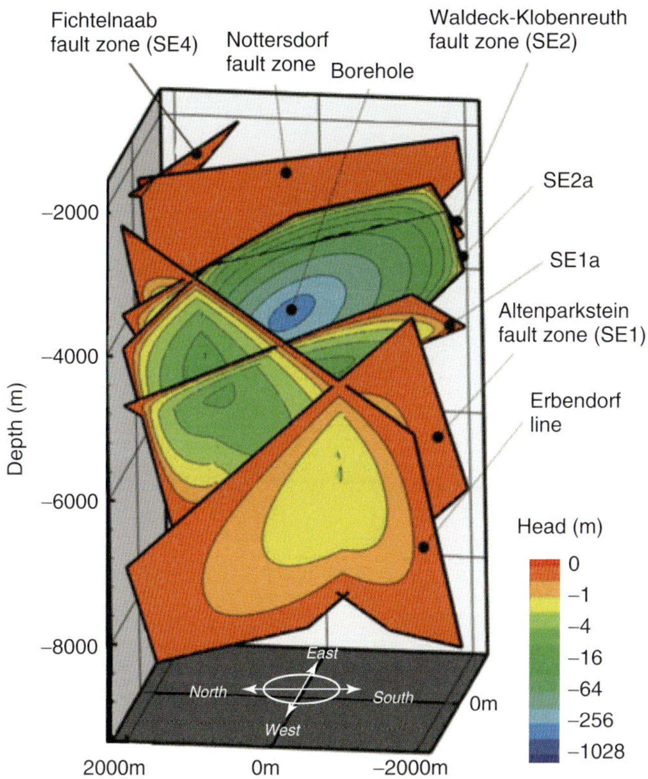

Figure 5.29 Conceptual deterministic fracture network model of the shear zones, illustration of the geomechanical facies concept [McDermott *et al.*, 2006a]. (This figure also appears on page 287.)

Index

Geothermal Energy Systems. Edited by Ernst Huenges

ISBN: 978-3-527-40831-3

l

o

p